Interdisciplinary Cancer Research

Volume 3

The "Interdisciplinary Cancer Research" series publishes comprehensive volumes on different cancers and presents the most updated and peer-reviewed articles on human cancers. Over the past decade, increased cancer research has greatly improved our understanding of the nature of cancerous cells which has led to the development of more effective therapeutic strategies to treat cancers. This translational series is of special value to researchers and practitioners working on cell biology, immunology, hematology, biochemistry, genetics, oncology and related fields.

Nima Rezaei
Editor

Hepatobiliary Cancers: An Interdisciplinary Approach

Springer

Editor
Nima Rezaei
Department of Clinical Immunology
Karolinska Institutet
Stockholm, Sweden

Cancer Immunology Project (CIP)
Universal Scientific Education and
Research Network (USERN)
Stockholm, Sweden

ISSN 2731-4561 ISSN 2731-457X (electronic)
Interdisciplinary Cancer Research
ISBN 978-3-031-44527-9 ISBN 978-3-031-44528-6 (eBook)
https://doi.org/10.1007/978-3-031-44528-6

This Springer imprint is published by the registered company Springer Nature Switzerland AG
The registered company address is: Gewerbestrasse 11, 6330 Cham, Switzerland

Paper in this product is recyclable.

Preface

Hepatobiliary cancers, although are not as common as other gastrointestinal cancers, have high rate of mortality and low survival rate. Hepatocellular carcinoma and biliary tract cancers are two major malignancies of hepatobiliary system.

The Interdisciplinary Cancer Research series publishes comprehensive volumes on different cancers. It plans to present the most updated and peer-reviewed interdisciplinary chapters on cancers. This interdisciplinary book series is of special value to researchers and practitioners working on cell biology, immunology, hematology, biochemistry, genetics, oncology and related fields. This is the main concept of Cancer Immunology Project (CIP) and Network of Immunity in Infection, Malignancy and Autoimmunity (NIIMA), which are two active interest groups of the Universal Scientific Education and Research Network (USERN).

The third volume of the book, entitled *Hepatobiliary Cancers: Interdisciplinary Approach*, starts with an introduction on hepatobiliary cancers, which requires interdisciplinary approach. The immune system from beginning to progression in liver cancer are discussed in Chap. 3. Tumor microenvironment in hepatocellular carcinoma is explained in Chap. 4, metabolic alterations of hepatocellular cancer stem cells are discussed in Chap. 5. The role of senescence in non-alcoholic steatohepatitis-related hepatocellular carcinoma is the subjects of Chap. 6, while regulators of genetic risk for the progression of non-alcoholic fatty liver disease to hepatocellular carcinoma are presented in Chap. 7. Chapter 8 presents tumor microenvironment and immunotherapy in advanced biliary tract cancers. After discussion on fibrosis and immunotherapy in hepatocellular carcinoma in Chap. 9, curative potential of nanomedicine in liver cancer is explained in Chap. 10. Locoregional therapies and drug resistance in hepatocellular carcinoma are discussed in Chaps. 11 and 12. Finally, liver stereotactic body radiotherapy is discussed in Chap. 13.

I hope that this interdisciplinary book will be comprehensible, cogent and of special value for researchers, oncologists and gastroenterologists who wish to extend their knowledge on hepatobiliary cancer and its treatment.

Stockholm, Sweden Nima Rezaei

Contents

Interdisciplinary Approach in Hepatobiliary Cancers 1
Vahid Mansouri and Nima Rezaei

Liver Cancer: Interdisciplinary Approach 41
Haiwei Zhang, Juan Zhao, Wei Yang, Zheng Li, Li Gong, and Yongsheng Li

The Immune System in Liver Cancer: From Beginning to Progression 85
Alex José de Melo Silva, Juliana Ellen de Melo Gama, Cleonilde Maria de Nascimento, Jessica Paula Lucena, Cicero Jadson da Costa, Camila Juliet Barbosa Fernandes, Danielle Maria Nascimento Moura, Helotonio Carvalho, and Sheilla Andrade de Oliveira

The Tumor Microenvironment in Hepatocellular Carcinoma 107
Katsuya Nagaoka, Yasuhito Tanaka, and Okio Hino

Metabolic Alterations of Hepatocellular Cancer Stem Cells 139
Jan Caira David, Marta Bento Afonso, and Cecília Maria Pereira Rodrigues

The Role of Senescence in NASH-Related HCC 167
Lampros Chrysavgis, Grigorios Papadopoulos, and Antonios Chatzigeorgiou

Regulators of Genetic Risk for the Progression of Non-alcoholic Fatty Liver Disease to Hepatocellular Carcinoma: Reconstruction of Transcriptional Network and Signature-Based Metabolic Profiling 193
Dora Lucía Vallejo-Ardila, Marco A. De Bastiani, and Diego A. Salazar

Tumor Microenvironment and Immunotherapy in Advanced Biliary Tract Cancers 229
Mustafa Korkmaz and Mehmet Artaç

Fibrosis and Immunotherapy in Hepatocellular Carcinoma 255
Sarah B. White and Dilip Rajasekhar Maddirela

Liver Cancer and the Curative Potential of Nanomedicine 283
Adrian Kuzmanović, Cheng Lin, and Matthias Bartneck

Locoregional Therapies for Hepatocellular Carcinoma 307
Alexander E. Hare and Mina S. Makary

Drug Resistance in Hepatocellular Carcinoma 325
Xinxin Chen, Jin Li, Yuhong Huang, and Chao Liang

Liver Stereotactic Body Radiotherapy (SBRT) 349
Olwen Leaman Alcibar, Fernando López Campos, José Antonio Blanco, Patricia Tavera Pomata, and Carmen Rubio

Index 363

About the Editor

Nima Rezaei, MD, PhD Professor Nima Rezaei gained his medical degree (MD) from Tehran University of Medical Sciences and subsequently obtained an MSc in Molecular and Genetic Medicine and a PhD in Clinical Immunology and Human Genetics from the University of Sheffield, UK. He also spent a short-term fellowship of Pediatric Clinical Immunology and Bone Marrow Transplantation in the Newcastle General Hospital. Professor Rezaei is now the Full Professor of Immunology and Vice Dean of Research and Technologies, School of Medicine, Tehran University of Medical Sciences, and the Co-founder and Head of the Research Center for Immunodeficiencies. He is also the Founder of Universal Scientific Education and Research Network (USERN). Prof. Rezaei has already been the Director of more than 100 research projects and has designed and participated in several international collaborative projects. Prof. Rezaei is the editor, editorial assistant, or editorial board member of more than 40 international journals. He has edited more than 50 international books, has presented more than 500 lectures/posters in congresses/meetings, and has published more than 1200 scientific papers in the international journals.

Interdisciplinary Approach in Hepatobiliary Cancers

Vahid Mansouri and Nima Rezaei

Abstract

Hepatobiliary cancers are one the most common and emerging cancers worldwide, causing a high burden, especially in endemic regions. The effective management of these cancers greatly depends on the successful implementation of screening and surveillance programs, accurate diagnosis, selection of best treatment, and careful follow-ups. The optimal management of these cancers could be more accessible through discussion between several providers with different specialties related to each step of management. The complex management, together with advancements in therapeutic options and promising results of combination therapies, necessitates close collaboration and communication between these specialties. Given the appealing outcomes of this multidisciplinary care, it has been getting increasingly integrated into practice, in the form of multidisciplinary tumor boards, or centralized clinics. Multidisciplinary care has been shown to result in higher patient satisfaction, improved receipt of timely

V. Mansouri
Network of Immunity in Infection, Malignancy and Autoimmunity (NIIMA), Universal Scientific Education and Research Network (USERN), Tehran, Iran

Gene Therapy Research Center, Digestive Disease Research Institute, Shariati Hospital, Tehran University of Medical Sciences, Tehran, Iran

N. Rezaei (✉)
Research Center for Immunodeficiencies, Children's Medical Center, Tehran University of Medical Sciences, Tehran, Iran

Department of Immunology, School of Medicine, Tehran University of Medical Sciences, Tehran, Iran

Cancer Immunology Project (CIP), Universal Scientific Education and Research Network (USERN), Stockholm, Sweden
e-mail: rezaei_nima@tums.ac.ir

Interdisciplinary Cancer Research, https://doi.org/10.1007/16833_2022_70
Published online: 19 November 2022

evidence-based care, and increased overall survival, indicating that it should be considered the standard of care for the management of hepatobiliary cancers.

Keywords

Digestive system diseases · Interdisciplinary communication · Neoplasms · Patient care team

1 Introduction

Hepatobiliary cancer refers to primary malignancies of the liver, such as hepatocellular carcinoma (HCC) or biliary tract carcinomas (BTCs) that arise from the intra- and extrahepatic biliary ductal system (Duffy and Greten 2017). Around 95% of hepatobiliary cancers are either HCC (80%) or intrahepatic cholangiocarcinoma (ICC) (15%) (McGlynn et al. 2015). In 2018, primary liver cancer was the sixth most prevalent cancer diagnosis and the fourth main cause of cancer death worldwide (Bray et al. 2018). It is anticipated to surpass colorectal cancer to become the third leading cause of cancer death in the United States by 2040, after lung and pancreatic cancer (Rahib et al. 2021).

Recent advances in the screening, diagnosis, treatment, and monitoring of hepatobiliary cancers make achieving optimal management complex. The majority of hepatocellular carcinoma (HCC) cases result from cirrhosis; hence, gastroenterologists play a crucial role in the initial screening procedure (Frenette et al. 2019). Furthermore, based on the current guidelines, the roles of diagnostic radiologists and pathologists in accurate radiographic and tissue evaluation are vital in diagnosis and staging (Ayuso et al. 2018). The best treatment for HCC is decided by a complex interaction of parameters such as tumor size, number, and location, as well as patient characteristics such as underlying liver function and performance status (Salgia and Mendiratta 2021). The emergence of new therapeutics, e.g., immunotherapies, targeted therapies, locoregional therapies, and oncolytic virus therapy, makes therapeutic decision-making even more diverse and complex (Luo et al. 2021). Consequently, establishing the most efficient treatment plan is frequently challenging, requiring knowledge of many therapeutic modalities from a variety of specialties, including hepatology, hepatobiliary surgery, transplant surgery, radiation oncology, medical oncology, interventional radiology, and palliative care specialists, among others (Salgia and Mendiratta 2021). In this chapter, we reviewed the interdisciplinary approach in hepatobiliary cancers, its importance, current practice models, and clinical outcomes, along with its challenges and future perspectives.

2 Epidemiology

Liver cancer has affected more than 900,000 cases (9.5 per 100,000 person-years) worldwide with approximately 830,000 deaths (8.7 deaths per 100,000 person-years) in 2020, according to the Global Cancer Observatory statistics (Sung et al. 2021; Fund 2022). It is predicted that liver cancer would affect more than a million individuals by 2025 (Llovet et al. 2021a, b, c). Liver cancers affect men more than women (Howlader et al. 2020). The prognosis of liver cancers is poor, with an average 5-year relative survival of about 20% (2011–2017), with a range from 35% for individuals who are diagnosed in early stages to 12% for metastatic cases (Howlader et al. 2020; Siegel et al. 2022). As most patients are identified with HCC at a late stage, the median survival time is between 6 and 20 months (Golabi et al. 2017). There is a geographical variation in the prevalence of liver cancers, from 5.1 per 100,000 person-years in Europe to 17.7 per 100,000 person-years in eastern Asia (Ferlay et al. 2018). The rate of liver cancer is rising at an alarming rate in western countries, while its trend is decreasing in Asian countries, given regional differences in the frequency of risk factors (Dasgupta et al. 2020). The main risk factors for the development of HCC are cirrhosis and viral hepatitis (Byrd et al. 2021). More than 90% and 60% of HCC cases occurred in patients with a background of cirrhosis and hepatitis, respectively (Yang et al. 2019; Llovet et al. 2021a, b, c). The growing number of HCC patients is generally caused by an increasing number of chronic hepatitis B and C, as well as emerging cases of nonalcoholic steatohepatitis (NASH) and nonalcoholic fatty liver disease (NAFLD) associated with obesity and diabetes (Kulik and El-Serag 2019; Llovet et al. 2021a, b, c).

There is more limited amount of information about ICC, compared with the HCC. Cholangiocarcinoma refers to cancers that originated from the epithelium of either intrahepatic or extrahepatic biliary tracts (Rodriguez-Pascual et al. 2018). Evaluation of available data is difficult given that intrahepatic and extrahepatic tumors are categorized separately (Rodriguez-Pascual et al. 2018). Intrahepatic BTCs are typically classified as primary liver tumors, and extrahepatic BTCs instead of being distinct entities usually were grouped with gallbladder cancers (Rodriguez-Pascual et al. 2018). Gallbladder and other extrahepatic bile duct cancers are estimated to account for 12,130 new cases and about 4400 deaths in the United States in 2022 (Siegel et al. 2022). BTCs are more prevalent in elderly people, with the average age of patients diagnosed with intrahepatic and extrahepatic BTC of 70 and 72, respectively (Siegel et al. 2022). BTCs generally are incurable and progressive, with a few cases surviving the first year (Rodriguez-Pascual et al. 2018). Common risk factors of BTCs include cirrhosis, viral hepatitis, primary sclerosing cholangitis, and liver fluke infections, the latter of which is responsible for the higher prevalence of BTC in western countries (Khan et al. 2019; Clements et al. 2020).

3 Management of Hepatobiliary Cancers

Management of hepatobiliary cancers primarily involves prevention and screening, early diagnosis, and hepatic resection, or transplantation (Mazzanti et al. 2016). Sensible management could prevent a great number of deaths attributed to liver cancer (Centers 2019). Sometimes patients with more advanced cancers have undergone combination therapies to be downstaged and become operable. Other modalities would be used to help patient longevity and to improve the function of liver with certain underlying conditions (Mazzanti et al. 2016; Dimitroulis et al. 2017).

The management of HCC is complex, given the multifaceted nature of these cancers (Personeni and Rimassa 2017). Public health measures and effective screening strategies should be used for the prevention and early detection of patients (Hainaut et al. 2019). Besides, clinical outcomes would greatly depend on patients' initial condition at diagnosis and proposed treatment modalities (Golabi et al. 2017). Most of the patients diagnosed with hepatobiliary cancers have some underlying liver disorders, which along with the cancer status itself should be considered when selecting the treatment approaches (Mazzanti et al. 2016). Additionally, given the nonspecific symptoms, many patients were not diagnosed until the advanced stages, which made their treatment more difficult and challenging (Golabi et al. 2017). Furthermore, although the prognosis of these patients is generally poor, the various management strategies could result in different rates of mortality, recurrence, and survival, which showed the necessity for an individualized treatment strategy (Golabi et al. 2017; Personeni and Rimassa 2017).

The management of BTC is also complex, given the various origins of cancer which could bias the diagnosis. Also, the unspecific initial presentation of the BTC results in latency in diagnosis. Similar to HCC, many of the BTCs are just diagnosed in advanced stages. Moreover, given the blockage effect of BTC on bile flow, opening the tract is usually needed via insertion of stents before the performance of diagnostic procedures, which itself could have considerable morbidity and mortality. Furthermore, although there was not a well-evidenced management strategy for locally advanced ones, there are not effective enough treatments for more disseminated ones (Rodriguez-Pascual et al. 2018).

Individualized treatment selection needs thorough clinical, radiological, and pathological assessments (Pascual et al. 2016). Personalized treatment is determined by liver function, tumor size and number, tumor morphology, biomarkers, macrovascular invasion, extrahepatic disease spread, donor availability, the severity of underlying liver disease, response to previous treatments, and other patient characteristics (Marrero et al. 2018; Rodriguez-Pascual et al. 2018; Mehta 2020).

3.1 Prevention and Screening

The rationale of HCC prevention is focused on decreasing the incidence of HCC risk factors. The main strategy which has been suggested for HCC prevention consists of the universal vaccination – mainly for hepatitis B virus (HBV) – and treatment of viral hepatitis, e.g., direct-acting antiviral (DAA) therapy for hepatitis C virus (HCV) and antiviral for HBV (Papatheodoridis et al. 2015; Kanwal et al. 2017). These two interventions played a significant role in decreasing the incidence of HCC. Also, there are other suggested means of HCC prevention, among which aspirin use and coffee consumption have stronger evidence than others. Using aspirin could decrease the cumulative incidence of HCC from 8% to 4% after a median follow-up of 8 years, based on a Swedish nationwide study (Simon et al. 2020). Coffee consumption was also suggested as a preventative strategy for patients with chronic liver disease by the European Association for the Study of the Liver (EASL) (Liver 2018).

As said before, many HCC patients were only diagnosed at advanced stages. It is well-known that the prognosis of patients diagnosed at early stages was remarkably better than those diagnosed in late stages (Villanueva 2019). For example, an early-stage HCC cancer that can be treated with curative modalities would have a 5-year survival of more than 70%, while a symptomatic advanced-stage HCC has a median survival of 1.5 years (Llovet et al. 2021a, b, c). Hence, there are growing recommendations for HCC surveillance. There is accumulating evidence of the superior prognosis of the at-risk group with surveillance compared to ones without HCC surveillance (Zhang et al. 2004; Lederle and Pocha 2012; Singal et al. 2014). Also, decision analysis models showed that the implementation of the surveillance system is cost-effective for the high-risk population, i.e., when the expected HCC risk exceeds 0.2% and 1.5% in patients affected by HBV and HCV, respectively (Andersson et al. 2008; Centers 2019). The American Association for the Study of Liver Diseases (AASLD) discussed the HCC risk – the estimated incidence of HCC – in its latest practice guideline on HCC (Marrero et al. 2018). Although the beneficial effect of incorporating HCC surveillance in the primary healthcare system is almost clear, the best screening modality is still under debate. The ideal screening tool should be accurate, easy to use in different situations, and make a reproducible result, while it should not be operator dependent which could cause inconsistency in findings. The most common recommended modality for HCC surveillance is abdominal sonography, although it is operator dependent and its performance is under question for at-risk patients with NASH or obesity (Atiq et al. 2017). Meanwhile, the biochemical tests yield more consistent results. The only biomarker approved for the HCC screening is alpha-fetoprotein (AFP) (Singal et al. 2014). Combining abdominal sonography with AFP significantly increases the sensitivity of HCC screening from 45% to 63%; therefore their usage could be considered, especially in communities with a higher prevalence of HCC (Tzartzeva et al. 2018). Currently, AASLD recommends 6-month interval sonography with or without AFP for HCC surveillance in cirrhotic patients (Marrero et al. 2018). Finally, imaging modalities like CT and MRI, however, provide higher sensitivity but, given the higher cost, risk of radiation and contrast exposure, and less accessibility to such equipment in developing countries, could be reserved for HCC high burden areas (Tzartzeva et al. 2018; Llovet et al. 2021a, b, c). Therefore, successful prevention and screening of HCC need an interdisciplinary collaboration of primary healthcare system, radiologists, and gastroenterologists.

There is not any recommended screening strategy for cholangiocarcinoma (CC), because the current methods of evaluation including imaging modalities or laboratory tests could not detect the tumor at early stages. However, recently, liquid biopsies are demonstrated to be capable of identifying the early-stage CCs; they are usually used for the diagnosis of cancer. Only there is a recommendation of using ultrasonography or MRI/MRCP and CA19-9 every 9–12 months as a screening modality in individuals aged higher than 40 years old in endemic areas and high-risk patients (e.g., primary sclerosing cholangitis) (Sungkasubun et al. 2016; Ahn and Yang 2019). Moreover, there are not any known ways to prevent the CC, because its identified risk factors, e.g., age, ethnicity, and bile duct abnormalities, are not changeable. Only recently, low-dose aspirin has been shown to reduce the incidence of CC (Choi et al. 2016; Shen and Shen 2021).

3.2 Diagnosis

Given the positive background of liver disease, i.e., cirrhosis or hepatitis in most of the patients with HCC, most cases could be identified through surveillance (Singal et al. 2020). However, given that the surveillance program would not implement in all clinical settings – especially in developing countries – and also the limitation of screening only to at-risk population, it may be possible to detect HCC as an enlarged liver mass. Patients could be asymptomatic with an incidental mass in their liver detected during imaging for other reasons or symptomatic showing nonspecific symptoms of HCC, i.e., abdominal pain, weight loss, and liver dysfunction (Llovet et al. 2021a, b, c).

The radiological studies remain the mainstay of HCC diagnosis. The suspicious individuals during the screening – having a liver nodule in sonography or having a high level of AFP – would be referred for imaging studies. The radiological hallmark of HCC is "arterial enhancement and delayed washout" which could be seen in CT scans with arterial contrast (Marrero et al. 2018). However, in about 10% of suspicious HCC cases, this diagnostic hallmark could not be detected (89% sensitivity and 96% specificity (van der Pol et al. 2019)). These cases with atypical presentation would need sampling to be histologically evaluated. The described histological features of HCC include stromal invasion, increased cell density, intratumoral portal tracts, unpaired arteries, pseudo-glandular pattern, and diffuse fatty changes (Kojiro et al. 2009). It is worth mentioning that given the difficulties of sampling procedure of small nodules and pathological interpretations, there would be still a trivial chance of missing the tumor (Forner et al. 2008). Along with serial repetitions of evaluations, the role of liquid biopsies is getting noticed (Xu et al. 2017). A study showed that a combined assessment of circulating tumor DNA with AFP could be successful in the detection of HCC with 100% sensitivity and 95% specificity (Qu et al. 2019).

HCC management is considered according to tumor staging, the patient's preserved liver function, and performance status (Marrero et al. 2018). EASL and AASLD both recommend the Barcelona Clinic Liver Cancer (BCLC) staging system

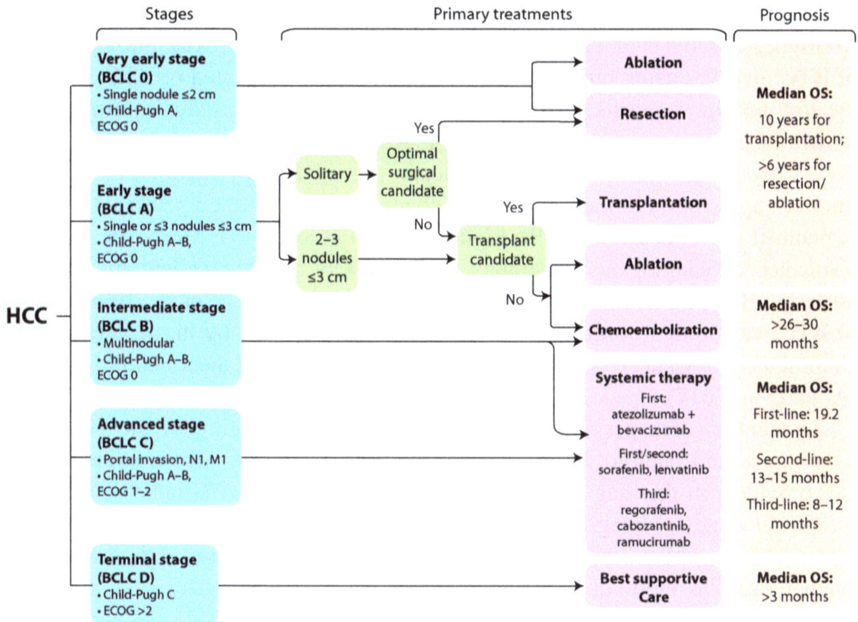

Fig. 1 Management of hepatocellular cancer based on the Barcelona Clinic Liver Cancer (BCLC) staging system. Patients should be staged into five categories according to disease extension, liver function, and performance status. Asymptomatic patients with low tumor burden and good liver function (BCLC 0/A) should be treated with local curative treatments (resection, ablation, or transplantation, which provides the highest survival). Those with asymptomatic multinodular disease and sufficient liver function (BCLC B) should be treated with chemoembolization, whereas patients with portal thrombosis or extrahepatic dissemination (BCLC C) should be treated with systemic treatments. *ECOG* Eastern Cooperative Oncology Group, *HCC* hepatocellular carcinoma, *M1* distant metastasis, *N1* lymph node metastasis, *OS* overall survival

for HCC staging and treatment allocation (Llovet et al. 1999a, b; Liver 2018; Marrero et al. 2018). BCLC incorporates tumor extension, liver function, and patient performance status to classify them into five stages (Cillo et al. 2006). Based on BCLC, local curative treatments could be used in patients with early-stage HCC tumors, while transarterial chemoembolization (TACE) and systemic therapies would be assigned to patients with intermediate and advanced stages, respectively (Liver 2018; Marrero et al. 2018; Llovet et al. 2021a, b, c) (Fig. 1). Although the main factor for choosing appropriate treatment is BCLC staging, the number of tumors in the imaging study and the remained liver function should also take into account. Given the hepatic disease in many HCC patients, the resection and systemic therapies could be administered for patients with adequate baseline liver function (Zhang et al. 2004; Mazzanti et al. 2016; Marrero et al. 2018). Management of patients according to BCLC staging has improved the survival of patients when compared with untreated patients (Llovet et al. 2021a, b, c).

Diagnostic modalities used for BTC were comprehensively reviewed by Rodriguez-Pascual et al. elsewhere (Rodriguez-Pascual et al. 2018). Initial diagnosis of BTC involves using ultrasonography in individuals suspected of BTC; however, the low accuracy and its operator-dependent findings necessitate confirmation of diagnosis with other modalities like CT scan or MRI (Hennedige et al. 2014). Previously ERCP was used as a diagnostic procedure; however, its invasiveness limits its application to the time when a concurrent EUS or therapeutic intervention is planned (Joshi et al. 2014; Rodriguez-Pascual et al. 2018). Moreover, magnetic resonance cholangiopancreatography (MRCP) acts as an excellent modality for assessing BTC, given its safety and high resolution (Hennedige et al. 2014). Staging of BTC was done according to the TNM staging provided by the American Joint Committee on Cancer (AJCC)/Union for International Cancer Control (UICC) (Amin et al. 2017).

As mentioned in previous lines, the diagnosis and staging of the HCC need radiologists and pathologists along with the GI specialists. Also, radiologists are key providers in the diagnosis of CCs. The role of radiologists is becoming more prominent as newer imaging equipment gives higher accuracy for detecting smaller tumors, diminishing the need for interventional sampling and pathological evaluations (Cassinotto et al. 2017).

3.3 Treatment of HCC

HCC treatment has progressed substantially during the last decades. Treatment options include curative approaches, including surgical resection, liver transplantation, and ablation, or non-curative approaches including TACE, transarterial radioembolization (TARE), radiotherapy, and systemic therapies. Also, combination treatments, given sequentially or concurrently, could be used to downstage the tumor making it resectable or achieve better outcomes (Llovet et al. 2021a, b, c). Successful treatment relies on the expertise and effective communication of hepatobiliary surgeon, medical oncologists and/or hepatologists, interventional radiologist, and radiation oncologist (Salgia and Mendiratta 2021). Also, patient-centered evidence-based decision-making was positively associated with better outcomes, highlighting the importance of patient-clinician(s) communication.

Hepatic resection is considered the treatment of choice in patients with a single liver nodule without metastasis whose post-operation remained liver function is not a concern. According to AASLD and EASL guidelines, hepatic resection could be done for patients with a single tumor, with well-compensated liver function (Child-Pugh A with total bilirubin <1 mg/dl), with the absence of clinically relevant portal hypertension (no varices or ascites), as well as without any performance restriction according to Eastern Cooperative Oncology Group score (ECOG score $= 0$) (Liver 2018; Marrero et al. 2018). Carefully applying these criteria results in perioperative mortality of $<3\%$ and 5-year and 10-year survival of approximately 70% and 7–15%, respectively (Zhou et al. 2001; Franssen et al. 2014). However, hepatic resection was applied for patients with multinodular tumors but with well-preserved liver function (Yin et al. 2014). Tumor recurrence is the main drawback of hepatic resections, affecting more than 70% of the patients within 5 years (Roayaie et al. 2013). Tumor recurrence could be either early (< 2 years), as a result of micrometastasis during resection procedures, or late (> 2 years) due to de novo carcinogenesis (Imamura et al. 2003). To this date, there was not any effective strategy to prevent tumor recurrence except the antiviral therapies against HBV and HCV, which showed a slight benefit (Yin et al. 2013; Waziry et al. 2017; Singal et al. 2019).

Liver transplantation has been associated with the best clinical outcomes and best survival for HCC patients (Tabrizian et al. 2019). However, the selection of appropriate candidates for transplantation was in debate for years. The Milan criteria, defined as a single tumor ≤5 cm or two to three tumors ≤3 cm without vascular invasion, are the most common advocated criteria (Mazzaferro et al. 1996). Also, there is an opportunity to downstage tumors to Milan criteria, to be able to be eligible for transplantation (Mazzaferro et al. 2009). The highest burden of tumor which could be considered for downstaging is defined based on the University of California San Francisco (UCSF) criteria which are also adopted by AASLD (Yao et al. 2007; Marrero et al. 2018). Although transplantation results in better survival and less recurrence compared to hepatic resection (recurrence rates of 10–15% at 5 years and 5-year and 10-year survival of 70% and 50%, respectively (Tabrizian et al. 2019)), long waiting lists and lack of decent donors limit its applicability and cause the patient dropout due to tumor progression (Llovet et al. 1999a, b). Neoadjuvant therapies, e.g., TACE or ablation (bridging therapy) during waiting for transplantation, could decrease the rates of dropouts and diminish the posttransplantation recurrence rates (Kulik et al. 2018). Furthermore, evaluation of patients regarding the response to neoadjuvant therapies by modified Response Evaluation Criteria In Solid Tumors (mRECIST), as a predictor of posttransplantation mortality, could guide the clinician in choosing the transplantation candidates (Cucchetti et al. 2020). In addition, further genetic investigations for identifying molecularly aggressive HCCs, which have a higher risk of progression – and subsequent waiting list dropout – and recurrence, could help to prioritize the patients on the waiting list (Miltiadous et al. 2015; Schulze et al. 2015).

However, over 70% of patients are not surgical candidates at the time of presentation; thus alternative treatment options must be considered for them (Salgia and Mendiratta 2021). In addition, the role of adjuvants following surgical treatments or combination treatments in lowering recurrences was assessed in various studies (Breitenstein et al. 2009; Wu et al. 2012; Kubo et al. 2013; Yin et al. 2013; Bruix et al. 2014).

Ablation techniques could be performed for patients with early-stage small tumors or as an alternative for patients with tumor up to 4 cm or two to three tumors ≤3 cm who are contraindicated for surgery or transplantation (e.g., due to liver dysfunction). The ablation could be done through chemical (i.e., ethanol injection) (Lencioni et al. 2003), thermal (i.e., radiofrequency ablation (RFA) (Xu et al. 2018), microwave ablation (MWA), cryoablation (CRA) (Wang et al. 2015), and laser interstitial thermotherapy (Di Costanzo et al. 2015)), and electrical (i.e., irreversible electroporation (IRE) (Cheng et al. 2015)) methods. RFA is considered the frontline ablation treatment by EASL and AASLD (Liver 2018; Marrero et al. 2018), with comparable outcomes to hepatic resection (Ng et al. 2017; Xu et al. 2018; Izumi et al. 2019). The median overall survival using RFA was reported to be approximately 60 months with a 5-year recurrence rate of 50–70% (Breen and Lencioni 2015; Liver 2018). MWA is widely used in medicine and easy to implement to achieve a larger ablation area (Llovet et al. 2021a, b, c); however, trials and meta-analysis did not reach any superiority for MWA compared with RFA (Yu et al. 2017; Hu et al. 2019). Other types of ablations, although had promising results, were not entered into clinical practice recommendations yet.

Radiotherapy is recommended for patients with unresectable tumors whose liver function is preserved (Child-Pugh A) (Llovet et al. 2021a, b, c). It could be used for HCC tumors with various numbers and different stages. Stereotactic radiotherapy was safely applied to tumors restricted to the liver, while it could be used for tumors with vascular invasion, as well as metastatic lesions (Tse et al. 2008; Bujold et al. 2013; Hong et al. 2016). Selected patients with well-preserved liver function had a median survival of 17 months following the radiotherapy (Bujold et al. 2013). Further studies would be needed for determining the modalities of choice for HCC radiotherapy.

Transarterial chemoembolization (TACE) is prescribed for patients with intermediate HCC. The positive effects of TACE on HCC were demonstrated by several studies (Llovet et al. 2002; Lo et al. 2002; Llovet and Bruix 2003), showing a range of median survival from 19.4 to 37 months when combined with tyrosine kinase inhibitors (TKI) (Okusaka et al. 2009; Kudo et al. 2014; Lencioni et al. 2016; Meyer et al. 2017a, b). TACE was also shown to be safe causing a treatment-related mortality of 0.6% (Lencioni et al. 2016). Response to prior treatment according to mRECIST was a predictor of a better survival following TACE (Vincenzi et al. 2015). TARE, another transarterial treatment, involves using glass or resin microspheres embedded by yttrium. The comparable (or even better) outcomes of TARE compared to TACE (Hilgard et al. 2010; Salem et al. 2010; Salem et al. 2016) caused it to be recommended by AASLD and EASL for intermediate-stage HCC (Liver 2018, Marrero et al. 2018).

Systemic therapies generally are used in advanced HCC cases. Following the discovery and approval of sorafenib, it became the first-line treatment for advanced-stage HCC based on the BCLC staging system (Llovet et al. 2008). However, from that time several other medications were assessed for their superior outcomes. Table 1 shows the studies corresponding to these novel therapeutics. Given the outcome and side effects, the atezolizumab plus bevacizumab regimen is more effective with manageable side effects than sorafenib (Finn et al. 2020, 2021). Hence, the current standard of care recommends using the combination regimen of atezolizumab (anti-PDL1 antibody) plus bevacizumab (anti-VEGF antibody), as the first-line treatment. Furthermore, as bevacizumab bears a potential risk of bleeding, upper gastrointestinal endoscopies would be needed for ruling out the untreated varices (Llovet et al. 2021a, b, c). There are several drugs used as a second line after HCC progression on sorafenib. Regorafenib, cabozantinib, and ramucirumab are

Table 1 Summary of clinical and radiological outcomes and adverse events among systemic therapies approved for advanced HCC

Study name	Year	No of participants	Study design	Treatment regimen	Median OS (months)	Median PFS (months)	ORR mRECIST; RECIST	Grade III–IV adverse events	TrM	TRAE-associated drug withdrawal	References
First-line treatments											
SHARP	2005–2008	502	Global, multicenter, open-label, phase 3 randomized trial	Sorafenib	13.4	3.7–4.3	NA; 2%	45%	NA	11%	Llovet et al. (2008)
REFLECT	2013–2015	954	Multicenter, randomized, open-label, noninferiority trial, phase 3	Lenvatinib	13.6	7.4	24.1%; 18.8%	57%	2%	9%	Kudo et al. (2018)
IMbrave150	2018–2020	558	Global, multicenter, randomized, open-label, phase 3	Atezolizumab + bevacizumab	19.2	6.8	35.4%; 29.8%[a]	36%	2%	16%	Finn et al. (2020)
Second-line treatments											
RESORCE	2013–2015	573	Randomized, double-blind, placebo-controlled, phase 3	Regorafenib	10.6	3.1	11%; 7%	50%	2%	10%	Bruix et al. (2017)
REACH-2	2015–2017	292	Randomized, double-blind, placebo-controlled, phase 3	Ramucirumab	8.5	2.8	NA; 5%	NR	2%	11%	Zhu et al. (2019)
CELESTIAL	2013–2021	707	Randomized, double-blind, controlled, phase 3	Cabozantinib	10.2	5.2	NA; 4%	68%[a]	1%	16%	Abou-Alfa et al. (2018)

(continued)

Table 1 (continued)

Study name	Year	No of participants	Study design	Treatment regimen	Median OS (months)	Median PFS (months)	ORR mRECIST; RECIST	Grade III–IV adverse events	TrM	TRAE-associated drug withdrawal	References
CheckMate 459	2016–2017	743	Randomized, multicenter, phase 3	Nivolumab	16.4	3.7	NA; 15%	22%	1%	4%	Yau et al. (2019), Yau et al. (2022)
KEYNOTE-224	2016–2017	104	Non-randomized, multicenter, open-label, phase 2	Pembrolizumab	13.2	4.9	19%; 17%	25%	1%	5%	Zhu et al. (2018) and Kudo et al. (2022)
CheckMate 040	2016	148	Multicenter, open-label, multicohort, phase 1/2	Nivolumab plus ipilimumab	22.8	N/A	NA; 32%	53%	2%	18%	Yau et al. (2020)

Abbreviations: *ORR* overall response rate, *PFS* progression-free survival, *mRECIST* modified RECIST, *NR* not reported, *RECIST* Response Evaluation Criteria In Solid Tumors, *OS* overall survival, *TRAE* treatment-related adverse events currently approved to be used for advanced-stage HCC as second-line treatment (Bruix et al. 2017; Abou-Alfa et al. 2018; Zhu et al. 2019), while three other regimens – nivolumab, pembrolizumab, and nivolumab plus ipilimumab – have been approved by the FDA based on the promising results in phase 1b/2 (Kudo et al. 2022; Yau et al. 2019, 2020, 2022; Zhu et al. 2018), and further investigations are currently on the way (Table 1).

3.3.1 Novel Therapeutics

Other agents are also used for the treatment of HCC (Liu et al. 2015). Immunotherapy using immune checkpoint pathway inhibitors in combination with antiangiogenic inhibitors showed promising results (Zappasodi et al. 2018). Cytotoxic T lymphocytes (CTLs) could be used against identified tumor-associated antigens (TAAs), e.g., AFP, human telomerase reverse transcriptase (hTERT), and melanoma antigen gene A (MAGE-A) (Mizukoshi et al. 2011; Liang et al. 2013). Autologous dendritic cells pulsed with specific TAA (DC vaccines) and siRNA/miR against oncogenes also had positive results (Palmer et al. 2009; Mansouri et al. 2021).

Furthermore, oncolytic viruses, which could selectively replicate in tumor tissues, were used against HCC. The selective multi-mechanistic effects of oncolytic viruses were discussed elsewhere (Russell et al. 2012). Pexastimogene devacirepvec (Pexa-Vec, JX-594), an oncolytic vaccinia virus armed with granulocyte-macrophage colony-stimulating factor (GM-CSF)-expressing genes, was successfully used in a phase 2 trial for the treatment of HCC (Park et al. 2008; Heo et al. 2013). Telomelysin (OBP-301), a telomerase-specific oncolytic adenovirus armed with telomerase reverse transcriptase (hTERT) promoter, which allows the replication of the virus in tissues with high telomerase expression, has been used in preclinical studies and entered the first phase trial for HCC (Lin et al. 2013).

3.3.2 Supportive Care

Quality of life in HCC patients is greatly affected, given the frequent symptoms which put a high burden on patients' daily lives (Gill et al. 2018). The most common symptoms of HCC patients include sleep disorders (50–65%), depression (60%), fatigue, malnutrition, anorexia and cachexia, pain, muscle cramps, and psychological issues (Colagrande et al. 2016). Carefully addressing these symptoms and treatment side effects should be considered as a part of HCC management. Supportive care is exactly defined for this purpose, i.e., the multi-professional attention to the individual's overall physical, psychological, spiritual, and cultural needs, available at all stages of the illness, for patients of all ages, and regardless of the current intention of any anticancer treatment (Centers 2019). Although the best supportive care (BSC) is still under debate, there are several recommendations on the importance of implementing such supportive (palliative) care and its positive results (Kumar and Panda 2014; Laube et al. 2021). Multiple caregivers with different fields of expertise, e.g., nutrition, mental and psychological consultation, pain management, etc., would be needed for successful supportive care.

3.4 Management of Cholangiocarcinoma

Management of CCs consists of a curative approach, i.e., tumor resection and liver transplantation, and some non-curative strategies, i.e., radiotherapy and systemic chemotherapy. Resection of BTC is challenging, and many proximal tumors failed to be resected with appropriate free margin (60–80%) (Tompkins et al. 1990; Schoenthaler et al. 1994). Besides, the use of surgery is limited because eligible patients should not have any distant or lymph node metastasis or invasion to adjacent organs or adjacent arteries and portal vein (Kurosaki et al. 1999). The difficulty of resection of the tumor with free margin often results in low long-term survival and frequent relapses (Burke et al. 1998; Sakamoto et al. 1998). Similarly, more restrictive criteria for transplantation candidates (absence of metastasis, acceptable liver function, inability of margin-free resection, and availability of compatible donor) have made it a rarely used strategy (Rosen et al. 2010). Furthermore, no clear guideline exists about the follow-up and monitoring after the surgery, but some recommend performing serial CA19-9 and CT scans (or MRI or PET) for detection of locoregional relapses (Rodriguez-Pascual et al. 2018).

The role of radiotherapy for BTC is largely unknown. However, for locally unresectable tumors, radiotherapy could be used as palliative care. The role of radiotherapy as a neoadjuvant therapy before surgery or after surgery as a combination therapy is still not sufficiently studied (Kim et al. 2002); however, the combination of radiotherapy with chemotherapy could be a reasonable choice for locally advanced tumors (Borghero et al. 2008). TACE or TARE, as described for HCC, could be used here, especially through the hepatic artery which feeds the tumor mass, although there is limited evidence for it (Rodriguez-Pascual et al. 2018). Chemotherapy with or without radiotherapy could be used after curative resection, especially for patients who are node-positive or have a margin-positive resection (Rodriguez-Pascual et al. 2018). Systemic therapies have been also used for CCs according to patient remained liver function, tumor location, previous treatment, and patient performance status. Gemcitabine-based regimen (usually in combination with cisplatin) is considered first-line treatment; however, the choice for second-line treatments extremely lacked evidence (Valle et al. 2010). The incorporation of genetic predisposition in management is largely neglected; however, molecular targeted therapies are under evaluation based on the favorable genes with marked different expressions in BTC (Nakamura et al. 2015; Farshidfar et al. 2017) (Table 2).

3.5 Response to Treatment and Monitoring

Some predictors have been assessed regarding their prediction of treatment response. The most investigated predictor is AFP, in which higher baseline levels could predict poor prognosis (Konopka 1984; Galle et al. 2019). On the other hand, the changes in AFP would be another predictor of the prognosis of HCC patients, in which a decline

Table 2 Summary of clinical and radiological outcomes and adverse events among targeted therapies for cholangiocarcinoma

Drug	Mechanism	Year	No of participants	Design	Median OS (months)	Median PFS (months)	ORR	Grade III–IV adverse events	TrM	Findings	References
Approved drugs											
Infigratinib (BGJ398)	Selective pan-FGFR kinase inhibitor	2014–2021	61	Multicenter, open-label, phase 2	N/A	5.8	14.8%	41%	0	Promising clinical activity and a manageable adverse event for previously treated patients with FGFR2 gene fusions or rearrangements	Javle et al. (2018)
Pemigatinib	Pan-FGFR inhibitor	2017–2019	146	Multicenter, open-label, single-arm, multicohort, phase 2	21.1	6.9	35.5%	64%	0	Potential benefit of pemigatinib in previously treated patients with FGFR2 fusions or rearrangements	Abou-Alfa et al. (2020)
Ivosidenib (AG-120)	Selective inhibitor of mutant IDH1	2017–2020	187	Multicenter, randomized, double-blind, placebo-controlled, phase 3	10.3	3.8	2.7	50%	0	FDA approval for locally advanced or metastatic CC with an isocitrate dehydrogenase 1 (IDH1) mutation	Administration (2021) and Zhu et al. (2021)

(continued)

Table 2 (continued)

Drug	Mechanism	Year	No of participants	Design	Median OS (months)	Median PFS (months)	ORR	Grade III–IV adverse events	TrM	Findings	References
Under investigation											
FGFR-directed agents											
Derazantinib (ARQ 087)	FGFR 1–3 inhibitor	2009–2016	17	Open-label multicenter phase 1/2 dose escalation	N/A	N/A	47%	N/A	N/A	Derazantinib is safe with manageable mild side effects	Mazzaferro et al. (2016)
Derazantinib (ARQ 087)	FGFR 1–3 inhibitor	2014–2017	29	Multicenter, open-label study, phase 1/2	7	5.7	20.7%	27.6%	0	Favorable safety profile of derazantinib	Mazzaferro et al. (2019)
Derazantinib (ARQ 087)	FGFR 1–3 inhibitor	2017–2022	28	Phase 2 (FIDES-01)	N/A	7.3	8.7%	N/A	N/A	Promising effects of derazantinib for FGFR2 $^{M/A+}$ patients	Javle et al. (2022)
Debio 1347	Selective FGFR kinase inhibitor	2013–2020	18 (5 CC)	Phase 1 extension study	N/A	N/A	40%	0	0	Acceptable safety and promising activity of Debio 1347 for CC	Cleary et al. (2020)
TAS-120	Irreversible pan-FGFR inhibitor	2014–2018	45	Phase 1 dose-escalation study	N/A	N/A	25%	51%	0	Acceptable safety profile of TAS-120	Meric-Bernstam et al. (2018)
Ponatinib	Pan-FGFR inhibitor	2014–2019	12	Multicenter, single institution, pilot study	15.7	2.4	8.3%	0	0	Safety and limited clinical activity of ponatinib for altered FGFR BTC	Ahn et al. (2022)

Erdafitinib	Selective pan-FGFR kinase inhibitor	2016–2018	14	Open-label, multicenter, phase 2a	N/A	5.59	50%	50%	0	Encouraging efficacy and acceptable safety of erdafitinib	Park et al. (2019)
EGFR-directed agents											
Erlotinib	EGFR inhibitor	2009–2012	266	Multicenter, open-label, randomized, phase 3	9.5	5.8	29.62%	4%	0	No significant difference in overall and progression-free survival compared with standard regimen	Lee et al. (2012)
Erlotinib	EGFR inhibitor	2010–2011	40	Noncontrolled, phase 2	6	2	35%	15%	2%	The combination did not have promising results (study terminated)	El-Khoueiry et al. (2014)
Others											
Tivantinib (ARQ 197)	c-MET inhibitor	2009–2011	74	Phase 1, dose-escalation study	N/A	4.2	20%	73%	3%	Monotherapy and combination therapy of tivantinib are safe and tolerable	Pant et al. (2014)
Vemurafenib	Inhibitor of the BRAF V600 kinase	2012–2014	122 (8 CC)	Histology-independent, flexible, early phase 2 "basket" study	N/A	N/A	12%	88%	0	Vemurafenib could be used in *BRAF V600* mutation-positive cancers	Hyman et al. (2015)

(continued)

Table 2 (continued)

Drug	Mechanism	Year	No of participants	Design	Median OS (months)	Median PFS (months)	ORR	Grade III–IV adverse events	TrM	Findings	References
Dabrafenib/ trametinib	BRAF inhibitor/ MEK inhibitor	2014–2018	43	A phase 2, open-label, single-arm, multicenter basket trial	14	9	47–51%	21%	0	Promising efficacy and manageable side effects for BRAFV600E-mutated biliary tract cancer	Subbiah et al. (2020)
Selumetinib (AZD6244, ARRY-142886)	Selective MEK1/ 2 inhibitor	2012–2013	13	Phase 1 dose escalation	N/A	6.4	25%	25%	0	Safe dose of selumetinib is determined	Bridgewater et al. (2016)
Cabozantinib	VEGFR2 and HGF inhibitor	2013–2014	19	Single-arm phase 2, terminated	5.2	1.8	0	89%	0	Limited efficacy and significant toxicity of cabozantinib	Goyal et al. (2017)
Ceritinib	ALK/ROS1 inhibitor	2015–2018	38 (7 CC)	Phase 1, dose escalation	13.7–29.1	3.4–4.8	14.28%	N/A	N/A	Ceritinib and gemcitabine-based could be further considered for ALK/ROS1-positive CCs	Fountzilas et al. (2020)

Abbreviations: *N/A* not accessible, *VEGFR2* vascular endothelial growth factor receptor 2, *HGF* hepatocyte growth factor, *BTC* biliary tract cancer, *TrM* treatment-related mortality, *CC* cholangiocarcinoma, *IDH* isocitrate dehydrogenase, *FGFR* fibroblast growth factor receptor, *OS* overall survival, *PFS* progression-free survival, *EGFR* epidermal growth factor receptor, *ORR* overall response rate would be associated with longer PFS and OS and an increase would be associated with tumor progression (Chau et al. 2018; Galle et al. 2019).

Some genes were shown to be associated with tumor progression and poor prognosis, e.g., PKM, SPP1, and TPM1 for cholangiocarcinoma (Huang et al. 2016) and SPP1, KIF20A, HMMR, TPX2, LAPTM4B, TTK, MAGEA6, ANX10, LECT2, CYP2C9, RDH16, and LCAT for HCC (Ouyang et al. 2020).

Evaluation of response to treatment is another vital step for consideration of a good treatment strategy. Biochemical and radiological follow-ups of HCC patients have been used to assess the treatment effectiveness. RECIST has been used as a standard means of radiological response evaluation in cancer treatment. However, there were several limitations for using RECIST for HCC (Lencioni and Llovet 2010; Lencioni 2013), which in turn has resulted in the definition and application of a modified version of RECIST (mRECIST) for HCC (Llovet and Lencioni 2020). Response rates based on mRECIST were associated with better survival (Lencioni et al. 2017; Meyer et al. 2017a, b).

As mentioned in previous parts, quality of life in HCC patients is severely affected. The frequent underlying liver disease, along with side effects of treatment and tumor progressions, makes it a high burden disease (Llovet et al. 2021a, b, c). The frequency of treatment-related adverse events for systemic therapies is presented in Table 1 (Table 1). Treatment side effects mainly include immune side effects which potentially could affect any organs. The emergence of new drugs like immune checkpoint inhibitors with a good safety profile helped lower the treatment-related adverse events (Jacome et al. 2021). On the other hand, fatigue, pain, insomnia, anorexia, weight loss, and sexual dysfunction were the most common symptoms reported by patients (Llovet et al. 2021a, b, c). The most common tool for evaluation of the quality of life in HCC patients is the self-reported European Organization for Research and Treatment of Cancer Quality of Life Questionnaire Core 30 (EORTC QLQ-C30) as well as its HCC-specific questionnaire (EORTC QLQ-HCC18) (Blazeby et al. 2004). The score of these questionnaires greatly helped for cost-beneficial drug decision-making and patient-centered care delivery (Dong et al. 2021).

4 Importance and Rationale of a Multidisciplinary Approach

The dynamic and still-evolving nature of cancers showed the importance of the involvement of several providers with different specialties greater than before (Bruix et al. 2016). The traditional approach of cancer management, i.e., sequential separate visits by multiple disciplines, demonstrated several inefficiencies (Plsek and Greenhalgh 2001). Hence, recently, multidisciplinary management of cancers was at the center of attention as the new standard of care.

The HCC management, from the first steps of screening and surveillance to management and monitoring, is complex, given the underlying chronic liver disease and the inherent risk of mortality. However, early diagnosis, careful staging, and selecting the best treatment options have been shown to considerably reduce the disease burden and increase survival and response rate (Singal et al. 2014; Siddique et al. 2017).

As mentioned before, HCC screening and surveillance need the cooperation of primary healthcare systems advocated by gastroenterologists (Centers 2019; Llovet et al. 2021a, b, c). Also, HCC diagnosis could be best made by enterprise radiologists and skilled pathologists. The role of radiologists in HCC early diagnosis and pathologists, especially regarding suspicious samples in a cirrhotic liver background, has been highlighted by the American Association for the Study of Liver Diseases and the International Consensus Group for Hepatocellular Neoplasia, respectively (Kojiro et al. 2009; Marrero et al. 2018). Also, the assessment of underlying liver disease is usually made according to the Child-Pugh classification and the model for end-stage liver disease (MELD) score (Cucchetti et al. 2006; Garcia-Tsao and Lim 2009). Given the complex genetic and pathological factors and various available therapeutic options for HCC, as well as the occurrence of side effects during treatment, e.g., varices, ascites, spontaneous bacterial peritonitis, hepatorenal syndrome, and hepatic encephalopathy, the best treatment should be determined by the joint contribution of gastroenterologists, hepatologists, surgeons, diagnostic and interventional radiologists, oncologists, and transplant teams (Sherman et al. 2011; Gish et al. 2012; Guy et al. 2012; Schulze et al. 2015; Siddique et al. 2017). The role of radiologists in disease monitoring and progression, along with the useful biomarkers, profoundly helped in cases where a change in treatment is needed. Besides, the evolving nature of the disease and numerous problems affecting patients' quality of life necessitate the involvement of some other caregivers with expertise in nutrition, mental and psychological consultation, pain management, special nursing, and social working (Naugler et al. 2015; Siddique et al. 2017; Centers 2019). Management of HCC by a multidisciplinary team (MDT), consisting of the aforementioned experts, was recommended by different guidelines (Sherman et al. 2011; Gish et al. 2012; Liver 2018; Marrero et al. 2018; Centers 2019). Only with a multidisciplinary approach, it could be assured that diagnosis and treatment of HCC are accurate, well-timed, evidence-based, and patient-centered and reached by expert consensus, which is the definition of optimal management (Taylor et al. 2010).

5 Suggested Models for Interdisciplinary Management

The MDT is designed to successfully implement the ideal management of HCC. Several systems navigate the management of HCC including the Cancer of the Liver Italian Program, BCLC, the American Joint Committee on Cancer tumor-node-metastasis (TNM), and the China liver cancer staging system (Daniele et al. 2007; Kim et al. 2019; Xie et al. 2020; Llovet et al. 2021a, b, c). Although there are differences in their performance in different populations, the common thing is their multidisciplinary approach to HCC (Cho et al. 2008; Den Winkel et al. 2012).

The multidisciplinary tumor models were extensively reviewed by Byrd et al. (2021). There are multiple different MDT models, from classical tumor boards to the centralized group clinic models, in which patients attend a co-located clinic. Historically tumor boards are used for the presentation of challenging tumor cases, usually in a retrospective manner. The use of tumor boards is currently more widespread (Byrd et al. 2021), and the Commission on Cancer recommends the presentation of cases in tumor boards, preferentially in a prospective manner (Cancer 2020). The structure of an HCC multidisciplinary tumor board involves a core member of experts from different fields including hepatologists, transplant surgeons, hepatobiliary surgeons, interventional radiologists, and medical oncologists. Also, pathologists and translational researchers could be integrated given the progressions in new molecular therapies. Other experts with specialties like dermatology, endocrinology, and rheumatology could have a role in MDT, especially in the management of the treatment side effects. The last but not the least important are the specially trained nurses, social workers, and managers which could bring the MDT with higher patient engagement and adherence to treatment and, moreover, could facilitate the quality-of-life considerations. The multidisciplinary tumor board involves a real-time interview with the patient, followed by reviewing documents, radiology images, pathology findings, etc. to reach an optimal decision for treatment. This model of MDT improves the accessibility of patients to providers in a cost-effective way (Gadsden and Kaplan 2019; Byrd et al. 2021; Salgia and Mendiratta 2021).

Another model of MDT involves a centralized co-location of experts in which patients could visit necessary providers in one place. This structure has two major differences in comparison with the previous model, the formation of an actual team and a higher level of patient engagement. This structure could also be implemented via virtual meetings, through which patients could be sequentially attended by the providers (Byrd et al. 2021). This model provides the opportunity for patient-centered up-to-date multifaceted care that brings higher patient satisfaction and quality of life, provider satisfaction, and lower costs (Elrashidi et al. 2018) (Fig. 2).

6 Clinical Outcomes of Multidisciplinary Management

Since the introduction of multidisciplinary management of cancers in the 2000s, various tumors were discussed in MTD with promising results (Osarogiagbon et al. 2015; Freytag et al. 2020). Similarly, management of HCC through MDT has not been an exception, either for decision-making between providers or for patient satisfaction. It has been shown that MDT could promote communication between providers and decrease the cumulative time to reach optimal management (Yopp et al. 2014). Also, in case of any treatment failure, changing treatments could be more agile and timely. Besides, it profoundly improves the consideration of curative treatment or combination therapies for all the patients, which were shown to have better outcomes compared with other conventional treatments (Byrd et al. 2021). Moreover, having multiple opinions from different specialists could help to reach a more holistic view of patient conditions and preferences, which in turn could result in better acceptance of treatment regimen and more long-term adherence to

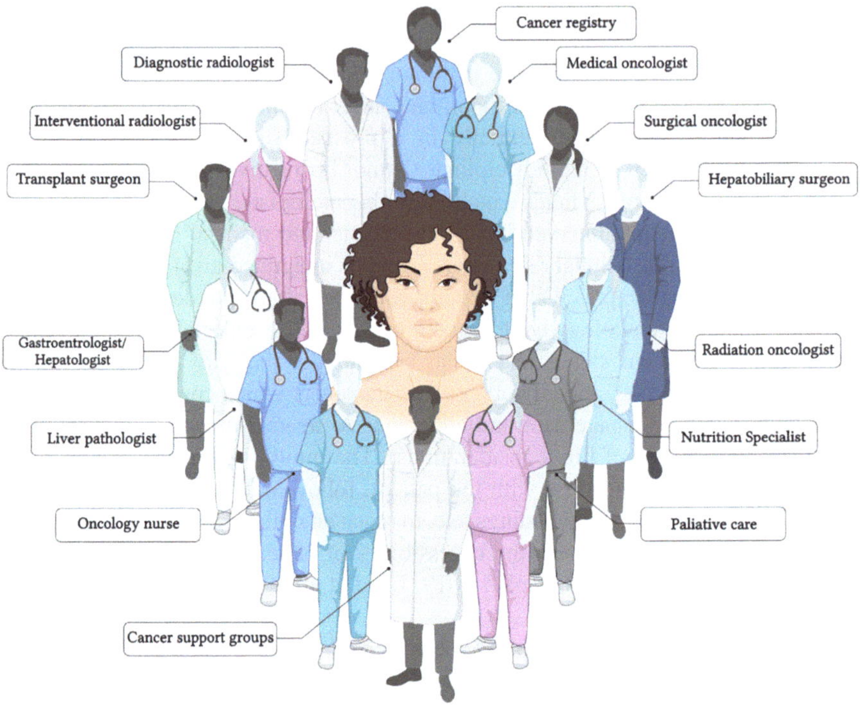

Fig. 2 Multidisciplinary liver tumor board for the management of hepatobiliary cancers

treatment, and higher patient overall satisfaction (Van Cleave et al. 1999; Chang et al. 2008; Burak and Kneteman 2010; Zhang et al. 2013; Gashin et al. 2014; Agarwal et al. 2017; Serper et al. 2017; Duininck et al. 2019; Sinn et al. 2019). Other outcomes of interdisciplinary management of HCC are presented in Table 3.

7 Challenges of Multidisciplinary Care

The multidisciplinary care for patients has also faced some challenges. Although there are several advantages for multidisciplinary care, the successful implementation of these models demands overcoming some issues. First of all, MDT needs strong leadership, through which it is assured that the treatment decision-making actually takes place based on the opinions of different disciplines and expertise and is not governed by some specialties (Soukup et al. 2018). Administrative issues regarding the management of tumor boards or centralized clinics would be another challenge. Planned visit flows – who and the order of visiting providers – and incorporating clinical coordinators, who would be in close touch with patients, could be problem-solving strategies in this regard (Lamb et al. 2011). Moreover, clinics need a minimum necessary size to include the essential disciplines, which

Table 3 The outcomes of HCC treatment through a multidisciplinary care

Study	Country	Institute(s)	Providers	Period	No of patients	Outcomes	References
Yopp et al.	USA	Parkland Memorial Health and Hospital System (PHHS), the safety-net hospital system for Dallas County, Texas, UT Southwestern Medical Center	Surgical oncology, transplant hepatology, interventional radiology, diagnostic radiology, radiation oncology, medical oncology, and palliative care	Pre: 2006–2010 Post: 2010–2011	Pre: 250 Post: 105	Shorter time to treatment Reduced stage-adjusted mortality	Yopp et al. (2014)
Serper et al.	USA	Multicentric, 128 Veterans Affairs centers	Hepatologists, gastroenterologists, surgeons, oncologists	2008–2014	Total: 3,988 MDT: 1,366 Multidisciplinary care: 2,155	Increased treatment receipt Reduced mortality	Serper et al. (2017)
Chang et al.	USA	San Francisco Veterans Affairs Medical Center	Surgeons, radiologists, medical oncologists, and gastroenterologists	Pre: 2000–2003 Post: 2003–2006	Pre: 62 Post: 121	Increased treatment receipt Reduced mortality	Chang et al. (2008)
Sinn et al.	Korea	Samsung Medical Center, Seoul	Hepatologists, surgeons, diagnostic radiologists, interventional radiologists	Pre: 2000–2005 Post: 2005–2013	Pre: 5,881 Post: 738	Reduced mortality	Sinn et al. (2019)
Agarwal et al.	USA	University of Wisconsin	Transplant hepatologists, medical oncologists, hepatobiliary and transplant surgeons, pathologists, diagnostic and interventional radiologists	Pre: 2002–2011 Post: 2007–2011	Pre: 349 Post: 306	Increased treatment receipt Reduced mortality	Agarwal et al. (2017)

(continued)

Table 3 (continued)

Study	Country	Institute(s)	Providers	Period	No of patients	Outcomes	References
Gashin et al.	USA	Beth Israel Deaconess Medical Center	Five hepatologists, three oncologists, one radiation oncologist, three interventional radiologists, one pathologist, three surgeons, three radiologists, and five mid-level staff including nurses, nurse practitioners, and physician assistants	2009–2010	137	Increased treatment receipt Reduced mortality Not following MDT decision was a negative prognostic factor	Gashin et al. (2014)
Duininck et al.	USA	Grady Memorial Hospital	Surgical oncologist, interventional radiologist, hepatologist, medical oncologists, radiation oncologists, and internal medicine physicians treating hepatitis C	Pre: 2009–2012 Post: 2013–2016	Pre: 70 Post: 134	Increased treatment receipt Reduced mortality	Duininck et al. (2019)
Zhang et al.	USA	Johns Hopkins Liver Tumor Center	Surgical oncologist, medical oncologist, radiation oncologist, radiologist, pathologist, interventional radiologist, hepatologist, and transplant surgeon	2009–2012	343	Alterations to imaging and pathology Interpretation for diagnosis Changes in the management plan	Zhang et al. (2013)

Charriere et al.	France	Toulouse University Hospital	Hepatology, oncology, hepatobiliary surgery, transplantation, and radiology	2006–2013	387	Not following MDT decision was a negative prognostic factor	Charriere et al. (2017)
Cohen et al.	USA	Temple University Hospital	Hepatologists, interventional radiologists, transplant surgeons, oncologists, residents, mid-level providers, and support staff	2013	About 50 pts	Higher patients' confidence in individualized treatment Higher survival and quality of life	Cohen and Black (2013)
Van Cleave et al.	USA	Philadelphia Veterans Affairs Medical Center	Physicians (radiology, surgical oncology, radiation oncology, medical oncology), nurses, pharmacists, social workers, and a chaplain	1995–1998	MDT: 18 pts. Non-MDT: 4 pts	Higher patient satisfaction Patient-centered care	Van Cleave et al. (1999)
Bunnell et al.	USA	Dana-Farber Cancer Institute	Surgeons, medical oncologists, radiation oncologists, dermatologists, nurse practitioners, and physician assistants	1997–2010	141 pts	High patient satisfaction A comprehensive, coordinated, and appropriate care	Bunnell et al. (2010)

Abbreviations: *MDT* multidisciplinary tumor board

made it difficult to be implemented, especially in small settings. In a smaller community environment, gathering specialists together may be difficult as they should travel some distance to be in the same location. Additionally, the clinical revenue of providers could be affected, because of the fewer number of patients a provider can visit. Finally, the legal rules governing these MDT have yet to be issued, which could help in making MDT members more committed and responsible for the consensus decision. Taken together, these centralized clinics often need so many resources, funding, workforce, and of course time, which may limit their widespread implementation across different healthcare systems. The limited number of these MDTs and prolonged time of reaching a consensus on patient-centered treatment – compared to the traditional single-person decision-making – significantly have threatened the coverage of multidisciplinary care (Byrd et al. 2021, Salgia and Mendiratta 2021).

8 Future Perspectives

The implementation of interdisciplinary and multidisciplinary care for hepatobiliary cancers should be further assessed in different settings regarding its effects on patient satisfaction, patients' knowledge about their care, the burden of disease, and also its burden on providers. The importance of multidisciplinary care has led to its inclusion in different guidelines as a common point.

Although the multidisciplinary approach is inherently costly and time-consuming and needs a decent specialized workforce, strategies are underway to make it more feasible, especially in low-income areas. Furthermore, the lack of a comprehensive team of regional specialists or the necessity of traveling a long way to form centralized MDT opens a horizon on virtual tumor boards and virtual MDT (vMDT). vMDT could significantly decrease the costs and increase the feasibility, thus accessibility of patients to multidisciplinary care. It also enables to expand the interdisciplinary care not only to new cancer cases but also to existing cases who need revision of their treatment. vMDT could also ease the contribution of more disciplines from anywhere, as needed, for example, management of treatment side effects which possibly need providers with other specialties.

The management of patients in MDTs also could provide a good opportunity to do the follow-ups more frequently and closely by various providers. For example, a patient may be planned to be attended by a nutritionist, while the medical oncologist could also do a quick unplanned follow-up if it is necessary.

9 Conclusions

Hepatobiliary cancers are a group of heterogeneous cancers, the most investigated of which is hepatocellular cancer. HCC management is based on a combination of tumor burden, degree of liver dysfunction, and patient performance status. According to the tumor stage, various treatment options could be selected, ranging from surgical therapies to ablations and systemic therapies. The optimal management would be achieved through discussion between several clinicians with different specialties related to each step of management, e.g., screening, diagnosis, and treatment. Furthermore, the increasing number of trials evaluating combination therapies, as well as the increasing number of transitions between therapies, necessitates close collaboration and communication between these specialties. As a result, multidisciplinary care has become more popular for HCC management, though it can take various forms, such as multidisciplinary tumor boards, or centralized clinics. Multidisciplinary care has been shown in studies to result in higher patient satisfaction, improved receipt of timely evidence-based care, and increased overall survival, indicating that it should be considered the standard of care for the management of hepatobiliary cancers.

References

Abou-Alfa GK, Meyer T, Cheng AL, El-Khoueiry AB, Rimassa L, Ryoo BY, Cicin I, Merle P, Chen Y, Park JW, Blanc JF, Bolondi L, Klumpen HJ, Chan SL, Zagonel V, Pressiani T, Ryu MH, Venook AP, Hessel C, Borgman-Hagey AE, Schwab G, Kelley RK (2018) Cabozantinib in patients with advanced and progressing hepatocellular carcinoma. N Engl J Med 379:54–63

Abou-Alfa GK, Sahai V, Hollebecque A, Vaccaro G, Melisi D, Al-Rajabi R, Paulson AS, Borad MJ, Gallinson D, Murphy AG, Oh DY, Dotan E, Catenacci DV, Van Cutsem E, Ji T, Lihou CF, Zhen H, Feliz L, Vogel A (2020) Pemigatinib for previously treated, locally advanced or metastatic cholangiocarcinoma: a multicentre, open-label, phase 2 study. Lancet Oncol 21: 671–684

Administration USFaD (2021) FDA approves ivosidenib for advanced or metastatic cholangiocarcinoma | FDA. Retrieved 6/4, 2022, from https://www.fda.gov/drugs/resources-information-approved-drugs/fda-approves-ivosidenib-advanced-or-metastatic-cholangiocarcinoma

Agarwal PD, Phillips P, Hillman L, Lucey MR, Lee F, Mezrich JD, Said A (2017) Multidisciplinary Management of Hepatocellular Carcinoma Improves Access to Therapy and Patient Survival. J Clin Gastroenterol 51:845–849

Ahn JC, Yang JD (2019) Screening indications and treatments for cholangiocarcinoma. Curr Hepatol Rep 18:408–416

Ahn DH, Uson Junior PLS, Masci P, Kosiorek H, Halfdanarson TR, Mody K, Babiker H, DeLeon T, Sonbol MB, Gores G, Smoot R, Bekaii-Saab T, Mahipal A, Mansfield A, Tran NH, Hubbard JM, Borad MJ (2022) A pilot study of Pan-FGFR inhibitor ponatinib in patients with FGFR-altered advanced cholangiocarcinoma. Investig New Drugs 40:134–141

Amin MB, Edge SB, Greene FL, Schilsky RL, Brookland RK, Washington MK, Gershenwald JE, Compton CC, Hess KR, Sullivan DC, Jessup MJ, Brierley JD, Gaspar LE, Balch CM, Winchester DP, Asare EA, Madera M, Gress DM, Meyer LR (2017) American Joint Committee on Cancer (AJCC). AJCC Cancer Staging Manual

Andersson KL, Salomon JA, Goldie SJ, Chung RT (2008) Cost effectiveness of alternative surveillance strategies for hepatocellular carcinoma in patients with cirrhosis. Clin Gastroenterol Hepatol 6:1418–1424

Atiq O, Tiro J, Yopp AC, Muffler A, Marrero JA, Parikh ND, Murphy C, McCallister K, Singal AG (2017) An assessment of benefits and harms of hepatocellular carcinoma surveillance in patients with cirrhosis. Hepatology 65:1196–1205

Ayuso C, Rimola J, Vilana R, Burrel M, Darnell A, Garcia-Criado A, Bianchi L, Belmonte E, Caparroz C, Barrufet M, Bruix J, Bru C (2018) Diagnosis and staging of hepatocellular carcinoma (HCC): current guidelines. Eur J Radiol 101:72–81

Blazeby JM, Currie E, Zee BC, Chie WC, Poon RT, Garden OJ, Group EQoL (2004) Development of a questionnaire module to supplement the EORTC QLQ-C30 to assess quality of life in patients with hepatocellular carcinoma, the EORTC QLQ-HCC18. Eur J Cancer 40:2439–2444

Borghero Y, Crane CH, Szklaruk J, Oyarzo M, Curley S, Pisters PW, Evans D, Abdalla EK, Thomas MB, Das P, Wistuba II, Krishnan S, Vauthey JN (2008) Extrahepatic bile duct adenocarcinoma: patients at high-risk for local recurrence treated with surgery and adjuvant chemoradiation have an equivalent overall survival to patients with standard-risk treated with surgery alone. Ann Surg Oncol 15:3147–3156

Bray F, Ferlay J, Soerjomataram I, Siegel RL, Torre LA, Jemal A (2018) Global cancer statistics 2018: GLOBOCAN estimates of incidence and mortality worldwide for 36 cancers in 185 countries. CA Cancer J Clin 68:394–424

Breen DJ, Lencioni R (2015) Image-guided ablation of primary liver and renal tumours. Nat Rev Clin Oncol 12:175–186

Breitenstein S, Dimitroulis D, Petrowsky H, Puhan MA, Mullhaupt B, Clavien PA (2009) Systematic review and meta-analysis of interferon after curative treatment of hepatocellular carcinoma in patients with viral hepatitis. Br J Surg 96:975–981

Bridgewater J, Lopes A, Beare S, Duggan M, Lee D, Ricamara M, McEntee D, Sukumaran A, Wasan H, Valle JW (2016) A phase 1b study of Selumetinib in combination with Cisplatin and Gemcitabine in advanced or metastatic biliary tract cancer: the ABC-04 study. BMC Cancer 16:153

Bruix J, Takayama T, Mazzaferro V, Chau G-Y, Yang J, Kudo M, Cai J, Poon R T-P, Han K-H, Tak W-Y, Lee HC, Song T, Roayaie S, Bolondi L, Lee KS, Makuuchi M, Souza F, Le Berre M-A, Meinhardt G, Llovet JM, STORM Investigators (2014) J Clin Oncol 32(15_suppl):4006

Bruix J, Reig M, Sherman M (2016) Evidence-based diagnosis, staging, and treatment of patients with hepatocellular carcinoma. Gastroenterology 150:835–853

Bruix J, Qin S, Merle P, Granito A, Huang YH, Bodoky G, Pracht M, Yokosuka O, Rosmorduc O, Breder V, Gerolami R, Masi G, Ross PJ, Song T, Bronowicki JP, Ollivier-Hourmand I, Kudo M, Cheng AL, Llovet JM, Finn RS, LeBerre MA, Baumhauer A, Meinhardt G, Han G, Investigators R (2017) Regorafenib for patients with hepatocellular carcinoma who progressed on sorafenib treatment (RESORCE): a randomised, double-blind, placebo-controlled, phase 3 trial. Lancet 389:56–66

Bujold A, Massey CA, Kim JJ, Brierley J, Cho C, Wong RK, Dinniwell RE, Kassam Z, Ringash J, Cummings B, Sykes J, Sherman M, Knox JJ, Dawson LA (2013) Sequential phase I and II trials of stereotactic body radiotherapy for locally advanced hepatocellular carcinoma. J Clin Oncol 31:1631–1639

Bunnell CA, Weingart SN, Swanson S, Mamon HJ, Shulman LN (2010) Models of multidisciplinary cancer care: physician and patient perceptions in a comprehensive cancer center. J Oncol Pract 6:283–288

Burak KW, Kneteman NM (2010) An evidence-based multidisciplinary approach to the management of hepatocellular carcinoma (HCC): the Alberta HCC algorithm. Can J Gastroenterol 24: 643–650

Burke EC, Jarnagin WR, Hochwald SN, Pisters PW, Fong Y, Blumgart LH (1998) Hilar cholangiocarcinoma: patterns of spread, the importance of hepatic resection for curative operation, and a presurgical clinical staging system. Ann Surg 228:385–394

Byrd K, Alqahtani S, Yopp AC, Singal AG (2021) Role of multidisciplinary Care in the Management of hepatocellular carcinoma. Semin Liver Dis 41:1–8

Cancer ACoSCo (2020) Optimal resources for cancer care (2020 standards). Chicago, American College of Surgeons Commission on Cancer

Cassinotto C, Aube C, Dohan A (2017) Diagnosis of hepatocellular carcinoma: An update on international guidelines. Diagn Interv Imaging 98:379–391

Centers AoCC (2019) Multidisciplinary hepatocellular carcinoma care. Association of Community Cancer Centers (ACCC)

Chang TT, Sawhney R, Monto A, Davoren JB, Kirkland JG, Stewart L, Corvera CU (2008) Implementation of a multidisciplinary treatment team for hepatocellular cancer at a veterans affairs medical center improves survival. HPB (Oxford) 10:405–411

Charriere B, Muscari F, Maulat C, Bournet B, Bonnet D, Bureau C, Otal P, Peron JM, Suc B (2017) Outcomes of patients with hepatocellular carcinoma are determined in multidisciplinary team meetings. J Surg Oncol 115:330–336

Chau I, Park JO, Ryoo BY, Yen CJ, Poon R, Pastorelli D, Blanc JF, Kudo M, Pfiffer T, Hatano E, Chung HC, Kopeckova K, Phelip JM, Brandi G, Ohkawa S, Li CP, Okusaka T, Hsu Y, Abada PB, Zhu AX (2018) Alpha-fetoprotein kinetics in patients with hepatocellular carcinoma receiving ramucirumab or placebo: an analysis of the phase 3 REACH study. Br J Cancer 119:19–26

Cheng RG, Bhattacharya R, Yeh MM, Padia SA (2015) Irreversible electroporation can effectively ablate hepatocellular carcinoma to complete pathologic necrosis. J Vasc Interv Radiol 26:1184–1188

Cho YK, Chung JW, Kim JK, Ahn YS, Kim MY, Park YO, Kim WT, Byun JH (2008) Comparison of 7 staging systems for patients with hepatocellular carcinoma undergoing transarterial chemoembolization. Cancer 112:352–361

Choi J, Ghoz HM, Peeraphatdit T, Baichoo E, Addissie BD, Harmsen WS, Therneau TM, Olson JE, Chaiteerakij R, Roberts LR (2016) Aspirin use and the risk of cholangiocarcinoma. Hepatology 64:785–796

Cillo U, Vitale A, Grigoletto F, Farinati F, Brolese A, Zanus G, Neri D, Boccagni P, Srsen N, D'Amico F, Ciarleglio FA, Bridda A, D'Amico DF (2006) Prospective validation of the Barcelona Clinic Liver Cancer staging system. J Hepatol 44:723–731

Cleary JM, Iyer G, Oh D-Y, Mellinghoff IK, Goyal L, Ng MCH, Meric-Bernstam F, Matos I, Chao T-Y, Ait Sarkouh R, Cretegny K, Nicolas-Metral V, Pokorska-Bocci A, Vaslin A, Zanna C, Zubel A, Tabernero J, Flaherty K, Hyman DM (2020) Final results from the phase I study expansion cohort of the selective FGFR inhibitor Debio 1,347 in patients with solid tumors harboring an FGFR gene fusion. J Clin Oncol 38:3603–3603

Clements O, Eliahoo J, Kim JU, Taylor-Robinson SD, Khan SA (2020) Risk factors for intrahepatic and extrahepatic cholangiocarcinoma: a systematic review and meta-analysis. J Hepatol 72:95–103

Cohen GS, Black M (2013) Multidisciplinary management of hepatocellular carcinoma: a model for therapy. J Multidiscip Healthc 6:189–195

Colagrande S, Inghilesi AL, Aburas S, Taliani GG, Nardi C, Marra F (2016) Challenges of advanced hepatocellular carcinoma. World J Gastroenterol 22:7645–7659

Cucchetti A, Ercolani G, Vivarelli M, Cescon M, Ravaioli M, La Barba G, Zanello M, Grazi GL, Pinna AD (2006) Impact of model for end-stage liver disease (MELD) score on prognosis after hepatectomy for hepatocellular carcinoma on cirrhosis. Liver Transpl 12:966–971

Cucchetti A, Serenari M, Sposito C, Di Sandro S, Mosconi C, Vicentin I, Garanzini E, Mazzaferro V, De Carlis L, Golfieri R, Spreafico C, Vanzulli A, Buscemi V, Ravaioli M, Ercolani G, Pinna AD, Cescon M (2020) Including mRECIST in the Metroticket 2.0 criteria improves prediction of hepatocellular carcinoma-related death after liver transplant. J Hepatol 73:342–348

Daniele B, Annunziata M, Barletta E, Tinessa V, Di Maio M (2007) Cancer of the Liver Italian Program (CLIP) score for staging hepatocellular carcinoma. Hepatol Res 37(Suppl 2):S206–S209

Dasgupta P, Henshaw C, Youlden DR, Clark PJ, Aitken JF, Baade PD (2020) Global trends in incidence Rates of primary adult Liver cancers: a systematic review and meta-analysis. Front Oncol 10:171

Den Winkel MO, Nagel D, Den Winkel PO, Lamerz R, Zech CJ, Straub G, Nickel T, Rentsch M, Stieber P, Kolligs FT (2012) Prognosis of patients with hepatocellular carcinoma. Validation and ranking of established staging-systems in a large western HCC-cohort. PLoS One 7:e45066

Di Costanzo GG, Tortora R, D'Adamo G, De Luca M, Lampasi F, Addario L, Galeota Lanza A, Picciotto FP, Tartaglione MT, Cordone G, Imparato M, Mattera S, Pacella CM (2015) Radiofrequency ablation versus laser ablation for the treatment of small hepatocellular carcinoma in cirrhosis: a randomized trial. J Gastroenterol Hepatol 30:559–565

Dimitroulis D, Damaskos C, Valsami S, Davakis S, Garmpis N, Spartalis E, Athanasiou A, Moris D, Sakellariou S, Kykalos S, Tsourouflis G, Garmpi A, Delladetsima I, Kontzoglou K, Kouraklis G (2017) From diagnosis to treatment of hepatocellular carcinoma: An epidemic problem for both developed and developing world. World J Gastroenterol 23:5282–5294

Dong Y, Wong JSL, Sugimura R, Lam KO, Li B, Kwok GGW, Leung R, Chiu JWY, Cheung TT, Yau T (2021) Recent advances and future prospects in immune checkpoint (ICI)-based combination therapy for advanced HCC. Cancers (Basel) 13:1949

Duffy AG, Greten TF (2017) Treating hepatobiliary Cancer: the immunologic approach. Dig Dis 35:390–396

Duininck G, Lopez-Aguiar AG, Lee RM, Miller L, Dariushnia S, Wu C, Alese OB, Lin JY, Wedd J, Adams A, Maithel SK, Russell MC (2019) Optimizing cancer care for hepatocellular carcinoma at a safety-net hospital: the value of a multidisciplinary disease management team. J Surg Oncol 120:1365–1370

El-Khoueiry AB, Rankin C, Siegel AB, Iqbal S, Gong IY, Micetich KC, Kayaleh OR, Lenz HJ, Blanke CD (2014) S0941: a phase 2 SWOG study of sorafenib and erlotinib in patients with advanced gallbladder carcinoma or cholangiocarcinoma. Br J Cancer 110:882–887

Elrashidi MY, Mohammed K, Bora PR, Haydour Q, Farah W, DeJesus R, Murad MH, Ebbert JO (2018) Co-located specialty care within primary care practice settings: a systematic review and meta-analysis. Healthc (Amst) 6:52–66

Farshidfar F, Zheng S, Gingras MC, Newton Y, Shih J, Robertson AG, Hinoue T, Hoadley KA, Gibb EA, Roszik J, Covington KR, Wu CC, Shinbrot E, Stransky N, Hegde A, Yang JD, Reznik E, Sadeghi S, Pedamallu CS, Ojesina AI, Hess JM, Auman JT, Rhie SK, Bowlby R, Borad MJ, Cancer Genome Atlas N, Zhu AX, Stuart JM, Sander C, Akbani R, Cherniack AD, Deshpande V, Mounajjed T, Foo WC, Torbenson MS, Kleiner DE, Laird PW, Wheeler DA, McRee AJ, Bathe OF, Andersen JB, Bardeesy N, Roberts LR, Kwong LN (2017) Integrative genomic analysis of cholangiocarcinoma identifies distinct IDH-mutant molecular profiles. Cell Rep 18:2780–2794

Ferlay J, Ervik M, Lam F, Colombet M, Mery L, Piñeros M, Znaor A, Soerjomataram I, Bray F (2018) Global cancer observatory: cancer today. International Agency for Research on Cancer, Lyon, pp 1–6

Finn RS, Qin S, Ikeda M, Galle PR, Ducreux M, Kim TY, Kudo M, Breder V, Merle P, Kaseb AO, Li D, Verret W, Xu DZ, Hernandez S, Liu J, Huang C, Mulla S, Wang Y, Lim HY, Zhu AX, Cheng AL, Investigators IM (2020) Atezolizumab plus Bevacizumab in Unresectable Hepatocellular Carcinoma. N Engl J Med 382:1894–1905

Finn RS, Qin S, Ikeda M, Galle PR, Ducreux M, Kim T-Y, Lim HY, Kudo M, Breder VV, Merle P, Kaseb AO, Li D, Verret W, Shao H, Liu J, Li L, Zhu AX, Cheng A-L (2021) IMbrave150: updated overall survival (OS) data from a global, randomized, open-label phase III study of atezolizumab (atezo) + bevacizumab (bev) versus sorafenib (sor) in patients (pts) with unresectable hepatocellular carcinoma (HCC). J Clin Oncol 39:267–267

Forner A, Vilana R, Ayuso C, Bianchi L, Solé M, Ayuso JR, Boix L, Sala M, Varela M, Llovet JM (2008) Diagnosis of hepatic nodules 20 mm or smaller in cirrhosis: prospective validation of the noninvasive diagnostic criteria for hepatocellular carcinoma. Hepatology 47:97–104

Fountzilas C, Adjei A, Opyrchal M, Evans R, Attwood K, Goey A, Wilton J, Ma WW, Iyer R (2020) Abstract CT139: Ceritinib (Cer) in combination with gemcitabine (Gem)-based chemotherapy in patients (pts) with advanced solid tumors, a phase I study. Cancer Res 80:CT139

Franssen B, Jibara G, Tabrizian P, Schwartz ME, Roayaie S (2014) Actual 10-year survival following hepatectomy for hepatocellular carcinoma. HPB (Oxford) 16:830–835

Frenette CT, Isaacson AJ, Bargellini I, Saab S, Singal AG (2019) A practical guideline for hepatocellular carcinoma screening in patients at risk. Mayo Clin Proc Innov Qual Outcomes 3:302–310

Freytag M, Herrlinger U, Hauser S, Bauernfeind FG, Gonzalez-Carmona MA, Landsberg J, Buermann J, Vatter H, Holderried T, Send T, Schumacher M, Koscielny A, Feldmann G, Heine M, Skowasch D, Schafer N, Funke B, Neumann M, Schmidt-Wolf IGH (2020) Higher number of multidisciplinary tumor board meetings per case leads to improved clinical outcome. BMC Cancer 20:355

Fund WCR (2022) Liver cancer statistics. World Cancer Research Fund International

Gadsden MM, Kaplan DE (2019) Multidisciplinary approach to HCC management: how can this be done? Dig Dis Sci 64:968–975

Galle PR, Foerster F, Kudo M, Chan SL, Llovet JM, Qin S, Schelman WR, Chintharlapalli S, Abada PB, Sherman M, Zhu AX (2019) Biology and significance of alpha-fetoprotein in hepatocellular carcinoma. Liver Int 39:2214–2229

Garcia-Tsao G, Lim JK, Members of Veterans Affairs Hepatitis CRCP (2009) Management and treatment of patients with cirrhosis and portal hypertension: recommendations from the Department of Veterans Affairs Hepatitis C Resource Center Program and the National Hepatitis C Program. Am J Gastroenterol 104:1802–1829

Gashin L, Tapper E, Babalola A, Lai KC, Miksad R, Malik R, Cohen E (2014) Determinants and outcomes of adherence to recommendations from a multidisciplinary tumour conference for hepatocellular carcinoma. HPB (Oxford) 16:1009–1015

Gill J, Baiceanu A, Clark PJ, Langford A, Latiff J, Yang PM, Yoshida EM, Kanavos P (2018) Insights into the hepatocellular carcinoma patient journey: results of the first global quality of life survey. Future Oncol 14:1701–1710

Gish RG, Lencioni R, Di Bisceglie AM, Raoul JL, Mazzaferro V (2012) Role of the multidisciplinary team in the diagnosis and treatment of hepatocellular carcinoma. Expert Rev Gastroenterol Hepatol 6:173–185

Golabi P, Fazel S, Otgonsuren M, Sayiner M, Locklear CT, Younossi ZM (2017) Mortality assessment of patients with hepatocellular carcinoma according to underlying disease and treatment modalities. Medicine (Baltimore) 96:e5904

Goyal L, Zheng H, Yurgelun MB, Abrams TA, Allen JN, Cleary JM, Knowles M, Regan E, Reardon A, Khachatryan A, Jain RK, Nardi V, Borger DR, Duda DG, Zhu AX (2017) A phase 2 and biomarker study of cabozantinib in patients with advanced cholangiocarcinoma. Cancer 123:1979–1988

Guy J, Kelley RK, Roberts J, Kerlan R, Yao F, Terrault N (2012) Multidisciplinary management of hepatocellular carcinoma. Clin Gastroenterol Hepatol 10:354–362

Hainaut P, Amadou A, Gormally E (2019) Cancer prevention and control: hepatocellular carcinoma. Ecancermedicalscience 13:949

Hennedige TP, Neo WT, Venkatesh SK (2014) Imaging of malignancies of the biliary tract- an update. Cancer Imaging 14:14

Heo J, Chao Y, Jonker DJ, Baron AD, Habersetzer F, Burke J, Breitbach C, Patt RH, Lencioni R, Homerin M, Limacher J-M, Lusky M, Hickman T, Longpre L, Kirn DH (2013) Phase IIb randomized trial of Pexa-Vec (pexastimogene devacirepvec; JX-594), a targeted oncolytic vaccinia virus, plus best supportive care (BSC) versus BSC alone in patients with advanced hepatocellular carcinoma who have failed sorafenib treatment (TRAVERSE). J Clin Oncol 31 (15_suppl):TPS4161

Hilgard P, Hamami M, Fouly AE, Scherag A, Muller S, Ertle J, Heusner T, Cicinnati VR, Paul A, Bockisch A, Gerken G, Antoch G (2010) Radioembolization with yttrium-90 glass microspheres in hepatocellular carcinoma: European experience on safety and long-term survival. Hepatology 52:1741–1749

Hong TS, Wo JY, Yeap BY, Ben-Josef E, McDonnell EI, Blaszkowsky LS, Kwak EL, Allen JN, Clark JW, Goyal L, Murphy JE, Javle MM, Wolfgang JA, Drapek LC, Arellano RS, Mamon HJ, Mullen JT, Yoon SS, Tanabe KK, Ferrone CR, Ryan DP, DeLaney TF, Crane CH, Zhu AX (2016) Multi-institutional phase II study of high-dose Hypofractionated proton beam therapy in patients with localized, Unresectable hepatocellular carcinoma and intrahepatic cholangiocarcinoma. J Clin Oncol 34:460–468

Howlader N, Noone AM, Krapcho M, Miller D, Brest A, Yu M, Ruhl J, Tatalovich Z, Mariotto A, Lewis DR, Chen HS, Feuer EJ, Cronin KA (eds) (2020) SEER cancer statistics review, 1975–2017. National Cancer Institute, Bethesda. https://seer.cancer.gov/csr/1975_2017/, based on November 2019 SEER data submission, posted to the SEER web site, April 2020

Hu J, Chen S, Wang X, Lin N, Yang J, Wu S (2019) Image-guided percutaneous microwave ablation versus Cryoablation for hepatocellular carcinoma in high-risk locations: intermediate-term results. Cancer Manag Res 11:9801–9811

Huang QX, Cui JY, Ma H, Jia XM, Huang FL, Jiang LX (2016) Screening of potential biomarkers for cholangiocarcinoma by integrated analysis of microarray data sets. Cancer Gene Ther 23: 48–53

Hyman DM, Puzanov I, Subbiah V, Faris JE, Chau I, Blay JY, Wolf J, Raje NS, Diamond EL, Hollebecque A, Gervais R, Elez-Fernandez ME, Italiano A, Hofheinz RD, Hidalgo M, Chan E, Schuler M, Lasserre SF, Makrutzki M, Sirzen F, Veronese ML, Tabernero J, Baselga J (2015) Vemurafenib in multiple nonmelanoma cancers with BRAF V600 mutations. N Engl J Med 373:726–736

Imamura H, Matsuyama Y, Tanaka E, Ohkubo T, Hasegawa K, Miyagawa S, Sugawara Y, Minagawa M, Takayama T, Kawasaki S, Makuuchi M (2003) Risk factors contributing to early and late phase intrahepatic recurrence of hepatocellular carcinoma after hepatectomy. J Hepatol 38:200–207

Izumi N, Hasegawa K, Nishioka Y, Takayama T, Yamanaka N, Kudo M, Shimada M, Inomata M, Kaneko S, Baba H, Koike K, Omata M, Makuuchi M, Matsuyama Y, Kokudo N (2019) A multicenter randomized controlled trial to evaluate the efficacy of surgery vs. radiofrequency ablation for small hepatocellular carcinoma (SURF trial). J Clin Oncol 37(15_suppl):4002

Jacome AA, Castro ACG, Vasconcelos JPS, Silva M, Lessa MAO, Moraes ED, Andrade AC, Lima FMT, Farias JPF, Gil RA, Prolla G, Garicochea B (2021) Efficacy and safety associated with immune checkpoint inhibitors in unresectable hepatocellular carcinoma: a meta-analysis. JAMA Netw Open 4:e2136128

Javle M, Lowery M, Shroff RT, Weiss KH, Springfeld C, Borad MJ, Ramanathan RK, Goyal L, Sadeghi S, Macarulla T, El-Khoueiry A, Kelley RK, Borbath I, Choo SP, Oh DY, Philip PA, Chen LT, Reungwetwattana T, Van Cutsem E, Yeh KH, Ciombor K, Finn RS, Patel A, Sen S, Porter D, Isaacs R, Zhu AX, Abou-Alfa GK, Bekaii-Saab T (2018) Phase II study of BGJ398 in patients with FGFR-altered advanced cholangiocarcinoma. J Clin Oncol 36:276–282

Javle MM, Abou-Alfa GK, Macarulla T, Personeni N, Adeva J, Bergamo F, Malka D, Vogel A, Knox JJ, Evans TRJ, Dimova-Dobreva M, Harris WP, Saulay M, Engelhardt M, Braun S, Droz Dit Busset M, Borad MJ (2022) Efficacy of derazantinib in intrahepatic cholangiocarcinoma patients with FGFR2 mutations or amplifications: interim results from the phase 2 study FIDES-01. J Clin Oncol 40:427–427

Joshi A, Rajpal K, Kakadiya K, Bansal A (2014) Role of CT and MRCP in evaluation of biliary tract obstruction. Curr Radiol Rep 2:72

Kanwal F, Kramer J, Asch SM, Chayanupatkul M, Cao Y, El-Serag HB (2017) Risk of hepatocellular Cancer in HCV patients treated with direct-acting antiviral agents. Gastroenterology 153: 996–1005 e1001

Khan SA, Tavolari S, Brandi G (2019) Cholangiocarcinoma: epidemiology and risk factors. Liver Int 39(Suppl 1):19–31

Kim S, Kim SW, Bang YJ, Heo DS, Ha SW (2002) Role of postoperative radiotherapy in the management of extrahepatic bile duct cancer. Int J Radiat Oncol Biol Phys 54:414–419

Kim IG, Hu XG, Wang HJ, Kim BW, Hong SY, Shen XY (2019) The 7th/8th American joint committee on Cancer and the modified Union for International Cancer Control staging system for hepatocellular carcinoma. Yonsei Med J 60:140–147

Kojiro M, Wanless I, Alves V (2009) International Consensus Group for Hepatocellular Neoplasia. The International Consensus Group for Hepatocellular N. Pathologic diagnosis of early hepatocellular carcinoma: a report of the international consensus group for hepatocellular neoplasia. Hepatology 49:658–664

Konopka A (1984) Is the information content of DNA evolutionarily significant? J Theor Biol 107: 697–704

Kubo S, Takemura S, Sakata C, Urata Y, Uenishi T (2013) Adjuvant therapy after curative resection for hepatocellular carcinoma associated with hepatitis virus. Liver Cancer 2:40–46

Kudo M, Han G, Finn RS, Poon RT, Blanc JF, Yan L, Yang J, Lu L, Tak WY, Yu X, Lee JH, Lin SM, Wu C, Tanwandee T, Shao G, Walters IB, Dela Cruz C, Poulart V, Wang JH (2014) Brivanib as adjuvant therapy to transarterial chemoembolization in patients with hepatocellular carcinoma: a randomized phase III trial. Hepatology 60:1697–1707

Kudo M, Finn RS, Qin S, Han KH, Ikeda K, Piscaglia F, Baron A, Park JW, Han G, Jassem J, Blanc JF, Vogel A, Komov D, Evans TRJ, Lopez C, Dutcus C, Guo M, Saito K, Kraljevic S, Tamai T, Ren M, Cheng AL (2018) Lenvatinib versus sorafenib in first-line treatment of patients with unresectable hepatocellular carcinoma: a randomised phase 3 non-inferiority trial. Lancet 391: 1163–1173

Kudo M, Finn RS, Edeline J, Cattan S, Ogasawara S, Palmer DH, Verslype C, Zagonel V, Fartoux L, Vogel A, Sarker D, Verset G, Chan SL, Knox J, Daniele B, Yau T, Gurary EB, Siegel AB, Wang A, Cheng AL, Zhu AX, Investigators K (2022) Updated efficacy and safety of KEYNOTE-224: a phase II study of pembrolizumab in patients with advanced hepatocellular carcinoma previously treated with sorafenib. Eur J Cancer 167:1–12

Kulik L, El-Serag HB (2019) Epidemiology and Management of Hepatocellular Carcinoma. Gastroenterology 156:477–491 e471

Kulik L, Heimbach JK, Zaiem F, Almasri J, Prokop LJ, Wang Z, Murad MH, Mohammed K (2018) Therapies for patients with hepatocellular carcinoma awaiting liver transplantation: a systematic review and meta-analysis. Hepatology 67:381–400

Kumar M, Panda D (2014) Role of supportive care for terminal stage hepatocellular carcinoma. J Clin Exp Hepatol 4:S130–S139

Kurosaki I, Hatakeyama K, Tsukada K (1999) Long-term survival of patients with biliary tract cancers with lymph node involvement. J Hepato-Biliary-Pancreat Surg 6:399–404

Lamb BW, Sevdalis N, Arora S, Pinto A, Vincent C, Green JS (2011) Teamwork and team decision-making at multidisciplinary cancer conferences: barriers, facilitators, and opportunities for improvement. World J Surg 35:1970–1976

Laube R, Sabih AH, Strasser SI, Lim L, Cigolini M, Liu K (2021) Palliative care in hepatocellular carcinoma. J Gastroenterol Hepatol 36:618–628

Lederle FA, Pocha C (2012) Screening for liver cancer: the rush to judgment. Ann Intern Med 156: 387–389

Lee J, Park SH, Chang HM, Kim JS, Choi HJ, Lee MA, Jang JS, Jeung HC, Kang JH, Lee HW, Shin DB, Kang HJ, Sun JM, Park JO, Park YS, Kang WK, Lim HY (2012) Gemcitabine and oxaliplatin with or without erlotinib in advanced biliary-tract cancer: a multicentre, open-label, randomised, phase 3 study. Lancet Oncol 13:181–188

Lencioni R (2013) New data supporting modified RECIST (mRECIST) for hepatocellular carcinoma. Clin Cancer Res 19:1312–1314

Lencioni R, Llovet JM (2010) Modified RECIST (mRECIST) assessment for hepatocellular carcinoma. Semin Liver Dis 30:52–60

Lencioni RA, Allgaier HP, Cioni D, Olschewski M, Deibert P, Crocetti L, Frings H, Laubenberger J, Zuber I, Blum HE, Bartolozzi C (2003) Small hepatocellular carcinoma in cirrhosis: randomized comparison of radio-frequency thermal ablation versus percutaneous ethanol injection. Radiology 228:235–240
Lencioni R, de Baere T, Soulen MC, Rilling WS, Geschwind JF (2016) Lipiodol transarterial chemoembolization for hepatocellular carcinoma: a systematic review of efficacy and safety data. Hepatology 64:106–116
Lencioni R, Montal R, Torres F, Park JW, Decaens T, Raoul JL, Kudo M, Chang C, Rios J, Boige V, Assenat E, Kang YK, Lim HY, Walters I, Llovet JM (2017) Objective response by mRECIST as a predictor and potential surrogate end-point of overall survival in advanced HCC. J Hepatol 66:1166–1172
Liang J, Ding T, Guo ZW, Yu XJ, Hu YZ, Zheng L, Xu J (2013) Expression pattern of tumour-associated antigens in hepatocellular carcinoma: association with immune infiltration and disease progression. Br J Cancer 109:1031–1039
Lin WH, Yeh SH, Yang WJ, Yeh KH, Fujiwara T, Nii A, Chang SS, Chen PJ (2013) Telomerase-specific oncolytic adenoviral therapy for orthotopic hepatocellular carcinoma in HBx transgenic mice. Int J Cancer 132:1451–1462
Liu CY, Chen KF, Chen PJ (2015) Treatment of Liver Cancer. Cold Spring Harb Perspect Med 5: a021535
Liver EAFTSOT (2018) EASL clinical practice guidelines: management of hepatocellular carcinoma. J Hepatol 69:182–236
Llovet JM, Bruix J (2003) Systematic review of randomized trials for unresectable hepatocellular carcinoma: chemoembolization improves survival. Hepatology 37:429–442
Llovet JM, Lencioni R (2020) mRECIST for HCC: performance and novel refinements. J Hepatol 72:288–306
Llovet JM, Brú C, Bruix J (1999a) Prognosis of hepatocellular carcinoma: the BCLC staging classification. In: Seminars in liver disease. © 1999 by Thieme Medical Publishers, Inc
Llovet JM, Fuster J, Bruix J (1999b) Intention-to-treat analysis of surgical treatment for early hepatocellular carcinoma: resection versus transplantation. Hepatology 30:1434–1440
Llovet JM, Real MI, Montana X, Planas R, Coll S, Aponte J, Ayuso C, Sala M, Muchart J, Sola R, Rodes J, Bruix J, Barcelona Liver Cancer G (2002) Arterial embolisation or chemoembolisation versus symptomatic treatment in patients with unresectable hepatocellular carcinoma: a randomised controlled trial. Lancet 359:1734–1739
Llovet JM, Ricci S, Mazzaferro V, Hilgard P, Gane E, Blanc JF, de Oliveira AC, Santoro A, Raoul JL, Forner A, Schwartz M, Porta C, Zeuzem S, Bolondi L, Greten TF, Galle PR, Seitz JF, Borbath I, Haussinger D, Giannaris T, Shan M, Moscovici M, Voliotis D, Bruix J, Group SIS (2008) Sorafenib in advanced hepatocellular carcinoma. N Engl J Med 359:378–390
Llovet JM, De Baere T, Kulik L, Haber PK, Greten TF, Meyer T, Lencioni R (2021a) Locoregional therapies in the era of molecular and immune treatments for hepatocellular carcinoma. Nat Rev Gastroenterol Hepatol 18:293–313
Llovet JM, Kelley RK, Villanueva A, Singal AG, Pikarsky E, Roayaie S, Lencioni R, Koike K, Zucman-Rossi J, Finn RS (2021b) Hepatocellular carcinoma. Nat Rev Dis Primers 7:6
Llovet JM, Villanueva A, Marrero JA, Schwartz M, Meyer T, Galle PR, Lencioni R, Greten TF, Kudo M, Mandrekar SJ, Zhu AX, Finn RS, Roberts LR, HCC APoEoTDi (2021c) Trial design and endpoints in hepatocellular carcinoma: AASLD consensus conference. Hepatology 73 Suppl 1:158–191
Lo CM, Ngan H, Tso WK, Liu CL, Lam CM, Poon RT, Fan ST, Wong J (2002) Randomized controlled trial of transarterial lipiodol chemoembolization for unresectable hepatocellular carcinoma. Hepatology 35:1164–1171
Luo XY, Wu KM, He XX (2021) Advances in drug development for hepatocellular carcinoma: clinical trials and potential therapeutic targets. J Exp Clin Cancer Res 40:172

Mansouri V, Beheshtizadeh N, Gharibshahian M, Sabouri L, Varzandeh M, Rezaei N (2021) Recent advances in regenerative medicine strategies for cancer treatment. Biomed Pharmacother 141:111875

Marrero JA, Kulik LM, Sirlin CB, Zhu AX, Finn RS, Abecassis MM, Roberts LR, Heimbach JK (2018) Diagnosis, staging, and Management of Hepatocellular Carcinoma: 2018 Practice Guidance by the American Association for the Study of Liver Diseases. Hepatology 68:723–750

Mazzaferro V, Regalia E, Doci R, Andreola S, Pulvirenti A, Bozzetti F, Montalto F, Ammatuna M, Morabito A, Gennari L (1996) Liver transplantation for the treatment of small hepatocellular carcinomas in patients with cirrhosis. N Engl J Med 334:693–699

Mazzaferro V, Llovet JM, Miceli R, Bhoori S, Schiavo M, Mariani L, Camerini T, Roayaie S, Schwartz ME, Grazi GL, Adam R, Neuhaus P, Salizzoni M, Bruix J, Forner A, De Carlis L, Cillo U, Burroughs AK, Troisi R, Rossi M, Gerunda GE, Lerut J, Belghiti J, Boin I, Gugenheim J, Rochling F, Van Hoek B, Majno P, Metroticket Investigator Study G (2009) Predicting survival after liver transplantation in patients with hepatocellular carcinoma beyond the Milan criteria: a retrospective, exploratory analysis. Lancet Oncol 10:35–43

Mazzaferro V, Shaib W, Rimassa L, Harris W, Personeni N, El-Rayes B, Tolcher A, Hall T, Wang Y, Schwartz B, Kazakin J, Droz Dit Busset M, Cotsoglou C, Papadopoulos K (2016) PD-019 ARQ 087, an oral pan- fibroblast growth factor receptor (FGFR) inhibitor, in patients (pts) with advanced and/or metastatic intrahepatic cholangiocarcinoma (iCCA). Ann Oncol 27: ii109

Mazzaferro V, El-Rayes BF, Droz Dit Busset M, Cotsoglou C, Harris WP, Damjanov N, Masi G, Rimassa L, Personeni N, Braiteh F, Zagonel V, Papadopoulos KP, Hall T, Wang Y, Schwartz B, Kazakin J, Bhoori S, de Braud F, Shaib WL (2019) Derazantinib (ARQ 087) in advanced or inoperable FGFR2 gene fusion-positive intrahepatic cholangiocarcinoma. Br J Cancer 120:165–171

Mazzanti R, Arena U, Tassi R (2016) Hepatocellular carcinoma: where are we? World J Exp Med 6: 21–36

McGlynn KA, Petrick JL, London WT (2015) Global epidemiology of hepatocellular carcinoma: an emphasis on demographic and regional variability. Clin Liver Dis 19:223–238

Mehta N (2020) Hepatocellular carcinoma-how to determine therapeutic options. Hepatol Commun 4:342–354

Meric-Bernstam F, Arkenau H, Tran B, Bahleda R, Kelley R, Hierro C, Ahn D, Zhu A, Javle M, Winkler R, He H, Huang J, Goyal L (2018) Efficacy of TAS-120, an irreversible fibroblast growth factor receptor (FGFR) inhibitor, in cholangiocarcinoma patients with FGFR pathway alterations who were previously treated with chemotherapy and other FGFR inhibitors. Ann Oncol 29:v100

Meyer T, Fox R, Ma YT, Ross PJ, James MW, Sturgess R, Stubbs C, Stocken DD, Wall L, Watkinson A, Hacking N, Evans TRJ, Collins P, Hubner RA, Cunningham D, Primrose JN, Johnson PJ, Palmer DH (2017a) Sorafenib in combination with transarterial chemoembolisation in patients with unresectable hepatocellular carcinoma (TACE 2): a randomised placebo-controlled, double-blind, phase 3 trial. Lancet Gastroenterol Hepatol 2:565–575

Meyer T, Palmer DH, Cheng AL, Hocke J, Loembe AB, Yen CJ (2017b) mRECIST to predict survival in advanced hepatocellular carcinoma: analysis of two randomised phase II trials comparing nintedanib vs sorafenib. Liver Int 37:1047–1055

Miltiadous O, Sia D, Hoshida Y, Fiel MI, Harrington AN, Thung SN, Tan PS, Dong H, Revill K, Chang CY, Roayaie S, Byrne TJ, Mazzaferro V, Rakela J, Florman S, Schwartz M, Llovet JM (2015) Progenitor cell markers predict outcome of patients with hepatocellular carcinoma beyond Milan criteria undergoing liver transplantation. J Hepatol 63:1368–1377

Mizukoshi E, Nakamoto Y, Arai K, Yamashita T, Sakai A, Sakai Y, Kagaya T, Yamashita T, Honda M, Kaneko S (2011) Comparative analysis of various tumor-associated antigen-specific t-cell responses in patients with hepatocellular carcinoma. Hepatology 53:1206–1216

Nakamura H, Arai Y, Totoki Y, Shirota T, Elzawahry A, Kato M, Hama N, Hosoda F, Urushidate T, Ohashi S, Hiraoka N, Ojima H, Shimada K, Okusaka T, Kosuge T, Miyagawa S, Shibata T (2015) Genomic spectra of biliary tract cancer. Nat Genet 47:1003–1010

Naugler WE, Alsina AE, Frenette CT, Rossaro L, Sellers MT (2015) Building the multidisciplinary team for management of patients with hepatocellular carcinoma. Clin Gastroenterol Hepatol 13: 827–835

Ng KKC, Chok KSH, Chan ACY, Cheung TT, Wong TCL, Fung JYY, Yuen J, Poon RTP, Fan ST, Lo CM (2017) Randomized clinical trial of hepatic resection versus radiofrequency ablation for early-stage hepatocellular carcinoma. Br J Surg 104:1775–1784

Okusaka T, Kasugai H, Shioyama Y, Tanaka K, Kudo M, Saisho H, Osaki Y, Sata M, Fujiyama S, Kumada T, Sato K, Yamamoto S, Hinotsu S, Sato T (2009) Transarterial chemotherapy alone versus transarterial chemoembolization for hepatocellular carcinoma: a randomized phase III trial. J Hepatol 51:1030–1036

Osarogiagbon RU, Freeman RK, Krasna MJ (2015) Implementing effective and sustainable multidisciplinary clinical thoracic oncology programs. Transl Lung Cancer Res 4:448–455

Ouyang G, Yi B, Pan G, Chen X (2020) A robust twelve-gene signature for prognosis prediction of hepatocellular carcinoma. Cancer Cell Int 20:207

Palmer DH, Midgley RS, Mirza N, Torr EE, Ahmed F, Steele JC, Steven NM, Kerr DJ, Young LS, Adams DH (2009) A phase II study of adoptive immunotherapy using dendritic cells pulsed with tumor lysate in patients with hepatocellular carcinoma. Hepatology 49:124–132

Pant S, Saleh M, Bendell J, Infante JR, Jones S, Kurkjian CD, Moore KM, Kazakin J, Abbadessa G, Wang Y, Chen Y, Schwartz B, Camacho LH (2014) A phase I dose escalation study of oral c-MET inhibitor tivantinib (ARQ 197) in combination with gemcitabine in patients with solid tumors. Ann Oncol 25:1416–1421

Papatheodoridis GV, Chan HL, Hansen BE, Janssen HL, Lampertico P (2015) Risk of hepatocellular carcinoma in chronic hepatitis B: assessment and modification with current antiviral therapy. J Hepatol 62:956–967

Park B-H, Hwang T, Liu T-C, Sze DY, Kim J-S, Kwon H-C, Oh SY, Han S-Y, Yoon J-H, Hong S-H (2008) Use of a targeted oncolytic poxvirus, JX-594, in patients with refractory primary or metastatic liver cancer: a phase I trial. Lancet Oncol 9:533–542

Park JO, Feng Y-H, Chen Y-Y, Su W-C, Oh D-Y, Shen L, Kim K-P, Liu X, Bai Y, Liao H, Nie J, Qing M, Ji Q, Li J, Zhao M, De Porre P, Monga M (2019) Updated results of a phase IIa study to evaluate the clinical efficacy and safety of erdafitinib in Asian advanced cholangiocarcinoma (CCA) patients with FGFR alterations. J Clin Oncol 37:4117–4117

Pascual S, Herrera I, Irurzun J (2016) New advances in hepatocellular carcinoma. World J Hepatol 8:421–438

Personeni N, Rimassa L (2017) Hepatocellular carcinoma: a global disease in need of individualized treatment strategies. J Oncol Pract 13:368–369

Plsek PE, Greenhalgh T (2001) Complexity science: the challenge of complexity in health care. BMJ 323:625–628

Qu C, Wang Y, Wang P, Chen K, Wang M, Zeng H, Lu J, Song Q, Diplas BH, Tan D, Fan C, Guo Q, Zhu Z, Yin H, Jiang L, Chen X, Zhao H, He H, Wang Y, Li G, Bi X, Zhao X, Chen T, Tang H, Lv C, Wang D, Chen W, Zhou J, Zhao H, Cai J, Wang X, Wang S, Yan H, Zeng YX, Cavenee WK, Jiao Y (2019) Detection of early-stage hepatocellular carcinoma in asymptomatic HBsAg-seropositive individuals by liquid biopsy. Proc Natl Acad Sci U S A 116:6308–6312

Rahib L, Wehner MR, Matrisian LM, Nead KT (2021) Estimated projection of US Cancer incidence and death to 2040. JAMA Netw Open 4:e214708

Roayaie S, Obeidat K, Sposito C, Mariani L, Bhoori S, Pellegrinelli A, Labow D, Llovet JM, Schwartz M, Mazzaferro V (2013) Resection of hepatocellular cancer </=2 cm: results from two Western centers. Hepatology 57:1426–1435

Rodriguez-Pascual J, Babbitt MT, Sanz E, Cubillo A (2018) Multidisciplinary approach of malignant tumors of the biliary tree. In: Topics in the surgery of the biliary tree. IntechOpen

Rosen CB, Heimbach JK, Gores GJ (2010) Liver transplantation for cholangiocarcinoma. Transpl Int 23:692–697

Russell SJ, Peng KW, Bell JC (2012) Oncolytic virotherapy. Nat Biotechnol 30:658–670

Sakamoto E, Nimura Y, Hayakawa N, Kamiya J, Kondo S, Nagino M, Kanai M, Miyachi M, Uesaka K (1998) The pattern of infiltration at the proximal border of hilar bile duct carcinoma: a histologic analysis of 62 resected cases. Ann Surg 227:405–411

Salem R, Lewandowski RJ, Mulcahy MF, Riaz A, Ryu RK, Ibrahim S, Atassi B, Baker T, Gates V, Miller FH, Sato KT, Wang E, Gupta R, Benson AB, Newman SB, Omary RA, Abecassis M, Kulik L (2010) Radioembolization for hepatocellular carcinoma using Yttrium-90 microspheres: a comprehensive report of long-term outcomes. Gastroenterology 138:52–64

Salem R, Gordon AC, Mouli S, Hickey R, Kallini J, Gabr A, Mulcahy MF, Baker T, Abecassis M, Miller FH, Yaghmai V, Sato K, Desai K, Thornburg B, Benson AB, Rademaker A, Ganger D, Kulik L, Lewandowski RJ (2016) Y90 Radioembolization significantly prolongs time to progression compared with chemoembolization in patients with hepatocellular carcinoma. Gastroenterology 151:1155–1163 e1152

Salgia R, Mendiratta V (2021) The multidisciplinary Management of Hepatocellular Carcinoma. Clin Liver Dis (Hoboken) 17:405–408

Schoenthaler R, Phillips TL, Castro J, Efird JT, Better A, Way LW (1994) Carcinoma of the extrahepatic bile ducts. The University of California at San Francisco experience. Ann Surg 219: 267–274

Schulze K, Imbeaud S, Letouze E, Alexandrov LB, Calderaro J, Rebouissou S, Couchy G, Meiller C, Shinde J, Soysouvanh F, Calatayud AL, Pinyol R, Pelletier L, Balabaud C, Laurent A, Blanc JF, Mazzaferro V, Calvo F, Villanueva A, Nault JC, Bioulac-Sage P, Stratton MR, Llovet JM, Zucman-Rossi J (2015) Exome sequencing of hepatocellular carcinomas identifies new mutational signatures and potential therapeutic targets. Nat Genet 47:505–511

Serper M, Taddei TH, Mehta R, D'Addeo K, Dai F, Aytaman A, Baytarian M, Fox R, Hunt K, Goldberg DS, Valderrama A, Kaplan DE, Group VS (2017) Association of Provider Specialty and Multidisciplinary Care with Hepatocellular Carcinoma Treatment and Mortality. Gastroenterology 152:1954–1964

Shen X, Shen X (2021) A potential role for aspirin in the prevention and treatment of cholangiocarcinoma. Int J Cancer 148:1323–1330

Sherman M, Burak K, Maroun J, Metrakos P, Knox JJ, Myers RP, Guindi M, Porter G, Kachura JR, Rasuli P, Gill S, Ghali P, Chaudhury P, Siddiqui J, Valenti D, Weiss A, Wong R (2011) Multidisciplinary Canadian consensus recommendations for the management and treatment of hepatocellular carcinoma. Curr Oncol 18:228–240

Siddique O, Yoo ER, Perumpail RB, Perumpail BJ, Liu A, Cholankeril G, Ahmed A (2017) The importance of a multidisciplinary approach to hepatocellular carcinoma. J Multidiscip Healthc 10:95–100

Siegel RL, Miller KD, Fuchs HE, Jemal A (2022) Cancer statistics, 2022. CA Cancer J Clin 72:7–33

Simon TG, Duberg AS, Aleman S, Chung RT, Chan AT, Ludvigsson JF (2020) Association of Aspirin with hepatocellular carcinoma and Liver-related mortality. N Engl J Med 382:1018–1028

Singal AG, Pillai A, Tiro J (2014) Early detection, curative treatment, and survival rates for hepatocellular carcinoma surveillance in patients with cirrhosis: a meta-analysis. PLoS Med 11:e1001624

Singal AG, Rich NE, Mehta N, Branch AD, Pillai A, Hoteit M, Volk M, Odewole M, Scaglione S, Guy J, Said A, Feld JJ, John BV, Frenette C, Mantry P, Rangnekar AS, Oloruntoba O, Leise M, Jou JH, Bhamidimarri KR, Kulik L, Ioannou GN, Huang A, Tran T, Samant H, Dhanasekaran R, Duarte-Rojo A, Salgia R, Eswaran S, Jalal P, Flores A, Satapathy SK, Kagan S, Gopal P, Wong R, Parikh ND, Murphy CC (2019) Direct-acting antiviral therapy for hepatitis C virus infection is associated with increased survival in patients with a history of hepatocellular carcinoma. Gastroenterology 157:1253–1263 e1252

Singal AG, Lampertico P, Nahon P (2020) Epidemiology and surveillance for hepatocellular carcinoma: new trends. J Hepatol 72:250–261

Sinn DH, Choi GS, Park HC, Kim JM, Kim H, Song KD, Kang TW, Lee MW, Rhim H, Hyun D, Cho SK, Shin SW, Jeong WK, Kim SH, Yu JI, Ha SY, Lee SJ, Lim HY, Kim K, Ahn JH, Kang W, Gwak GY, Paik YH, Choi MS, Lee JH, Koh KC, Joh JW, Lim HK, Paik SW (2019) Multidisciplinary approach is associated with improved survival of hepatocellular carcinoma patients. PLoS One 14:e0210730

Soukup T, Lamb BW, Arora S, Darzi A, Sevdalis N, Green JS (2018) Successful strategies in implementing a multidisciplinary team working in the care of patients with cancer: an overview and synthesis of the available literature. J Multidiscip Healthc 11:49–61

Subbiah V, Lassen U, Elez E, Italiano A, Curigliano G, Javle M, de Braud F, Prager GW, Greil R, Stein A, Fasolo A, Schellens JHM, Wen PY, Viele K, Boran AD, Gasal E, Burgess P, Ilankumaran P, Wainberg ZA (2020) Dabrafenib plus trametinib in patients with BRAF (V600E)-mutated biliary tract cancer (ROAR): a phase 2, open-label, single-arm, multicentre basket trial. Lancet Oncol 21:1234–1243

Sung H, Ferlay J, Siegel RL, Laversanne M, Soerjomataram I, Jemal A, Bray F (2021) Global Cancer statistics 2020: GLOBOCAN estimates of incidence and mortality worldwide for 36 cancers in 185 countries. CA Cancer J Clin 71:209–249

Sungkasubun P, Siripongsakun S, Akkarachinorate K, Vidhyarkorn S, Worakitsitisatorn A, Sricharunrat T, Singharuksa S, Chanwat R, Bunchaliew C, Charoenphattharaphesat S, Molek R, Yimyaem M, Sornsamdang G, Soonklang K, Wittayasak K, Auewarakul CU, Mahidol C (2016) Ultrasound screening for cholangiocarcinoma could detect premalignant lesions and early-stage diseases with survival benefits: a population-based prospective study of 4,225 subjects in an endemic area. BMC Cancer 16:346

Tabrizian P, Holzner M, Halazun K, Agopian VG, Busuttil RW, Yao F, Roberts J, Florman SS, Schwartz ME, Emond JC (2019) A US multicenter analysis of 2529 HCC patients undergoing liver transplantation: 10-year outcome assessing the role of down-staging to within Milan criteria. Hepatology (Wiley) 72(6):2014–2028

Taylor C, Munro AJ, Glynne-Jones R, Griffith C, Trevatt P, Richards M, Ramirez AJ (2010) Multidisciplinary team working in cancer: what is the evidence? BMJ 340:c951

Tompkins RK, Saunders K, Roslyn JJ, Longmire WP Jr (1990) Changing patterns in diagnosis and management of bile duct cancer. Ann Surg 211:614–620; discussion 620-611

Tse RV, Hawkins M, Lockwood G, Kim JJ, Cummings B, Knox J, Sherman M, Dawson LA (2008) Phase I study of individualized stereotactic body radiotherapy for hepatocellular carcinoma and intrahepatic cholangiocarcinoma. J Clin Oncol 26:657–664

Tzartzeva K, Obi J, Rich NE, Parikh ND, Marrero JA, Yopp A, Waljee AK, Singal AG (2018) Surveillance imaging and alpha fetoprotein for early detection of hepatocellular carcinoma in patients with cirrhosis: a meta-analysis. Gastroenterology 154:1706–1718 e1701

Valle J, Wasan H, Palmer DH, Cunningham D, Anthoney A, Maraveyas A, Madhusudan S, Iveson T, Hughes S, Pereira SP, Roughton M, Bridgewater J, Investigators ABCT (2010) Cisplatin plus gemcitabine versus gemcitabine for biliary tract cancer. N Engl J Med 362: 1273–1281

Van Cleave J, Devine P, Odom-Ball P (1999) Multidisciplinary care of hepatocellular carcinoma. Cancer Pract 7:302–308

van der Pol CB, Lim CS, Sirlin CB, McGrath TA, Salameh JP, Bashir MR, Tang A, Singal AG, Costa AF, Fowler K, McInnes MDF (2019) Accuracy of the Liver imaging reporting and data system in computed tomography and magnetic resonance image analysis of hepatocellular carcinoma or overall malignancy-a systematic review. Gastroenterology 156:976–986

Villanueva A (2019) Hepatocellular Carcinoma. N Engl J Med 380:1450–1462

Vincenzi B, Di Maio M, Silletta M, D'Onofrio L, Spoto C, Piccirillo MC, Daniele G, Comito F, Maci E, Bronte G, Russo A, Santini D, Perrone F, Tonini G (2015) Prognostic relevance of objective response according to EASL criteria and mRECIST criteria in hepatocellular carcinoma patients treated with loco-regional therapies: a literature-based meta-analysis. PLoS One 10:e0133488

Wang C, Wang H, Yang W, Hu K, Xie H, Hu KQ, Bai W, Dong Z, Lu Y, Zeng Z, Lou M, Wang H, Gao X, Chang X, An L, Qu J, Li J, Yang Y (2015) Multicenter randomized controlled trial of percutaneous cryoablation versus radiofrequency ablation in hepatocellular carcinoma. Hepatology 61:1579–1590

Waziry R, Hajarizadeh B, Grebely J, Amin J, Law M, Danta M, George J, Dore GJ (2017) Hepatocellular carcinoma risk following direct-acting antiviral HCV therapy: a systematic review, meta-analyses, and meta-regression. J Hepatol 67:1204–1212

Wu C-Y, Chen Y-J, Ho HJ, Hsu Y-C, Kuo KN, Wu M-S, Lin J-T (2012) Association between nucleoside analogues and risk of hepatitis B virus–related hepatocellular carcinoma recurrence following liver resection. JAMA 308:1906–1913

Xie DY, Ren ZG, Zhou J, Fan J, Gao Q (2020) 2019 Chinese clinical guidelines for the management of hepatocellular carcinoma: updates and insights. Hepatobiliary Surg Nutr 9:452–463

Xu RH, Wei W, Krawczyk M, Wang W, Luo H, Flagg K, Yi S, Shi W, Quan Q, Li K, Zheng L, Zhang H, Caughey BA, Zhao Q, Hou J, Zhang R, Xu Y, Cai H, Li G, Hou R, Zhong Z, Lin D, Fu X, Zhu J, Duan Y, Yu M, Ying B, Zhang W, Wang J, Zhang E, Zhang C, Li O, Guo R, Carter H, Zhu JK, Hao X, Zhang K (2017) Circulating tumour DNA methylation markers for diagnosis and prognosis of hepatocellular carcinoma. Nat Mater 16:1155–1161

Xu XL, Liu XD, Liang M, Luo BM (2018) Radiofrequency ablation versus hepatic resection for small hepatocellular carcinoma: systematic review of randomized controlled trials with meta-analysis and trial sequential analysis. Radiology 287:461–472

Yang JD, Hainaut P, Gores GJ, Amadou A, Plymoth A, Roberts LR (2019) A global view of hepatocellular carcinoma: trends, risk, prevention and management. Nat Rev Gastroenterol Hepatol 16:589–604

Yao FY, Xiao L, Bass NM, Kerlan R, Ascher NL, Roberts JP (2007) Liver transplantation for hepatocellular carcinoma: validation of the UCSF-expanded criteria based on preoperative imaging. Am J Transplant 7:2587–2596

Yau T, Park JW, Finn RS, Cheng AL, Mathurin P, Edeline J, Kudo M, Han KH, Harding JJ, Merle P, Rosmorduc O, Wyrwicz L, Schott E, Choo SP, Kelley RK, Begic D, Chen G, Neely J, Anderson J, Sangro B (2019) CheckMate 459: a randomized, multi-center phase III study of nivolumab (NIVO) vs sorafenib (SOR) as first-line (1L) treatment in patients (pts) with advanced hepatocellular carcinoma (aHCC). Ann Oncol 30:v874–v875

Yau T, Kang YK, Kim TY, El-Khoueiry AB, Santoro A, Sangro B, Melero I, Kudo M, Hou MM, Matilla A, Tovoli F, Knox JJ, Ruth He A, El-Rayes BF, Acosta-Rivera M, Lim HY, Neely J, Shen Y, Wisniewski T, Anderson J, Hsu C (2020) Efficacy and safety of Nivolumab plus Ipilimumab in patients with advanced hepatocellular carcinoma previously treated with Sorafenib: the CheckMate 040 randomized clinical trial. JAMA Oncol 6:e204564

Yau T, Park JW, Finn RS, Cheng AL, Mathurin P, Edeline J, Kudo M, Harding JJ, Merle P, Rosmorduc O, Wyrwicz L, Schott E, Choo SP, Kelley RK, Sieghart W, Assenat E, Zaucha R, Furuse J, Abou-Alfa GK, El-Khoueiry AB, Melero I, Begic D, Chen G, Neely J, Wisniewski T, Tschaika M, Sangro B (2022) Nivolumab versus sorafenib in advanced hepatocellular carcinoma (CheckMate 459): a randomised, multicentre, open-label, phase 3 trial. Lancet Oncol 23: 77–90

Yin J, Li N, Han Y, Xue J, Deng Y, Shi J, Guo W, Zhang H, Wang H, Cheng S, Cao G (2013) Effect of antiviral treatment with nucleotide/nucleoside analogs on postoperative prognosis of hepatitis B virus-related hepatocellular carcinoma: a two-stage longitudinal clinical study. J Clin Oncol 31:3647–3655

Yin L, Li H, Li AJ, Lau WY, Pan ZY, Lai EC, Wu MC, Zhou WP (2014) Partial hepatectomy vs. transcatheter arterial chemoembolization for resectable multiple hepatocellular carcinoma beyond Milan criteria: a RCT. J Hepatol 61:82–88

Yopp AC, Mansour JC, Beg MS, Arenas J, Trimmer C, Reddick M, Pedrosa I, Khatri G, Yakoo T, Meyer JJ, Shaw J, Marrero JA, Singal AG (2014) Establishment of a multidisciplinary hepatocellular carcinoma clinic is associated with improved clinical outcome. Ann Surg Oncol 21: 1287–1295

Yu J, Yu XL, Han ZY, Cheng ZG, Liu FY, Zhai HY, Mu MJ, Liu YM, Liang P (2017) Percutaneous cooled-probe microwave versus radiofrequency ablation in early-stage hepatocellular carcinoma: a phase III randomised controlled trial. Gut 66:1172–1173

Zappasodi R, Merghoub T, Wolchok JD (2018) Emerging concepts for immune checkpoint blockade-based combination therapies. Cancer Cell 33:581–598

Zhang BH, Yang BH, Tang ZY (2004) Randomized controlled trial of screening for hepatocellular carcinoma. J Cancer Res Clin Oncol 130:417–422

Zhang J, Mavros MN, Cosgrove D, Hirose K, Herman JM, Smallwood-Massey S, Kamel I, Gurakar A, Anders R, Cameron A, Geschwind JF, Pawlik TM (2013) Impact of a single-day multidisciplinary clinic on the management of patients with liver tumours. Curr Oncol 20:e123–e131

Zhou XD, Tang ZY, Yang BH, Lin ZY, Ma ZC, Ye SL, Wu ZQ, Fan J, Qin LX, Zheng BH (2001) Experience of 1000 patients who underwent hepatectomy for small hepatocellular carcinoma. Cancer 91:1479–1486

Zhu AX, Finn RS, Edeline J, Cattan S, Ogasawara S, Palmer D, Verslype C, Zagonel V, Fartoux L, Vogel A, Sarker D, Verset G, Chan SL, Knox J, Daniele B, Webber AL, Ebbinghaus SW, Ma J, Siegel AB, Cheng AL, Kudo M, investigators K (2018) Pembrolizumab in patients with advanced hepatocellular carcinoma previously treated with sorafenib (KEYNOTE-224): a non-randomised, open-label phase 2 trial. Lancet Oncol 19:940–952

Zhu AX, Kang YK, Yen CJ, Finn RS, Galle PR, Llovet JM, Assenat E, Brandi G, Pracht M, Lim HY, Rau KM, Motomura K, Ohno I, Merle P, Daniele B, Shin DB, Gerken G, Borg C, Hiriart JB, Okusaka T, Morimoto M, Hsu Y, Abada PB, Kudo M, investigators R-s (2019) Ramucirumab after sorafenib in patients with advanced hepatocellular carcinoma and increased alpha-fetoprotein concentrations (REACH-2): a randomised, double-blind, placebo-controlled, phase 3 trial. Lancet Oncol 20:282–296

Zhu AX, Macarulla T, Javle MM, Kelley RK, Lubner SJ, Adeva J, Cleary JM, Catenacci DVT, Borad MJ, Bridgewater JA, Harris WP, Murphy AG, Oh DY, Whisenant JR, Lowery MA, Goyal L, Shroff RT, El-Khoueiry AB, Chamberlain CX, Aguado-Fraile E, Choe S, Wu B, Liu H, Gliser C, Pandya SS, Valle JW, Abou-Alfa GK (2021) Final overall survival efficacy results of Ivosidenib for patients with advanced cholangiocarcinoma with IDH1 mutation: the phase 3 randomized clinical ClarIDHy trial. JAMA Oncol 7:1669–1677

Liver Cancer: Interdisciplinary Approach

Haiwei Zhang, Juan Zhao, Wei Yang, Zheng Li, Li Gong, and Yongsheng Li

Abstract

Liver cancer is currently the fifth most common malignant tumor and the second leading cause of tumor deaths. Its pathogenesis is extremely complex, involving multiple signaling pathways such as VEGFR, EGFR, Raf/MEK/ERK, PI3K/AKT, and Wnt. Chronic hepatitis B virus infection is the most common cause. Combining the high-risk factors, imaging features and serological molecular markers of liver cancer, biopsy or surgical resection of liver space-occupying lesions or extrahepatic metastases, the clinical diagnosis of liver cancer is made

H. Zhang
Phase I Clinical Trial Center, Chongqing University Cancer Hospital, Chongqing, China

Chongqing Key Laboratory of Translational Research for Cancer Metastasis and Individualized Treatment, Chongqing University Cancer Hospital, Chongqing, China

School of Medicine, Chongqing University, Chongqing, China
e-mail: haiweizh@cqu.edu.cn

J. Zhao · W. Yang · Z. Li · L. Gong
Phase I Clinical Trial Center, Chongqing University Cancer Hospital, Chongqing, China

Chongqing Key Laboratory of Translational Research for Cancer Metastasis and Individualized Treatment, Chongqing University Cancer Hospital, Chongqing, China
e-mail: ZJde2021@163.com; cqchyang_wei@163.com; fengqiyunyonglong@163.com; gongli@tom.com

Y. Li (✉)
Phase I Clinical Trial Center, Chongqing University Cancer Hospital, Chongqing, China

Chongqing Key Laboratory of Translational Research for Cancer Metastasis and Individualized Treatment, Chongqing University Cancer Hospital, Chongqing, China

School of Medicine, Chongqing University, Chongqing, China

Department of Medical Oncology, Chongqing University Cancer Hospital, Chongqing, China
e-mail: lys@cqu.edu.cn

Interdisciplinary Cancer Research, https://doi.org/10.1007/16833_2022_96
Published online: 28 December 2022

through histopathological and/or cytological examinations. Among the treatment options for liver cancer, surgical treatment is the most important means for long-term survival of liver cancer patients, including liver resection and transplantation. In the past, anti-vascular targeted therapy was the fundamental treatment for advanced liver cancer. Combination immunotherapy has opened a new gate, and the treatment of immunotherapy plus anti-angiogenesis has driven liver cancer treatment into the era of Immune 2.0. Local therapy and systemic therapy are indispensable to each other, and chemotherapy, targeted therapy, and immunotherapy play a synergistic role, indicating that the spring of interdisciplinary diagnosis and therapy for liver cancer has arrived. This chapter systematically reviews the pathogenic factors, pathogenesis, diagnosis, and treatment of liver cancer, and highlight the future direction of liver cancer diagnosis and treatment.

Keywords

Diagnosis · Hepatitis B virus · Hepatocellular carcinoma · Immunotherapy · Liver cancer · Targeted therapy

1 Introduction

Over 70% of all liver cancers occur in Asia, with more men than women suffering from the disease. The primary liver cancer has three pathological types: intrahepatic cholangiocarcinoma (ICC), hepatocellular carcinoma (HCC), and combined hepatocellular and cholangiocarcinoma (cHCC-CCA) (Bray et al. 2018). Among those with primary liver cancer, HCC accounts for 75–85%, ICC accounts for 10–15%, and cHCC-CCA accounts for less than 10%. HCC ranked sixth in the number of cancers and third in the number of cancer-related deaths (Siegel et al. 2022).

The factors that contribute to liver cancer include hepatitis B virus (HBV) infection, hepatitis C virus (HCV) infection, heavy drinking, steatohepatitis without alcohol, other causes of cirrhosis, family history of liver cancer, especially men >40 years of age, obesity, and type 2 diabetes, (Siegel et al. 2022; Vogel et al. 2018). Liver cancer is diagnosed through histological analysis and contrast-enhanced imaging. The therapy approaches of HCC are interdisciplinary, including hepatectomy, liver transplantation, ablation, transarterial chemoembolisation (TACE), radiotherapy, and systemic antitumor therapy (Vogel et al. 2018). In this chapter, we review the recent advances in HCC from a multidisciplinary perspective, including risk factors, signal pathways, diagnosis, and treatments.

2 Risk Factors for Liver Cancer

Literature has reported many risk factors contributing to liver cancer as listed below (Table 1).

Table 1 Risk factors of liver cancer

Risk factor	References	Study subjects	Main findings	Conclusion
HBV	Chen and Yang et al. (2011)	11,893 men from Chinese mainland without hepatocellular carcinoma	The relative risk of HCC is 9.6 in HBsAg positive individuals (95% CI, 6.0 to 15.2) and 60.2 (95% CI, 35.5 to 102.1) HBeAg-positive individuals, respectively.	Infection with HBV increases the risk of hepatocellular carcinoma.
HBV	Geier et al. (2002)	23,820 participants from seven townships in the Taiwan region	The risk of developing HCC was highly correlated with serum HBV DNA levels and HBV genotypes.	With chronic hepatitis B, HBV DNA levels and HBV genotype C are associated with HCC.
HBV	Llovet et al. (2003)	145 patients with newly diagnosed HCC from the Northern Territory of Australia	The annual incidence of HCC among aborigines was 22.7 per 100 000. HBV infection is present in 8% of aboriginals over 40 years of age, and it is responsible for 60% of all cases of HCC.	The most common HCC causative factor among natives is HBV infection.
HCV	Park et al. (2015)	18,031 HCC patients from 14 countries	In Europe, North America, parts of Central Asia, Japan, North Africa, and the Middle East, especially Egypt, the main virus-related cause of HCC is HCV infection.	HCV infection is an important cause of developing HCC.
HCV	Tsukuma et al. (1993)	Adult patients with cirrhosis ($n = 240$) or chronic hepatitis ($n = 677$)	Patients with hepatitis C antibodies positive had a 4-fold increase in liver cancer risk (RR, 4.09; 95% CI, 1.30 to 12.85).	The risk of liver cancer in patients with HCV infection is greatly increased.

(continued)

Table 1 (continued)

Risk factor	References	Study subjects	Main findings	Conclusion
Cirrhosis	West et al. (2017)	3,107 cirrhotic patients from the UK	Based on the etiology of HCC in 3107 patients with cirrhosis, the predicted 10-year cumulative incidence is as follows:1.2% for alcohol, 4.0% for chronic viral hepatitis, 3.2% for autoimmune or metabolic diseases, and 1.1% for cryptogenic causes.	A 10-year cumulative HCC incidence of 4% is estimated in cirrhotic patients. The risk varies depending on the etiology of the cirrhosis.
NAFLD	Younossi et al. (2015)	4,929 HCC patients and 14,937 HCC control population from the United States	HCC was associated with NAFLD in 14.1% of all HCC cases. The number of NAFLD-HCCs increased by 9% per year between 2004 and 2009.	NAFLD is emerging as a major cause of HCC development in America.
Aflatoxin	Sun et al. (1999)	Prospective cohort study in Chinese population ($n = 18, 244$)	In the HBsAg-negative population, the increased risk of HCC caused by AFB1 exposure alone was 1.9-fold (95% CI, 5 to 7.5) higher than in nonexposed subjects.	There is a risk of HCC development associated with aflatoxin.
Aristolochic acid	Ng et al. (2017)	1,400 HCC patients from different geographical regions	47% of Chinese HCC patients, 29% of Southeast Asian HCC patients, 13% of Korean HCC patients, and 2.7% of Japanese HCC patients showed AA features.	Aristolochic acid exposure is geographically ubiquitous, especially in the Chinese region, being more widely affected.

(continued)

Table 1 (continued)

Risk factor	References	Study subjects	Main findings	Conclusion
Alcohol	Donato et al. (2002)	464 patients with HCC and 824 controls without liver disease from Italy	Alcohol intake was positively associated with the incidence of HCC. In the absence of hepatitis virus infection, alcohol consumption affected the development of HCC as well.	A risk factor for developing HCC is alcohol consumption.
Tobacco	Liu et al. (2018)	2011 liver cancer patients and 7,933 healthy individuals from Chinese mainland.	A significant association was found between age at first smoking, smoking duration, and the number of cigarettes smoked.	The risk of liver cancer is increased in smokers.
Diabetes	Schlesinger et al. (2013)	Prospective cohort study in European patients with diabetes ($n = 363{,}426$)	HCC was diagnosed in 176 patients during 8.5 years of follow-up. And the diabetes-related HCC risk was 2.17 (RR, 1.36 to 3.47; 95% CI, 1.36 to 3.47).	Diabetes mellitus diagnosis increases the risk of developing HCC.
Obesity	Batty et al. (2005)	Prospective cohort study in London population ($n = 18{,}403$)	Hepatocellular cancer risk is higher in obese individuals (RR, 3.76; 95% CI, 1.36–10.4).	Obesity is associated with HCC development.

2.1 HBV

Chronic infection with HBV is the leading cause of liver cancer worldwide, accounting for 50–55% of all HCC cases (Plummer et al. 2016). By causing chronic necroinflammatory disease, HBV induces hepatocyte mutations that eventually lead to HCC (Humans 2012a). The lifetime risk of HCC in HBV-infected populations is 10–25% (McGlynn et al. 2015). Several meta-analysis have shown that the relative risk of HCC in HBV-infected populations is approximately 15–20 times higher than that in uninfected populations (Donato et al. 1998; Shi et al. 2005). In the HBV carrier population, cofactors that increase the risk of HCC include demographic characteristics (e.g., male, older age, family history of HCC, Asian or African ancestry), viral infection (e.g., high serum HBV DNA levels, long duration of infection, HBV genotype C, coinfection with HCV, HIV), and environment or lifestyle (e.g., exposure to aflatoxin, chronic heavy alcohol consumption, smoking) (El-Serag 2012). A large study in the Taiwan region showed the correlation of HBV replication with the risk of HCC in men (Geier et al. 2002). Several population-based studies have shown that HBV genotype C is more likely to cause HCC than genotypes A2, Ba, Bj, and D (McMahon 2009).

Vaccination against HBV reduces the risk of chronic infection and liver cancer (Goldstein et al. 2005). A national neonatal HBV immunization program initiated in the Taiwan region in July 1984 successfully reduced the prevalence of HBV, the incidence of HCC, and mortality from fulminant hepatitis in infants in the vaccination cohort (Chiang et al. 2013). One study predicted that neonatal widespread hepatitis B vaccination would reduce HCC associated with HBV infection by 70–85% (Goldstein et al. 2005). With the development of methods to target HBV minichromosomal cccDNA (Ding and Robek 2014), new drugs will soon be available.

2.2 HCV

Infection with HCV is another leading cause of liver cancer worldwide (Plummer et al. 2016). HCV causes liver cancer mainly by inducing fibrosis and cirrhosis (Tong et al. 1997). Studies have shown that individuals with HCV have an increased risk of developing HCC (Tsukuma et al. 1993), and the risk of HCC in patients infected with HCV is 15- to 20-fold higher compared to those without. (El-Serag 2012). People with active HCV infection are at greater risk of developing HCC due to certain factors, such as male, older, Hispanic, longer infection duration, HIV/HBV coinfection, insulin resistance, obesity, diabetes, heavy drinking, and smoking, especially HCV genotype 1b (Chang et al. 2013). According to a meta-analysis, patients infected with HCV genotype 1b have an approximately twofold increased risk of developing HCC compared with patients infected with other genotypes (Raimondi et al. 2009). A meta-analysis involving 32 studies indicated that HBV and HCV had a synergistic effect, coinfection significantly increases the risk of HCC (Shi et al. 2005).

Prevention of transmission of transfused blood supplies can prevent hepatitis C virus infection, but further efforts are needed to prevent infection in high-risk groups (Llovet et al. 2003). It has been shown that HCV eradication reduces HCC risk by 81% in HCV-infected patients with advanced fibrosis or cirrhosis (van der Meer et al. 2012). And the risk of HCC among high-risk HCV patients has been reduced by routine use of highly potent antivirals for HCV infection in Olmsted County (Hosaka et al. 2013; Hsu et al. 2015). Moreover, the disease burden of HCV-associated HCC has steadily declined as antiviral agents against HCV have become more widely available in the last decade (Yang et al. 2011).

2.3 Cirrhosis

Most HCC patients have cirrhosis as a background, which accounts for 85–95% of all cases (Forner et al. 2018), indicating that cirrhosis is a very important factor that induces HCC (Befeler and Di Bisceglie 2002). In cirrhosis patients infected with HCV and HBV, the risk of developing HCC is estimated to be between 10% and 37% (El-Serag 2002), and patients with HBV/HCV-related cirrhosis have a 2-3-fold greater lifetime risk of HCC than patients with alcoholic cirrhosis (Ganne-Carrie et al. 2018). Cirrhosis patients with diabetes mellitus and the family history of liver cancer have a higher HCC risk. Furthermore, other factors contribute to this process, such as older, male, and a combination of low platelet count ($<100\times10^9$L) and esophageal varices (Kim et al. 2018).

2.4 NAFLD

Nonalcoholic fatty liver disease (NAFLD) plays an increasingly important role in HCC development (Altekruse et al. 2009). In a study involving approximately 130,000 patients, NAFLD patients developed HCC at a rate of 3.51 times greater than the healthy individuals (Alexander et al. 2019). NAFLD was associated with 14.1% of HCC patients between 2004 and 2009, according to a US study using the SEER database, and the number of HCC patients associated with NAFLD increased by 9% per year (Younossi et al. 2015).

2.5 Aflatoxin

Aflatoxin is often found in *Aspergillus flavus*-contaminated cereals and rape seeds, and it was certified as class I carcinogen by the International Agency for Research on Cancer (IARC) because of its potent hepatocarcinogenic effects (Wild et al. 2015). Among aflatoxin types, AFB1 is the most dangerous in liver carcinogenesis (Sun et al. 2013; Yang and Roberts 2010), and AFB1 exposure is one of the most important causes of HCC in sub-Saharan Africa, Southeast Asia, and rural China (Ross et al. 1992). A case-control study found that in a hepatitis B surface antigen-negative population, AFB1 exposure alone increases the risk of HCC by 1.9-fold (95% CI: 5–7.5), while combination with chronic HBV infection increased by 60.1-fold (95% CI:6.4–561.8) (Ross et al. 1992).

In the 1980s, the Chinese government implemented water and food conversion measures, and populations in high-incidence liver cancer areas showed a significant decrease in AFB1 exposure levels, and that the contribution of AFB1 to the etiology of HCC in China was diminishing (Sun et al. 2013). To reduce exposure to aflatoxins, the IARC suggested several intervention strategies in 2014, including selecting genetically resistant seeds, improving post-harvest process methods, and using mycotoxin-capturing enterosorbents. And in the population with these interventions, the aflatoxin exposure level was significantly reduced (Wild et al. 2015).

2.6 Aristolochic Acid

Aristolochic acid (AA) is mainly found in the *Aristolochia* or Chinese wild ginger, which is often used as a traditional Chinese herbal medicine (Arlt et al. 2002). The IARC classified plants containing aristolochic acid as a class I carcinogen in 2012 (Humans 2012b). Mutational characteristics of AA exposure in cell, animal, and human culture models suggest that AA-mediated carcinogenesis is associated with the formation and persistence of aristolochic acid lactam (AL)-DNA (Sidorenko et al. 2012). A large study of HCC in different regions showed that the SNP features of AA are present in 78% population of the Taiwan region, 47% of Chinese mainland, 29% of Southeast Asia, 13% of Korea, 4.8% of North America, 2.7% of Japan, and 1.7% of Europe (Ng et al. 2017). Although herbs containing AA have been banned in many countries and regions, the dose–response relationship between AA exposure and HCC has been well established in HBV-infected populations (Chen et al. 2018a), and it is recommended that HBV-infected populations who have taken these herbs should be closely followed up.

2.7 Alcohol

Alcohol is an independent risk factor that promotes HCC development. Compared to individuals without liver disease, heavy drinking has been shown to increase the risk of HCC, with a 6-fold increased risk of HCC in patients who drink 1500 g/year over their lifetime (Marrero et al. 2005). Alcohol may promote the development of HCC through genotoxicity and by promoting the development of cirrhosis. According to epidemiological studies, the proportion of alcoholics among HCC patients infected with HBV/HCV was much higher than that of non-alcoholics (60% vs 27%), suggesting that alcohol consumption significantly promotes the development of HCC and may play a synergistic role in HBV/HCV-related hepatocarcinogenesis (Brechot et al. 1996). Abstinence from alcohol decreased HCC risk by 6% to 7% per year, but it took 23 years for the risk to reach that of nondrinkers (95% CI 14–70 years) (Heckley et al. 2011).

2.8 Tobacco

Tobacco contains a variety of carcinogens, 11 of which are classified as class I carcinogens by the IARC (Humans 2012b). Several epidemiological studies and meta-analysis have revealed that current smoking increases liver cancer risk (RR: 1.55, 95% CI: 1.46–1.65). And smoking duration (years), frequency, and age at first smoking all significantly and positively correlated with liver cancer in a dose–response relationship (Liu et al. 2018). Moreover, a study reported a 17.9-fold increased risk of HCC among those who used both alcohol and tobacco (Mukaiya et al. 1998). Also, a study observed a synergistic effect between smoking and HBV infection (Liu et al. 2018). Therefore, limiting smoking is a key measure to control the occurrence of HCC, especially for HBV-infected individuals.

2.9 Diabetes and Obesity

Of the known risk factors for liver cancer, diabetes has been reported to have the highest attributable proportion in the American population (Welzel et al. 2013). One meta-analysis found that diabetes status increased HCC risk over two times (Schlesinger et al. 2013), consistent with two other studies (Wang et al. 2012a, b).

As its prevalence increases worldwide in last two decades, Obesity, another independent risk factor, is becoming one of the leading causes of HCC (Saitta et al. 2019). In Taiwan region, the risk of developing HCC was 2-fold in obese patients without viral infection, but 4-fold in obese HCV-positive patients, and the risk increased 100-fold when obese people with diabetes had HBV or HCV infections (Chen et al. 2008).

2.10 Gut Microbiota

Increasing evidence supported that the gut microbiota contributes to HCC development by distributing the liver environment (Yu and Schwabe 2017). Gut barrier failure and dysbiosis were reported to promote various liver diseases including liver fibrosis, alcoholic liver disease (ALD), NAFLD, which are the risk of HCC development (Schnabl and Brenner 2014). Moreover, targeting the gut-microbiota-liver axis has been shown to inhibit the HCC development induced by DEN–CCl_4 in animal models (Dapito et al. 2012).

3 Biomarkers in Liver Cancer

Omics, which includes genomics, transcriptomics, proteomics, metabolomics, and microbiomics, is one of the most common technical tools in oncology nowadays. Omics is used to sketch the expression profile and explore the functions of genes, genetic products, and metabolites of liver cancer. Based on multiomics data, several biomarkers have been revealed to diagnose and predict the prognosis of liver cancer (Jang et al. 2021).

3.1 Genomics

Genomics is the science of studying all the genetic material of an organism, analyzing and quantifying all the genes for a single individual or class of organisms, as well as studying the interrelationships and their effects on the organism. Currently, genomics research relies on all kinds of sequencing technologies such as RNA sequencing (RNA-seq), whole genome sequencing (WGS), and whole exon sequencing (WES). Utilizing second generation sequencing (NGS) technologies and expanded databases, such as Gene Expression Omnibus (GEO) and the Cancer Genome Atlas (TCGA) project, the genomic landscape of HCC has been profiled

(Zhang 2012), TP53 and CTNNB1 are high-frequency mutated genes that are associated with liver cancer development (Candia et al. 2020).

3.1.1 HBV/HCV-Associated HCC

Several omic studies have identified the liver cancer-related genes, CCNE1, MML4, and telomerase reverse transcriptase (TERT), as important targets for HBV integration in HCC (Jang et al. 2021). Recent studies have shown that HBV viral enhancers can induce overexpression of oncogenes (Pillai et al. 2020). Interestingly, gene rearrangements at HBV integration sites may also lead to changes in the copy number of cancer-driven genes, and the number of HBV integrations is an independent prognostic factor for HBV-related HCC (Peneau et al. 2022). Several studies have reported the integration of HBV into the CCNE1 and TERT genes and the role of ANGPT1 in the induction of HCC neoangiogenesis and disease progression (Fujimoto et al. 2016). ARID2 is also reported to involve in HCC in patients infected HCV (Li et al. 2011). And BRD7 expression can be suppressed by HCV infection, leading to dysregulation of HCC cell proliferation (Zhang et al. 2016). Also, it is more common for TERT promoter mutations to be associated with chronic HCV infection and excessive alcohol consumption than with HBV-associated HCC.

3.1.2 Aflatoxin-Associated HCC

In aflatoxin-associated HCC, TP53 is the most common mutation gene. Moreover, the expression of programmed cell death ligand 1 (PD-L1) and higher capillary density were both shown to be correlated with high frequency mutations in the adhesion G protein coupled receptor B1 (adgrb1) gene in tumor tissue. In China, 9.8% of liver cancer cases contained aflatoxin-associated genetic traits, a disproportionately high percentage compared to other regions. The pathogenesis of aflatoxin-associated HCC will be of interest to be elucidated (Zhang et al. 2017).

3.1.3 Alcohol-Associated HCC

As a consequence of persistent liver injury with alcoholism, telomeres of adult hepatocytes shorten due to inactivity of telomerase as a result of necrosis. However, the telomerase reactivates during hepatocarcinogenesis (Forbes et al. 2017). Moreover, patients who intake excessive alcohol often have mutations in ARID1A and CTNNB1 in their HCC cells (Candia et al. 2020).

3.2 Transcriptomics

Noncoding RNAs play a crucial role in numerous cellular, physiological, and pathological processes, as revealed by transcriptomics, and abundantly expressed in a variety of cancers. The increasing evidence has shown that ncRNAs are an important source of biomarkers and therapeutic targets for malignancies due to their abnormal expression and close association with carcinogenesis and metastasis.

3.2.1 MicroRNAs (miRNAs)

Many miRNAs have been evaluated as potential biomarkers for liver cancer diagnosis. In the serum of patients with early-stage HCC, the level of miR-21 was upregulated while miR-199-a downregulated, suggesting that miR-21 and miR-199-a may be used to detect early liver cancer (Mjelle et al. 2019; Guo et al. 2015). A combination of four miRNAs (miR-320a, miR-20a-5p, miR-375, and miR-324-3p) have also been identified to be biomarkers for diagnosing early liver cancer (Wen et al. 2015). Moreover, the expression of miR-10a, miR-122, miR-375, and miR-423 had significant increases in HCC compared to controls, which may help increase the efficiency of HCC diagnosis (An et al. 2018).

3.2.2 Long Noncoding RNAs (lncRNAs)

LncRNAs have been linked to the diagnosis of liver cancer and could be used as biomarkers for the diagnosis of HCC, such as lnc-PCDH9-13:1 from tissues, plasma, and saliva, and lncRNA-UCA1 and lncRNA-WRAP53 from serum (Xie et al. 2018). Studies have shown that combination with alpha-fetoprotein (AFP) can improve diagnostic efficiency of lncRNAs. Combining lncRNA-UCA1 and lncRNA-WRAP53 with AFP showed 100% sensitivity (Huang et al. 2020b).

3.2.3 Circular RNAs (circRNAs)

CircRNAs, members of the ncRNA family, also involved in the development of HCC, and their stability in circulation makes them an ideal candidate for HCC early screening because they are less susceptible to degradation by nucleic acid exonucleases. Several circRNAs, including circRNA-100338, circRNA SMARCA5, hsa_circ_0028502, may be useful as prognostic indicators and as diagnostic markers (Huang et al. 2020a; Matboli et al. 2018).

3.3 Epigenetics

HCC is caused by somatic mutations resulting in abnormal chromatin regulation and epigenetic features. Histone modifications and methylation disorders have been reported in cases of HBV infection. However, epigenetic alterations are rarely mapped genome-wide while most studies focus on specific genes or pathways.

3.3.1 DNA Methylation

Recently, the methylation patterns of nearly 200 liver cancer patients were analyzed by an array-based platform, revealing distinct cancer-specific DNA hypermethylation clusters (Cancer Genome Atlas Research Network. Electronic address and Cancer Genome Atlas Research, 2017). The DNA methylation clusters in HBV/HCV-associated liver cancer are significantly different from that in normal tissue (Song et al. 2016). Moreover, the DNA methylation in HCV-associated liver cancer is more common than that in HBV-associated liver cancer, and the tumor-promoting genes are upregulated in liver cancer due to hypermethylated DNA and promoters. For example, GSTP1, RASSF1A, DOK1, and CHRNA3 exhibit high levels of aberrant DNA hypermethylation in HCC, which can be predictive of poor prognosis when combined with clinicopathological data. In addition, global transcriptional reprogramming caused by hypermethylated enhancers of C/EBP-β promotes hepatocarcinogenicity (Xiong et al. 2019). And the significant correlation has been reported between the hypermethylation status of RASSF1A, APC, and GSTP1 and HCC. In patients with liver cancer, DNA methylation at six CpG loci was identified to distinguish liver cancer from healthy samples (Xiong et al. 2019). Furthermore, The CpG island 3 methylation in HBV-associated HCC is elevated compared to hepatitis and cirrhotic liver tissue (Zhang et al. 2019). Furthermore, HBV infection promotes DNA methyltransferase activity, methylating HBV-DNA and host CpG islands simultaneously.

3.3.2 RNA Methylation

Mammalian RNA is commonly modified with N6-methyladenosine (m6A). Recent studies show that the m6A modifications in HCC are aberrant compared with healthy individuals. Overexpression of METTL3, a m6A methylesterase, results in rapid degradation of SOCS2 and hepatocarcinogenesis (Chen et al. 2018b). Low expression of METTL14, another methylation transferase, implied unfavorable prognosis in nonrecurrent HCC patients. METTL14 can mediate miR-126 maturation, which involves in HCC metastases (Ma et al. 2017). Except for METTL3 and METTL14, other genes associated with m6A, especially YTHDF2, have been shown to be risk profiles through transcriptome sequencing in HCC. These m6A-modified regulators could therefore serve as prognostic biomarkers in HBV-associated HCC patients.

3.3.3 Histone Modifications

In HBV infected patients, HBV-cccDNA can form microchromatin by assembling with histones from host cells to promote viral gene expression, and this process is regulated by posttranslational modifications to histones (PTMs) dynamically. A range of histone modifications on HBV cccDNA has been revealed. H3K36me3, H3K27ac, H3K4me2, H3K4me3, and H3K9ac are reported to activate gene expression, while H3K9me3 silences gene expression (Tropberger et al. 2015). Moreover, H3K4me1 histone modification, which is enriched via KMT2c/d, is responsible for activating HBV transcription. Histone modification not only affects HBV transcription on the cccDNA microchromosome but also regulates host gene expression (Yang et al. 2020). A recent study reported that an assembly shift of the H3.3 from HIRA to HBV-cccDNA could activate H3K4me and HBV transcription (Alvarez-Astudillo et al. 2021). As shown by the first genome-wide PTM map generated by ChIP-Seq, high-level PTMs at specific sites of HBV genome are associated with HBV transcriptional activation, while low-level PTMs correlate with transcriptional repression, even silence. Therefore, HBV-cccDNA transcription may be regulated by PTM through chromatin modification, offering a new treatment approach for the chronic hepatitis B.

3.4 Proteomics

Proteomics can dynamically and comprehensively assess changes in protein quantity and properties during disease onset and progression, it can be used as an important tool for screening liver cancer markers and monitoring liver cancer metastasis. The tool has identified three proteins that may serve as liver cancer biomarkers (Du et al. 2020), while HBV-associated liver cancer is classified into three subgroups based on activation of key signaling pathways and metabolic reprogramming in liver-specific HBV-associated liver cancer (Gao et al. 2019). Zhang and colleagues observed liver-specific metabolic reprogramming that occurs in tumor neighborhood groups as tumor size increases (Zhang et al. 2021a). Furthermore, several prognostic markers for liver cancer metabolic reprogramming have been identified, which may facilitate a more effective treatment. The role of ADP1b3 in immune regulation and therapeutic strategies have been demonstrated in HCC progression (Lu et al. 2021). Based on Wang et al.'s findings, 406 RNA binding protein (RBP) transcripts and proteins were identified to differentially express between HCC tumors and adjacent tumors, and 11 proteins were proposed as HCC biomarkers to develop grouped risk assessment models for HCC (Wang et al. 2020b).

3.5 Metabolomics

Metabolomics focuses on metabolites, including amino acids, lipids, sugars, and hormones, which will reflect the nutritional status and disease conditions of cancer patients (Zhang et al. 2021a). As the most important metabolomic organ, the liver metabolomic profile provides unprecedented opportunities for early diagnosing and treating liver cancer, because the metabolism of lipids, glucose, and amino acids will be changed during liver cancer development and progression.

Luo et al. found that a combination of serum metabolic biomarkers, including phenylalanyl-tryptophan and glycocholic acid, had excellent diagnostic properties to detect early liver cancer in high-risk patients in a multicenter, large sample study (Luo et al. 2018). Lu et al. used liquid chromatography coupled with mass spectrometry to identify Pep (36:4) and Pep (40:6) in serum diminishing liver cancer from other liver diseases and may serve as early diagnostic biomarkers of liver cancer. A recent study identified glycocholic acid 2-deoxychelerythrone, glycine deoxycholate coupling, and taurocholic acid as potential early clinical diagnostic markers (Lu et al. 2018). Based on high-resolution mass spectrometry (HRMS), a prospective cohort study assessing metabolic disturbances prior to the diagnosis of liver cancer identified changes in circulating levels of 92 metabolites associated with risk of liver cancer, 14 of which had high confidence (Stepien et al. 2021). These metabolites included N1-acetyl spermidine, isatin, p-hydroxyphenyl lactate, tyrosine, sphingosine, L-cyclic (leucyl-prolinyl), glycingoose deoxycholic acid. Liu et al. performed metabolomic profiling of changes in metabolites from portal blood and found the elevated levels of L-tryptophan, DL-3-phenyl lactate, glycocholic acid, and 1-methylnicotinamide metabolites in patients with HCC, suggesting hepatic impairment and poor survival.

The serum level of carbohydrates, which are energy sources of hepatocytes, are significantly reduced in HCC patients (Nishikawa et al. 2014). And similar with other cancer cells, glycolysis is the main source of energy in HCC, but not oxidative phosphorylation, which causes the reduced carbohydrate profile in HCC (Gao et al. 2015). Besides, amino acid metabolism is also dysregulated in HCC development. The profile of serum amino acids in patients with HCC is significantly changed due to protein requirement of malignant tumor cells (Dennis and Norris 2015). And the level of fatty acids in HCC patients significantly increase to sustain energy supply and membrane synthesis during the aggressive proliferation of HCC cells.

3.6 Microbiomics

Several studies have identified a key role for gut microbes in immunity, inflammatory responses, and liver cancer development (Ren et al. 2019). Behary et al. showed that NAFLD-associated liver cancer has a unique microbiota and metabolome profile that modulates peripheral immunity (Behary et al. 2021). Ponziani et al. observed that the abundance of *Ackermannia* and *Bifidobacterium* may influence hepatocarcinogenesis and progression by reducing the inflammatory response in the gut and liver (Ponziani et al. 2019). The increase of *Bacteroidetes spp.* and *Bacillariophyceae* was also found in the HCC group compared to non-HCC patients with cirrhosis. The microbiota structure in adults is specific, but recent studies have shown that although the microbiota structure differs in liver cancer patients from different regions, there may be a generalized microbial dysbiosis in liver cancer, suggesting the possibility of microbial markers as early screening markers (Reuter and Bajaj 2020). However, gut microbes are influenced by many factors, such as the host, the environment and diet, thus the microbiome network in HCC awaits further explored (Zmora et al. 2019).

3.7 Molecular Pathways in HCC

Several signaling pathways have been reported to be involved in the tumorigenesis and progression of HCC (Fig. 1).

3.7.1 Receptor Tyrosine Kinase (RTK) Pathway

RTKs act both as receptors and enzymes that bind to ligands and phosphorylate target proteins on tyrosine residues. The receptors that activate RTK signaling include EGFR, FGFR, HGF/c-MET, VEGFR, and PDGFR.

Aberrant epidermal growth factor receptor (EGFR) signal cascades have been found to be directly associated with liver cancer growth and development (Jin et al. 2021). And high expression of vascular endothelial growth factor (VEGF) family genes was significantly associated with poor outcome of HCC, and the function and

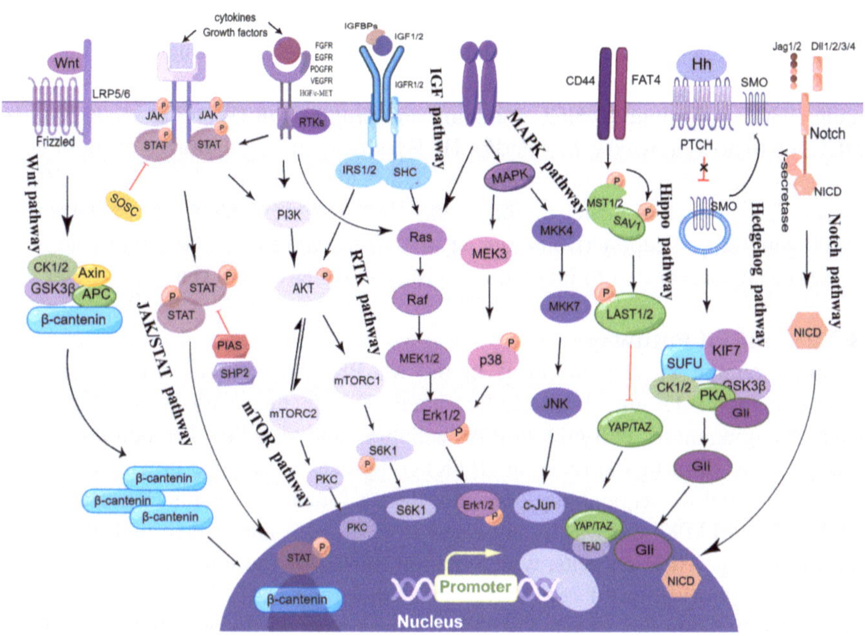

Fig. 1 The signal pathways involved in hepatocellular carcinoma (By Figdraw)

immunoinfiltrating properties of VEGF family genes are highly heterogeneous. Moreover, VEGF upregulation in serum of HCC patients has been linked to complications and poor survival. And VEGF can stimulate endothelial cell mitogenesis and chemotaxis and promotes angiogenesis and tumor growth once it interacts with its receptor (Melincovici et al. 2018).

As a highly conserved tyrosine kinase transmembrane receptor, fibroblast growth factor receptor (FGFR) branches several different signal pathways, which are involved in homeostasis maintenance and liver development. Overexpression of FGFR2, a member of FGFR family, predicts unfavorable prognosis in HCC. Aberrant overexpression of FGFR4, another member of FGFR family, was reported to be a progressive biomarker of HCC, and mutations on its kinase domain may induce acquired resistance to targeted FGFR4 inhibitors (Hatlen et al. 2019). Amplification of the FGFR4 locus (Chr. 11q13) occurs in up to 8% of HCC patients. FGFR selective blockade can prevent tumor growth in HCC models (Hagel et al. 2015). As a selective FGFR4 inhibitor, fisogatinib showed potential benefit in FGF19-overexpressing patients who were recruited in phase I trial (Kim et al. 2019). In addition, a non-inferiority phase III trial demonstrated that lenvatinib, a FGFR-blocking polytyrosine kinase inhibitor, has shown the first-line non-inferior clinical efficacy compared to sorafenib (Al-Salama et al. 2019). Additionally, FGF2 and FGF19 contribute to the maintenance of HCC stem cells (Twaroski et al. 2015; Wang et al. 2021).

c-Met, which can be activated by hepatocyte growth factor (HGF), is important during embryogenesis for tissue modeling and the epithelial-mesenchymal transition (Smith et al. 2017). Several experimental models have demonstrated that c-Met and/or HGF contribute to HCC initiation and progression (Wang et al. 2020a). In HCC, up to 50% showed c-Met and/or HGF pathway dysregulation as result of gene overexpression (50%), amplification (24%), and mutation (4%) (Goyal et al. 2013). Tepotinib, a selective c-Met inhibitor, has shown clinical benefit in patients with MET gene amplification (Bouattour et al. 2018). Moreover, the amplification of c-Met has been reported to involve in resistance to Lenvatinib (Fu et al. 2020).

3.7.2 MAPK Pathway

Serine-threonine kinases involved in the MAPK pathway convert extracellular signaling molecules into intracellular signals to influence various cell functions, such as angiogenesis, cell growth differentiation, and viability. The components of this pathway form ligand-receptor (RTK) complexes to activate downstream transcription signal by sequentially phosphorylating four key proteins (Ras/Raf/MEK/ERK). The MAPK pathway is activated in over 50% early-stage and almost all advanced-stage HCC patients, which mainly caused by the mRNA overexpression or overphosphorylation of core competents including RAF, MEK and ERK. The molecular therapies that targeted MAPK pathway have opened new field in the HCC treatment since sorafenib, a multikinase inhibitor against RAF kinase and receptors tyrosine kinase (i.e. VEGFR2/3, PDGFR etc.), was approved in 2007, more and more MAPK inhibitors enter clinical and preclinical studies of HCC treatment (Yamada et al. 2021).

3.7.3 JAK/STAT Pathway

JAK/STAT signaling pathway can be activated by various growth factors or cytokines during liver regeneration and injury repairment. The ligand-receptor interaction activates JAK, enabling it to phosphorylate tyrosine residues of the receptor, thereby providing a docking site for STATs and facilitating transcription of downstream target genes (Wen et al. 2022). Dysregulated JAK/STAT signaling has been reported to involve in many cancers including HCC (Luo and Balko 2019). For example, constitutive activation of STAT3 is present in more than 60% of HCC patients (He and Karin 2011). JAK1 missense mutations were found in 9% of HBV-related HCC cases, which activate phosphorylation of JAK1 and STAT3 in an independent-cytokine manner (Kan et al. 2013). Thus, the inhibitors of JAK/STAT signaling pathway have been explored for HCC treatment in the preclinical and clinical trials. According to the targets, the inhibitors divide into two class: the JAK inhibitors (ruxolitinib, cryptotanishinone, pacritinib, and WP1066), and the STAT inhibitors (napabucasin, AZD9150, OPB-111077, and OPB-31121). Some inhibitors show a potential therapeutic efficacy in HCC, although their adverse events and outcome need to be further clarified (Hin Tang et al. 2020).

3.7.4 mTOR Pathway

The mTOR pathway is composed of two complexes: mTORC1 and mTORC2, which are responsible for proliferation, aging, and metabolism. It has been estimated that the mTOR pathway activates in approximately 50% of HCC cases (Joechle et al. 2021). Both of the two mTOR complexes, and their downstream effectors (for example AKT and mTORC2/Rictor) involve in the development of various liver cancers.

3.7.5 Wnt/β-Catenin Pathway

The Wnt/β-catenin signaling pathway is responsible for various physiological and pathological activities. Over 50% HCC patients have been identified for mutations of APC, β-catenin and CK1/2, which continuously activate Wnt/β-catenin signaling. For example, activated mutations in CTNNB1 account for 11%–41% of liver cancers. Small molecular inhibitors that target CTNNB1 have been explored to treat liver cancer by inactivating Wnt pathway, such as γ-secretase inhibitors and sorafenib (He and Tang 2020).

3.7.6 Hippo Pathway

The Hippo signaling pathway involves in various liver-specific functions, including adult liver homeostasis, development, and liver regeneration after injury. Aberrant activation of the downstream transcriptional coactivators, Yes-associated protein 1 (YAP), and PDZ-binding pattern (TAZ) is associated with diverse chronic liver disorders (Liu et al. 2020). Hyperactivation of YAP/TAZ by binding TEAD promotes initiation and development of liver cirrhosis and HCC. Therefore, it has become increasingly interesting to develop pharmacological inhibitors of YAP/TEAD interaction. Visudyne, also known as verteporfin, has been identified to inhibit YAP-induced liver overgrowth by disrupting YAP/TEAD association. The effectiveness of this drug is, however, somewhat reduced by off-target effects and cytotoxicity (Wei et al. 2017). Also, tankyrase inhibitors eliminate aberrant cell proliferation by modulating YAP activation in HCC (Jia et al. 2017). To block the YAP/TEAD complex, a variety of promising therapeutic agents are currently being evaluated in preclinical studies (Song et al. 2018).

3.7.7 Hedgehog (Hh) Pathway

Hedgehog (Hh) signaling controls cell fate and tissue development during embryogenesis of insects and vertebrates. Hh ligands initiate the canonical Hh signaling by binding with Ptch receptor, then relieve the inhibition of Smo that translocates and accumulates at cilia, activates the downstream transcription factors Glis which enter the nucleus and regulate target genes (Gao et al. 2018). It has been shown that Hh signaling in HCC enhances cell proliferation, viability, and migration inappropriately (Machado and Diehl 2018). The majority of cells in the pathological specimen sections are Hh positive in cirrhosis and HCC, which implies reactivated Hh signaling (Matz-Soja et al. 2016). Moreover, HCC cell lines and tumor tissues from HCC patients showed higher levels of components, including Smo, Gli1/2, and SHH. Shh neutralizing antibodies or Smo antagonists specifically inhibit the growth of HCC cell lines (Huang et al. 2006). Targeting of Gli2, one of the Gli transcription factors that is predominantly involved in HCC cell proliferation, resulted in a decrease in cell growth among a variety of HCC cell lines (Kim et al. 2007).

3.7.8 IGF Pathway

Insulin-like growth factor (IGF) signaling has been implicated in the growth, development, and liver regeneration, dysfunction of IGF signaling will lead to liver dysfunction, even liver cancers (Adamek and Kasprzak 2018). The fundamental components of IGF signaling include multiple proteases, IGF1/2 (the signaling ligands), IGFR1/2 (IGF receptors), and IGFBPs (IGF binding proteins). In approximately 20% of HCC patients at the early stage, the IGF signaling is hyperactivated. The proliferation of HCC cells is inhibited by selectively blocking IGFR1. Gene expression analysis revealed that IGF2 was overexpressed in 15% of cases by over 20-fold in 228 human HCCs (Martinez-Quetglas et al. 2016). It appears that decreased serum levels of IGF1 are associated with HCC progression, while levels of circulating IGF2 are rather uncertain markers. In HCC, the IGF1 secretion decreases, and the IGFR1 expression increases in HCC tissue, so IGF1 and IGFR1 may be used as prognostic markers (Adamek and Kasprzak 2018).

3.7.9 Notch Pathway

The Notch signaling pathway regulates several fundamental cellular processes, including cell development and differentiation. It has been found that abnormal Notch signaling is associated with hepatocarcinogenesis (Kawaguchi and Kaneko 2021). According to gene expression and genomic analysis of liver cancer, Notch-related genes including JAG1/2 and DLL1/2/3/4 were associated with aggressiveness (Kawaguchi and Kaneko 2021). Over 30% of human HCCs exhibited abnormal Notch signaling. Moreover, the expression of JAG1, a key element of Notch signaling, has been demonstrated to be associated with HBx that is a hepatitis B virus protein, which indicates that Notch signaling involves in the HBV-induced liver cancer (Valizadeh et al. 2021). In summary, Notch signaling exerts multiple functions during liver cancer and may generate new targets for therapeutic manipulation of liver cancer.

3.7.10 Telomere Shortening

The telomere is a DNA-protein structure at the end of each chromosome, and it prevents DNA material from being lost during replication. The length of telomere is maintained by telomerase, a key enzyme of telomere synthesis complex which contains dyskerin, telomerase RNA component (TERC), and telomerase reverse transcriptase (TERT) (Nault et al. 2019). It has been shown that a lack of telomerase and telomere shortening is critical to the development of cirrhosis in hepatocytes, consequently leading to HCC.

In HCC patients, the single nucleotide polymorphisms (SNPs) in TERT or in other proteins involved in telomere synthesis complex have been reported to associate with prognosis. For example, the SNP panels including TERT, TRF1, TEP1, and

POT1 predicted poor outcome, and longer telomere length implied high risk of death in HBV-infected HCC patients (Nault et al. 2019). Moreover, HCV-infected HCC patients who have TERT overexpression or TERT promoter mutations are at a lower risk of recurrence-free survival (Chen et al. 2014; Yu et al. 2017). Normal cells do not reactivate telomerase, but cancer cells do. So, it is considered an attractive therapeutic target for HCC. The preclinical data have shown that imetelstat (GRN163L), a telomerase inhibitor of telomerase, can inhibit tumor cell proliferation and induce apoptosis (Shay and Wright 2011). Furthermore, imetelstat also sensitizes doxorubicin in the Hep3B cell line (Djojosubroto et al. 2005).

4 Staging

Staging of HCC patients is a crucial part of the management of the disease. Besides tumor stage, liver function impairment should also be considered when evaluating the prognosis. Additionally, cancer-related symptoms have consistently been shown to negatively impact survival. There have been several proposals to stratify patients according to expected outcomes on the basis of prognostic prediction. Different staging systems for liver cancers have been adopted by different countries or regions. For example, the AJCC (American Joint Committee on Cancer) TNM system is mostly used in the United States, and the Barcelona Clinic Liver Cancer (BCLC) staging system is worldwide applied, especially in Europe and America, while the Japan Society of Heptatology (JSH) staging system in Japan (Ayuso et al. 2018). Among these staging systems, the BCLC system is the most widely used guideline for HCC staging and treatment and has been evaluated by large multicenter clinical trials. Based on the results of recent investigations incorporating strong evidence, the BCLC system has been updated in 2022. In the 2022 version of BCLC system, the prognosis of HCC is evaluated according to three factors, which are tumor burden, liver function, and physical status, and is refined by other factors including the albumin-bilirubin (ALBI) score, the Child-Pugh score, AFP concentration, and model for end-stage liver disease (MELD) score (Reig et al. 2022). In China, the staging and treatment of HCC are mainly guided by the China liver cancer staging (CNLC) system. In 2022 version of CNLC (Fig. 2), six elements involved in evaluating HCC prognosis: general health status by performance score (PS), liver function by Child-Pugh, extrahepatic metastases, blood vessel invasion, tumor number, and tumor size. Specially, blood vessel invasion, as an independent factor, is included in the parameters of patient characterization for decision of individualized treatment approach (Commission 2022; Zhou et al. 2020).

5 Treatment

Treatment options for hepatocelluar carcinoma include surgery, transplantation, ablation, transarterial chemoembolisation (TACE), radiotherapy, and systemic therapy. The sequential combination of various treatment methods is the best choice for

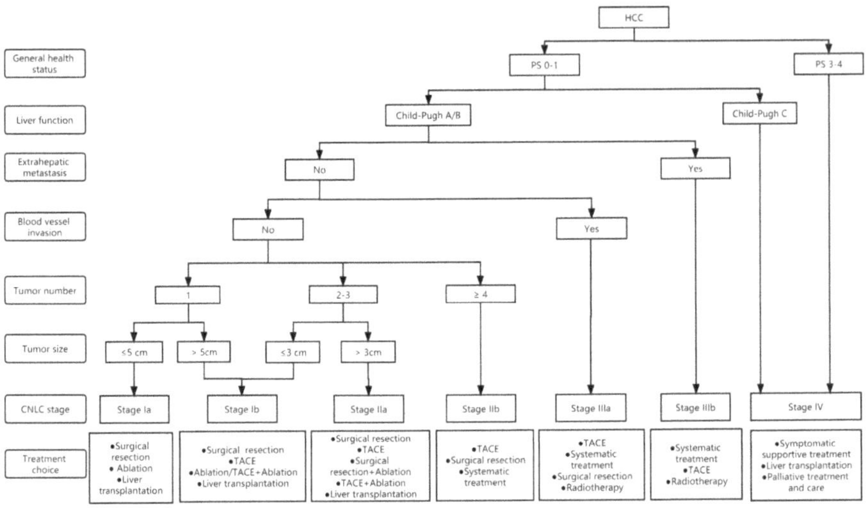

Fig. 2 The flowchart of China liver cancer staging and treatment. PS, performance status; TACE, transarterial chemoembolization; systematic treatment includes: (1) first-line treatment: atezolizumab plus bevacizumab, sintilimab plus bevacizumab analogs; donafenib, lenvatinib, sorafenib; FOLFOX4. (2) Second-line treatment: regorafenib, apatinib, camrelizumab, tislelizumab

each patient according to tumor stage, but there is a conflict between the current medicine system of subspecialty care based on different treatment modalities and the achievement of standardized combination therapies. Therefore, it is recommended that HCC patients should be treated by multidisciplinary teams that include surgeons, radiologists, pathologists, and oncologists, especially for difficult and complex cases, to avoid the limitations of single-disciplinary treatment, promote interdisciplinary communication, and improve the overall efficacy.

5.1 Surgical Intervention

Surgery is an important approach for patients with early HCC to achieve long-term survival, including resection and liver transplantation. Resection or transplantation depends on the patients' general health status, the presence and degree of portal hypertension, liver function and tumor features, and comorbidities. Staging criteria and surgical criteria for liver cancer vary from countries and regions, but a careful preoperative multidisciplinary assessment is considered as a crucial step in the surgical decision-making process for liver cancer.

5.1.1 Resection

Partial hepatectomy is the first treatment choice for early-stage HCC patients without cirrhosis, based on the principle of removing the tumor completely and preserving a sufficient volume of function. The common methods to assess liver function include Child-Pugh scores, indocyanine green clearance, and models for advanced liver disease. Generally, when HCC patients have no portal hypertension and their scores of Child-Pugh are class A, liver resection is considered an optimal procedure (Benson et al. 2021). Compared with traditional open hepatectomy, there are several advantages to laparoscopic hepatectomy (LH) including smaller incisions, less surgical trauma, less bleeding, shorter hospital stays, faster recovery from surgery, and fewer complications after operation. Studies have shown that the oncological outcome of LH is comparable to open hepatectomy in selected patients (Pan et al. 2021). However, LH lacks direct palpation of the mass, and the LH surgical plan is difficult to change intraoperatively. In addition, tumors located in segments VII and VIII of the liver are more difficult to resect than tumors in other parts of the liver in LH, resulting in longer operative time, more intraoperative bleeding, and an increased probability of conversion to open surgery (Lee et al. 2016; Teo et al. 2015). Despite the above-mentioned limitations of LH, the trend of minimally invasive surgery is inevitable. Surgical training, cautious preoperative selection and planning, and careful intraoperative surgical manipulation are required to overcome these shortcomings and ensure patients' safety. In addition, preoperative 3D visualization techniques allow for individualized liver volume calculations and virtual liver resection, facilitating to improve the accuracy of liver resection (Huettl et al. 2021).

Laparoscopic techniques are less invasive, less painful, and have faster recovery, but application of LH is limited by inherent flaws, such as low instrument freedom, amplifying hand tremor by long surgical instruments, and operator fatigue caused by ergonomic design. Robotic hepatectomy, a new surgical modality, is considered to overcome the limitations of LH due to advantages including a stable, clear, and magnified 3D field of view, filtering out operators' hand tremor, flexible and stable operation with multiple degrees of freedom, and expanding the application of minimally invasive hepatectomy. In recent years, robotic surgical systems have been increasingly used in liver surgery, and the safety and feasibility has been continuously proven. Studies have reported the benefits of robotic hepatectomy compared to open surgery in patients with postoperative and long-term overall survival (OS) (Sucandy et al. n.d.). To some extent, robotic hepatectomy has expanded the indications for LH. However, robotic hepatectomy is expensive, and future research will focus on its long-term efficacy, especially oncological outcome, improvement in surgical technique and method, and cost-benefit ratios.

To improve the outcome of hepatic resection, modified surgical technique has been commonly used. Anatomical resection (AR) systematically removes a segment of the liver and area innervated by the portal vein along with the arteries including lobes, segments, and subsegments of the liver. Non-anatomical resection (NAR) removes liver tissue within a certain distance of the tumor and its surrounding area, without regarding to the lobes and segments of the liver. Theoretically, AR removes more liver tissue and can minimize microvascular invasion (MVI) residuals resulting in a superior oncological outcome to that of NAR. Studies have shown that patients undergoing AR operation have a lower local recurrence rate than NAR in HCC cases with MVI, although there is no difference in OS (Hidaka et al. 2020).

In patients with early-stage HCC, hepatectomy offers a better outcome in terms of long-term recurrence rates and local recurrence than other modalities. Comparing with radiofrequency ablation (RFA), a potential alternative curative local therapy, the local recurrence rate after surgical resection is significantly lower, though no significant difference in treatment efficacy for ≤2cm HCC was observed (Feng et al. 2012; Liu et al. 2016). For recurrent HCC, the prognosis of surgical resection remains better than that of RFA (Xia et al. 2020). Conversion therapy, which converts unresectable HCC into resectable HCC, is one way to achieve radical resection and it offers long-term benefit for the intermediate to advanced HCC patients. The efficacy of multimodal, high-intensity anti-tumor treatment strategy alone or in combination to facilitate the translation therapy has been demonstrated (Zhu et al. 2021). Adjuvant therapy after surgery is primarily intended to reduce recurrence. Immunotherapy, targeted agents, and immunomodulators alone or in combination are being actively explored as postoperative adjuvant therapy for HCC (Brown et al. 2019; Zhang et al. 2021b).

5.1.2 Liver Transplantation

Liver transplantation is the curative treatment of liver cancer, especially for small lesions that cannot be surgically removed or ablated. The main international criteria for assessing suitability for liver transplantation are the University of California San Francisco (UCSF) criteria and the Milan criteria. The Milan criteria requires one lesion ≤5 cm in length, or no more than three lesions and a maximum length of ≤3 cm. In addition, the tumor should be free of large intrahepatic vascular invasion and distant metastases. However, the Milan criteria are restrictive in terms of size and number of liver lesions that many patients with HCC would be deprived of liver transplantation. Consequently, there are several new liver transplant recipient selection criteria that have been clinically validated, such as the criteria developed by UCSF, the criteria developed by Up-to-Seven, and the Hangzhou criteria developed in China. Clinical evidence showed that these criteria have not only expanded the recipient population but also achieved similar survival rates to the Milan criteria (Lei and Yan 2013; Lyu et al. 2022).

Liver transplantation has an excellent prognosis, which has 70% 5-year survival rate and 50% 10-year survival rate (Tabrizian et al. 2019). However, the shortage of liver tissue and the uncertain waiting list often result in tumor progression and dropping out, and eventually losing the opportunity of liver transplantation. Thus, bridge therapy is needed to prevent patients from losing the chance to transplant liver by controlling tumor progression while they wait for a donor liver. Moreover, some HCC patients whose tumor burden exceeds the criteria for liver transplantation may be eligible after downstaging therapy to reduce their tumor load. In a phase IIb/III randomized trial, patients who received downstaging therapy with local, surgical, or systemic treatment followed by liver transplantation showed an improved 5-year tumor-free survival (77% vs. 18%) and 5-year OS (78% vs. 31%) compared to those without downstaging therapy (Mazzaferro et al. 2020).

A variety of therapy strategies including TACE, ablation, radiotherapy, and systemic treatment are available for bridging or downstaging therapy. However, it is worth noting that there is limited evidence whether these treatments reduce the probability of recurrence after liver transplantation. In addition, the endpoint of successful downstaging prior to transplantation needs to be further validated (Lee et al. 2020; Yao and Fidelman 2016). Recurrence is the major challenge after transplantation. Studies have shown that immunosuppressive regimens that based on mTOR inhibitors (e.g. everolimus, rapamycin) can reduce tumor recurrence in liver transplant recipients and improve survival (Schnitzbauer et al. 2020). In particular, immune checkpoint inhibitors (ICIs) must be used with caution before and after liver transplantation for HCC (Nordness et al. 2020; Shi et al. 2021). Comprehensive treatment of postoperative recurrence of HCC based on multidisciplinary care may prolong patient survival.

5.2 Locoregional Treatment

Locoregional treatment aims to selectively induce tumor necrosis, including ablation, transarterial therapy, and radiotherapy. It is considered a sign of tumor necrosis when there is no contrast uptake in the treated tumor (Ho et al. 2011). Thus, effect evaluation can be made by comparing the change of contrast uptake in tumor site before and after treatment. Though RECIST is the standard method of tumor evaluation, and several research teams have recommended modified RECIST (mRECIST) as a more appropriate evaluation method for HCC, which is designed to evaluate differences in living tumor tissue rather than focusing only on tumor shrinkage (Llovet et al. 2021a; Zhou et al. 2020). mRECIST is reported to be more effective in identifying treatment response of patients receiving local treatment and systemic therapy (Zhou et al. 2021). AFP response after locoregional therapy is also considered to be a predictor of tumor response. Local treatment can create the possibility of surgical resection for patients with unresectable HCC, thereby increasing the survival benefit for patients. In addition, local treatment is also a bridging therapy option for patients in a waiting list, but there is limited evidence whether it reduces the probability of recurrence after liver transplantation.

5.2.1 Ablation

Although surgery has been adopted as the preferred treatment option in HCC patients at early stage, surgery is not suitable for patients with cirrhosis or intolerant of surgery. The limitations of surgery have led to the emergence of several alternative curative treatments. Image-guided ablation is an important alternative to surgery. Ablation treatment directly destroys the tumor cells by means of chemical, thermal, or electrical mechanism (Llovet et al. 2021a). Currently, the most common ablation techniques are radiofrequency ablation (RFA) and microwave ablation (MWA). Other ablation modalities have been investigated widely; however, they are less used clinically for lack of strong evidence, such as laser ablation, cryoablation, high intensity focused ultrasound ablation, and irreversible electroporation.

RFA

As one of the most commonly used ablative modalities, RFA is to conduct alternating current through radiofrequency ablation electrodes and induce resistive tissue heating, which is conducted by indirect heating of adjacent tissues and forms an ablation zone through coagulation necrosis. After applying RFA according to guideline, the 5-year survival rate for patients with hepatocellular carcinoma ranged from 40% to 68%, while the 10-year survival rate ranged from 27% to 32% (N'Kontchou et al. 2009; Shiina et al. 2012). In a multicenter RCT using RFA as opposed to surgical resection , disease-free survival and overall survival were similar or slightly worse, but complication rates were lower and hospital stays shorter (Kudo et al. 2021). There is evidence that RFA has similar efficacy to surgical resection for patients with HCC with one lesion ≤2 cm in diameter, especially for centrally located HCC (Peng et al. 2012). To achieve larger overlapping ablation zones, RFA can be conducted with multiple or expandable needle electrodes. A large multicenter case-matched study including 362 patients with HCC ≤5 cm showed that RFA with no-touch multipolar bipolar was more effective for primary ablation and experienced a more sustained local response than monopolar RFA, without increasing severe complication rate (Hocquelet et al. 2017). In general, RFA is convenient to operate, and has a short hospital stay and a controlled scope of ablation. RFA is particularly useful in patients with advanced age, other comorbidities, severe cirrhosis, and deep or central HCC tumors in the liver.

MWA

MWA, another common thermal ablation method, converts electromagnetic energy into thermal energy to heat tissue rapidly within a few minutes. The thermal energy induces tumor cells coagulative necrosis. Theoretically, MWA has a greater advantage over RFA in terms of short ablation times, ablation efficiency, and reduction of the "heat sink effect" associated with RFA. Nevertheless, the long-term survival and local efficacy between MWA and RFA are not statistically different (Tan et al. 2019; Violi et al. 2018). The complication rate is even higher in patients with HCC >3 cm for MWA (Glassberg et al. 2019). However, the variability between equipment and devices within each ablative modality complicates comparative studies between RFA and MWA. In clinical practice, the choice of ablation technique, RFA or MWA, depends on the tumor size, location, and other factors.

Percutaneous Ethanol Injection (PEI)

PEI is a seminal chemical ablation approach historically. PEI induces coagulative necrosis by injecting 99% dehydrated ethanol into the tumor, which causes "chemical burn" of HCC. PEI is considered as an effective, cost-effective, and minimally invasive technique. However, its clinical application has been limited by inhomogeneous distribution within tumors, inability to reach tumor margins, and poor retention. Though PEI has been largely replaced by RFA and MWA, it remains considered for tumors less than 2 cm, particularly if tumors are located nearby high-risk areas like major vessels, the gastrointestinal tract, or the bile ducts (Commission 2022; Llovet et al. 2021b).

Table 2 Advantages and limitations of various ablative modalities

Ablation technique	Advantage	Limitation	Level of evidence
RFA	Convenient operation; Less bleeding; Shorter hospital stay; Lower rate of postoperative complications than surgery	Heat sink effect; Limitation of ablation volume; Technically infeasible when the tumor location is conspicuous and dangerous	High
MWA	Higher ablation temperature; Larger ablation zone; Shorter ablation time; Lower risk of heat sink effects; No skin burns from grounding pads	Complications; such as mild pain, post-ablation syndrome, and abnormal pleural effusion	Moderate
PEI	Efficacious and cost-effective; Minimally invasive procedure	High rate of local recurrence; Intratumor inhomogeneous distributions; Repeated operation; Tumor margins unreachable; Poor retention at required site of action	Moderate
CRA	Visual ice ball formation from CRA; Activation of cellular immune function of the tumor; less damage to nearby large blood vessels or organs; Severe pain relief	Small ablation zone of each individual probe; High potential for residual tumor cells at the edge of the frozen area inducing risk of recurrence; High risk of serious complications such as organ dehiscence and "cold shock"	Low
LA	Minimal invasiveness; Efficient light penetration; Extremely well-tolerated; without severe complications.	Possible accompanied vasovagal reaction	Low
HIFU	Noninvasiveness	Long procedure time; Coustic shadowing; thermal injury of the overlying soft tissue	Low
IRE	Precise positioning; Without heat-sink effect; Enhancing efficacy nearby vessels.	Muscle contractions; Increased risk of cardiac arrhythmia; Inability to track the ablate	Low

Abbreviation: *RFA* Radiofrequency ablation, *MWA* Microwave ablation, *PEI* Percutaneous ethanol injection, *CRA* Cryoablation, *HIFU* High intensity focused ultrasound ablation, *LA* Laser ablation, *IRE* Irreversible electroporation

Other Ablative Modalities

Newer ablative techniques have been under investigation currently, such as CRA, LA, HIFU, and IRE. Mechanically, ablative modality consists of energy-based and non-energy-based ablation. Table 2 summarizes advantages and limitations of various ablative modalities. CRA causes cell death and ischemia based on repetitive freeze–thaw cycles by intracellular and intravascular ice crystal formation respectively. Initially, the metallic probe is cooled by high-pressure argon gas. Subsequently the probe is warmed by infusion of helium gas, leading to thawing of the tissue. Intracellular and intravascular ice crystal form through these cycles. Though CRA has comparable outcomes to RFA and MWA, the efficacy of CRA compared to RFA and MWA remains controversial and more clinical studies are required (Gupta et al. 2021; Luo et al. 2022). LA delivers high-energy laser radiation to target tissues through laser optical fibers, and cause tumor cells coagulative necrosis (Chai et al. 2019). HIFU focuses the ultrasound waves of high energy into one small spot to elevate tissue temperature and induce tumor cell coagulative necrosis (Hill et al. 1994). IRE induces apoptosis by increasing cell membrane permeability. In theory, IRE can avoid unnecessary thermal damage to significant structures, but require general anesthesia, which increases the complexity of this technology.

Although these new ablative modalities remain promising with theoretical advantages and early signals of efficacy, these advantages are compromised by corresponding limitations such as the need for general anesthesia, multi-probe techniques, complexity, insufficient safety, and efficacy information (Hui et al. 2021; Kim et al. 2014). Therefore, these techniques are not yet broadly recommended for clinical practice.

5.2.2 Transarterial Therapy

According to BCLC staging system, more than 50% of HCC patients were diagnosed at stage C or D in North America and China, and only 18% of HCC patients in North America are treated with radical surgery (Park et al. 2015). Transarterial therapy has been widely used for patients with inability or unwillingness to accept curative treatments. Approximately 1/4 of the blood supply to the liver is supplied by the hepatic artery and the other 3/4 is supplied by the portal vein, whereas liver cancer tissue is supplied by about 90% hepatic arterial blood and <10% portal venous blood. Thus, transarterial therapy has a scientific theoretical basis. The following transarterial treatments are currently available: hepatic artery infusion chemotherapy (HAIC), transarterial chemoembolization (TACE), transarterial embolization (TAE), and transarterial radioembolization (TARE) (Llovet et al. 2021a).

HAIC involves infusion of chemotherapeutic agents such as anthracyclines, platinum, and fluorouracil via the tumor supply artery. A number of different substances have been used in TAE in order to prevent arterial flow, including particles made of gelatin sponge or polyvinyl alcohol, and microspheres of polyacrylamide. Along with arterial blockage, TACE also delivers high concentration of chemotherapeutic drugs to tumor cells through drug-loaded lipiodol emulsions or microspheres, which are able to prolong embolic agents contact with cancer cells and minimize their systemic toxicity. According to the embolic agent, TACE can be classified as conventional-TACE or drug-eluting beads-TACE (DEB-TACE) (Kotsifa et al. 2022). HAIC has a large overlap of indications with cTACE, but HAIC shows the following advantages: (1) HAIC does not use any embolic agents, which can eliminate adverse events such as embolic syndrome and ectopic embolism; (2) HAIC usually does not cause adhesion between the tumor and adjacent organs such as the diaphragm, gallbladder, and gastrointestinal tract, thereby HAIC has little impact on subsequent surgical operations; (3) HAIC is easy to operate and can be performed according to uniform standards; (4) HAIC has a higher objective efficiency for giant HCC and surgery conversion rate (Gourd et al. 2020). DEB-TACE is a modified treatment modality developed on the basis of cTACE, which refers to the use of chemotherapy agents loaded eluting microspheres-based therapy. Recent studies indicated that the overall efficacy of DEB-TACE and cTACE is not significantly different, but DEB-TACE has an advantage in terms of treatment response (Liang et al. 2021). Therefore, TACE is considered as the mainstream method of vascular intervention for HCC, the use of HAIC and DEB-TACE may become increasingly widespread with the development of vascular interventional materials and improvement of chemotherapy regimens.

TARE with yttrium 90 (Y-90) has increasingly been used to treat HCC, which directly delivers Y-90 microspheres that are capable of emitting high doses of beta radiation into the tumor capillaries, thus avoiding radiation damage to normal liver tissue (Ibrahim et al. 2008). Currently, glass Y-90 microspheres and resin Y-90 microspheres have been approved for clinical TARE use. Studies have shown that the survival time and remission rate of unresectable HCC patients treated with TARE are similar with those of TACE (Lobo et al. 2016). However, patients with unresectable liver cancer treated with TARE showed longer time to progression (TTP), fewer adverse events, and less posttreatment pain compared to TACE. Additionally, TACE is an inpatient procedure, whereas TARE is usually an outpatient one. Though produced positive efficacy of TARE with Y-90 for intermediate HCC was reported in phase II investigations (Salem et al. 2013), As of yet, this recommendation has not been widely adopted by the guidelines. The comparison of TACE with TARE requires more controlled clinical studies before a high-level recommendation can be made. In the future, doctors will need to address the microsphere type and lack of individualized radiation methods associated with TARE in clinical practice.

Currently, transarterial therapy combined with other therapeutic strategies is being widely investigated, such as invasive ablation, external radiation therapy, surgical resection, antiviral therapy, and systemic therapy, with the goal of controlling HCC, improving quality of life and prolonging survival (Benson et al. 2021; Lewis et al. 2019). Studies have reported that TACE combined with RFA or MWA showed similar safety and efficacy, and combination therapy is more effective than other alone (Ginsburg et al. 2015; Maataoui et al. 2005). Moreover, the consequence and mode of combination have an impact on the efficacy. For example, TACE combined with sequential RFA within 30 days is more effective than delayed FRA in recurrent HCC (Liu et al. 2019).

5.2.3 Radiotherapy

For unresectable or inoperable HCC patients, radiotherapy, which includes internal radiation therapy (IRRT) and external beam radiation therapy (EBRT), can be used alone or in combination with other therapies to hinder tumor growth, and help relieve HCC-related pain and discomfort through local radiotherapy. EBRT is referred to the use of radiation (photons or particles) produced by radiotherapy equipment to irradiate tumors from outside the body. EBRT has been advanced in recent decade.

The most common EBRT techniques include image-guided radiotherapy (IGRT), intensity-modulated radiation therapy (IMRT), and stereotactic body radiation therapy (SBRT). Using IGRT and IMRT, enhanced radiation dose can be targeted to the tumor with increased accuracy while surrounding healthy tissue can be spared. SBRT, an advanced EBRT technique delivering high dose of radiation to the target area, is recommended by guidelines as the first-line treatment for patients with recurrent, unresectable, advanced HCC. EBRT can also be used as downstaging therapy for resection and bridging therapy for liver transplant (Jang et al. 2020; Sapisochin et al. 2017). Internal radiation treatment involves implanting radionuclides into the tumor with a needle or through the body duct. The efficacy of proton radiotherapy (PBT) for postoperative recurrent or residual HCC lesions (<3 cm in size and ≤ 2 in number) is similar to that of RFA (Kim et al. 2021). It should be noted, however, that studies comparing radiotherapy with other locoregional therapies are largely retrospective, with low evidence. Hence, well-designed randomized studies are needed before radiotherapy can be recommended by guideline.

5.3 Systemic Therapy

Systemic therapy mainly refers to anti-tumor therapy, including molecular targeted drug therapy, immunotherapy, chemotherapy, and Chinese herbal medicine. It also includes treatment for the underlying disease of liver cancer, such as antiviral therapy, liver and bile protection, and symptomatic treatment. Systemic therapy is the major treatment for intermediate to advanced HCC, because less than 30% of patients with liver cancer are suitable for radical treatment at the first diagnosis due to its insidious onset (Sonbol et al. 2020). In patients with advanced HCC, or those who have failed other treatments, systemic therapy has been shown to be beneficial.

5.3.1 First-Line Treatment

Atezolizumab, an immune checkpoint inhibitor (ICI), in combination with bevacizumab (Atezo-Bev) is the first-line treatment for unresectable HCC patients, as it confers better survival compared with sorafenib. According to the IMbrave150 study, atezolizumab plus bevacizumab significantly improved 12-month overall survival and median progression free survival compared with sorafenib (Finn and Zhu 2021). In China, a combination of Sindilizumab and a bevacizumab biosimilar is recommended as the first-line treatment for patients with advanced HCC that has not received to prior systemic treatment (Ren et al. 2021).

The understanding of the molecular events of tumorigenesis and progression has enabled the development of targeted therapies. Several agents have been approved by FDA during the last two decades. Sorafenib is the first effective first-line choice to propose evidence-based sequential therapy in HCC (Llovet et al. 2008). Lenvatinib is also recommended when the liver function score of advanced HCC patients is Child-Pugh grade A (Personeni et al. 2019). Furthermore, Donafinib has been approved for the treatment of unresectable HCC patients without prior systemic therapy in China. Compared with sorafenib, donafenib significantly prolonged median survival in advanced HCC, and reduced the risk of death by 17%, had a good safety and tolerability profile, but showed similar median PFS (Qin et al. 2021a). In CNLC (2022 version), FOLFOX4 is recommended to treat locally advanced or metastatic hepatocellular carcinoma that cannot be surgically resected or treated locally (Qin et al. 2014).

5.3.2 Second-Line Treatment

The current second-line treatment drugs for liver cancer are all tested in patients who have received sorafenib but failed or intolerant, and the choice of second-line drugs also lacks uniform standards and data of large-scale phase III clinical trials. However, multikinase inhibitors are clearly rational given the existing evidence of their efficacy. According to RESORCE study, Regorafenib has only been approved for HCC patients who previously received sorafenib but not suitable for patients who didn't receive tyrosine kinase inhibitors (TKIs). Its adverse effects are similar to those of sorafenib, therefore, it is also not suitable for those patients who cannot tolerate sorafenib (Bruix et al. 2017). Similarly, cabozantinib is also approved for patients with HCC after sorafenib. Among drugs available in this setting, only cabozantinib can be prescribed to all patients regardless of their baseline AFP levels or tolerance to sorafenib (Abou-Alfa et al. 2018). Apatinib, a small molecule targeted new drug developed in China, is approved for advanced HCC patients who have received at least first-line systemic treatment but have failed or intolerant (Qin et al. 2021b). In addition to the above drugs, other ICIs and TKIs are approved as second-line choice for advanced HCC. Ramucirumab is approved only for HCC patients with AFP higher than 400 ng/mL (Zhu et al. 2019). Pembrolizumab and nivolumab in combination with ipilimumab have been conditionally approved for HCC patients who have been previously treated with or are intolerant to sorafenib. Camrelizumab is approved to treat advanced HCC patients who have previously received systemic therapy with sorafenib or oxaliplatin. Currently, many studies are consistently being conducted to explore the combination of immune checkpoint inhibitors with targeted drugs, chemotherapy drugs, and local therapy for the second-line treatment of liver cancer.

6 Future Prospective

Over the last 20 years, HCC incidence has been on the rise worldwide, and this trend is expected to continue in the next decade. However, the etiologies in Asia-Pacific and Euro-American regions are significantly different. The main risk factors of HCC in eastern Asia and sub-Saharan Africa are chronic hepatitis B and aflatoxin B1 exposure, while hepatitis C, excessive alcohol intake in the Euro-American regions and Japan (Forner et al. 2018). The etiology of HCC in developed countries is increasingly becoming NASH and NAFLD, associated with metabolic syndrome or diabetes mellitus (Llovet et al. 2021b). Numerous studies try to explore the molecular networks in HCC and attempt to develop new drugs for HCC treatment.

However, there has been limited success with drugs targeting metastasis, cell proliferation, and metabolites, and in particular, no etiology-specific treatments have been initiated (Swamy et al. 2017). Optimal treatment for HCC depends on several factors, including the tumor size, tumor number, location, and patient's status which is evaluated by performance status (PS), liver function, etc. The treatment landscape continues to evolve in complexity with recent advances in liver-directed and systemic therapies. For patients with early-stage HCC, several curative treatment options are available, including liver transplants, surgical resections, and ablation. For patients with intermediate and locally advanced HCC, the main modalities are local therapy, such as TACE, ablation, TARE, and SEBT. For patients with advanced HCC, several systemic therapies, including ICIs alone or combination with TKIs or VEGF inhibitors, have been approved, such as sorafenib, atezolizumab plus bevacizumab, regorafenib, lenvatinib, cabozantinib, and ramucirumab (Benson et al. 2021; Commission 2022; Reig et al. 2022; Vogel et al. 2018).

The molecular cascades of signal transduction underlying hepatocarcinogenesis and metastasis have been discovered through recent advances in tumor biology, immunology, and molecular genetics, which help us to better understand characteristics of HCC and to develop selective agents for HCC treatment. A growing number of molecular targets are available for the development of anti-liver cancer drugs. MET inhibitors, including tivantinib, cabozantinib, and golvatinib, are currently being studied for advanced HCC after a locoregional or systemic treatment has failed, especially in patients with high MET expression (Qi et al. 2015). It has been reported that specific inhibitors of FGFR4 and FGF19 kinases are in phase I, such as BLU-554 (NCT02508467) and H3B-6527 (NCT02834780), while FGF401 (NCT02325739) is in phase II (Lu et al. 2019). Different pan-FGFR inhibitors, including ABSK-011 (NCT04906434), [225Ac]-FPI-1966 (NCT05363605), H3B-6527 (NCT02834780), and erdafitinib (NCT02421185), are in different stages of clinical development (Raja et al. 2019). Clinical trials of precision medicine have not yet been conducted for some molecular drivers, such as chromosomal aberrations and dysregulated signaling (MAPK, RTK, mTOR, JAK/STAT, Hippo, Hh, or IGF signaling) (Garcia-Lezana et al. 2021).

With advances in tumor immunology, the emerging immunotherapeutic approaches are being applied to treat patients with hepatocellular carcinoma, such as vaccines against cancer, chimeric antigen receptor (CAR), T-cells, DC cells, NK cells, various ICIs, cytokine-induced killer cells (CIK), and TILs-mediated immunotherapy and adoptive immunotherapy (Rai and Mukherjee 2022). Several components of the tumor microenvironment are also important in the development of drugs, including cytokines, immune components, and nonimmune components, such as cancer-associated fibroblasts (CAF), T lymphocytes, NK cells, macrophages in NASH and Fibrosis, macrophages in HCC, neutrophils, IL-6, and TGF-β. Novel immunotherapy of HCC about PD-L1 and CTLA4 inhibitor (NCT03298451, NCT04294498), CAR-T therapy (NCT03013712, NCT02729493, NCT0334925) are in the development stage (Sas et al. 2022).

Artificial intelligence (AI), machine learning and other technologies have been applied to diagnosis, prognosis analysis, efficacy prediction and drug resistance prediction of liver cancer. Artificial Intelligence Algorithm was used to assistance with diagnosis (Zhang et al. 2022) and predict sensitivity to immunotherapy (Chapiro 2022). Deep learning of histopathological images of HCC to automated diagnosis and somatic mutation prediction for HCC (Liao et al. 2020). Machine learning is also used to assist in liver transplantation (Briceno 2020; Khorsandi et al. 2021).

Additionally, metabolic reprogramming plays a key role in polarizing and activating tumor-associated macrophages (TAMs) and initiating liver cancer (Wu et al. 2020). Reprogramming of lipid metabolism is an important component of tumor cell tumorigenicity, especially during HCC initiation and invasion. In future studies, attention should be paid to the therapies targeting epigenetic and mTOR pathway, to develop novel therapeutic methods for HCC (Hu et al. 2020).

Multiple treatment modalities are commonly used sequentially or in concert to attain more durable outcomes or to downstage patients for transplant eligibility in clinical practice. Therefore, it will be challenging to determine which treatment plan will be most effective. This requires expertise in a wide variety of therapeutic modalities, including hepatologists, surgeons, oncologists, radiologists, clinical pharmacologist, oncology nutritionist, tumor biologist, and palliative care specialists. With the multidisciplinary team communicating frequently and thoroughly, the interdisciplinary approaches will lead to precise diagnosis and treatment for liver cancers.

Acknowledgements The authors thank the editors and anonymous reviewers for their efforts on improving this manuscript during the review process. This work was supported by the Major International (Regional) Joint Research Program of the National Natural Science Foundation of China (No. 81920108027), Chongqing Outstanding Youth Science Foundation (No. cstc2020jcyj-jqX0030), Chongqing Talent Innovation Leader Project (No. CQYC20200301111), Funding for Chongqing University Innovation Research Group, and the Fundamental Research Funds for the Central University (2019CDYGYB005).

Conflicts of Interest All the figures and tables in this chapter are original. There is no interest to declare.

References

Abou-Alfa GK, Meyer T, Cheng AL, El-Khoueiry AB, Rimassa L, Ryoo BY, Cicin I, Merle P, Chen Y, Park JW et al (2018) Cabozantinib in patients with advanced and progressing hepatocellular carcinoma. N Engl J Med 379:54–63

Adamek A, Kasprzak A (2018) Insulin-like growth factor (IGF) system in liver diseases. Int J Mol Sci 19:1308

Alexander M, Loomis AK, van der Lei J, Duarte-Salles T, Prieto-Alhambra D, Ansell D, Pasqua A, Lapi F, Rijnbeek P, Mosseveld M et al (2019) Risks and clinical predictors of cirrhosis and hepatocellular carcinoma diagnoses in adults with diagnosed NAFLD: real-world study of 18 million patients in four European cohorts. BMC Med 17:95

Al-Salama ZT, Syed YY, Scott LJ (2019) Lenvatinib: a review in hepatocellular carcinoma. Drugs 79:665–674

Altekruse SF, McGlynn KA, Reichman ME (2009) Hepatocellular carcinoma incidence, mortality, and survival trends in the United States from 1975 to 2005. J Clin Oncol 27:1485–1491

Alvarez-Astudillo F, Garrido D, Varas-Godoy M, Gutierrez JL, Villanueva RA, Loyola A (2021) The histone variant H3.3 regulates the transcription of the hepatitis B virus. Ann Hepatol 21: 100261

An Y, Gao S, Zhao WC, Qiu BA, Xia NX, Zhang PJ, Fan ZP (2018) Novel serum microRNAs panel on the diagnostic and prognostic implications of hepatocellular carcinoma. World J Gastroenterol 24:2596–2604

Arlt VM, Stiborova M, Schmeiser HH (2002) Aristolochic acid as a probable human cancer hazard in herbal remedies: a review. Mutagenesis 17:265–277

Ayuso C, Rimola J, Vilana R, Burrel M, Darnell A, Garcia-Criado A, Bianchi L, Belmonte E, Caparroz C, Barrufet M et al (2018) Diagnosis and staging of hepatocellular carcinoma (HCC): current guidelines. Eur J Radiol 101:72–81

Batty GD, Shipley MJ, Jarrett RJ, Breeze E, Marmot MG, Smith GD (2005) Obesity and overweight in relation to organ-specific cancer mortality in London (UK): findings from the original Whitehall study. Int J Obes (Lond) 29:1267–1274

Befeler AS, Di Bisceglie AM (2002) Hepatocellular carcinoma: diagnosis and treatment. Gastroenterology 122:1609–1619

Behary J, Amorim N, Jiang XT, Raposo A, Gong L, McGovern E, Ibrahim R, Chu F, Stephens C, Jebeili H et al (2021) Gut microbiota impact on the peripheral immune response in non-alcoholic fatty liver disease related hepatocellular carcinoma. Nat Commun 12:187

Benson AB, D'Angelica MI, Abbott DE, Anaya DA, Anders R, Are C, Bachini M, Borad M, Brown D, Burgoyne A et al (2021) Hepatobiliary cancers, version 2.2021, NCCN clinical practice guidelines in oncology. J Natl Compr Canc Netw 19:541–565

Bouattour M, Raymond E, Qin S, Cheng AL, Stammberger U, Locatelli G, Faivre S (2018) Recent developments of c-Met as a therapeutic target in hepatocellular carcinoma. Hepatology 67: 1132–1149

Bray F, Ferlay J, Soerjomataram I, Siegel RL, Torre LA, Jemal A (2018) Global cancer statistics 2018: GLOBOCAN estimates of incidence and mortality worldwide for 36 cancers in 185 countries. CA Cancer J Clin 68:394–424

Brechot C, Nalpas B, Feitelson MA (1996) Interactions between alcohol and hepatitis viruses in the liver. Clin Lab Med 16:273–287

Briceno J (2020) Artificial intelligence and organ transplantation: challenges and expectations. Curr Opin Organ Transplant 25:393–398

Brown ZJ, Greten TF, Heinrich B (2019) Adjuvant treatment of hepatocellular carcinoma: prospect of immunotherapy. Hepatology 70:1437–1442

Bruix J, Qin S, Merle P, Granito A, Huang YH, Bodoky G, Pracht M, Yokosuka O, Rosmorduc O, Breder V et al (2017) Regorafenib for patients with hepatocellular carcinoma who progressed on sorafenib treatment (RESORCE): a randomised, double-blind, placebo-controlled, phase 3 trial. Lancet 389:56–66

Cancer Genome Atlas Research Network. Electronic address, W.B.E., and Cancer Genome Atlas Research, N (2017) Comprehensive and integrative genomic characterization of hepatocellular carcinoma. Cell 169:1327–1341 e1323

Candia J, Bayarsaikhan E, Tandon M, Budhu A, Forgues M, Tovuu LO, Tudev U, Lack J, Chao A, Chinburen J et al (2020) The genomic landscape of Mongolian hepatocellular carcinoma. Nat Commun 11:4383

Chai W, Zhao Q, Song H, Cheng C, Tian G, Jiang Ta (2019) Treatment response and preliminary efficacy of hepatic tumour laser ablation under the guidance of percutaneous and endoscopic ultrasonography. World J Surg Oncol 17:133

Chang KC, Wu YY, Hung CH, Lu SN, Lee CM, Chiu KW, Tsai MC, Tseng PL, Huang CM, Cho CL et al (2013) Clinical-guide risk prediction of hepatocellular carcinoma development in chronic hepatitis C patients after interferon-based therapy. Br J Cancer 109:2481–2488

Chapiro J (2022) Translating artificial intelligence from code to bedside: the road towards AI-driven predictive biomarkers for immunotherapy of hepatocellular carcinoma. J Hepatol.
Chen CJ, Yang HI (2011) Natural history of chronic hepatitis B REVEALed. J Gastroenterol Hepatol 26:628–638
Chen CL, Yang HI, Yang WS, Liu CJ, Chen PJ, You SL, Wang LY, Sun CA, Lu SN, Chen DS et al (2008) Metabolic factors and risk of hepatocellular carcinoma by chronic hepatitis B/C infection: a follow-up study in Taiwan. Gastroenterology 135:111–121
Chen YL, Jeng YM, Chang CN, Lee HJ, Hsu HC, Lai PL, Yuan RH (2014) TERT promoter mutation in resectable hepatocellular carcinomas: a strong association with hepatitis C infection and absence of hepatitis B infection. Int J Surg 12:659–665
Chen CJ, Yang YH, Lin MH, Lee CP, Tsan YT, Lai MN, Yang HY, Ho WC, Chen PC, Health Data Analysis in Taiwan Research, G (2018a) Herbal medicine containing aristolochic acid and the risk of hepatocellular carcinoma in patients with hepatitis B virus infection. Int J Cancer 143: 1578–1587
Chen M, Wei L, Law CT, Tsang FH, Shen J, Cheng CL, Tsang LH, Ho DW, Chiu DK, Lee JM et al (2018b) RNA N6-methyladenosine methyltransferase-like 3 promotes liver cancer progression through YTHDF2-dependent posttranscriptional silencing of SOCS2. Hepatology 67:2254–2270
Chiang CJ, Yang YW, You SL, Lai MS, Chen CJ (2013) Thirty-year outcomes of the national hepatitis B immunization program in Taiwan. JAMA 310:974–976
Commission, G.O.o.N.H (2022) Standard for diagnosis and treatment of primary liver cancer (2022 edition). J Clinical Oncol 38:288
Dapito DH, Mencin A, Gwak GY, Pradere JP, Jang MK, Mederacke I, Caviglia JM, Khiabanian H, Adeyemi A, Bataller R et al (2012) Promotion of hepatocellular carcinoma by the intestinal microbiota and TLR4. Cancer Cell 21:504–516
Dennis EA, Norris PC (2015) Eicosanoid storm in infection and inflammation. Nat Rev Immunol 15:511–523
Ding S, Robek MD (2014) Cytidine deamination and cccDNA degradation: A new approach for curing HBV? Hepatology 60:2118–2121
Djojosubroto MW, Chin AC, Go N, Schaetzlein S, Manns MP, Gryaznov S, Harley CB, Rudolph KL (2005) Telomerase antagonists GRN163 and GRN163L inhibit tumor growth and increase chemosensitivity of human hepatoma. Hepatology 42:1127–1136
Donato F, Boffetta P, Puoti M (1998) A meta-analysis of epidemiological studies on the combined effect of hepatitis B and C virus infections in causing hepatocellular carcinoma. Int J Cancer 75: 347–354
Donato F, Tagger A, Gelatti U, Parrinello G, Boffetta P, Albertini A, Decarli A, Trevisi P, Ribero ML, Martelli C et al (2002) Alcohol and hepatocellular carcinoma: the effect of lifetime intake and hepatitis virus infections in men and women. Am J Epidemiol 155:323–331
Du Z, Liu X, Wei X, Luo H, Li P, Shi M, Guo B, Cui Y, Su Z, Zeng J et al (2020) Quantitative proteomics identifies a plasma multi-protein model for detection of hepatocellular carcinoma. Sci Rep 10:15552
El-Serag HB (2002) Hepatocellular carcinoma and hepatitis C in the United States. Hepatology 36: S74–S83
El-Serag HB (2012) Epidemiology of viral hepatitis and hepatocellular carcinoma. Gastroenterology 142(1264-1273):e1261
Feng K, Yan J, Li XW, Xia F, Ma KS, Wang SG, Bie P, Dong JH (2012) A randomized controlled trial of radiofrequency ablation and surgical resection in the treatment of small hepatocellular carcinoma. J Hepatol 57:794–802
Finn RS, Zhu AX (2021) Evolution of systemic therapy for hepatocellular carcinoma. Hepatology 73(Suppl 1):150–157
Forbes SA, Beare D, Boutselakis H, Bamford S, Bindal N, Tate J, Cole CG, Ward S, Dawson E, Ponting L et al (2017) COSMIC: somatic cancer genetics at high-resolution. Nucleic Acids Res 45:D777–D783

Forner A, Reig M, Bruix J (2018) Hepatocellular carcinoma. Lancet 391:1301–1314
Fu R, Jiang S, Li J, Chen H, Zhang X (2020) Activation of the HGF/c-MET axis promotes lenvatinib resistance in hepatocellular carcinoma cells with high c-MET expression. Med Oncol 37:24
Fujimoto A, Furuta M, Totoki Y, Tsunoda T, Kato M, Shiraishi Y, Tanaka H, Taniguchi H, Kawakami Y, Ueno M et al (2016) Whole-genome mutational landscape and characterization of noncoding and structural mutations in liver cancer. Nat Genet 48:500–509
Ganne-Carrie N, Chaffaut C, Bourcier V, Archambeaud I, Perarnau JM, Oberti F, Roulot D, Moreno C, Louvet A, Dao T et al (2018) Estimate of hepatocellular carcinoma incidence in patients with alcoholic cirrhosis. J Hepatol 69:1274–1283
Gao R, Cheng J, Fan C, Shi X, Cao Y, Sun B, Ding H, Hu C, Dong F, Yan X (2015) Serum metabolomics to identify the liver disease-specific biomarkers for the progression of hepatitis to hepatocellular carcinoma. Sci Rep 5:18175
Gao L, Zhang Z, Zhang P, Yu M, Yang T (2018) Role of canonical hedgehog signaling pathway in liver. Int J Biol Sci 14:1636–1644
Gao Q, Zhu H, Dong L, Shi W, Chen R, Song Z, Huang C, Li J, Dong X, Zhou Y et al (2019) Integrated proteogenomic characterization of HBV-related hepatocellular carcinoma. Cell 179: 561–577 e522
Garcia-Lezana T, Lopez-Canovas JL, Villanueva A (2021) Signaling pathways in hepatocellular carcinoma. Adv Cancer Res 149:63–101
Geier A, Gartung C, Dietrich CG (2002) Hepatitis B e antigen and the risk of hepatocellular carcinoma. N Engl J Med 347:1721–1722. Author reply 1721–1722
Ginsburg M, Zivin SP, Wroblewski K, Doshi T, Vasnani RJ, Van Ha TG (2015) Comparison of combination therapies in the management of hepatocellular carcinoma: transarterial chemoembolization with radiofrequency ablation versus microwave ablation. J Vasc Interv Radiol 26:330–341
Glassberg MB, Ghosh S, Clymer JW, Qadeer RA, Ferko NC, Sadeghirad B, Wright GWJ, Amaral JF (2019) Microwave ablation compared with radiofrequency ablation for treatment of hepatocellular carcinoma and liver metastases: a systematic review and meta-analysis. Oncotarget Ther 12:6407–6438
Goldstein ST, Zhou F, Hadler SC, Bell BP, Mast EE, Margolis HS (2005) A mathematical model to estimate global hepatitis B disease burden and vaccination impact. Int J Epidemiol 34:1329–1339
Gourd K, Lai C, Reeves C (2020) ESMO Virtual Congress 2020. Lancet Oncol 21:1403–1404
Goyal L, Muzumdar MD, Zhu AX (2013) Targeting the HGF/c-MET pathway in hepatocellular carcinoma. Clin Cancer Res 19:2310–2318
Guo W, Qiu Z, Wang Z, Wang Q, Tan N, Chen T, Chen Z, Huang S, Gu J, Li J, Yao M, Zhao Y, He X (2015) MiR-199a-5p is negatively associated with malignancies and regulates glycolysis and lactate production by targeting hexokinase 2 in liver cancer. Hepatology (Baltimore, Md.), 62(4):1132–1144. https://doi.org/10.1002/hep.27929
Gupta P, Maralakunte M, Kumar-M P, Chandel K, Chaluvashetty SB, Bhujade H, Kalra N, Sandhu MS (2021) Overall survival and local recurrence following RFA, MWA, and cryoablation of very early and early HCC: a systematic review and Bayesian network meta-analysis. EU Radiol 31:5400–5408
Hagel M, Miduturu C, Sheets M, Rubin N, Weng W, Stransky N, Bifulco N, Kim JL, Hodous B, Brooijmans N et al (2015) First selective small molecule inhibitor of FGFR4 for the treatment of hepatocellular carcinomas with an activated FGFR4 signaling pathway. Cancer Discov 5:424–437
Hatlen MA, Schmidt-Kittler O, Sherwin CA, Rozsahegyi E, Rubin N, Sheets MP, Kim JL, Miduturu C, Bifulco N, Brooijmans N et al (2019) Acquired on-target clinical resistance validates FGFR4 as a driver of hepatocellular carcinoma. Cancer Discov 9:1686–1695
He G, Karin M (2011) NF-kappaB and STAT3 – key players in liver inflammation and cancer. Cell Res 21:159–168

He S, Tang S (2020) WNT/beta-catenin signaling in the development of liver cancers. Biomed Pharmacother 132:110851

Heckley GA, Jarl J, Asamoah BO, U, G.G. (2011) How the risk of liver cancer changes after alcohol cessation: a review and meta-analysis of the current literature. BMC Cancer 11:446

Hidaka M, Eguchi S, Okuda K, Beppu T, Shirabe K, Kondo K, Takami Y, Ohta M, Shiraishi M, Ueno S et al (2020) Impact of anatomical resection for hepatocellular carcinoma with microportal invasion (vp1) a multi-institutional study by the Kyushu study group of liver surgery. Ann Surgery 271:339–346

Hill CR, Rivens I, Vaughan MG, ter Haar GR (1994) Lesion development in focused ultrasound surgery: a general model. Ultrasound Med Biol 20:259–269

Hin Tang JJ, Hao Thng DK, Lim JJ, Toh TB (2020) JAK/STAT signaling in hepatocellular carcinoma. Hepat Oncol 7:HEP18

Ho MH, Yu CY, Chung KP, Chen TW, Chu HC, Lin CK, Hsieh CB (2011) Locoregional therapy-induced tumor necrosis as a predictor of recurrence after liver transplant in patients with hepatocellular carcinoma. Ann Surg Oncol 18:3632–3639

Hocquelet A, Aube C, Rode A, Cartier V, Sutter O, Manichon AF, Boursier J, N'Kontchou G, Merle P, Blanc JF et al (2017) Comparison of no-touch multi-bipolar vs. monopolar radiofrequency ablation for small HCC. J Hepatol 66:67–74

Hosaka T, Suzuki F, Kobayashi M, Seko Y, Kawamura Y, Sezaki H, Akuta N, Suzuki Y, Saitoh S, Arase Y et al (2013) Long-term entecavir treatment reduces hepatocellular carcinoma incidence in patients with hepatitis B virus infection. Hepatology 58:98–107

Hsu YC, Wu CY, Lin JT (2015) Hepatitis C virus infection, antiviral therapy, and risk of hepatocellular carcinoma. Semin Oncol 42:329–338

Hu B, Lin JZ, Yang XB, Sang XT (2020) Aberrant lipid metabolism in hepatocellular carcinoma cells as well as immune microenvironment: A review. Cell Prolif 53:e12772

Huang S, He J, Zhang X, Bian Y, Yang L, Xie G, Zhang K, Tang W, Stelter AA, Wang Q et al (2006) Activation of the hedgehog pathway in human hepatocellular carcinomas. Carcinogenesis 27:1334–1340

Huang XY, Huang ZL, Huang J, Xu B, Huang XY, Xu YH, Zhou J, Tang ZY (2020a) Exosomal circRNA-100338 promotes hepatocellular carcinoma metastasis via enhancing invasiveness and angiogenesis. J Exp Clin Cancer Res 39:20

Huang Z, Zhou JK, Peng Y, He W, Huang C (2020b) The role of long noncoding RNAs in hepatocellular carcinoma. Mol Cancer 19:77

Huettl F, Saalfeld P, Hansen C, Preim B, Poplawski A, Kneist W, Lang H, Huber T (2021) Virtual reality and 3D printing improve preoperative visualization of 3D liver reconstructions-results from a preclinical comparison of presentation modalities and user's preference. Ann Transl Med 9:1074

Hui TCH, Kwan J, Pua U (2021) Advanced techniques in the percutaneous ablation of liver tumours. Diagnostics 11:585

Humans, I.W.G.o.t.E.o.C.R.t. (2012a) Biological agents. Volume 100 B. A review of human carcinogens. IARC Monogr Eval Carcinog Risks Hum 100:1–441

Humans, I.W.G.o.t.E.o.C.R.t. (2012b) Pharmaceuticals. Volume 100 A. A review of human carcinogens. IARC Monogr Eval Carcinog Risks Hum 100:1–401

Ibrahim SM, Lewandowski RJ, Sato KT, Gates VL, Kulik L, Mulcahy MF, Ryu RK, Omary RA, Salem R (2008) Radioembolization for the treatment of unresectable hepatocellular carcinoma: a clinical review. World J Gastroenterol 14:1664–1669

Jang WI, Bae SH, Kim MS, Han CJ, Park SC, Kim SB, Cho EH, Choi CW, Kim KS, Hwang S et al (2020) A phase 2 multicenter study of stereotactic body radiotherapy for hepatocellular carcinoma: safety and efficacy. Cancer 126:363–372

Jang JW, Kim HS, Kim JS, Lee SK, Han JW, Sung PS, Bae SH, Choi JY, Yoon SK, Han DJ et al (2021) Distinct patterns of HBV integration and TERT alterations between in tumor and non-tumor tissue in patients with hepatocellular carcinoma. Int J Mol Sci 22:7056

Jia J, Qiao Y, Pilo MG, Cigliano A, Liu X, Shao Z, Calvisi DF, Chen X (2017) Tankyrase inhibitors suppress hepatocellular carcinoma cell growth via modulating the Hippo cascade. PLoS One 12:e0184068

Jin H, Shi Y, Lv Y, Yuan S, Ramirez CFA, Lieftink C, Wang L, Wang S, Wang C, Dias MH et al (2021) EGFR activation limits the response of liver cancer to lenvatinib. Nature 595:730–734

Joechle K, Guenzle J, Hellerbrand C, Strnad P, Cramer T, Neumann UP, Lang SA (2021) Role of mammalian target of rapamycin complex 2 in primary and secondary liver cancer. World J Gastrointest Oncol 13:1632–1647

Kan Z, Zheng H, Liu X, Li S, Barber TD, Gong Z, Gao H, Hao K, Willard MD, Xu J et al (2013) Whole-genome sequencing identifies recurrent mutations in hepatocellular carcinoma. Genome Res 23:1422–1433

Kawaguchi K, Kaneko S (2021) Notch Signaling and Liver Cancer. Adv Exp Med Biol 1287:69–80

Khorsandi SE, Hardgrave HJ, Osborn T, Klutts G, Nigh J, Spencer-Cole RT, Kakos CD, Anastasiou I, Mavros MN, Giorgakis E (2021) Artificial intelligence in liver transplantation. Transplant Proc 53:2939–2944

Kim Y, Yoon JW, Xiao X, Dean NM, Monia BP, Marcusson EG (2007) Selective down-regulation of glioma-associated oncogene 2 inhibits the proliferation of hepatocellular carcinoma cells. Cancer Res 67:3583–3593

Kim BS, Seo YS, Kim YS, Lee CH, Lee HA, Um SH, Yoo JJ, Kim SG, Suh SJ, Jung YK et al (2018) Reduced risk of hepatocellular carcinoma by achieving a subcirrhotic liver stiffness through antiviral agents in hepatitis B virus-related advanced fibrosis or cirrhosis. J Gastroenterol Hepatol 33:503–510

Kim RD, Sarker D, Meyer T, Yau T, Macarulla T, Park JW, Choo SP, Hollebecque A, Sung MW, Lim HY et al (2019) First-in-human Phase I study of fisogatinib (BLU-554) validates aberrant FGF19 signaling as a driver event in hepatocellular carcinoma. Cancer Discov 9:1696–1707

Kim TH, Koh YH, Kim BH, Kim MJ, Lee JH, Park B, Park JW (2021) Proton beam radiotherapy vs. radiofrequency ablation for recurrent hepatocellular carcinoma: a randomized phase III trial. J Hepatol 74:603–612

Kotsifa E, Vergadis C, Vailas M, Machairas N, Kykalos S, Damaskos C, Garmpis N, Lianos GD, Schizas D (2022) Transarterial chemoembolization for hepatocellular carcinoma: why, when, how? J Pers Med 12:436

Kudo M, Hasegawa K, Kawaguchi Y, Takayama T, Izumi N, Yamanaka N, Shimada M, Inomata M, Kaneko S, Baba H et al (2021) A multicenter randomized controlled trial to evaluate the efficacy of surgery versus radiofrequency ablation for small hepatocellular carcinoma (SURF trial): Analysis of overall survival. J Clin Oncol 39(15_suppl):4093

Lee W, Han HS, Yoon YS, Cho JY, Choi Y, Shin HK, Jang JY, Choi H, Jang JS, Kwon SU (2016) Comparison of laparoscopic liver resection for hepatocellular carcinoma located in the posterosuperior segments or anterolateral segments: a case-matched analysis. Surgery 160: 1219–1226

Lee S, Kim KW, Song GW, Kwon JH, Hwang S, Kim KH, Ahn CS, Moon DB, Park GC, Lee SG (2020) The real impact of bridging or downstaging on survival outcomes after liver transplantation for hepatocellular carcinoma. Liver Cancer 9:721–733

Lei J, Yan L (2013) Outcome comparisons among the Hangzhou, Chengdu, and UCSF criteria for hepatocellular carcinoma liver transplantation after successful downstaging therapies. J Gastrointest Surg 17:1116–1122

Lewis AR, Padula CA, McKinney JM, Toskich BB (2019) Ablation plus transarterial embolic therapy for hepatocellular carcinoma larger than 3 cm: science, evidence, and future directions. Seminar Interven Radiol 36:303–309

Li M, Zhao H, Zhang X, Wood LD, Anders RA, Choti MA, Pawlik TM, Daniel HD, Kannangai R, Offerhaus GJ et al (2011) Inactivating mutations of the chromatin remodeling gene ARID2 in hepatocellular carcinoma. Nat Genet 43:828–829

Liang B, Makamure J, Shu SL, Zhang LJ, Sun T, Zheng CS (2021) Treatment response, survival, and safety of transarterial chemoembolization With callispheres(R) microspheres versus conventional transarterial chemoembolization in hepatocellular carcinoma: a meta-analysis. Front Oncol 11:576232

Liao H, Long Y, Han R, Wang W, Xu L, Liao M, Zhang Z, Wu Z, Shang X, Li X et al (2020) Deep learning-based classification and mutation prediction from histopathological images of hepatocellular carcinoma. Clin Transl Med 10:e102

Kim, Y.-s., Lim, H.K., Rhim, H., and Lee, M.W. (2014). Ablation of hepatocellular carcinoma. Best Pract & Res Clin Gastroenterol 28, 897-908.

Liu PH, Hsu CY, Hsia CY, Lee YH, Huang YH, Chiou YY, Lin HC, Huo TI (2016) Surgical resection versus radiofrequency ablation for single hepatocellular carcinoma <= 2 cm in a propensity score model. Ann Surg 263:538–545

Liu X, Baecker A, Wu M, Zhou JY, Yang J, Han RQ, Wang PH, Jin ZY, Liu AM, Gu X et al (2018) Interaction between tobacco smoking and hepatitis B virus infection on the risk of liver cancer in a Chinese population. Int J Cancer 142:1560–1567

Liu D, Liu M, Su L, Wang Y, Zhang X, Long H, Kuang M, Xie X, Lin M (2019) Transarterial chemoembolization followed by radiofrequency ablation for hepatocellular carcinoma: impact of the time interval between the two treatments on outcome. J Vasc Interv Radiol 30:1879–1886

Liu Y, Wang X, Yang Y (2020) Hepatic Hippo signaling inhibits development of hepatocellular carcinoma. Clin Mol Hepatol 26:742–750

Llovet JM, Burroughs A, Bruix J (2003) Hepatocellular carcinoma. Lancet 362:1907–1917

Llovet JM, Ricci S, Mazzaferro V, Hilgard P, Gane E, Blanc JF, de Oliveira AC, Santoro A, Raoul JL, Forner A et al (2008) Sorafenib in advanced hepatocellular carcinoma. N Engl J Med 359:378–390

Llovet JM, De Baere T, Kulik L, Haber PK, Greten TF, Meyer T, Lencioni R (2021a) Locoregional therapies in the era of molecular and immune treatments for hepatocellular carcinoma. Nat Rev Gastroenterol Hepatol 18:293–313

Llovet JM, Kelley RK, Villanueva A, Singal AG, Pikarsky E, Roayaie S, Lencioni R, Koike K, Zucman-Rossi J, Finn RS (2021b) Hepatocellular carcinoma. Nat Rev Dis Primers 7:6

Lobo L, Yakoub D, Picado O, Ripat C, Pendola F, Sharma R, ElTawil R, Kwon D, Venkat S, Portelance L et al (2016) Unresectable hepatocellular carcinoma: radioembolization versus chemoembolization: a systematic review and meta-analysis. Cardiovasc Interv Radiol 39: 1580–1588

Lu Y, Chen J, Huang C, Li N, Zou L, Chia SE, Chen S, Yu K, Ling Q, Cheng Q et al (2018) Comparison of hepatic and serum lipid signatures in hepatocellular carcinoma patients leads to the discovery of diagnostic and prognostic biomarkers. Oncotarget 9:5032–5043

Lu X, Chen H, Patterson AV, Smaill JB, Ding K (2019) Fibroblast growth factor receptor 4 (FGFR4) selective inhibitors as hepatocellular carcinoma therapy: advances and prospects. J Med Chem 62:2905–2915

Lu S, Cai S, Peng X, Cheng R, Zhang Y (2021) Integrative transcriptomic, proteomic and functional analysis reveals ATP1B3 as a diagnostic and potential therapeutic target in hepatocellular carcinoma. Front Immunol 12:636614

Luo N, Balko JM (2019) Role of JAK-STAT pathway in cancer signaling. In: Badve S, Kumar GL (eds) Predictive biomarkers in oncology: applications in precision medicine. Springer International Publishing, Cham, pp 311–319

Luo P, Yin P, Hua R, Tan Y, Li Z, Qiu G, Yin Z, Xie X, Wang X, Chen W et al (2018) A Large-scale, multicenter serum metabolite biomarker identification study for the early detection of hepatocellular carcinoma. Hepatology 67:662–675

Luo J, Dong Z, Xie H, Zhang W, An L, Yu Z, Chang X, Chen Y, Yang W, Hu K et al (2022) Efficacy and safety of percutaneous cryoablation for elderly patients with small hepatocellular carcinoma: a prospective multicenter study. Liver Int 42:918–929

Lyu SC, Wang J, Ren ZY, Cao D, He Q (2022) Prognosis of liver transplantation for hepatocellular carcinoma in terms of different criteria: a single center experience. Curr Med Sci 42:548–554

Ma JZ, Yang F, Zhou CC, Liu F, Yuan JH, Wang F, Wang TT, Xu QG, Zhou WP, Sun SH (2017) METTL14 suppresses the metastatic potential of hepatocellular carcinoma by modulating N(6) -methyladenosine-dependent primary MicroRNA processing. Hepatology 65:529–543

Maataoui A, Qian J, Vossoughi D, Khan MF, Oppermann E, Bechstein WO, Vogl TJ (2005) Transarterial chemoembolization alone and in combination with other therapies: a comparative study in an animal HCC model. Eur Radiol 15:127–133

Machado MV, Diehl AM (2018) Hedgehog signalling in liver pathophysiology. J Hepatol 68:550–562

Marrero JA, Fontana RJ, Fu S, Conjeevaram HS, Su GL, Lok AS (2005) Alcohol, tobacco and obesity are synergistic risk factors for hepatocellular carcinoma. J Hepatol 42:218–224

Martinez-Quetglas I, Pinyol R, Dauch D, Torrecilla S, Tovar V, Moeini A, Alsinet C, Portela A, Rodriguez-Carunchio L, Sole M et al (2016) IGF2 Is Up-regulated by epigenetic mechanisms in hepatocellular carcinomas and is an actionable oncogene product in experimental models. Gastroenterology 151:1192–1205

Matboli M, Shafei AE, Ali MA, Ashry AM, Kamal KM, Agag MA, Reda I, Tash EF, Ali M (2018) circRNAs (hsa_circ_00156, hsa_circ _000224, and hsa_circ _000520) are novel potential biomarkers in hepatocellular carcinoma. J Cell Biochem. https://doi.org/10.1002/jcb.28045

Matz-Soja M, Rennert C, Schonefeld K, Aleithe S, Boettger J, Schmidt-Heck W, Weiss TS, Hovhannisyan A, Zellmer S, Kloting N et al (2016) Hedgehog signaling is a potent regulator of liver lipid metabolism and reveals a GLI-code associated with steatosis. Elife 5:e13308

Mazzaferro V, Citterio D, Bhoori S (2020) Liver transplantation in hepatocellular carcinoma after tumour downstaging (XXL): a randomised, controlled, phase 2b/3 trial (vol 21, pg 947, 2020). Lancet Oncology 21:E373–E373

McGlynn KA, Petrick JL, London WT (2015) Global epidemiology of hepatocellular carcinoma: an emphasis on demographic and regional variability. Clin Liver Dis 19:223–238

McMahon BJ (2009) The natural history of chronic hepatitis B virus infection. Hepatology 49:S45–S55

Melincovici CS, Bosca AB, Susman S, Marginean M, Mihu C, Istrate M, Moldovan IM, Roman AL, Mihu CM (2018) Vascular endothelial growth factor (VEGF) – key factor in normal and pathological angiogenesis. Rom J Morphol Embryol 59:455–467

Mjelle R, Dima SO, Bacalbasa N, Chawla K, Sorop A, Cucu D, Herlea V, Saetrom P, Popescu I (2019) Comprehensive transcriptomic analyses of tissue, serum, and serum exosomes from hepatocellular carcinoma patients. BMC Cancer 19:1007

Mukaiya M, Nishi M, Miyake H, Hirata K (1998) Chronic liver diseases for the risk of hepatocellular carcinoma: a case-control study in Japan. Etiologic association of alcohol consumption, cigarette smoking and the development of chronic liver diseases. Hepatogastroenterology 45:2328–2332

Nault JC, Ningarhari M, Rebouissou S, Zucman-Rossi J (2019) The role of telomeres and telomerase in cirrhosis and liver cancer. Nat Rev Gastroenterol Hepatol 16:544–558

Ng AWT, Poon SL, Huang MN, Lim JQ, Boot A, Yu W, Suzuki Y, Thangaraju S, Ng CCY, Tan P et al (2017) Aristolochic acids and their derivatives are widely implicated in liver cancers in Taiwan and throughout Asia. Sci Transl Med 9:eaan6446

Nishikawa T, Bellance N, Damm A, Bing H, Zhu Z, Handa K, Yovchev MI, Sehgal V, Moss TJ, Oertel M et al (2014) A switch in the source of ATP production and a loss in capacity to perform glycolysis are hallmarks of hepatocyte failure in advance liver disease. J Hepatol 60:1203–1211

N'Kontchou G, Mahamoudi A, Aout M, Ganne-Carrie N, Grando V, Coderc E, Vicaut E, Trinchet JC, Sellier N, Beaugrand M et al (2009) Radiofrequency ablation of hepatocellular carcinoma: long-term results and prognostic factors in 235 Western patients with cirrhosis. Hepatology 50:1475–1483

Nordness MF, Hamel S, Godfrey CM, Shi CJ, Johnson DB, Goff LW, O'Dell H, Perri RE, Alexopoulos SP (2020) Fatal hepatic necrosis after nivolumab as a bridge to liver transplant for HCC: are checkpoint inhibitors safe for the pretransplant patient? Am J Transplant 20:879–883

Pan Y, Xia SJ, Cai JQ, Chen K, Cai XJ (2021) Efficacy of laparoscopic hepatectomy versus open surgery for hepatocellular carcinoma with cirrhosis: a meta-analysis of case-matched studies. Front Oncol 11:652272

Park JW, Chen M, Colombo M, Roberts LR, Schwartz M, Chen PJ, Kudo M, Johnson P, Wagner S, Orsini LS et al (2015) Global patterns of hepatocellular carcinoma management from diagnosis to death: the BRIDGE Study. Liver Int 35:2155–2166

Peneau C, Imbeaud S, La Bella T, Hirsch TZ, Caruso S, Calderaro J, Paradis V, Blanc JF, Letouze E, Nault JC et al (2022) Hepatitis B virus integrations promote local and distant oncogenic driver alterations in hepatocellular carcinoma. Gut 71:616–626

Peng ZW, Lin XJ, Zhang YJ, Liang HH, Guo RP, Shi M, Chen MS (2012) Radiofrequency ablation versus hepatic resection for the treatment of hepatocellular carcinomas 2 cm or smaller: a retrospective comparative study. Radiology 262:1022–1033

Personeni N, Pressiani T, Rimassa L (2019) Lenvatinib for the treatment of unresectable hepatocellular carcinoma: evidence to date. J Hepatocell Carcinoma 6:31–39

Pillai A, Ahn J, Kulik L (2020) Integrating genomics into clinical practice in hepatocellular carcinoma: the challenges ahead. Am J Gastroenterol 115:1960–1969

Plummer M, de Martel C, Vignat J, Ferlay J, Bray F, Franceschi S (2016) Global burden of cancers attributable to infections in 2012: a synthetic analysis. Lancet Glob Health 4:e609–e616

Ponziani FR, Bhoori S, Castelli C, Putignani L, Rivoltini L, Del Chierico F, Sanguinetti M, Morelli D, Paroni Sterbini F, Petito V et al (2019) Hepatocellular carcinoma is associated with gut microbiota profile and inflammation in nonalcoholic fatty liver disease. Hepatology 69:107–120

Qi XS, Guo XZ, Han GH, Li HY, Chen J (2015) MET inhibitors for treatment of advanced hepatocellular carcinoma: a review. World J Gastroenterol 21:5445–5453

Qin S, Cheng Y, Liang J, Shen L, Bai Y, Li J, Fan J, Liang L, Zhang Y, Wu G et al (2014) Efficacy and safety of the FOLFOX4 regimen versus doxorubicin in Chinese patients with advanced hepatocellular carcinoma: a subgroup analysis of the EACH study. Oncologist 19:1169–1178

Qin S, Bi F, Gu S, Bai Y, Chen Z, Wang Z, Ying J, Lu Y, Meng Z, Pan H et al (2021a) Donafenib versus Sorafenib in first-line treatment of unresectable or metastatic hepatocellular carcinoma: a randomized, open-label, parallel-controlled Phase II-III trial. J Clin Oncol 39:3002–3011

Qin S, Li Q, Gu S, Chen X, Lin L, Wang Z, Xu A, Chen X, Zhou C, Ren Z et al (2021b) Apatinib as second-line or later therapy in patients with advanced hepatocellular carcinoma (AHELP): a multicentre, double-blind, randomised, placebo-controlled, phase 3 trial. Lancet Gastroenterol Hepatol 6:559–568

Rai V, Mukherjee S (2022) Targets of immunotherapy for hepatocellular carcinoma: an update. World J Hepatol 14:140–157

Raimondi S, Bruno S, Mondelli MU, Maisonneuve P (2009) Hepatitis C virus genotype 1b as a risk factor for hepatocellular carcinoma development: a meta-analysis. J Hepatol 50:1142–1154

Raja A, Park I, Haq F, Ahn SM (2019) FGF19-FGFR4 signaling in hepatocellular carcinoma. Cells 8:536

Reig M, Forner A, Rimola J, Ferrer-Fabrega J, Burrel M, Garcia-Criado A, Kelley RK, Galle PR, Mazzaferro V, Salem R et al (2022) BCLC strategy for prognosis prediction and treatment recommendation: the 2022 update. J Hepatol 76:681–693

Ren Z, Li A, Jiang J, Zhou L, Yu Z, Lu H, Xie H, Chen X, Shao L, Zhang R et al (2019) Gut microbiome analysis as a tool towards targeted non-invasive biomarkers for early hepatocellular carcinoma. Gut 68:1014–1023

Ren Z, Xu J, Bai Y, Xu A, Cang S, Du C, Li Q, Lu Y, Chen Y, Guo Y et al (2021) Sintilimab plus a bevacizumab biosimilar (IBI305) versus sorafenib in unresectable hepatocellular carcinoma (ORIENT-32): a randomised, open-label, phase 2-3 study. Lancet Oncol 22:977–990

Reuter B, Bajaj JS (2020) Microbiome: emerging concepts in patients with chronic liver disease. Clin Liver Dis 24:493–520

Ross RK, Yuan JM, Yu MC, Wogan GN, Qian GS, Tu JT, Groopman JD, Gao YT, Henderson BE (1992) Urinary aflatoxin biomarkers and risk of hepatocellular carcinoma. Lancet 339:943–946

Saitta C, Pollicino T, Raimondo G (2019) Obesity and liver cancer. Ann Hepatol 18:810–815

Salem R, Mazzaferro V, Sangro B (2013) Yttrium 90 radioembolization for the treatment of hepatocellular carcinoma: biological lessons, current challenges, and clinical perspectives. Hepatology 58:2188–2197

Sapisochin G, Barry A, Doherty M, Fischer S, Goldaracena N, Rosales R, Russo M, Beecroft R, Ghanekar A, Bhat M et al (2017) Stereotactic body radiotherapy vs. TACE or RFA as a bridge to transplant in patients with hepatocellular carcinoma. An intention-to-treat analysis. J Hepatol 67:92–99

Sas Z, Cendrowicz E, Weinhauser I, Rygiel TP (2022) Tumor microenvironment of hepatocellular carcinoma: challenges and opportunities for new treatment options. Int J Mol Sci 23:3778

Schlesinger S, Aleksandrova K, Pischon T, Jenab M, Fedirko V, Trepo E, Overvad K, Roswall N, Tjonneland A, Boutron-Ruault MC et al (2013) Diabetes mellitus, insulin treatment, diabetes duration, and risk of biliary tract cancer and hepatocellular carcinoma in a European cohort. Ann Oncol 24:2449–2455

Schnabl B, Brenner DA (2014) Interactions between the intestinal microbiome and liver diseases. Gastroenterology 146:1513–1524

Schnitzbauer AA, Filmann N, Adam RP, Bachellier P, Bechstein WO, Becker T, Bhoori S, Bilbao I, Brockmann J, Burra P et al (2020) mTOR inhibition is most beneficial after liver transplantation for hepatocellular carcinoma in patients with active tumors. Ann Surg 272:855–862

Shay JW, Wright WE (2011) Role of telomeres and telomerase in cancer. Semin Cancer Biol 21: 349–353

Shi J, Zhu L, Liu S, Xie WF (2005) A meta-analysis of case-control studies on the combined effect of hepatitis B and C virus infections in causing hepatocellular carcinoma in China. Br J Cancer 92:607–612

Shi GM, Wang JP, Huang XW, Huang XY, He YF, Ji Y, Chen Y, Wu D, Lu JC, Sun QM et al (2021) Graft programmed death ligand 1 expression as a marker for transplant rejection following anti-programmed death 1 immunotherapy for recurrent liver tumors. Liver Transplan 27:444–449

Shiina S, Tateishi R, Arano T, Uchino K, Enooku K, Nakagawa H, Asaoka Y, Sato T, Masuzaki R, Kondo Y et al (2012) Radiofrequency ablation for hepatocellular carcinoma: 10-year outcome and prognostic factors. Am J Gastroenterol 107:569–577

Sidorenko VS, Yeo JE, Bonala RR, Johnson F, Scharer OD, Grollman AP (2012) Lack of recognition by global-genome nucleotide excision repair accounts for the high mutagenicity and persistence of aristolactam-DNA adducts. Nucleic Acids Res 40:2494–2505

Siegel RL, Miller KD, Fuchs HE, Jemal A (2022) Cancer statistics, 2022. CA Cancer J Clin 72:7–33

Smith MA, Licata T, Lakhani A, Garcia MV, Schildhaus HU, Vuaroqueaux V, Halmos B, Borczuk AC, Chen YA, Creelan BC et al (2017) MET-GRB2 signaling-associated complexes correlate with oncogenic MET signaling and sensitivity to MET kinase inhibitors. Clin Cancer Res 23: 7084–7096

Sonbol MB, Riaz IB, Naqvi SAA, Almquist DR, Mina S, Almasri J, Shah S, Almader-Douglas D, Uson Junior PLS, Mahipal A et al (2020) Systemic therapy and sequencing options in advanced hepatocellular carcinoma: a systematic review and network meta-analysis. JAMA Oncol 6:e204930

Song MA, Kwee SA, Tiirikainen M, Hernandez BY, Okimoto G, Tsai NC, Wong LL, Yu H (2016) Comparison of genome-scale DNA methylation profiles in hepatocellular carcinoma by viral status. Epigenetics 11:464–474

Song S, Xie M, Scott AW, Jin J, Ma L, Dong X, Skinner HD, Johnson RL, Ding S, Ajani JA (2018) A novel YAP1 inhibitor targets CSC-enriched radiation-resistant cells and exerts strong antitumor activity in esophageal adenocarcinoma. Mol Cancer Ther 17:443–454

Stepien M, Keski-Rahkonen P, Kiss A, Robinot N, Duarte-Salles T, Murphy N, Perlemuter G, Viallon V, Tjonneland A, Rostgaard-Hansen AL et al (2021) Metabolic perturbations prior to hepatocellular carcinoma diagnosis: findings from a prospective observational cohort study. Int J Cancer 148:609–625

Sucandy I, Shapera E, Syblis CC, Crespo K, Przetocki VA, Ross SB, Rosemurgy AS (n.d.) Propensity score matched comparison of robotic and open major hepatectomy for malignant liver tumors. Surgical Endoscopy and Other Interventional Techniques

Sun Z, Lu P, Gail MH, Pee D, Zhang Q, Ming L, Wang J, Wu Y, Liu G, Wu Y et al (1999) Increased risk of hepatocellular carcinoma in male hepatitis B surface antigen carriers with chronic hepatitis who have detectable urinary aflatoxin metabolite M1. Hepatology 30:379–383

Sun Z, Chen T, Thorgeirsson SS, Zhan Q, Chen J, Park JH, Lu P, Hsia CC, Wang N, Xu L et al (2013) Dramatic reduction of liver cancer incidence in young adults: 28 year follow-up of etiological interventions in an endemic area of China. Carcinogenesis 34:1800–1805

Swamy SG, Kameshwar VH, Shubha PB, Looi CY, Shanmugam MK, Arfuso F, Dharmarajan A, Sethi G, Shivananju NS, Bishayee A (2017) Targeting multiple oncogenic pathways for the treatment of hepatocellular carcinoma. Target Oncol 12:1–10

Tabrizian P, Holzner M, Halazun K, Agopian VG, Busuttil RW, Yao F, Roberts J, Florman SS, Schwartz ME, Emond JC et al (2019) A US multicenter analysis of 2529 HCC patients undergoing liver transplantation: 10-year outcome assessing the role of down-staging to within Milan criteria. Hepatology 70:10A–11A

Tan WC, Deng QW, Lin SY, Wang YH, Xu GL (2019) Comparison of microwave ablation and radiofrequency ablation for hepatocellular carcinoma: a systematic review and meta-analysis. Int J Hypertherm 36:264–272

Teo JY, Kam JH, Chan CY, Goh BKP, Wong JS, Lee VTW, Cheow PC, Chow PKH, Ooi LLP, Chung AYF et al (2015) Laparoscopic liver resection for posterosuperior and anterolateral lesions-a comparison experience in an Asian centre. Hepatobiliary Surg Nutr 4:379–390

Tong MJ, Lai LP, Murakami-Mori K (1997) Development of hepatocellular carcinoma after clearance of hepatitis C virus with interferon therapy. West J Med 167:103–105

Tropberger P, Mercier A, Robinson M, Zhong W, Ganem DE, Holdorf M (2015) Mapping of histone modifications in episomal HBV cccDNA uncovers an unusual chromatin organization amenable to epigenetic manipulation. Proc Natl Acad Sci U S A 112:E5715–E5724

Tsukuma H, Hiyama T, Tanaka S, Nakao M, Yabuuchi T, Kitamura T, Nakanishi K, Fujimoto I, Inoue A, Yamazaki H et al (1993) Risk factors for hepatocellular carcinoma among patients with chronic liver disease. N Engl J Med 328:1797–1801

Twaroski K, Mallanna SK, Jing R, DiFurio F, Urick A, Duncan SA (2015) FGF2 mediates hepatic progenitor cell formation during human pluripotent stem cell differentiation by inducing the WNT antagonist NKD1. Genes Dev 29:2463–2474

Valizadeh A, Sayadmanesh A, Asemi Z, Alemi F, Mahmoodpoor A, Yousefi B (2021) Regulatory roles of the notch signaling pathway in liver repair and regeneration: a novel therapeutic target. Curr Med Chem 28:8608–8626

van der Meer AJ, Veldt BJ, Feld JJ, Wedemeyer H, Dufour JF, Lammert F, Duarte-Rojo A, Heathcote EJ, Manns MP, Kuske L et al (2012) Association between sustained virological response and all-cause mortality among patients with chronic hepatitis C and advanced hepatic fibrosis. JAMA 308:2584–2593

Violi NV, Duran R, Guiu B, Cercueil JP, Aube C, Digklia A, Pache I, Deltenre P, Knebel JF, Denys A (2018) Efficacy of microwave ablation versus radiofrequency ablation for the treatment of hepatocellular carcinoma in patients with chronic liver disease: a randomised controlled phase 2 trial. Lancet Gastroenterol Hepatol 3:317–325

Vogel A, Cervantes A, Chau I, Daniele B, Llovet JM, Meyer T, Nault JC, Neumann U, Ricke J, Sangro B et al (2018) Hepatocellular carcinoma: ESMO Clinical Practice Guidelines for diagnosis, treatment and follow-up. Ann Oncol 29:iv238–iv255

Wang C, Wang X, Gong G, Ben Q, Qiu W, Chen Y, Li G, Wang L (2012a) Increased risk of hepatocellular carcinoma in patients with diabetes mellitus: a systematic review and meta-analysis of cohort studies. Int J Cancer 130:1639–1648

Wang P, Kang D, Cao W, Wang Y, Liu Z (2012b) Diabetes mellitus and risk of hepatocellular carcinoma: a systematic review and meta-analysis. Diabetes Metab Res Rev 28:109–122

Wang H, Rao B, Lou J, Li J, Liu Z, Li A, Cui G, Ren Z, Yu Z (2020a) The function of the HGF/c-Met axis in hepatocellular carcinoma. Front Cell Dev Biol 8:55
Wang L, Zhang Z, Li Y, Wan Y, Xing B (2020b) Integrated bioinformatic analysis of RNA binding proteins in hepatocellular carcinoma. Aging (Albany NY) 13:2480–2505
Wang J, Zhao H, Zheng L, Zhou Y, Wu L, Xu Y, Zhang X, Yan G, Sheng H, Xin R et al (2021) FGF19/SOCE/NFATc2 signaling circuit facilitates the self-renewal of liver cancer stem cells. Theranostics 11:5045–5060
Wei H, Wang F, Wang Y, Li T, Xiu P, Zhong J, Sun X, Li J (2017) Verteporfin suppresses cell survival, angiogenesis and vasculogenic mimicry of pancreatic ductal adenocarcinoma via disrupting the YAP-TEAD complex. Cancer Sci 108:478–487
Welzel TM, Graubard BI, Quraishi S, Zeuzem S, Davila JA, El-Serag HB, McGlynn KA (2013) Population-attributable fractions of risk factors for hepatocellular carcinoma in the United States. Am J Gastroenterol 108:1314–1321
Wen Y, Han J, Chen J, Dong J, Xia Y, Liu J, Jiang Y, Dai J, Lu J, Jin G et al (2015) Plasma miRNAs as early biomarkers for detecting hepatocellular carcinoma. Int J Cancer 137:1679–1690
Wen X, Zhang B, Wu B, Xiao H, Li Z, Li R, Xu X, Li T (2022) Signaling pathways in obesity: mechanisms and therapeutic interventions. Signal Transduct Target Ther 7:298
West J, Card TR, Aithal GP, Fleming KM (2017) Risk of hepatocellular carcinoma among individuals with different aetiologies of cirrhosis: a population-based cohort study. Aliment Pharmacol Ther 45:983–990
Wild CP, Miller JD, Groopman JD (2015) In: Wild CP, Miller JD, Groopman JD (eds) Mycotoxin control in low- and middle-income countries. IARC Working Group Report No 9, International Agency for Research on Cancer, Lyon
Wu L, Zhang X, Zheng L, Zhao H, Yan G, Zhang Q, Zhou Y, Lei J, Zhang J, Wang J et al (2020) RIPK3 orchestrates fatty acid metabolism in tumor-associated macrophages and hepatocarcinogenesis. Cancer Immunol Res 8:710–721
Xia Y, Li J, Liu GH, Wang K, Qian GJ, Lu ZH, Yang T, Yan ZL, Lei ZQ, Si AF et al (2020) Long-term effects of repeat hepatectomy vs percutaneous radiofrequency ablation among patients with recurrent hepatocellular carcinoma a randomized clinical trial. Jama Oncology 6:255–263
Xie Z, Zhou F, Yang Y, Li L, Lei Y, Lin X, Li H, Pan X, Chen J, Wang G et al (2018) Lnc-PCDH9-13:1 is a hypersensitive and specific biomarker for early hepatocellular carcinoma. EBioMedicine 33:57–67
Xiong L, Wu F, Wu Q, Xu L, Cheung OK, Kang W, Mok MT, Szeto LLM, Lun CY, Lung RW et al (2019) Aberrant enhancer hypomethylation contributes to hepatic carcinogenesis through global transcriptional reprogramming. Nat Commun 10:335
Yamada K, Oikawa T, Kizawa R, Motohashi S, Yoshida S, Kumamoto T, Saeki C, Nakagawa C, Shimoyama Y, Aoki K et al (2021) Unconventional secretion of PKCdelta exerts tumorigenic function via stimulation of ERK1/2 signaling in liver cancer. Cancer Res 81:414–425
Yang JD, Roberts LR (2010) Hepatocellular carcinoma: a global view. Nat Rev Gastroenterol Hepatol 7:448–458
Yang JD, Harmsen WS, Slettedahl SW, Chaiteerakij R, Enders FT, Therneau TM, Orsini L, Kim WR, Roberts LR (2011) Factors that affect risk for hepatocellular carcinoma and effects of surveillance. Clin Gastroenterol Hepatol 9(617–623):e611
Yang B, Li B, Jia L, Jiang Y, Wang X, Jiang S, Du S, Ji X, Yang P (2020) 3D landscape of hepatitis B virus interactions with human chromatins. Cell Discov 6:95
Yao FY, Fidelman N (2016) Reassessing the boundaries of liver transplantation for hepatocellular carcinoma: where do we stand with tumor down-staging? Hepatology 63:1014–1025
Younossi ZM, Otgonsuren M, Henry L, Venkatesan C, Mishra A, Erario M, Hunt S (2015) Association of nonalcoholic fatty liver disease (NAFLD) with hepatocellular carcinoma (HCC) in the United States from 2004 to 2009. Hepatology 62:1723–1730
Yu LX, Schwabe RF (2017) The gut microbiome and liver cancer: mechanisms and clinical translation. Nat Rev Gastroenterol Hepatol 14:527–539

Yu JI, Choi C, Ha SY, Park CK, Kang SY, Joh JW, Paik SW, Kim S, Kim M, Jung SH et al (2017) Clinical importance of TERT overexpression in hepatocellular carcinoma treated with curative surgical resection in HBV endemic area. Sci Rep 7:12258

Zhang Z (2012) Genomic landscape of liver cancer. Nature Genetics 44:1075–1077

Zhang Q, Wei L, Yang H, Yang W, Yang Q, Zhang Z, Wu K, Wu J (2016) Bromodomain containing protein represses the Ras/Raf/MEK/ERK pathway to attenuate human hepatoma cell proliferation during HCV infection. Cancer Lett 371:107–116

Zhang W, He H, Zang M, Wu Q, Zhao H, Lu LL, Ma P, Zheng H, Wang N, Zhang Y et al (2017) Genetic features of aflatoxin-associated hepatocellular carcinoma. Gastroenterology 153(249-262):e242

Zhang C, Huang C, Sui X, Zhong X, Yang W, Hu X, Li Y (2019) Association between gene methylation and HBV infection in hepatocellular carcinoma: a meta-analysis. J Cancer 10:6457–6465

Zhang F, Wang Y, Chen G, Li Z, Xing X, Putz-Bankuti C, Stauber RE, Liu X, Madl T (2021a) Growing human hepatocellular tumors undergo a global metabolic reprogramming. Cancers (Basel) 13:1980

Zhang W, Zhang B, Chen XP (2021b) Adjuvant treatment strategy after curative resection for hepatocellular carcinoma. Front Med 15:155–169

Zhang Y, Cui J, Wan W, Liu J (2022) Multimodal imaging under artificial intelligence algorithm for the diagnosis of liver cancer and its relationship with expressions of EZH2 and p57. Comput Intell Neurosci 2022:4081654

Zhou J, Sun H, Wang Z, Cong W, Wang J, Zeng M, Zhou W, Bie P, Liu L, Wen T et al (2020) Guidelines for the diagnosis and treatment of hepatocellular carcinoma (2019 Edition). Liver Cancer 9:682–720

Zhou M, Zhang C, Nie J, Sun Y, Xu Y, Wu F, Huang Y, Li S, Wang Y, Zhou Y et al (2021) Response evaluation and survival prediction following PD-1 inhibitor in patients with advanced hepatocellular carcinoma: comparison of the RECIST 1.1, iRECIST, and mRECIST criteria. Front Oncol 11:764189

Zhu AX, Kang YK, Yen CJ, Finn RS, Galle PR, Llovet JM, Assenat E, Brandi G, Pracht M, Lim HY et al (2019) Ramucirumab after sorafenib in patients with advanced hepatocellular carcinoma and increased alpha-fetoprotein concentrations (REACH-2): a randomised, double-blind, placebo-controlled, phase 3 trial. Lancet Oncol 20:282–296

Zhu XD, Huang C, Shen YH, Ji Y, Ge NL, Qu XD, Chen LL, Shi WK, Li ML, Zhu JJ et al (2021) Downstaging and resection of initially unresectable hepatocellular carcinoma with tyrosine kinase inhibitor and anti-PD-1 antibody combinations. Liver Cancer 10:320–329

Zmora N, Suez J, Elinav E (2019) You are what you eat: diet, health and the gut microbiota. Nat Rev Gastroenterol Hepatol 16:35–56

The Immune System in Liver Cancer: From Beginning to Progression

Alex José de Melo Silva, Juliana Ellen de Melo Gama, Cleonilde Maria de Nascimento, Jessica Paula Lucena, Cicero Jadson da Costa, Camila Juliet Barbosa Fernandes, Danielle Maria Nascimento Moura, Helotonio Carvalho, and Sheilla Andrade de Oliveira

Abstract

Hepatocellular carcinoma is one of the deadliest cancers in the world; it is a disease that develops due to different etiologies. As a result of progressive inflammation, this tumor can be progressive and immunotolerant to the immune system. Moreover, several immune cells as CD4+ and CD8+ T cell lymphocytes and interleukin (IL)-10 and IL-1b are involved in its tolerance, once their activities are reduced, and the presence of an efficient immunosuppressive microenvironment that gives subsidies to this process is an important factor. Several cells as Kupffer cells act to keep on the tissue homeostasis; however, other mechanisms such as signaling pathways, for instance, phosphatidylinositol 3-kinase (PI3K)/AKT/mammalian target of rapamycin (mTOR), receptor tyrosine pathways, Wnt/β-catenin, and Janus kinase (JAK)-signal transducer and activator of transcription factor (STAT) (JAK/STAT), are involved in this cancer, helping it to develop properly and growing up effectively in the microenvironment. The immune system is essential for cancer progression inhibition; however, it can be inefficient once this tumor has abilities to evade it due to the action of several factors such as the TGF-β1. Therefore, this chapter provides a general overview of the immune system function and hepatocellular carcinoma, either from its beginning or its progression into the tumor microenvironment.

A. J. de Melo Silva (✉) · J. E. de Melo Gama · C. M. de Nascimento · J. P. Lucena · C. J. da Costa · C. J. B. Fernandes · D. M. N. Moura · S. A. de Oliveira
Department of Immunology, Aggeu Magalhães Institute, Recife, PE, Brazil
e-mail: ajmsufpe@hotmail.com

H. Carvalho
Department of Immunology, Aggeu Magalhães Institute, Recife, PE, Brazil

Department of Biophysics, Federal University of Pernambuco, Recife, PE, Brazil

Interdisciplinary Cancer Research, https://doi.org/10.1007/16833_2022_37
Published online: 18 November 2022

Keywords

Hepatocellular carcinoma · Immune system · Immunotolerance · Liver cancer · Microenvironment

1 Introduction

Hepatocellular carcinoma (HCC) accounts for at least 80–90% of the primary liver cancers resulting from a chronic inflammation process (Ringelhan et al. 2018). Its progression, as well as the therapeutic approach, is affected by the continuous inflammation of the tumor microenvironment leading to a tumor immune microenvironment (TIME) (Grivennikov et al. 2010; Shalapour and Karin 2019; Hou et al. 2020). It has been considered a high immunotolerant site due to the interactions that occur in it, among the cells such as the liver-resident cells and other, for instance, peripheral leukocytes, the decreased or incomplete activation of CD4+ and CD8+ T cell lymphocytes, and an immunosuppressive environment that is mediated by some molecules, such as interleukin (IL)-10 and IL-1b (Robinson et al. 2016). The Kupffer cells (KCs), which maintain the liver homeostasis through their capacity to present antigens and phagocytosis, also play an important role in immunotolerance (Mossanen et al. 2019). Several signaling pathways are involved in the HCC pathogenesis, and their identification provides an alternative to designing novel targets, which may be utilized as a therapeutic approach (Dimri and Satyanarayana 2020). The understanding of HCC biology has provided evidence regarding alterations observed in several molecular pathways such as the phosphatidylinositol 3-kinase (PI3K)/AKT/mammalian target of rapamycin (mTOR), receptor tyrosine pathways, Wnt/β-catenin, Janus kinase (JAK)-signal transducer and activator of transcription factor (STAT) (JAK/STAT), Wnt/β-catenin, Ras mitogen-activated protein kinase (Ras/Raf/MAPK), Hedgehog, Notch (Dotto 2009), and transforming growth factor (TGF)-β signaling pathway (Fabregat and Caballero-Díaz 2018), which are extremely essential for the development of cancer. Several evasion mechanisms have been described in HCC pathogenesis, which are provided by diverse factors, such as TGF-β1, one of the most important molecules in this process to trigger metastasis (Fu et al. 2019), which can also cooperate with other cells such as fibroblasts to promote the spreading of cancer cells (Sachdeva et al. 2015). Thus, in this chapter, we focus on describing the main immune cells and their activities, as well as the inflammation process and progress to HCC, and signaling pathways and metastasis, providing an overview regarding how all of these mechanisms address HCC development.

2 The Function of the Immune System in HCC

The most common causes of HCC include alcohol abuse, cholestatic liver disease (Wilson disease), chronic infection with hepatitis B virus (HBV) (which remains the leading cause of HCC worldwide), and chronic infection with hepatitis C virus

(HCV), being currently the leading cause of end-stage liver disease and HCC in the Western world (Chiu et al. 2017). In general, 90% of HCCs develop due to underlying chronic liver inflammation, the induction of fibrosis, and/or subsequent cirrhosis. The ability of the liver to repair itself after acute damage is the key to understanding the pathophysiology. Hepatocytes are able to serve as their own main source of replacement. However, in chronic “necroinflammation,” constant cell death, compensatory regeneration, and activation of non-parenchymal cells, together with an altered immune response, promote liver fibrosis and tumorigenesis (Chiu et al. 2017; Kubes and Jenne 2018).

The liver also plays a central role in the immunoregulation of HCC, because the immune microenvironment in the liver is a multicomponent system consisting of hepatocytes, immune cell subsets, immune receptors and ligands, cytokines and chemokines, extracellular matrix, and other elements. In a healthy liver, this complex network is delicately regulated to maintain a dynamic balance between tolerance and immunity. When chronic inflammation is induced by some infectious or pathogenic agents, both the innate and adaptive immune systems can be altered (Hou et al. 2020; Chiu et al. 2017; Kubes and Jenne 2018).

Natural killer (NK) cells are found in the greatest concentration within the liver. In an effort to evade the immune response, tumor cells downregulate antigen-presenting molecules, such as major histocompatibility complex (MHC) I; this absence of self is recognized by NK cells that then eliminate the suspect cells through secretion of granules containing perforin, granzymes (proteases), and interferon (IFN)-γ that result in cell lysis and, moreover, protection against damages that lead to HCC through various mechanisms, such as clearance of hepatitis virus-infected cells and exerting antifibrotic properties (Lawal et al. 2021; Liu and Zeng 2012).

The reduction of NK cell numbers and their activity is associated with HCC development (Woo et al. 2015; Juengpanich et al. 2019; Li et al. 2015; Lin et al. 2019). KCs are macrophages that reside in the liver’s sinusoids, and their main function is to detect and phagocytose pathogens that have been marked as foreign through complement-mediated opsonization (Woo et al. 2015). There is a population of macrophages that live in or around the tumor microenvironment and are known as tumor-associated macrophages (TAMs). They can be classified into two subsets: anti-tumorigenic (M1 phenotype) and protumorigenic (M2 phenotype), which their profile changes according to the expression of chemokines, cytokines, and growth factors. In HCC and other tumors, TAMs differ predominantly from the M2 subset in a process called macrophage polarization, facilitated in part by colony-stimulating factor 1 (CSF-1) and C-C motif ligand 2 (CCL2). M2 macrophages produce cytokines, which promote HCC tumor growth (IL-5 and TGF-β), invasion (IL-6), and metastasis (stromal cell-derived factor 1 [SDF-1]) (Lawal et al. 2021). In contrast, M1 macrophages secrete a different set of cytokines that aid in tumor control, including IL-12 (Kubes and Jenne 2018; Lawal et al. 2021; Liu et al. 2018a, b).

Tumor-associated neutrophils (TANs) are neutrophils that infiltrate into the tumor microenvironment activated and modulated by various molecules, such as IL-17, IFN-β, CXCL1, CXCL2, and CXCL5. Similar to TAMs, TANs can be classified as anti-tumor N1 or protumor N2 phenotypes (Woo et al. 2015). Myeloid-derived suppressor cells (MDSCs) are a subset of inflammatory monocytes with the ability to downregulate adaptive immune responses. MDSCs in HCC have been shown to directly suppress the cytotoxic function of NK cells and are known to promote tumor angiogenesis and are upregulated in HCC, partly due to the hypoxic tumor microenvironment (Mikulak et al. 2019; Prieto et al. 2015).

Dendritic cells (DCs) link the innate and adaptive immunity through a spectrum of functions, not only limited to be DCs and phagocytosis of tumor cells but also act as professional antigen-presenting cells (APCs) that elicit a robust T cell response. DC subsets have an important role in balancing immune activation and tolerance in the liver. A cytotoxic subset of DCs (NK1.1+ DCs) stimulates an inflammatory Th1-predominant (Th1 cells secrete pro-inflammatory cytokines, including IL-2 and IFN-γ, which have potent anti-tumor effects) response that helps with tumor control. Conversely, other DC populations in the liver have immunosuppressive properties. Furthermore, DCs that are present in the liver elicit a weaker T cell response compared to those living in other tissues (Ringelhan et al. 2018; Wang et al. 2021).

Representing the adaptive immunity, T and B lymphocytes also play important roles in mediating immunity in HCC. The most important T cell subtype in determining clinical outcomes in HCC is the regulatory T cell (Treg). Tregs are a subpopulation of T cells with a suppressive regulatory influence on the immune system. They are activated by the interaction between T cell receptors (TCRs) and their ligands, as well as the IL-10 and TGF-β signaling pathways (Woo et al. 2015; Yu et al. 2019). B lymphocytes exert important phagocytic functions, presenting antigens and activating T lymphocytes, in addition to inducing the secretion of specific antibodies with anti-tumor properties by several mechanisms that are still being elucidated. Balanced immunoregulation is the key in response to HCC, as it can characterize the difference between tumor progression and control (Yu et al. 2019).

2.1 Immune Cells in Hepatocarcinogenesis

Continuous liver injury is associated with intrahepatic inflammation and excessive accumulation of extracellular matrix (ECM), which progresses to hepatic fibrosis (Llovet et al. 2016). There is an increase in cell renewal and exacerbated tissue regeneration progressing to liver cirrhosis. In addition, immune cells which migrate to lesion sites secrete some cytokines and growth factors. These changes trigger the proliferation and expansion of abnormal growth cells (Schattenberg et al. 2011). Hepatocarcinogenesis is a multifactorial process and can be in two different ways: benign (hepatic adenomas) and malignant (HCC) (Zaki et al. 2021). HCC is the most prevalent, accounting for more than 80% of cases (Llovet et al. 2016). Among the cells in the liver tissue, hepatocytes, hepatic stellate cells (HSCs), and sinusoidal endothelial cells have immunological functions, for example, antigen presentation, homeostasis for immunotolerance, ECM production, and activation of the immune system (Zaki et al. 2021; Barry et al. 2020). In addition, there is a rich population of the immune cells such as natural killer T (NKT) and natural killer (NK) cells; T and

B cells; KCs, which are resident macrophages in the liver; and resident lymphocytes or just a transitory cell in the liver (Shetty et al. 2018; Strick-Marchand et al. 2004). The innate immune cells that make up the inflammatory microenvironment are KCs, granulocytes such as neutrophils, and DCs (Tumanov et al. 2009).

Once occur an injury in the liver tissue, KCs are activated and release pro-inflammatory cytokines that attract neutrophils, monocytes, and other immune cells to the damaged site, resulting in the activation of HSCs (Zaki et al. 2021). As a result of chronicity, the KCs can still be activated by intracellular structures (damage-associated molecular patterns [DAMPs]) resulting from destroyed hepatocytes (Woo et al. 2015). These macrophages that surround or are internalized are the TAMs. The TAMs produce IL-6, induce immunosuppression, impair immune cells that have anti-tumor activity, and promote the reproduction of hepatocytes in order to compensate the tissue damage (Garcia-Martinez et al. 2016).

Neutrophil infiltration is associated with activation of HSCs resulting in the development of liver fibrosis, and increasing cell growth, angiogenesis, tumorigenesis, and metastasis (Qian and Pollard 2010). Despite the protective effect of NKT cells, it has been described that these cells are also able to activate the HSCs. These activated HSCs start to produce and secrete ECM, in addition to releasing pro-inflammatory cytokines that activate KCs, leading to positive feedback (Ibusuki et al. 2013; Wehr et al. 2013). The NK cells have been involved with a positive prognosis of HCC, due to their ability to fight against viral infections and other liver damage, and elimination of active HSCs (Syn et al. 2012; Gur et al. 2012). Taking into account the cells of the adaptive immune system, during the liver chronic injury, the activated T cells synthesize and secret some cytokines, such as lymphotoxins. These cytokines induce the activation and transformation of hepatocytes (Liu et al. 2018a, b). However, CD8+ and CD4+ T lymphocytes are associated with anti-tumor effector functions, leading to a greater prognosis in some patients (Haybaeck et al. 2009; Flecken et al. 2014).

The functions of CD8+ T cells, as well as CD4+ T lymphocytes in the hepatic tumor microenvironment, are not yet well evidenced. The presence and proliferation of Treg cells in the patients have been associated with a low rate of patients' survival, due to its immunosuppressive action on T and NK cells. Thus, the increase in this population correlates with a high rate of spreading and recurrence of the tumor (Ringelhan et al. 2018). Regarding B cells, these are described with controversial functions. While B cells that infiltrate the tumor can trigger the angiogenesis process, other populations associated with T cells are involved in a positive prognosis of the individuals (Keenan et al. 2019). Some studies indicate an anti-tumor role of B cells, stimulating the death of tumor cells and controlling tumor expansion (Garnelo et al. 2017).

3 Inflammation in HCC

HCC development is a common carcinogenesis process linked to inflammation, due to its association with chronic damage and the inflammation resulting from it within the liver (Nakagawa and Maeda 2012). Inflammation in an advanced and chronic stage is characterized by the presence of different immune cells, such as immature myeloid cells and macrophages, besides disturbing the cytokines' synthesis being linked to being important events to trigger the HCC development and its progression. During a stage of premalignancy, several pathways associated with inflammation signaling are chronically activated resulting in nitrogen oxygen species (NOS) and reactive oxygen species (ROS) formation. Furthermore, associated with this activation, some inflammatory cells such as stromal cells, in the tumor microenvironment, originate an ample array of some molecules, to illustrate proangiogenic and growth factors, and cytokines and chemokines (Wei et al. 2011), being these lasts two indicated as molecules linked to being a "bridge" between the inflammation process and carcinogenesis (Refolo et al. 2020). Some processes are associated with hepatocarcinogenesis, and one of the primaries is the death of epithelial cells due to their capacity to trigger inflammation. Furthermore, signaling pathways are directly linked to inflammation-mediated hepatocarcinogenesis. These pathways included in this process are innate and adaptive immunity and others such as cytokine signaling (as seen in the following topics) (Chen et al. 2018).

The immune response against HCC includes several cell responses and mechanisms. The communication among the tumor, its microenvironment, and the immune system is a consequence of an interaction between factors with pro- and anti-tumor activity, being an unstable phenomenon changing between eradication and its progress to a potential metastasis. Indeed, the tumor could be recognized as an immunogenic process, once it can generate a T cell response through the host immune system, and this response once recognized cancer can eliminate the carcinogenic cells (Yang et al. 2019). Some molecules can negatively regulate the immune response amplitude against cancer, being present in the peripheral self-tolerance. These molecules include some checkpoint molecules such as programmed cell death protein 1 (PD-1) and the cytotoxic T-lymphocyte associated protein-4 (CTLA-4). The presence of different pathways can be used by cancer to upregulate these checkpoint molecules. This overexpression can lead to the inhibition of some activities such as that developed by the T cells either CD4+ helpers or CD8+ effectors, which combines the mechanisms of innate and adaptive immunity. Moreover, this establishes a subsequent inhibition of immunity against tumors, which is the most significant mechanism of immune resistance. Besides that, the interaction that occurs in the tumor microenvironment, tumoral cells, and immune components also highly contributes in a significant way to this resistance (Chen and Mellman 2013).

3.1 Inflammation: Pro- and Anti-Inflammatory Cytokines in Liver Cancer

The inflammatory response plays a key role in the development and progression of liver cancer, inducing proliferation, angiogenesis, and metastasis (Subbarao 2021; Zappavigna et al. 2020). The main trigger of inflammation associated with hepatocarcinogenesis in the cytokine-mediated signaling is one of the major causes of this process, and the death of the epithelial cells. This involves especially tumor necrosis factor α (TNF-α), IL-6, nuclear factor kappa B (NF-κB), c-Jun N-terminal kinase (JNK), and the signal transducer and activator of transcription 3 (STAT3) (Yang et al. 2019). Many inflammatory and carcinogenic agents can activate the transcription factor, NF-kB, which is able to induce the production of pro-inflammatory cytokines, which, in turn, attract immune system cells increasing tissue inflammation levels (Subbarao 2021; Zappavigna et al. 2020).

The Akt signaling pathway, which acts on the progression of the inflammatory response and apoptosis in several types of cancer, has a mechanism that induces the activation of NF-κB and is also linked to the death promoter associated with Bcl-2 (Bad) for the progression of liver cancer. When activated, the Akt pathway promotes cell survival and growth in response to extracellular signals, mediating growth, proliferation, migration, and angiogenesis responses through the phosphorylation of a series of intracellular proteins (Tang et al. 2016; Chagpar et al. 2010). Furthermore, it can activate the IκB kinase IKK-α, resulting in degradation of IκB and nuclear translocation of NF-κB, promoting the expression of cytokines such as IL-1β, IL-18, IL-6, and TNF-α, which accelerates the inflammatory response (Chagpar et al. 2010). TNF-α is one of the most well-characterized protumorigenic cytokines in hepatocarcinogenesis (Ringelhan et al. 2018) with a significant role in the stages of angiogenesis and metastasis (Landskron et al. 2014; Patil et al. 2019). Although it is pro-inflammatory, it can play a dual role in carcinogenesis, being able to induce an anti-tumor response, when at high levels, (Landskron et al. 2014) or generate a tumor phenotype, inducing the generation of ROS and nitrogen (Micheau and Tschopp 2003), especially by activating the NF-κB and JNK signaling pathways (Ringelhan et al. 2018).

Another important promoter of hepatocarcinogenesis is the activation of STAT3 mediated by IL-6, which drives the repair and replication of hepatocytes (Yang et al. 2019). Some studies have shown the association of increased levels of these markers with HCC, being an important pathway for initiation and progression of this type of cancer (Yang et al. 2019; Aleksandrova et al. 2014; Wang et al. 2011; He et al. 2013; Gao et al. 2007). IL-6 plays a fundamental role in cell proliferation and inhibition of apoptosis and may involve in the activation of JAK (in addition to STAT3) by binding to IL-6Rα receptor and gp130 co-receptor (glycoprotein 130), and activation of IL-6-induced tumorigenesis can also occur through hypermethylation of tumor suppressor genes (Landskron et al. 2014).

As for anti-inflammatory cytokines related to carcinogenic processes, it is cited that TGF-β, a pleiotropic cytokine with multifaceted roles in liver cancer, in the early stages, can act as a tumor suppressor, inhibiting the cell cycle progression and promoting apoptosis. In more advanced stages, it promotes malignant progression, inducing the epithelial-mesenchymal transition (EMT) and increasing the survival of cancer cells, promoting migration, invasion, and metastasis (Fabregat and Caballero-Díaz 2018; Landskron et al. 2014; Dituri et al. 2019; Tsui et al. 2020). Activated TGF-β signaling in HCC is associated with liver fibrosis, inflammation, and cancer development (Tu et al. 2019).

Another powerful anti-inflammatory cytokine is IL-10, able to provide pro- and anti-tumor effects (Iyer and Cheng 2012). Its anti-tumor effect may occur through the inhibition of NF-kB, which leads to a decrease in the expression of pro-inflammatory cytokines such as IL-1α, IL-1β, IL-6, IL-8, IL-12, and TNF-α (Iyer and Cheng 2012; Wei et al. 2011). Its protumorigenic effect is played by the mechanism of induction of phosphorylation and sustained activation of STAT3 (Iyer and Cheng 2012), which may be involved in the progression of chronic inflammation that leads to HCC (El-Emshaty et al. 2015). Furthermore, due to its immunosuppressive effect on DCs and macrophages, IL-10 can attenuate antigen presentation, cell maturation, and differentiation, facilitating hepatocarcinogenesis by suppressing immune surveillance mechanisms (Iyer and Cheng 2012; Taylor et al. 2006).

Chemotactic factors are also involved in liver cancer promotion. In this sense, chemokines are cited, which can act as growth factors and are responsible for controlling the infiltration of leukocytes into the tumor and regulating immune angiogenesis. In addition, they stimulate cell growth and metastasis in various tumors, in addition to stimulating tumor cells and leukocytes' migration into inflammatory sites (Subbarao 2021; Zappavigna et al. 2020).

4 Oncogenic Signaling Cascade in HCC

There are several signaling pathways associated with HCC: TGF-β, Wnt/β-catenin, Hedgehog, Notch, endothelial growth factor (EGF), hepatocyte growth factor (HGF), VEFG, JAK/STAT, ERK/MAPK, and PI3K/AKT/mTOR, among others, which means many things can go wrong and ultimately lead to HCC. With respect specifically to oncogenes or tumor suppressor genes, the tumor suppressor gene, TP53, which codifies p53 protein, is commonly mutated in HCC. According to data from the cBioPortal for Cancer Genomics (Taylor et al. 2006; Cerami et al. 2012), 31% of the tumors have TP53 mutations and 26% have mutations on the CTNNB1 oncogene, which codifies β-catenin. The MDM2-p53 seems to play an important role in HCC since mutations in TP53 induced by hepatitis viruses cause its silencing, and overexpression of MDM2, which increases p53 degradation, ultimately leading to genomic instability and tumor development (Gao et al. 2013). These events are the result of compromising checkpoint control, and DNA repair mechanisms as well as apoptosis, processes normally controlled by p53.

In the case of the TGF-β pathway, binding to its receptor activates Smad2/3 by phosphorylation which associates with Smad4 and translocates to the nucleus, resulting in Snail upregulation, E-cadherin downregulation, EMT, and metastasis (Fabregat and Caballero-Díaz 2018). The TGF-β pathway also induces HCC proliferation and vascular endothelial growth factor (VEGF) expression with the recruitment of endothelial cells (Cao et al. 2020). Besides this, FOXP3 expression is related to a better prognosis in HCC patients and does that by inhibiting the TGF-β Smad2/3 pathway (Sun et al. 2012). The importance of the Wnt/β-catenin pathway is highlighted by the high mutation frequency of the β-catenin transcription factor. However, the β-catenin mutation is associated with a better prognosis for the disease

(Shi et al. 2017). Wnt/β-catenin is involved in angiogenesis in HCC acting through Ribosomal protein S15a (Wang et al. 2015). The leucine-rich repeat-containing G (LGR5) receptor connects the Wnt/β-catenin pathway with HCC metastasis by inducing EMT (Liu et al. 2017a, b, c, d), and its overexpression is associated with poor prognosis in HCC patients (Ko et al. 2019). Besides this, repression of p53 mediated by LGR5, in a process dependent on death receptor PDCD5, is associated with HCC resistance to doxorubicin chemotherapy (Ma et al. 2019).

The Hedgehog pathway is another important component of HCC. Activation of the SHH receptor triggers phosphorylation and nuclear translocation of the transcription factor Gli, inducing the expression of cell cycle proteins like Cyclin D and c-Myc, and MMP proteins, associated with cell invasion (Guo et al. 2018a, b). Besides this, Gli is also related to angiogenesis through increased expression of VEGF (Jeng et al. 2019). Hedgehog signaling's role in cellular invasion and metastasis is dependent on the ERK pathway since ERK1/2 inhibitors impaired these events in the HCC Bel-7,402 cell line (Pinter et al. 2013). The Notch pathway is controlled by Notch receptor activation, which controls the expression of genes associated with cell proliferation, invasion, apoptosis-like p53, and the cyclin-D and c-Myc oncogenes (Dotto 2009). Notch pathway, however, shows cross talk with several other signaling pathways, like Wnt, SHH, PI3K, and VEGF, and depending on whether each cross talk is in place, the effect observed will be cell proliferation, cancer stem cell maintenance, or angiogenesis, respectively (Lu et al. 2012).

Additionally, the Notch3 receptor suppresses p53 posttranscriptional modification through cyclin-G1 expression, Mdm-2, and miR-221 (Giovannini et al. 2016). The EGF pathway is involved in the activation of Akt, the MAPKs ERK, p38, and NF-κB, and high expression of EGF is often associated with HCC (Giovannini et al. 2014). However, EGF mutations are rare in HCC. HGF signaling pathway, on the other hand, has implications for most of the relevant aspects of HCC, like cell proliferation, invasion, and angiogenesis. Binding of HGF to its c-met receptor induces activation of PI3K, ERK, and JNK/STAT3 signaling pathways (Su et al. 2006), and it has been demonstrated that microRNAs, especially miR-17-5p and miR-20a-5p, have an essential role in reducing HCC metastasis by inhibiting the HGF/ERBB3-NF-κB axis.

Despite the role of other growth factors in inducing angiogenesis in HCC, VEGF is the main pathway involved in this process, also taking part in the activation of the PI3K/Akt pathway. In fact, inhibition of VEGF signaling by using a monoclonal antibody against it has proved as a potential therapy for HCC by inducing apoptosis in tumor cells since VEGF has a crucial role as an oncogene (Wang et al. 2020). JAK/STAT3 is another pathway that can be activated in HCC. Normally stimulated by pro-inflammatory cytokines, the STAT3 transcription factor can lead to cancer cell proliferation, angiogenesis, and metastasis (Liu et al. 2020). The ERK pathway is activated in response to several different growth factors, and it has been described that the Ras/Raf/MEK/ERK pathway is constitutively activated in HCC, which suggests its role in tumor development (Hwang et al. 2004). In fact, micro-RNAs are known to play a role in inhibiting this pathway and suppressing tumor progression and invasion (Zhou et al. 2017a, b; Guo et al. 2018a, b).

Despite the high frequency of mutations in the Ras and Raf oncogenes in mice, these mutations are rarely observed in humans (Delire and Stärkel 2015). Finally, the PI3K/AKT/mTOR is described to be constitutively activated in several types of tumors, and the insulin-like growth factor (IGF) and IGF receptor are upregulated in HCC (Alexia et al. 2004). Besides this, the pathway is essential to induce HCC in mice and humans in a process dependent on the c-Myc protein, codified by its corresponding oncogene (Xu et al. 2019). c-Myc is also reported to be overexpressed in HCC, which is generally associated with a poor prognosis (Liu et al. 2017a, b, c, d).

5 Liver Cancer and Immune System Surveillance

The liver's defense mechanism is mediated by hepatic immune cells, including KCs, NK cells, NKT cells, and γδ T cells. Both are responsible for controlling the presence and eradication of harmful substances in the region; however, variables such as excessive alcohol consumption, presence of viruses, and accumulation of lipids require a greater response in activating this defense system (Yang et al. 2019; Zhang et al. 2018). Innate lymphoid cells (ILCs) act directly or indirectly in the establishment of HCC. ILC1s produce IFN-γ that activates NK cells; ILC3s participate by releasing IL-17 and IL-22 which can promote angiogenesis and in turn is contributed to tumor growth (Dadi et al. 2016; Abe et al. 2014).

Chronic inflammation causes greater recruitment of cells of the innate immune system, since there is no correct distribution of the O2 supply, a hypoxic microenvironment is created, and the cells of this system activate the hepatic stellate cells (HSCs) that induce a greater collagen deposition contributing to the formation of fibrosis and worsening of tissue hypoxia. Consequently, this finding affects the functioning of immune cells, as it induces the infiltration and accumulation of immunosuppressive innate immune cells, such as myeloid progenitors (MDSCs) in this region (Yuen and Wong 2020). In the tumor microenvironment, MDSCs rapidly expand and potentiate the immunosuppressive action, as they do not differentiate into mature cells, which in turn, in this undifferentiated state, suppress the TCR signaling pathway under generating high ROS levels. In addition, expression of arginine-1 (ARG1), inducible nitric oxide synthase (iNOS), TGF-β, IL-10, COX-2, and indoleamine 2,3-dioxygenase (IDO) impacts the response of other components of the immune system (Yuen and Wong 2020).

Tumor metabolism requires greater energy input in the form of adenosine triphosphate (ATP), so they no longer obtain it through the usual pathway, oxidative phosphorylation (OXPHOS), and gain access through aerobic glycolysis. Such a change, despite being less efficient since the production of ATP by glycolysis generates only two molecules of ATP and that of OXPHOS provides molecules of ATP (Syn et al. 2012; Gur et al. 2012; Liu et al. 2018a, b), skipping the OXPHOS stage together makes use of a series of biomolecules that enable the synthesis of new structures and cell proliferation. This shift caused by cancer cells to the glycolytic shift impacts several pathways, including the hypoxia-inducible factor Myc, PI3K/Akt/mTOR, p53, and Ras (Piñeiro Fernández et al. 2019).

Once oxygen levels are decreased and TGF-β increased, it can polarize NK cell differentiation into a proangiogenic phenotype. Furthermore, recruitment of endothelial cells and adhesion to the tumor site by expression of agrin regulate levels of vascular endothelial growth receptor (VEGFR)2 and angiogenesis (Njah et al. 2019). The acquisition of new blood vessels decreases the hypoxic load of this environment and allows the uncontrolled growth of the tumor (Piñeiro Fernández et al. 2019).

6 Mechanism of Immunotolerance in Liver Cancer

The continuous exposition of the liver to self and foreign antigens has created a specialized mechanism of tolerance to avoid immune system overactivation. This mechanism has been related to specific cell types that express anti-inflammatory mediators and inhibitory cell surface ligands for T cell activation. The types involved in the process include non-parenchymal and parenchymal cells, such as resident DCs, stellate hepatic cells (HSCs), KCs, liver sinusoidal endothelial cells (LSECs), the hepatocytes, and immune cells, like natural killer cells (NKs, NKTs) and regulatory T cells (Tregs), with the latter being residents or recruited to the liver (Roth and Decaens 2017; Jenne and Kubes 2013).

The non- and parenchymal cells' ability to modulate the immunotolerance mechanism relies on their functions as APCs. Some of them such as DCs and LSECs have this recognized capacity, while KCs, HSCs, and even hepatocytes can develop this capacity through interaction with other cells and molecules within the liver microenvironment. These conventional and nonconventional APCs cause an initial activation of T cells that follows a state of nonresponsiveness in CD4+ and CD8+ T cells, and altogether they can contribute to the conversion of conventional T cells to Tregs expressing high levels of CD25 and Foxp3 (Horst et al. 2016; Doherty 2019). The specific mechanisms by which activation followed by exhaustion of CD4+ and CD8+ T lymphocytes are complex and occur in a variety of ways (Ficht and Iannacone 2020). In general, it includes immunomodulation via the regulation of the expression of cell surface molecules (MHCI and II) and co-stimulatory molecules, such as CD80 and CD86, secretion of soluble mediators, and a complex cytokine milieu that includes pro-inflammatory cytokines such as IL-2, IL-12, and IFN-γ, and anti-inflammatory cytokines (IL-10 and TGF-β). In this context, secretion of TGF-β and IL-10 from different cell sources, but mainly from LSECs and KCs, plays a pivotal in T cell exhaustion and antigen-presentation inhibition (Ringelhan et al. 2018). The overexpression of other regulators known as immune checkpoint molecules, such as CTLA-4, T cell immunoglobulin mucin-3 (TIM3), and the axis PD-1/programmed death-ligand 1(PD-L1), also contributes to T cell exhaustion and tolerance (Wen et al. 2019).

In a healthy liver, hepatic immunotolerance is a highly controlled protective mechanism; however, the deregulation of this system can lead to the establishment of chronic liver diseases and carcinogenesis. Therefore, in an oncogenic context, the exacerbation of the hepatic tolerogenic effect or the escape of immune surveillance helps the development and progression of the HCC, since this immunosuppressive environment has an important effect on the innate and adaptive immune cell response (Zheng and Tian 2019). The immunotolerance also seems to contribute to the metastasis of several tumors such as metastasis of malignant melanoma or colon carcinoma to the liver, probably due to the evasion of immune surveillance (Sällberg and Pasetto 2020).

The immunotolerance of the liver has been explored as a target for immunotherapy since the stimulation of the latent immune system can be achieved by interfering with some of the immunosuppressive mechanisms and activating anticancer immunity (Liu et al. 2021). Different approaches have been tested in HCC patients, including the use of cytokines and chemokines, inhibitors for immune checkpoints, vaccines, and cellular therapy (Wang and Wang 2019). The application of monoclonal antibodies against PD-1, PD-L1, and CTLA-4 seems promising (Uson Junior et al. 2021). For cellular therapy, strategies using adoptive cellular transfusion of cytokine-induced killer cells, genetically modified cells (engineered chimeric antigen receptor (CAR)-T cells and TCR-T cells), and stem cells are explored. Other potential targets under investigation include Tregs, microRNAs, and NKs. Multiple preclinical and clinical trials, at different stages, are in progress to verify the safety and efficacy of these new treatments, with some of them studying the combination of therapies; however, the outcomes have been variable. Although challenging, mostly due to tumor heterogeneity and lack of biomarkers in HCC, the studies of new immunotherapies for this type of cancer represent an opportunity to find a more successful way to treat these patients, considering the few therapeutic options currently available (Llovet et al. 2021).

7 HCC and Immune System Evasion

HCC has several evasion mechanisms, many of which are related to an induced-anti-inflammatory microenvironment. MDSCs, TAMs, TANs, and tumor-infiltrating lymphocytes are known sources of immunosuppressive cytokines (Huang et al. 2018). TGF-β1 plays a major role in immune evasion due to angiogenesis and EMT promotion that supports invasion and metastasis (Yoshida et al. 2018) via TGF/Smads (Moreno-Cubero and Larrubia 2016). Also, it has a suppressive response on the innate immune system, stimulating CD8+ Treg cell infiltration and leading to HCC evasion (Zongyi and Xiaowu 2020). Moreover, it was shown that during tumor progression, IL-10 can impair MHCII response by downregulation and NK cell cytotoxicity (Zhou et al. 2017a, b).

Tumor microenvironment changes mechanisms that can be presented in antigen/marker content. Immune evasion checkpoints, such as PD-1 and CTLA-4, are inhibitory molecules involved in impairing immune response (Ruiz de Galarreta et al. 2019). Particularly, PD-1 acts as a stimulus to apoptosis in immune cells, via β-catenin (Sun et al. 2021), and it is associated with tumor relapse (Galle et al. 2019). From all abnormal antigens, α-fetoprotein is a pro-proliferative protein mainly produced by the fetal liver and downregulated afterward. It is one of the most common prognostic serum markers (Sun et al. 2019). Some other markers have been seen in higher levels in the tumorigenic environment, like CD96 and CD155, being associated with NK cell impairment and poor outcomes (Ye et al. 2018). Besides, it was seen that tumor cell stimuli can promote macrophage polarization, from M1 to M2, presenting upregulation of IL-10, Arginase-1, and Fizz-1 (Sükei et al. 2021). The ECM content is mainly composed of collagen I, III, and IV and fibronectin. It promotes cell survival by integrin binding and PI3K signaling (Keenan et al. 2019).

Due to its multiple origins and frequently concomitant diseases, HCC treatment has always been a challenge. The therapy mainly included resection, radiation, or transplant, depending on the etiology and its stage (Xia et al. 2020). Sorafenib and other tyrosine kinase inhibitors were the standard procedure (Cui et al. 2019). However, its antivascular outcome drives HCC resistance due to hypoxia by IL-10/STAT3 signaling (Cui et al. 2019). Immunotherapy has emerged as an interesting treatment strategy for some types of cancer, like breast cancer and non-small cell lung cancer. And currently, some antibodies were approved by the Food and Drug Administration (FDA) for the treatment of HCC, targeting the PD-1/PL-1 axis regulation (Mocan et al. 2019), acting directly at the PD-1 receptor on malignant cells (El Dika et al. 2019).

8 Immunity and HCC Metastasis

HCC is associated with a recurrence rate of around 65% after treatment dressing with liver resection or percutaneous tumor ablation (Cammà et al. 2005; Lencioni et al. 2003; Colombo and Iavarone 2014). This may be occurred due to metastases from the primary tumor and where malignant cells infiltrate the surrounding parenchyma and enter the circulation by intravasation of blood vessels. Disseminated tumor cells migrate to distant areas where, after arrest, they leak out of the circulation and enter new tissues (McAllister and Weinberg 2014; Zeeshan and Mutahir 2017). Primary liver cancer can be hampered or triggered by some metabolic and immunological changes necessary for the adaptation of the hepatic microenvironment. In liver homeostasis, there is a cooperation between immune cells (lymphocytes and myeloid cells) and nonimmune cells (endothelial cells, pericytes, and fibroblasts) (Safarzadeh et al. 2021). Thus, in normal liver tissue, leukocytes or myeloid cells maintain the immunosuppressive orientation in liver disease; however, with the persistence of the aggressive stimulus, cellular changes activate the innate immune response (with the participation of KCs, neutrophils, DCs, and monocytes) and adaptive immune responses (Safarzadeh et al. 2021). In this process, the tumor microenvironment secretes factors such as extracellular vesicles, cytokines, chemokines, and other components (Li et al. 2021), which will promote the immune system's response against tumor cells through the regulation of inhibitory and costimulatory responses (Marin-Acevedo et al. 2018; Eggert and Greten 2017).

Importantly, the liver contains specialized endothelial cells (liver sinusoidal endothelial cells) and specialized fibroblasts (HSCs), which perform immune functions (Eggert and Greten 2017). HCC cells can reciprocally stimulate the

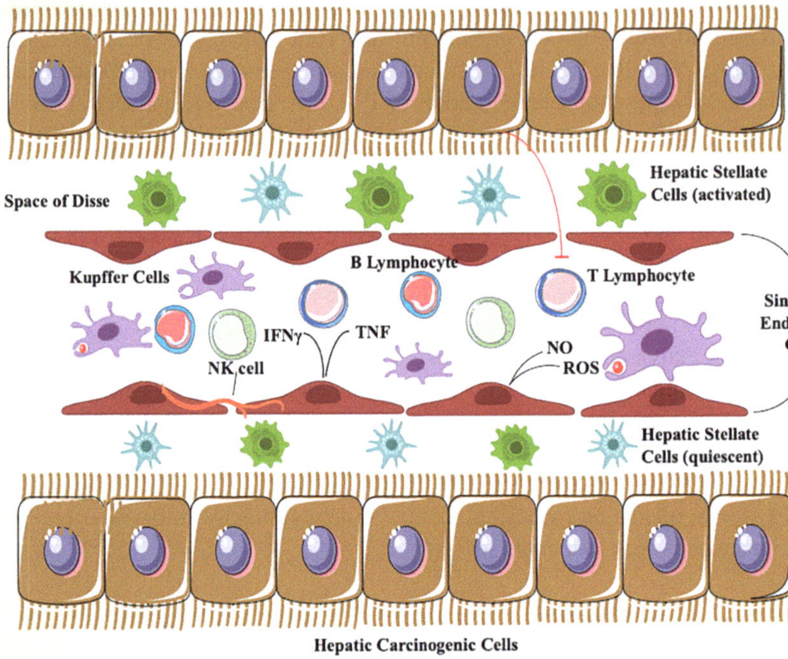

Fig. 1 Different cell functions in the tumor microenvironment. Sinusoidal endothelial cells secrete molecules such as TNF, IFN-γ, NO, and ROS. Moreover, Kupffer cells (KCs) act as an anti-tumor immune response besides phagocytosis. NK cells act in a neoangiogenic process in the tumor microenvironment, while the hepatic tumor cells inhibit the T lymphocytes' activities

proliferation of these tumor-associated fibroblasts, demonstrating the participation of fibroblasts in the growth, extravasation, and metastatic spread of cells (Sachdeva et al. 2015). In liver cancer tissue, cells act in different ways to kill tumor cells: sinusoidal endothelial cells secrete TNF, IFN-γ, nitric oxide (NO), and ROS; KCs carry out phagocytosis and recruit T cells to the liver to exert an anti-tumor immune response along with B cells and NK cells. Importantly, in the tumor microenvironment, NK cells promote the formation of hepatic blood vessels, and the tumor cells by themselves also support angiogenesis and inhibit CD8+ T cell cytotoxicity (Fig. 1) (Safarzadeh et al. 2021). The tumor microenvironment and immune response will differ between primary liver tumors, which reside in inflamed liver tissue, and the tumor environment resulting from liver metastasis from non-liver tumors, which are found in healthy liver tissue.

9 Conclusion

Despite many effective advances that have been achieved in cancer therapy, mechanisms to inhibit it, and the possibility to impair metastasis, it is still challenging for the clinical and research fields. Moreover, several immune cells participate in the process and drive an important role in the liver as well as the tumor microenvironment. The inflammation process in HCC drives an essential mechanism, taking count of many molecules such as cytokines with the function of pro- and anti-inflammatory helping the system to eradicate the tumorigenic cell. The signaling pathways, once evidenced, might be an attractive target to design a novel approach and a reliable strategy to eliminate cancer and attack all fronts. Finally, the mechanism of escape and metastasis stands out due to some different immune molecules, which can provide a future target to possible therapeutic alternatives for those patients affected by HCC, and in addition, present a promising approach in HCC cases, improving the patients' clinical prognosis achieving a positive and effective approach.

References

Abe H, Kimura A, Tsuruta S, Fukaya T, Sakaguchi R, Morita R et al (2014) Aryl hydrocarbon receptor plays protective roles in ConA-induced hepatic injury by both suppressing IFN-γ expression and inducing IL-22. Int Immunol 26(3):129–137

Aleksandrova K, Boeing H, Nöthlings U, Jenab M, Fedirko V, Kaaks R et al (2014) Inflammatory and metabolic biomarkers and risk of liver and biliary tract cancer. Hepatology 60(3):858–871

Alexia C, Fallot G, Lasfer M, Schweizer-Groyer G, Groyer A (2004) An evaluation of the role of insulin-like growth factors (IGF) and of type-I IGF receptor signalling in hepatocarcinogenesis and in the resistance of hepatocarcinoma cells against drug-induced apoptosis. Biochem Pharmacol 68(6):1003–1015

Barry AE, Baldeosingh R, Lamm R, Patel K, Zhang K, Dominguez DA et al (2020) Hepatic stellate cells and Hepatocarcinogenesis. Front Cell Dev Biol 8:709

Cammà C, Di Marco V, Orlando A, Sandonato L, Casaril A, Parisi P et al (2005) Unità Interdipartimentale Neoplasie Epatiche (U.I.N.E) group. Treatment of hepatocellular carcinoma in compensated cirrhosis with radio-frequency thermal ablation (RFTA): a prospective study. J Hepatol 42(4):535–540

Cao H, Chen X, Wang Z, Wang L, Xia Q, Zhang W (2020) The role of MDM2-p53 axis dysfunction in the hepatocellular carcinoma transformation. Cell Death Discovery 6:53

Cerami E, Gao J, Dogrusoz U, Gross BE, Sumer SO, Aksoy BA et al (2012) The cBio cancer genomics portal: an open platform for exploring multidimensional cancer genomics data. Cancer Discov 2(5):401–404

Chagpar RB, Links PH, Pastor CM, Furber LA, Hawrysh AD, Chamberlain MD, Anderson DH (2010) Direct positive regulation of PTEN by the p85 subunit of phosphatidylinositol 3-kinase. PNAS 107(12):5471–5476

Chen DS, Mellman I (2013) Oncology meets immunology: the cancer-immunity cycle. Immunity 39(1):1–10

Chen HJ, Hu MH, Xu FG, Xu HJ, She JJ, Xia HP (2018) Understanding the inflammation-cancer transformation in the development of primary liver cancer. Hepatoma Res 4:29

Chiu DK, Tse AP, Xu IM, Di Cui J, Lai RK, Li LL et al (2017) Hypoxia inducible factor HIF-1 promotes myeloid-derived suppressor cells accumulation through ENTPD2/CD39L1 in hepatocellular carcinoma. Nat Commun 8(1):517

Colombo M, Iavarone M (2014) Role of antiviral treatment for HCC prevention. Best Pract Res Clin Gastroenterol 28(5):771–781

Cui C, Fu K, Yang L, Wu S, Cen Z, Meng X et al (2019) Hypoxia-inducible gene 2 promotes the immune escape of hepatocellular carcinoma from nature killer cells through the interleukin-10-STAT3 signaling pathway. J Exp Clin Cancer Res 38(1):229

Dadi S, Chhangawala S, Whitlock BM, Franklin RA, Luo CT, Oh SA et al (2016) Cancer Immunosurveillance by tissue-resident innate lymphoid cells and innate-like T cells. Cell 164(3):365–377
Delire B, Stärkel P (2015) The Ras/MAPK pathway and hepatocarcinoma: pathogenesis and therapeutic implications. Eur J Clin Investig 45(6):609–623
Dimri M, Satyanarayana A (2020) Molecular signaling pathways and therapeutic targets in hepatocellular carcinoma. Cancers (Basel) 12(2):491
Dituri F, Mancarella S, Cigliano A, Chieti A, Giannelli G (2019) TGF-β as multifaceted orchestrator in HCC progression: signaling, EMT, immune microenvironment, and novel therapeutic perspectives. Semin Liver Dis 39(1):53–69
Doherty DG (2019) Antigen-specific immune tolerance in the liver. Nat Biomed Eng 3(10): 763–765
Dotto GP (2009) Crosstalk of notch with p53 and p63 in cancer growth control. Nat Rev Cancer 9(8):587–595
Eggert T, Greten TF (2017) Tumor regulation of the tissue environment in the liver. Pharmacol Ther 173:47–57
El Dika I, Khalil DN, Abou-Alfa GK (2019) Immune checkpoint inhibitors for hepatocellular carcinoma. Cancer 125(19):3312–3319
El-Emshaty HM, Nasif WA, Mohamed IE (2015) Serum cytokine of IL-10 and IL-12 in chronic liver disease: the immune and inflammatory response. Dis Markers 2015:707254
Fabregat I, Caballero-Díaz D (2018) Transforming growth factor-β-induced cell plasticity in liver fibrosis and Hepatocarcinogenesis. Front Oncol 8:357
Ficht X, Iannacone M (2020) Immune surveillance of the liver by T cells. Sci Immunol 5(51):2351
Flecken T, Schmidt N, Hild S, Gostick E, Drognitz O, Zeiser R et al (2014) Immunodominance and functional alterations of tumor-associated antigen-specific CD8+ T-cell responses in hepatocellular carcinoma. Hepatology 59(4):1415–1426
Fu Y, Liu S, Zeng S, Shen H (2019) From bench to bed: the tumor immune microenvironment and current immunotherapeutic strategies for hepatocellular carcinoma. J Exp Clin Cancer Res 38(1):396
Galle PR, Foerster F, Kudo M, Chan SL, Llovet JM, Qin S et al (2019) Biology and significance of alpha-fetoprotein in hepatocellular carcinoma. Liver Int 39(12):2214–2229
Gao SP, Mark KG, Leslie K, Pao W, Motoi N, Gerald WL et al (2007) Mutations in the EGFR kinase domain mediate STAT3 activation via IL-6 production in human lung adenocarcinomas. J Clin Investig 117(12):3846–3856
Gao J, Aksoy BA, Dogrusoz U, Dresdner G, Gross B, Sumer SO et al (2013) Integrative analysis of complex cancer genomics and clinical profiles using the cBioPortal. Sci Signal 6(269):l1
Garcia-Martinez I, Santoro N, Chen Y, Hoque R, Ouyang X, Caprio S et al (2016) Hepatocyte mitochondrial DNA drives nonalcoholic steatohepatitis by activation of TLR9. J Clin Investig 126(3):859–864
Garnelo M, Tan A, Her Z, Yeong J, Lim CJ, Chen J et al (2017) Interaction between tumour-infiltrating B cells and T cells controls the progression of hepatocellular carcinoma. Gut 66(2): 342–351
Giovannini C, Minguzzi M, Baglioni M, Fornari F, Giannone F, Ravaioli M et al (2014) Suppression of p53 by Notch3 is mediated by cyclin G1 and sustained by MDM2 and miR-221 axis in hepatocellular carcinoma. Oncotarget 5(21):10607–10620
Giovannini C, Bolondi L, Gramantieri L (2016) Targeting Notch3 in hepatocellular carcinoma: molecular mechanisms and therapeutic perspectives. Int J Mol Sci 18(1):56
Grivennikov SI, Greten FR, Karin M (2010) Immunity, inflammation, and cancer. Cell 140:883–899
Guo P, Wang Y, Dai C, Tao C, Wu F, Xie X et al (2018a) Ribosomal protein S15a promotes tumor angiogenesis via enhancing Wnt/β-catenin-induced FGF18 expression in hepatocellular carcinoma. Oncogene 37(9):1220–1236

Guo C, Zhao D, Zhang Q, Liu S, Sun MZ (2018b) miR-429 suppresses tumor migration and invasion by targeting CRKL in hepatocellular carcinoma via inhibiting Raf/MEK/ERK pathway and epithelial-mesenchymal transition. Sci Rep 8(1):2375

Gur C, Doron S, Kfir-Erenfeld S, Horwitz E, Abu-Tair L, Safadi R, Mandelboim O (2012) NKp46-mediated killing of human and mouse hepatic stellate cells attenuates liver fibrosis. Gut 61(6): 885–893

Haybaeck J, Zeller N, Wolf MJ, Weber A, Wagner U, Kurrer MO et al (2009) A lymphotoxin-driven pathway to hepatocellular carcinoma. Cancer Cell 16(4):295–308. Erratum in: Cancer Cell, 16(5), pp. 447

He G, Dhar D, Nakagawa H, Font-Burgada J, Ogata H, Jiang Y et al (2013) Identification of liver cancer progenitors whose malignant progression depends on autocrine IL-6 signaling. Cell 155(2):384–396

Horst AK, Neumann K, Diehl L, Tiegs G (2016) Modulation of liver tolerance by conventional and nonconventional antigen-presenting cells and regulatory immune cells. Cell Mol Immunol 13(3):277–292

Hou J, Zhang H, Sun B, Karin M (2020) The immunobiology of hepatocellular carcinoma in humans and mice: basic concepts and therapeutic implications. J Hepatol 72(1):167–182

Huang J, Qiu M, Wan L, Wang G, Huang T, Chen Z et al (2018) TGF-β1 promotes hepatocellular carcinoma invasion and metastasis via ERK pathway-mediated FGFR4 expression. Cell Physiol Biochem 45(4):1690–1699

Hwang YH, Choi JY, Kim S, Chung ES, Kim T, Koh SS et al (2004) Over-expression of c-raf-1 proto-oncogene in liver cirrhosis and hepatocellular carcinoma. Hepatol Res 29(2):113–121

Ibusuki R, Uto H, Arima S, Mawatari S, Setoguchi Y, Iwashita Y et al (2013) Transgenic expression of human neutrophil peptide-1 enhances hepatic fibrosis in mice fed a choline-deficient, L-amino acid-defined diet. Liver Int 33(10):1549–1556

Iyer SS, Cheng G (2012) Role of interleukin 10 transcriptional regulation in inflammation and autoimmune disease. Crit Rev Immunol 32(1):23–63

Jeng KS, Jeng CJ, Jeng WJ, Sheen IS, Li SY, Leu CM et al (2019) Sonic hedgehog signaling pathway as a potential target to inhibit the progression of hepatocellular carcinoma. Oncol Lett 18(5):4377–4384

Jenne CN, Kubes P (2013) Immune surveillance by the liver. Nat Immunol 14(10):996–1006

Juengpanich S, Shi L, Iranmanesh Y, Chen J, Cheng Z, Khoo AK et al (2019) The role of natural killer cells in hepatocellular carcinoma development and treatment: a narrative review. Transl Oncol 12(8):1092–1107

Keenan BP, Fong L, Kelley RK (2019) Immunotherapy in hepatocellular carcinoma: the complex interface between inflammation, fibrosis, and the immune response. J Immunother Cancer 7(1): 267

Ko CJ, Li CJ, Wu MY, Chu PY (2019) Overexpression of LGR-5 as a predictor of poor outcome in patients with hepatocellular carcinoma. Int J Environ Res Public Health 16(10):1836

Kubes P, Jenne C (2018) Immune responses in the liver. Annu Rev Immunol 36:247–277

Landskron G, De la Fuente M, Thuwajit P, Thuwajit C, Hermoso MA (2014) Chronic inflammation and cytokines in the tumor microenvironment. J Immunol Res 2014:149185

Lawal G, Xiao Y, Rahnemai-Azar AA, Tsilimigras DI, Kuang M, Bakopoulos A, Pawlik TM (2021) The immunology of hepatocellular carcinoma. Vaccines (Basel) 9(10):1184

Lencioni RA, Allgaier HP, Cioni D, Olschewski M, Deibert P, Crocetti L et al (2003) Small hepatocellular carcinoma in cirrhosis: randomized comparison of radio-frequency thermal ablation versus percutaneous ethanol injection. Radiology 228(1):235–240

Li S, Yang F, Ren X (2015) Immunotherapy for hepatocellular carcinoma. Drug Discoveries Therapeutics 9(5):363–371

Li X, Ramadori P, Pfister D, Seehawer M, Zender L, Heikenwalder M (2021) The immunological and metabolic landscape in primary and metastatic liver cancer. Nat Rev Cancer 21:541–557

Lin Y, Xu J, Lan H (2019) Tumor-associated macrophages in tumor metastasis: biological roles and clinical therapeutic applications. J Hematol Oncol 12(1):76

Liu Y, Zeng G (2012) Cancer and innate immune system interactions: translational potentials for cancer immunotherapy. J Immunother 35(4):299–308
Liu P, Ge M, Hu J, Li X, Che L, Sun K et al (2017a) A functional mammalian target of rapamycin complex 1 signaling is indispensable for c-Myc-driven hepatocarcinogenesis. Hepatology 66(1):167–181
Liu L, Qin S, Zheng Y, Han L, Zhang M, Luo N et al (2017b) Molecular targeting of VEGF/VEGFR signaling by the anti-VEGF monoclonal antibody BD0801 inhibits the growth and induces apoptosis of human hepatocellular carcinoma cells in vitro and in vivo. Cancer Biol Ther 18(3):166–176
Liu J, Yu GZ, Cheng XK, Li XD, Zeng XT, Ren XQ (2017c) LGR5 promotes hepatocellular carcinoma metastasis through inducting epithelial-mesenchymal transition. Oncotarget 8(31): 50896–50903
Liu Q, Zhang H, Jiang X, Qian C, Liu Z, Luo D (2017d) Factors involved in cancer metastasis: a better understanding to "seed and soil" hypothesis. Mol Cancer 16(1):176
Liu Z, Chen D, Ning F, Du J, Wang H (2018a) EGF is highly expressed in hepatocellular carcinoma (HCC) and promotes motility of HCC cells via fibronectin. J Cell Biochem 119(5):4170–4183
Liu P, Chen L, Zhang H (2018b) Natural killer cells in liver disease and hepatocellular carcinoma and the NK cell-based immunotherapy. J Immunol Res 2018:1206737
Liu DL, Lu LL, Dong LL, Liu Y, Bian XY, Lian BF et al (2020) miR-17-5p and miR-20a-5p suppress postoperative metastasis of hepatocellular carcinoma via blocking HGF/ERBB3-NF-κB positive feedback loop. Theranostics 10(8):3668–3683
Liu L, Dai B, Li R, Liu Z, Zhang Z (2021) Intravital molecular imaging reveals the restrained capacity of CTLs in the killing of tumor cells in the liver. Theranostics 11(1):194–208
Llovet JM, Zucman-Rossi J, Pikarsky E, Sangro B, Schwartz M, Sherman M, Gores G (2016) Hepatocellular carcinoma. Nat Rev Dis Primers 2:16018
Llovet JM, Castet F, Heikenwalder M, Maini MK, Mazzaferro V, Pinato DJ et al (2021) Immunotherapies for hepatocellular carcinoma. Nat Rev Clin Oncol 11
Lu JT, Zhao WD, He W, Wei W (2012) Hedgehog signaling pathway mediates invasion and metastasis of hepatocellular carcinoma via ERK pathway. Acta Pharmacol Sin 33(5):691–700
Ma Z, Guo D, Wang Q, Liu P, Xiao Y, Wu P et al (2019) Lgr5-mediated p53 repression through PDCD5 leads to doxorubicin resistance in hepatocellular carcinoma. Theranostics 9(10): 2967–2983
Marin-Acevedo JA, Dholaria B, Soyano AE, Knutson KL, Chumsri S, Lou Y (2018) Next generation of immune checkpoint therapy in cancer: new developments and challenges. J Hematol Oncol 11(1):39
McAllister SS, Weinberg RA (2014) The tumour-induced systemic environment as a critical regulator of cancer progression and metastasis. Nat Cell Biol 16(8):717–727
Micheau O, Tschopp J (2003) Induction of TNF receptor I-mediated apoptosis via two sequential signaling complexes. Cell 114(2):181–190
Mikulak J, Bruni E, Oriolo F, Di Vito C, Mavilio D (2019) Hepatic natural killer cells: organ-specific sentinels of liver immune homeostasis and physiopathology. Front Immunol 10:946
Mocan T, Sparchez Z, Craciun R, Bora CN, Leucuta DC (2019) Programmed cell death protein-1 (PD-1)/programmed death-ligand-1 (PD-L1) axis in hepatocellular carcinoma: prognostic and therapeutic perspectives. Clin Transl Oncol 21(6):702–712
Moreno-Cubero E, Larrubia JR (2016) Specific CD8(+) T cell response immunotherapy for hepatocellular carcinoma and viral hepatitis. World J Gastroenterol 22(28):6469–6483
Mossanen JC, Kohlhepp M, Wehr A, Krenkel O, Liepelt A, Roeth AA et al (2019) CXCR6 inhibits Hepatocarcinogenesis by promoting natural killer T- and CD4+ T-cell-dependent control of senescence. Gastroenterology 156(6):1877–1889.e4
Nakagawa H, Maeda S (2012) Inflammation- and stress-related signaling pathways in hepatocarcinogenesis. World J Gastroenterol 18(31):4071–4081

Njah K, Chakraborty S, Qiu B, Arumugam S, Raju A, Pobbati AV et al (2019) A role of Agrin in maintaining the stability of vascular endothelial growth factor Receptor-2 during tumor angiogenesis. Cell Rep 28(4):949–965.e7
Patil KR, Mahajan UB, Unger BS, Goyal SN, Belemkar S, Surana SJ et al (2019) Animal models of inflammation for screening of anti-inflammatory drugs: implications for the discovery and development of Phytopharmaceuticals. Int J Mol Sci 20(18):4367
Piñeiro Fernández J, Luddy KA, Harmon C, O'Farrelly C (2019) Hepatic tumor microenvironments and effects on NK cell phenotype and function. Int J Mol Sci 20(17):4131
Pinter M, Sieghart W, Schmid M, Dauser B, Prager G, Dienes HP et al (2013) Hedgehog inhibition reduces angiogenesis by downregulation of tumoral VEGF-A expression in hepatocellular carcinoma. United European Gastroenterol J 1(4):265–275
Prieto J, Melero I, Sangro B (2015) Immunological landscape and immunotherapy of hepatocellular carcinoma. Nat Rev Gastroenterol Hepatol 12:681–700
Qian BZ, Pollard JW (2010) Macrophage diversity enhances tumor progression and metastasis. Cell 141(1):39–51
Refolo MG, Messa C, Guerra V, Carr BI, D'Alessandro R (2020) Inflammatory mechanisms of HCC development. Cancers (Basel) 12(3):641
Ringelhan M, Pfister D, O'Connor T, Pikarsky E, Heikenwalder M (2018) The immunology of hepatocellular carcinoma. Nat Immunol 19(3):222–232
Robinson MW, Harmon C, O'Farrelly C (2016) Liver immunology and its role in inflammation and homeostasis. Cell Mol Immunol 13(3):267–276
Roth GS, Decaens T (2017) Liver immunotolerance and hepatocellular carcinoma: Pathophysiological mechanisms and therapeutic perspectives. Eur J Cancer 87:101–112
Ruiz de Galarreta M, Bresnahan E, Molina-Sánchez P, Lindblad KE, Maier B, Sia D et al (2019) β-Catenin activation promotes immune escape and resistance to anti-PD-1 therapy in hepatocellular carcinoma. Cancer Discov 9(8):1124–1141
Sachdeva M, Chawla YK, Arora SK (2015) Immunology of hepatocellular carcinoma. World J Hepatol 7(17):2080–2090
Safarzadeh A, Alizadeh M, Beyranvand F, Falavand Jozaaee R, Hajiasgharzadeh K, Baghbanzadeh A et al (2021) Varied functions of immune checkpoints during cancer metastasis. Cancer Immunol Immunotherapy 70(3):569–588
Sällberg M, Pasetto A (2020) Liver, tumor and viral hepatitis: key players in the complex balance between tolerance and immune activation. Front Immunol 11:552
Schattenberg JM, Schuchmann M, Galle PR (2011) Dysregulation of apoptosis signaling pathways. J Gastroenterol Hepatol 6(s1):213–219
Shalapour S, Karin M (2019) Pas de Deux: control of anti-tumor immunity by cancer-associated inflammation. Immunity 51(1):15–26
Shetty S, Lalor PF, Adams DH (2018) Liver sinusoidal endothelial cells - gatekeepers of hepatic immunity. Nat Rev Gastroenterol Hepatol 15(9):555–567
Shi JY, Ma LJ, Zhang JW, Duan M, Ding ZB, Yang LX et al (2017) FOXP3 is a HCC suppressor gene and acts through regulating the TGF-β/Smad2/3 signaling pathway. BMC Cancer 17(1): 648
Strick-Marchand H, Morosan S, Charneau P, Kremsdorf D, Weiss MC (2004) Bipotential mouse embryonic liver stem cell lines contribute to liver regeneration and differentiate as bile ducts and hepatocytes. Proc Nat Acad Sci U S A 101(22):8360–8365
Su MC, Hsu C, Kao HL, Jeng YM (2006) CD24 expression is a prognostic factor in intrahepatic cholangiocarcinoma. Cancer Lett 235(1):34–39
Subbarao SA (2021) Cancer vs. SARS-CoV-2 induced inflammation, overlapping functions, and pharmacological targeting. Inflammopharmacology 29(2):343–366
Sükei T, Palma E, Urbani L (2021) Interplay between cellular and non-cellular components of the tumour microenvironment in hepatocellular carcinoma. Cancers (Basel) 13(21):5586

Sun XT, Yuan XW, Zhu HT, Deng ZM, Yu DC, Zhou X, Ding YT (2012) Endothelial precursor cells promote angiogenesis in hepatocellular carcinoma. World J Gastroenterol 18(35): 4925–4933

Sun H, Huang Q, Huang M, Wen H, Lin R, Zheng M et al (2019) Human CD96 correlates to natural killer cell exhaustion and predicts the prognosis of human hepatocellular carcinoma. Hepatology 70(1):168–183

Sun Y, Wu L, Zhong Y, Zhou K, Hou Y, Wang Z et al (2021) Single-cell landscape of the ecosystem in early-relapse hepatocellular carcinoma. Cell 184(2):404–421.e16

Svinka J, Mikulits W, Eferl R (2014) STAT3 in hepatocellular carcinoma: new perspectives. Hepatic Oncol 1(1):107–120

Syn WK, Agboola KM, Swiderska M, Michelotti GA, Liaskou E, Pang H et al (2012) NKT-associated hedgehog and osteopontin drive fibrogenesis in non-alcoholic fatty liver disease. Gut 61(9):1323–1329

Tang B, Tang F, Wang Z, Qi G, Liang X, Li B et al (2016) Upregulation of Akt/NF-κB-regulated inflammation and Akt/Bad-related apoptosis signaling pathway involved in hepatic carcinoma process: suppression by carnosic acid nanoparticle. Int J Nanomedicine 11:6401–6420

Taylor A, Verhagen J, Blaser K, Akdis M, Akdis CA (2006) Mechanisms of immune suppression by interleukin-10 and transforming growth factor-beta: the role of T regulatory cells. Immunology 117(4):433–442

Tsui YM, Chan LK, Ng IO (2020) Cancer stemness in hepatocellular carcinoma: mechanisms and translational potential. Br J Cancer 122(10):1428–1440

Tu S, Huang W, Huang C, Luo Z, Yan X (2019) Contextual regulation of TGF-β signaling in liver cancer. Cell 8(10):1235

Tumanov AV, Koroleva EP, Christiansen PA, Khan MA, Ruddy MJ, Burnette B et al (2009) T cell-derived lymphotoxin regulates liver regeneration. Gastroenterology 136(2):694–704.e4

Uson Junior PLS, Nagalo BM, Ahn DH, Bekaii-Saab T, Borad MJ (2021) Combination immunotherapy for hepatocellular carcinoma: where are we currently? Semin Liver Dis 41(2):136–141

Wang L, Wang FS (2019) Clinical immunology and immunotherapy for hepatocellular carcinoma: current progress and challenges. Hepatol Int 13(5):521–533

Wang H, Lafdil F, Wang L, Park O, Yin S, Niu J et al (2011) Hepatoprotective versus oncogenic functions of STAT3 in liver tumorigenesis. Am J Pathol 179(2):714–724

Wang Z, Sheng YY, Gao XM, Wang CQ, Wang XY, Lu XU et al (2015) β-Catenin mutation is correlated with a favorable prognosis in patients with hepatocellular carcinoma. Mol Clin Oncol 3(4):936–940

Wang H, Rao B, Lou J, Li J, Liu Z, Li A et al (2020) The function of the HGF/c-met Axis in hepatocellular carcinoma. Front Cell Dev Biol 8:55

Wang SZ, Lee SD, Sarkar D, Lee HM, Khan A, Bhati C et al (2021) Immunological characterization of hepatocellular carcinoma. Hepatoma Res 7:6

Wehr A, Baeck C, Heymann F, Niemietz PM, Hammerich L, Martin C et al (2013) Chemokine receptor CXCR6-dependent hepatic NK T cell accumulation promotes inflammation and liver fibrosis. J Immunol 190(10):5226–5236

Wei YG, Liu F, Li B, Chen X, Ma Y, Yan LN et al (2011) Interleukin-10 gene polymorphisms and hepatocellular carcinoma susceptibility: a meta-analysis. World J Gastroenterol 17(34): 3941–3947

Wen L, Xin B, Wu P, Lin CH, Peng C, Wang G et al (2019) An efficient combination immunotherapy for primary liver cancer by harmonized activation of innate and adaptive immunity in mice. Hepatology 69(6):2518–2532

Woo SR, Corrales L, Gajewski TF (2015) Innate immune recognition of cancer. Annual Rev Immunol 33:445–474

Xia S, Pan Y, Liang Y, Xu J, Cai X (2020) The microenvironmental and metabolic aspects of sorafenib resistance in hepatocellular carcinoma. EBioMedicine 51:102610

Xu Z, Xu M, Liu P, Zhang S, Shang R, Qiao Y et al (2019) The mTORC2-Akt1 Cascade is crucial for c-Myc to promote Hepatocarcinogenesis in mice and humans. Hepatology 70(5):1600–1613

Yang YM, Kim SY, Seki E (2019) Inflammation and liver cancer: molecular mechanisms and therapeutic targets. Semin Liver Dis 39(1):26–42
Ye Y, Xu Y, Lai Y, He W, Li Y, Wang R et al (2018) Long non-coding RNA cox-2 prevents immune evasion and metastasis of hepatocellular carcinoma by altering M1/M2 macrophage polarization. J Cell Biochem 119(3):2951–2963
Yoshida K, Matsuzaki K, Murata M, Yamaguchi T, Suwa K, Okazaki K (2018) Clinico-pathological importance of TGF-β/Phospho-Smad signaling during human hepatic Fibrocarcinogenesis. Cancers (Basel) 10(6):183
Yu SJ, Ma C, Heinrich B, Brown ZJ, Sandhu M, Zhang Q et al (2019) Targeting the crosstalk between cytokine-induced killer cells and myeloid-derived suppressor cells in hepatocellular carcinoma. J Hepatol 70(3):449–457
Yuen VW, Wong CC (2020) Hypoxia-inducible factors and innate immunity in liver cancer. J Clin Investig 130(10):5052–5062
Zaki MYW, Mahdi AK, Patman GL, Whitehead A, Maurício JP, McCain MV et al (2021) Key features of the environment promoting liver cancer in the absence of cirrhosis. Sci Rep 11(1): 16727
Zappavigna S, Cossu AM, Grimaldi A, Bocchetti M, Ferraro GA, Nicoletti GF et al (2020) Anti-inflammatory drugs as anticancer agents. Int J Mol Sci 21(7):2605
Zeeshan R, Mutahir Z (2017) Cancer metastasis - tricks of the trade. Bosn J Basic Med Sci 17(3): 172–182
Zhang Q, Lou Y, Bai XL, Liang TB (2018) Immunometabolism: a novel perspective of liver cancer microenvironment and its influence on tumor progression. World J Gastroenterol 24(31): 3500–3512
Zheng M, Tian Z (2019) Liver-mediated adaptive immune tolerance. Front Immunol 10:2525
Zhou K, Luo X, Wang Y, Cao D, Sun G (2017a) MicroRNA-30a suppresses tumor progression by blocking Ras/Raf/MEK/ERK signaling pathway in hepatocellular carcinoma. Biomed Pharmacother 93:1025–1032
Zhou G, Sprengers D, Boor PPC, Doukas M, Schutz H, Mancham S et al (2017b) Antibodies against immune checkpoint molecules restore functions of tumor-infiltrating T cells in hepatocellular carcinomas. Gastroenterology 153(4):1107–1119.e10
Zongyi Y, Xiaowu L (2020) Immunotherapy for hepatocellular carcinoma. Cancer Lett 470:8–17

The Tumor Microenvironment in Hepatocellular Carcinoma

Katsuya Nagaoka, Yasuhito Tanaka, and Okio Hino

Abstract

The tumor microenvironment refers to the noncancerous tissue surrounding cancer cells. The tumor microenvironment consists of stromal cells and normal parenchymal cells that surround the cancer cells. In the tumor microenvironment, dynamic interactions between immunocompetent cells and cancer cells occur, as described by the cancer immune-editing theory and the cancer immune cycle. At the same time, changes in the surrounding environment such as tumor angiogenesis, reprogramming of cancer energy metabolism, and fibrosis associated with chronic inflammation are observed. When the immune balance of the liver is disrupted by chronic hepatitis or cirrhosis, the risk of developing hepatocellular carcinoma (HCC) increases. HCC forms a unique tumor microenvironment based on the peculiarities of the liver's vascular control and immune system. This chapter outlines the microenvironment in HCC, with a particular focus on immunity, angiogenesis, and metabolism.

While the recent medical environment regarding HCC has changed drastically due to the introduction of new drugs and changes in worldwide lifestyle, the recent development of single-cell analysis technology has led to dramatic progress in the detailed analysis of the tumor microenvironment in HCC. In this chapter, we will also introduce the latest information on clinical applications of

K. Nagaoka · Y. Tanaka (✉)
Department of Gastroenterology and Hepatology, Graduate School of Medical Sciences, Kumamoto University, Kumamoto, Japan
e-mail: ytanaka@kumamoto-u.ac.jp

O. Hino
Department of Pathology and Oncology, Faculty of Medicine, Juntendo University, Tokyo, Japan

Nitobe Inazo Memorial Center, Tokyo, Japan

Keisen Jogakuen, Tokyo, Japan

Interdisciplinary Cancer Research, https://doi.org/10.1007/16833_2022_66
Published online: 18 November 2022

the microenvironment of HCC, including diagnosis and treatment. In addition, we will discuss recent topics such as immune cycle and immune checkpoint molecules in HCC, the concept of hot and cold tumors, tumor microenvironment and tumor macroenvironment, and especially organ-specific metastasis and metastatic niche.

Keywords

Angiogenesis · Cancer immunity · Cancer metabolism · Fibrosis · Hepatocellular carcinoma · Immune checkpoint · Tumor microenvironment

1 Introduction

Most primary liver cancers are either hepatocellular carcinomas or intrahepatic cholangiocarcinomas. Hepatocellular carcinomas account for about 90%. Hepatocellular carcinoma (HCC) has the fifth highest incidence rate and is the fourth leading cause of death worldwide (Global Burden of Disease Cancer et al. 2019; Yang et al. 2019). The causes of HCC are known to include metabolic diseases, such as steatohepatitis, obesity, and diabetes, alcohol, chronic hepatitis caused by hepatitis B and C viruses, hemochromatosis, and toxins produced by molds such as aflatoxin (Yang et al. 2019).

The combination of atezolizumab, an anti-programmed death-ligand 1 (PD-L1) antibody, and bevacizumab, an anti-vascular endothelial growth factor (VEGF) antibody, (combination therapy with atezolizumab plus bevacizumab), was recently shown to be effective in the treatment of unresectable advanced HCC (Finn et al. 2020a). Atezolizumab plus bevacizumab demonstrated superiority, in both overall survival (OS) and progression-free survival (PFS), over the previous first-line tyrosine kinase inhibitor (TKI), sorafenib, making it a new first-line therapy for HCC patients with preserved liver function (Finn et al. 2020a). Atezolizumab is the first immunotherapy agent to be adopted as first-line therapy for the treatment of HCC. It has also been reported that bevacizumab acts synergistically by altering the immune responsiveness of the tumor immune microenvironment (Hegde et al. 2018). With the widespread use of immunotherapy for HCC, interest in the microenvironment of HCC, including tumor immunity, is increasing among physicians and liver disease researchers.

Cancer is caused by genetic abnormalities. Cancers caused by monogenic disorders, in which the responsible gene has been clearly identified, tend to be familial tumors, manifesting at young ages, and occur in 5–10% of all cancers. On the other hand, the remaining 90% of cancers are explained by the concept of multistep accumulations of genetic alterations (Vogelstein and Kinzler 1993; Weiss 2004). In recent years, however, it has been suggested that changes in the microenvironment surrounding tissue stem cells may occur in parallel with the accumulation of genetic mutations, leading to the development of cancer (Meylan et al. 2020; Kobayashi et al. 2019).

Research into the relationship between cancer cells and the microenvironment dates back to the nineteenth century. An English surgeon, Stephen Paget, proposed his "seed and soil" theory in 1889. This is the idea that cancer cells (seeds) do not form metastases anywhere in the body by chance but have an affinity for the environment (soil) of a particular organ for their growth (Paget 1989). This suggests that the surrounding environment at the cellular level, i.e., the tumor microenvironment, is very important for the growth of cancer cells.

The tumor microenvironment plays an important role, not only in metastasis but also in the proliferation of cancer cells in the primary tumor. In their review "Hallmarks of cancer" published in Cell (Hanahan and Weinberg 2000; Hanahan and Weinberg 2011), Hanahan and Weinberg identified the following ten malignant traits that cancer cells have acquired: (1) sustained proliferative signaling, (2) evasion of growth suppressors, (3) enabled replication, (4) genome instability and mutation, (5) resistance to cell death, (6) deregulation of cellular energetics, (7) sustained angiogenesis, (8) tumor-promoting inflammation, (9) avoidance of immune destruction, and (10) tissue invasion and metastasis. Of these, items 5–10 are directly related to the interaction between the cancer cells themselves, surrounding host cells, neoplastic blood vessels, and cancer immunity. Cancer cells are generated from normal cells and express incomplete non-self-antigens. The struggle between the immune cells of the self- and the non-self-cancer cells, to eliminate the nonself, is complex and combined with changes in immune responsiveness due to angiogenesis. In particular, most hepatocellular carcinomas are caused by chronic inflammation of the liver tissue, and the blood flow to the liver also is unique, which makes the HCC microenvironment more or less complicated compared to the general cancer microenvironment.

This chapter introduces the microenvironment as it affects tumors, in terms of immunity, angiogenesis, and metabolism, and provides an overview of the latest information on the cancer microenvironment, including phenomena specific to HCC.

2 The Microenvironment of Cancer

Normal tissues are classified anatomically into parenchyma and stroma. In the liver, the parenchyma is composed of hepatocytes, while the stroma is composed of the endothelial cells of the vasculatures (arteries, veins, portal veins, and lymphatic vessels), bile duct epithelial cells, connective tissue, immunocompetent cells, and inflammatory cells. The cancer microenvironment refers to the noncancerous tissues surrounding the cancer cells and comprises stromal cells surrounding cancer and normal parenchymal cells. In general, the characteristics of the cancer microenvironment overlap with those of the cancer stroma, but because HCC is often triggered by chronic liver diseases, including hepatitis and cirrhosis, hepatocytes, which are the site of inflammation around cancer, have a significant impact on the cancer microenvironment as well as the cancer stroma.

The diversity of cancer is one of the most important characteristics of cancerous diseases. The environment at the cellular level, or microenvironment, surrounding cancer is also extremely diverse. However, without exception, proliferating cancers have an interrelationship with their surrounding microenvironment, and certain common denominators have been found among these diverse characteristics; this knowledge has been accumulating in recent years.

The components of the microenvironment and the related mechanisms are diverse, and even in the case of HCC, a variety of components and underlying mechanisms can be assumed (Hernandez-Gea et al. 2013) (Kurebayashi et al. 2021). In this section, we briefly introduce some of these topics that have recently attracted attention to clinical practice, such as cancer immunity, tumor angiogenesis, cancer energy metabolism, and chronic inflammation and fibrosis.

2.1 Cancer Immunity

2.1.1 Cancer Immunoediting

The concept of "cancer immunosurveillance," in which the immune system eliminates not only non-self-derived entities such as pathogens but also mutated cancer cells derived from self, has been debated since around 1950 (Burnet 1970), but until recently, it had not been proven whether it really exists. In 2001, Schreiber et al. showed that mice lacking the *RAG* gene, which is involved in the activation of recombination of immunoglobulin genes related to the diversity of the immune response, have an increased incidence of cancer (Shankaran et al. 2001). Schreiber et al. proposed the hypothesis of cancer immunoediting, which is a further development of the concept of cancer immunosurveillance and explains the process from carcinogenesis to cancer progression, despite the existence of cancer immunosurveillance (Dunn et al. 2002; Schreiber et al. 2011). Cancer immunoediting consists of three phases: elimination, equilibrium, and escape. Elimination is the process by which mutant cells (cancer cells) are removed by various immune cells. CD4+ T cells and CD8+ T cells are thought to act as important effectors in the elimination of cancer. In the escape phase, the cancer cells build a favorable environment for their own proliferation (cancer microenvironment) by acquiring a mechanism that suppresses the antitumor immune response. Equilibrium is the "dormant" state of the tumor, intermediate between elimination and escape, and is a state of "dormancy" in tumors; it is a process in which mutant cells with low immunogenicity (edited by the immune system), and which could not be eliminated in the elimination phase, do not grow over time. This is explained as a process of maintaining equilibrium with the attack of the immune system.

2.1.2 The Cancer Immunity Cycle

The "cancer immunity cycle" proposed by Chen et al. in 2013 divides the immune response to cancer into seven steps (Table 1) (Chen and Mellman 2013, 2017). First, cancer cells are destroyed by immune cells, such as phagocytes and lymphocytes, and cancer antigens are released (Step 1). Next, dendritic cells phagocytose the

Table 1 The cancer immunity cycle

Step	Events	Cells	Organs	References
1	Release of cancer cell antigens		Tumor tissues	Ferguson et al. (2011)
2	Cancer antigen presentation	Dendritic cells/APCs	Tumor tissues/lymph nodes	Lippitz (2013) and Ruiz de Galarreta et al. (2019)
3	Priming and activation	APCs/T cells	Lymph nodes	Franciszkiewicz et al. (2012) and Lee et al. (2022)
4	Trafficking of T cells to tumors	CTLs	Blood vessels	Franciszkiewicz et al. (2012) and Peng et al. (2012)
5	Infiltration of T cells into tumors	CTLs, endothelial cells	Blood vessels	Franciszkiewicz et al. (2013) and Kurebayashi et al. (2021)
6	Recognition of cancer cells by T cells	CTLs, cancer cells	Tumor tissues	Mellman et al. (2011)
7	Killing of cancer cells	Immune and cancer cells	Tumor tissues	Chen et al. (2012) and Giraud et al. (2021)

released cancer antigens and degrade them so that they can be recognized by T cells (processing) and presented (Step 2). The antigen-presenting dendritic cells migrate to the lymph nodes, where naïve T cells are activated by the presentation of cancer antigens (priming) and proliferate (Step 3). Activated effector T cells are induced by various chemokines to migrate through the blood vessels toward the cancer cells (Step 4). Activated effector T cells pass through the gaps in the vascular endothelium, infiltrate cancer (Step 5), specifically recognize and bind to the cancer cells expressing the antigen presented in Step 3 (Step 6), and kill the target cancer cells (Step 7). These cancer antigens are exposed in the process of cancer cell injury, and "Step 1" of the next cycle begins. Then, more cancer cells are eliminated in successive cycles (epitope spreading). Thus, when the cycles are successful, cancer can be continuously eliminated, but, on the other hand, in cancerous tissues, multiple steps are found frequently to be impaired and the cancer cells escape from the immunosurveillance.

2.2 Tumor Angiogenesis

Angiogenesis is an important physiological function that creates a network of new blood vessels, as required, such as in wound healing. Hypoxia induces VEGF through the activation of hypoxia-inducible factor (HIF) (Rey and Semenza 2010). VEGF is produced by vascular endothelial cells and various other types of cells, including fibroblasts, platelets, macrophages, and keratinocytes (Schultz et al. 2003). VEGF acts mainly on vascular endothelial cells in the process of angiogenesis, promoting vascular endothelial cell proliferation, lumen formation, and endothelial cell migration, as well as increasing vascular permeability (Senger et al. 1983;

Ke et al. 1996). Under physiological conditions, these processes of angiogenesis are correctly regulated. On the other hand, angiogenesis is also actively induced in cancer tissues, upon ischemia. The cancer cells themselves secrete VEGF (Kim et al. 1993), and angiogenic factors such as VEGF are abundant in the cancer microenvironment. The difference between physiological angiogenesis in normal tissues and pathological angiogenesis in cancerous tissues is that pathological angiogenesis in cancer does not stop, even after the provision of a sufficient blood supply to the tissues.

In addition to the disorderly angiogenesis in cancerous tissues, it has been pointed out that the traits of vascular endothelial cells are altered after normal vascular endothelial cells invade cancerous tissues under the influence of the surrounding tumor microenvironment. Tumor endothelial cells may express specific factors that induce malignant transformation of cancer, such as metastasis, drug resistance (Ohga et al. 2012), and so on. In addition, tumor endothelial cells are immature structures with high vascular permeability because of the weak adhesion between endothelial cells and pericytes. Increased vascular permeability leads to edema of the tumor tissue and increased intratumor pressure. This induces impaired reflux and hypoxia in tumor blood vessels (Boucher and Jain 1992), which triggers further activation of VEGF, inhibits the delivery of small molecules such as anticancer agents, and prevents the migration of immunocompetent cells from the vascular lumen to the interstitial tissue.

Thus, tumor angiogenesis leads not only to the pathological proliferation of vascular endothelial cells but also to the disruption of tumor immunity. By normalizing immature tumor blood vessels, VEGF inhibitors have been shown to reduce the malignant effects of cancer in the tumor microenvironment and improve the invasion of immunocompetent cells into cancerous tissues (Fukumura et al. 2018), and this other effect of VEGF inhibitors is attracting attention in the treatment of liver cancer.

2.3 Cancer Energy Metabolism

Many cancer cells undergo metabolic reprogramming to adapt to the tumor microenvironment and maintain their proliferative potential. One of these strategies is a metabolic mode called the Warburg effect (or aerobic glycolysis) (Warburg 1956; Cairns et al. 2011). In an aerobic environment, the amount of adenosine triphosphate (ATP) obtained from a single molecule of glucose is much higher through mitochondrial respiration than through glycolysis alone, and, therefore, glycolytic enhancement in normal cells generally occurs under anaerobic conditions (anaerobic glycolysis). In cancer cells, ATP and lactate production by the glycolytic system are enhanced, regardless of the oxygen conditions (aerobic glycolysis). Interestingly, this phenomenon disrupts physiological energy homeostasis, allowing the cells to adapt to the cancer microenvironment of hypoxia and hypotrophy. In addition, the accumulation of lactate in tumor tissues is also associated with an immunosuppressive phenotype of immune cells infiltrating the tumor, represented by a change in the polarity of M2-type macrophages (Colegio et al. 2014; Ohashi et al. 2017; Noe et al. 2021).

2.4 Chronic Inflammation and Fibrosis

Chronic inflammation of organs can lead to fibrosis. Fibrosis due to chronic inflammation is not localized, such as the scarring that occurs after physical damage to tissues but occurs as widespread fibrosis that forms slowly, as in cirrhosis, chronic pancreatitis, chronic glomerulonephritis, and pulmonary fibrosis, which are generally associated with progressive organ dysfunction. The pathogenesis of fibrotic diseases is explained as organ damage caused by the accumulation of collagenous extracellular matrix due to persistent tissue damage such as chronic inflammation. The extracellular matrix is an insoluble substance that fills the extracellular space; it is produced by fibroblasts in the stroma and is abundant in connective tissue.

Normal fibroblasts are present in large numbers in the stroma in a disordered manner, while myofibroblasts exist in various organs such as the liver, pancreas, kidney, and lung under normal conditions in a specific cell form (stellate form). Tissue organ fibrosis is thought to be a state in which active myofibroblasts emerge during tissue repair and regeneration and increase excessive fibroid degeneration. Myofibroblasts are given the name "myo" because, when inflammation becomes chronic and they are repeatedly subjected to stress, their ability to produce fibroids is enhanced, and, at the same time, the expression of α-smooth muscle-type myosin (SMA) is markedly increased. Hepatic stellate cells (Tsuchida and Friedman 2017), pancreatic stellate cells, and intraglomerular mesangial cells also express α-SMA and act as myofibroblast-like cells when they are activated (Zhao and Burt 2007). The origin of myofibroblasts is diverse, and cases have been reported in which tissue-resident fibroblasts and pericytes were activated to become myofibroblasts (Mack and Yanagita 2015).

The fibroblasts that proliferate in the cancer stroma are called cancer-associated fibroblasts (CAFs) and are similar or nearly identical to myofibroblasts, for which α-SMA is a marker. Mesenchymal stem cells, tissue-resident fibroblasts, and some pericytes have been postulated as sources of CAFs (Kobayashi et al. 2019; Sahai et al. 2020).

3 The Diversity and Peculiarity of the Liver Cancer Microenvironment

The liver has two blood supply channels: the portal vein and the hepatic artery. Three-quarters of the blood flow is supplied by the portal vein, with rich nutrients and relatively high levels of oxygen, and the remaining quarter is supplied by the hepatic artery, with oxygen-rich blood (Rappaport 1980). The capillaries of the liver differ in their properties and shape from those of the rest of the body and are called "the sinusoid." The blood flowing through the portal vein and hepatic artery joins and mixes in the sinusoid and flows out into the hepatic veins. The sinusoid is lined by sinusoidal endothelial cells, and the lumen contains Kupffer cells (liver-resident macrophages) and Pit cells (liver-specific natural killer (NK) cells). The hepatic stellate cells reside in the "space of Disse," located between the laminae of the hepatocytes and the sinusoids. The hepatic stellate cells, like the pericytes, surround the sinusoidal endothelial cells (Blomhoff and Wake 1991). When fibrosis occurs in the liver, hepatic stellate cells are activated, and collagen accumulates in the space of Disse (Tsuchida and Friedman 2017), causing hypoxia of the hepatocytes in this region.

3.1 The Liver Tissue as an Immune System

The liver has a powerful capacity to manage the influx, via the portal vein, of a wide variety of microorganisms, microbial compounds (e.g., lipopolysaccharides), and microbial-associated molecules from the intestine. The liver contains the largest number of resident macrophages (Kupffer cells) in the body, accounting for about 90% of all tissue macrophages. These play a central role in trapping pathogens, such as eliminating bacteremia and recruiting immune cells to the liver. In addition, liver tissue contains a high density of NK cells, natural killer T cells (NKT cells), hepatic migratory and resident lymphocytes, and a greater ratio of CD8+ and CD4+ T cells than in the periphery (Jenne and Kubes 2013); these are centrally involved in the detection and clearance of infectious organisms from the blood. Furthermore, liver sinusoidal endothelial cells (LSECs), hepatic stellate cells (HSCs), and hepatocytes have also been reported to be involved in pathogen detection and capture, antigen presentation, and cytokine production (Jenne and Kubes 2013; Protzer et al. 2012; Robinson et al. 2016).

3.2 Chronic Liver Injury and Immune Imbalance

The liver plays an important immunomodulatory role through its ability to maintain immune tolerance of non-pathological or homeostatic inflammatory stimuli. This prevents liver injury and induces systemic tolerance. Immune tolerance in the liver results from the interaction of resident cells (e.g., Kupffer cells, LSECs, dendritic cells, and hepatocytes) with peripheral leukocytes. This immunosuppressive environment is maintained by a balance of basal pro-inflammatory cytokines (interleukin (IL)-2, IL-7, IL-12, IL-15, and interferon (IFN)-γ) and anti-inflammatory cytokines (IL-10, IL-13, and transforming growth factor (TGF)-β) (Robinson et al. 2016).

This balance can be disrupted by persistent elevation of inflammatory signals due to liver injury from chronic infections (e.g., hepatitis B virus (HBV) and hepatitis C virus (HCV) infections), fat accumulation (in nonalcoholic steatohepatitis (NASH)), and alcoholic liver disease. During HBV infection, strong antiviral CD8+ T-cell responses to the viral surface and core antigens facilitate the clearance of HBV (Knolle and Thimme 2014). On the other hand, if a load of circulating HBV antigens is too great, this leads to T-cell exhaustion (Lopes et al. 2008) and contributes to the development of chronic hepatitis. HCV evades detection by the immune system through its high rates of replication and mutation (Hedegaard et al. 2017). HCV also activates specific inflammasomes in macrophages and monocytes (e.g., the NLR family pyrin domain containing 3 (NLRP3) inflammasome, a protein complex involved in the induction of inflammation) (Chattergoon et al. 2014) without inducing type I IFN. Ongoing inflammation caused by chronic alcohol consumption also activates NLRP3 inflammasomes and amplifies inflammatory signals. Although NASH partially shares histopathological features with alcoholic steatohepatitis, the immunological pathogenic factors involved are distinctly different. The difference between NASH and "aseptic inflammation" such as alcoholic liver disease is that inflammasome activation occurs not only in macrophages but also in hepatocytes and other immune cells (Szabo and Petrasek 2015). Antibody-mediated inhibition of CD8+ T cells suppresses NASH liver injury; therefore, intrahepatic CD8+ T cells are thought to play a major role in NASH liver injury (Wolf et al. 2014).

3.3 The Immune Microenvironment in Hepatocellular Carcinoma

Similar to the situation in chronic infections, sustained overstimulation of T cells in tumor localities leads to T-cell exhaustion. According to Yarchoan et al., immunohistochemical staining of resected HCC and noncancerous liver tissues showed decreased expression of CD8 in the tumor and decreased expression of CD163 at the tumor interface. 83% of patients had positive clusters of PD-L1 expression. In addition, in 65% of cases, the tumor-infiltrating lymphocytes expressed LAG-3, a CD4 analog that is an auxiliary receptor for helper T (Th) cells. Both PD-L1 and LAG-3 were upregulated in tumor-infiltrating lymphocytes, compared to the liver background (Yarchoan et al. 2017).

The detailed analysis of the relationship between the tumor microenvironment and the immune microenvironment has been dramatically advanced by the recent development of technology for single-cell analysis. The following are some of the important reports on the interaction between HCC cells and the surrounding immunocompetent cells.

3.3.1 Similarities Between the Tumor and Fetal Liver Microenvironments

Sharma et al. performed comprehensive gene expression profiling by single-cell RNA-seq using single-cell cDNA libraries obtained from human fetal liver, HCC, and adjacent noncancerous liver tissue, and from mouse models. It was found that the tumor microenvironment is markedly reprogrammed like that of a fetus, especially in HCC, with the reappearance of fetal endothelial cells and fetal tumor-associated macrophages and that VEGF/NOTCH signaling plays a role in maintaining the immunosuppressive fetal ecosystem in HCC. In addition, Sharma et al. performed a highly multiplexed assay for digital spatial profiling using formalin-fixed paraffin-embedded tumor samples and found that most regions of the tumor and fetal liver expressed immunosuppressive marker (FOXP3, CTLA-4, LAG-3, BATF3)-positive T cells and contained fetal-like (FOLR2+) macrophages and fetal-associated (PLVAP+) epithelial cells, and were classified into the same cluster, while the expression profiles from the adjacent noncancerous liver tissues were classified into different clusters. These results from single-cell analysis supported the similarity between the tumor microenvironment and the fetal liver microenvironment (Sharma et al. 2020).

3.3.2 The Heterogeneous Nature of the Immune Microenvironment in Multiple Liver Cancers

47 surgical specimens from 15 patients who underwent resection of multiple hepatocellular carcinomas, including intrahepatic metastases and multicentric occurrence, for curative treatment, were analyzed using integrated analysis of whole-exome/transcriptome sequencing, multiple immunostainings, immunopeptidome, and T-cell receptor (TCR) analysis. The study showed spatial heterogeneity of the immune microenvironment, for example, intrahepatic metastases had less T-cell infiltration and more M2 macrophages than multicentric occurrence, while multicentric occurrence had higher expression of inhibitory immune checkpoints than intrahepatic metastases (Dong et al. 2020).

3.3.3 The Characteristics of Immunocompetent Cells in Early Recurrent Hepatocellular Carcinoma

Gene expression profiling, by single-cell RNA-seq, of 18 primary and early relapse hepatocellular carcinomas within 2 years of surgery showed that early recurrent tumors had decreased levels of regulatory T cells and increased numbers of dendritic cells and infiltrating CD8+ T cells, compared to primary tumors. The results suggested the existence of a potential immune evasion mechanism in recurrent tumor cells that weakens antigen presentation by dendritic cells and recruits innate CD8+ T cells to recurrent tumors (Sun et al. 2021).

3.3.4 NASH-Associated Hepatocellular Carcinoma and Immunotherapy

Analysis of CD8+, programmed death-1 (PD-1) + T cells using flow cytometry, fluorescence immunostaining, mass spectrometry, and single-cell RNA-seq in a mouse model of NASH-derived hepatocarcinoma showed an increase in CD8+ PD-1+ T cells in the NASH liver compared to the normal liver. The number of CD8+ T cells infiltrating the tumor tissue in NASH-HCC was also increased following anti-PD-1 antibody treatment. Furthermore, these CD8+ PD-1+ T cells lacked immunosurveillance and had hepatotoxic potential, suggesting that anti-PD-1 antibody treatment may contribute to the worsening and increase of HCC. Analysis of human NAFLD/NASH livers showed that, as in mice, CD8+ PD-1+ T cells were also increased in number and that the median overall survival after immunotherapy alone for HCC in patients with NAFLD was shorter than that in patients with chronic viral liver disease, suggesting the possibility of a poor response to checkpoint inhibitors (Pfister et al. 2021).

4 Clinical Application of the Hepatocarcinoma Microenvironment (Diagnosis)

Here we present a clinical example of the use of the characteristics of the tumor microenvironment for the diagnosis of HCC and prediction of the treatment effect.

4.1 Positron Emission Tomography (PET)

Positron emission tomography (PET) is a clinical application of the Warburg effect (Ohashi et al. 2017), described above. Because 18F-FDG is taken up by cells in vivo in the same way as glucose, it can be used for detecting and imaging some body areas of high glucose consumption. Theoretically, it is thought that cancer cells are exposed to hypoxia until they acquire angiogenic potential during the process of carcinogenesis; therefore, they proliferate by shifting their energy production from mitochondrial oxidative phosphorylation to glycolysis in order to adapt to the hypoxic conditions. Because the glycolysis of cancer cells is enhanced in the presence or absence of oxygen, PET scanning is used to detect the presence and location of cancer in clinical practice. Poorly differentiated HCC and recurrent HCC tend to have a higher accumulation of 18F-FDG than well-differentiated and moderately differentiated HCC.

4.2 Companion Diagnostics (PD-L1 Protein Immunostaining Test)

Immunohistochemical staining with anti-PD-L1 antibodies has already been used to evaluate the expression of PD-L1 protein in tumor tissue to determine the indication for anti-PD-L1 antibody agents (Doroshow et al. 2021). There have also been attempts to use the PD-L1 test as a biomarker to predict therapeutic efficacy; however, the clinical significance of the biomarker may vary with the primary tumor. For example, in non-squamous non-small cell carcinomas of the lung (mostly adenocarcinomas), the OS of patients bearing lung tumors with PD-L1 expression of less than 1%, and treated with nivolumab, an anti-PD-L1 antibody agent, is similar to that of patients treated with docetaxel (conventional taxane chemotherapy that inhibits cell division). Therefore, the National Comprehensive Cancer Network (NCCN) guidelines for the treatment of lung cancer recommend that patients with non-squamous cell carcinoma who are found to have a PD-L1 expression rate of less than 1% should be prioritized for antineoplastic agents other than nivolumab, such as docetaxel (Ettinger et al. 2021; Borghaei et al. 2015). On the other hand, in a comparative study of nivolumab in patients with gastric cancer after the third-line treatment, there was a trend toward a greater effect of nivolumab in patients with a PD-L1 expression rate of 1% or higher, while, even in patients with a PD-L1 expression rate of less than 1%, the nivolumab group was significantly better than the placebo group (Kang et al. 2017). Therefore, nivolumab is recommended for the third-line treatment of gastric cancer, regardless of the PD-L1 expression rate.

In HCC, the combination of atezolizumab (an anti-PD-L1 antibody) and bevacizumab (an anti-VEGF antibody) is recommended as the first-line treatment for patients who are eligible for systemic chemotherapy (Finn et al. 2020a). For the atezolizumab + bevacizumab combination therapy, the association between response and PD-L1 expression in cancer tissues also has been verified in an international phase Ib clinical trial (GO30140 study) (Lee et al. 2020) but, at present, the association with response has not been fully found in the analysis of PD-L1 expression alone.

4.3 M2BPGi and CSF1R

Analysis of sera from patients with chronic hepatitis C and liver cirrhosis identified two glycan markers for HCC, Mac-2 binding protein glycosylation isomer (M2BPGi) (also known as Wisteria Floribunda agglutinin (WFA) + -M2BP) (Kuno et al. 2013) and WFA + - colony-stimulating factor 1 receptor (CSF1R) (Ocho et al. 2014). M2BPGi is a glycosylation isomer of Mac-2 binding protein (M2BP). M2BP binds to Mac-2 (or galectin 3), which is a cytokine expressed from macrophages and Kupffer cells. M2BP is thought to be secreted by hepatic stellate cells and may also induce inflammation in the extracellular matrix of liver tissue; it promotes the progression of fibrosis (Bekki et al. 2017; Shirabe et al. 2018; Gantumur et al. 2021). In addition, detection of M2BP in sera is useful for predicting liver fibrosis (Ichikawa et al. 2017). Serum M2BPGi levels are also associated with the risk of developing HCC (Ichikawa et al. 2017) due to HBV and HCV infection.

WFA + -CSF1R is a membrane protein that has functions related to cell proliferation and macrophage differentiation. The CSF1R ligand, colony-stimulating factor 1 (or M-CSF), may be involved in the progression of hepatocarcinogenesis, VEGF production from hepatic macrophages, and angiogenesis in liver tumors in mouse models of liver cancer (Hara et al. 2014). Moreover, analysis of sera from patients with chronic liver disease type C and HCC showed that WFA + -CSF1R could be a useful prognostic biomarker for liver carcinogenesis and cirrhosis (Iio et al. 2016).

5 The Clinical Application of the Hepatocarcinoma Microenvironment (Treatment)

International phase III trials of primary chemotherapy for patients with advanced HCC showed prolonged survival with sorafenib therapy, a receptor TKI (TKI, also called multi-kinase inhibitor) (Llovet et al. 2008), in 2009. In 2018, lenvatinib therapy showed non-inferiority to sorafenib for OS (Kudo et al. 2018). In 2020, the atezolizumab-bevacizumab combination showed superiority over sorafenib for survival (IMbrave150 study) (Finn et al. 2020a). Based on these results, the guidelines of the Japanese Society of Hepatology, the American Society of Hepatology, the American Society of Clinical Oncology, and the European Society of Clinical Oncology recommend atezolizumab-bevacizumab combination therapy as the first-line treatment for advanced HCC. Transcatheter arterial chemoembolization (TACE) is also used for HCC, within a limited number and size range (Llovet et al. 2021), as described below. TKIs, cancer immunotherapy, and TACE have all been shown to be related to the cancer microenvironment in their treatment mechanisms.

5.1 Transcatheter Arterial Chemoembolization (TACE)

Transcatheter arterial chemoembolization (TACE) is a procedure in which a catheter is inserted into an artery, often the femoral artery, through a puncture and then advanced through the celiac and hepatic arteries to the vessels feeding the liver tumor. Contrast media and/or anticancer drugs are injected and the feeding vessels are embolized. The non-tumor area of the liver receives a dual blood supply from the portal vein and arteries (portal vein, artery is 3:1) (Rappaport 1980), while the tumor area receives predominantly an arterial blood supply (artery: portal vein is 9:1) (Breedis and Young 1954; Takayasu et al. 1995). TACE is a drug delivery system specific for HCC and takes advantage of this difference in blood supply between the tumor and noncancerous areas. The tumor reduction effect of TACE is superior to systemic therapies, such as TKIs and immune checkpoint inhibitors (ICIs) when the size and number of tumors are within a certain range, but a problem is that repeated treatment results in decreased liver function. Because decreased liver function reduces the therapeutic effect of TKIs and ICIs, appropriate consideration of the limitations of TACE is important (Arizumi et al. 2015; Ueshima et al. 2019).

5.2 Lenvatinib

Lenvatinib, an orally available TKI, inhibits vascular endothelial growth factor receptor (VEGFR) 1–3, fibroblast growth factor receptor (FGFR) 1–4, platelet-derived growth factor receptor (PDGFR)-α, rearranged during transfection tyrosine kinase (RET), and stem cell factor receptor (SCFR, c-Kit) (Yamamoto et al. 2014). As for the previously approved TKI, sorafenib, lenvatinib inhibits both the malignant transformation of HCC cells and tumor angiogenesis of its related cells (vascular endothelial cells, etc.). Based on the analysis of mouse models, compared to sorafenib, lenvatinib is thought to have a stronger antitumor effect, especially by more strongly inhibiting tumor angiogenesis induced by VEGF and fibroblast growth factor (FGF) (Tohyama et al. 2014).

5.3 Sequential Therapy with Lenvatinib and TACE

Lenvatinib reportedly improves the delivery of antitumor agents and reduces tumor-stromal pressure by normalizing tumor blood vessels, due to its VEGF-inhibitory effect (Tohyama et al. 2014); it is expected that the tumors in patients that are judged to be "TACE unsuitable" may be altered to benefit from TACE after lenvatinib therapy. In this situation, the addition of "on-demand TACE" to selective residual disease is expected to be more effective than lenvatinib alone while reducing the loss of liver reserve (Kudo et al. 2020).

5.4 Inhibition of the VEGF-VEGF Receptor (VEGFR) Pathway

In addition to multi-kinase inhibitors such as lenvatinib, antibody-based drugs targeting the VEGF pathway are used for the treatment of HCC. Bevacizumab, an antibody used in atezolizumab-bevacizumab combination therapy, which is the first-line treatment for HCC, binds to VEGF-A (Presta et al. 1997) and inhibits its binding to its receptors (VEGFR-1, VEGFR-2, and neuropilin-1) (Pan et al. 2007). Besides HCC, it is used to treat colorectal cancer, non-small cell lung cancer, ovarian cancer, malignant glioma, (Kurkjian and Kim 2012), and so on. Other than cancer, it is expected to have a therapeutic effect on age-related macular degeneration (The 2018). In all these diseases, bevacizumab is administered for the prevention of disease progression through its anti-angiogenic effects and suppression of inflammation and anti-vascular permeability by VEGF inhibition.

Ramucirumab is an antibody against VEGFR-2, and the US Food and Drug Administration (FDA) has approved ramucirumab as a single agent for use in patients with advanced HCC who have been previously treated with sorafenib and whose tumor marker serum alpha-fetoprotein (AFP) levels are elevated above a certain threshold (Zhu et al. 2019). Lenvatinib is a candidate for first-line therapy in patients who are unable to use the atezolizumab-bevacizumab combination, while ramucirumab is a second- and subsequent-line therapy. Although there are no direct comparative studies, clinically, the therapeutic effect of ramucirumab is more moderate than that of lenvatinib. On the other hand, ramucirumab has fewer side effects and is better tolerated (Kudo et al. 2018, 2021). The ligands for VEGFR-2 are VEGF-A, VEGF-C, and VEGF-D, and bevacizumab specifically inhibits VEGF-A. Plasma VEGF-A levels reportedly increase in response to bevacizumab (Siegel et al. 2008), suggesting that the VEGF-VEGFR pathway is reactivated in response to bevacizumab failure. Therefore, ramucirumab, which suppresses the VEGF-VEGFR pathway using a different target from bevacizumab (Lin et al. 2018), is theoretically a good candidate as second- and subsequent-line therapy for patients who fail atezolizumab-bevacizumab combination therapy.

Bevacizumab and ramucirumab are well tolerated with relatively few side effects when used alone. The main adverse events are hypertension, proteinuria, and bleeding. VEGF induces nitric oxide (NO) production and lowers blood pressure by NO synthase (NOS). VEGF inhibition inhibits NOS, and hypertension is thought to be caused by decreased NO expression and increased vascular resistance due to vessel rarefaction in vascular endothelial cells (Izzedine et al. 2009).

5.5 Immunotherapy for Hepatocellular Carcinoma

PD-1 is a receptor expressed on the surfaces of activated T lymphocytes and acts as a checkpoint in excessive immune and inflammatory responses because its expression increases with the stimulation of inflammatory mediators and cell activation to suppress and regulate immune activity downstream (Donisi et al. 2020). The ligands of PD-1 are PD-L1 and PD-L2. PD-L1 is usually expressed on antigen-presenting cells, and PD-L2 is expressed on the surfaces of activated antigen-presenting cells and macrophages (Taube et al. 2014). These cells express PD-L1 and are thought to suppress the immune response against self by suppressing immune activity via PD-1 receptors on surrounding activated T lymphocytes (Donisi et al. 2020). Cancer cells hijack this function and suppress the proliferative and injury functions of T cells by constantly expressing PD-L1 and PD-L2, thereby slipping through the immunosurveillance system (Chen and Mellman 2017).

Nivolumab and pembrolizumab are antibody drugs against PD-1 expressed on the surface of activated T lymphocytes. Nivolumab showed a positive trend in the CheckMate-459 study in the first-line treatment of unresectable HCC compared to sorafenib but did not achieve a significant increase in OS, the primary endpoint (D'Alessio et al. 2021). On the other hand, pembrolizumab showed improvement in OS and PFS compared to placebo in the randomized, double-blind phase III KEYNOTE-240 trial in HCC patients treated with sorafenib; however, this did not meet the pre-specified level of statistical significance (Finn et al. 2020b). Based on these results, both nivolumab monotherapy and pembrolizumab monotherapy are not used as first-line agents for advanced HCC. Exceptionally, pembrolizumab can be administered if high-frequency microsatellite instability (MSI-high) is found in cancer-derived genomic testing from liver tumor tissue or liquid biopsy (Le et al. 2015, 2017). However, MSI-high is less common in HCC patients (Cortes-Ciriano et al. 2017; Eso et al. 2020). In the international phase II KEYNOTE-224 trial, which preceded the phase III KEYNOTE-240 trial of pembrolizumab therapy, patients with PD-L1-expressing tumors had better response rates and PFS than patients with PD-L1-non-expressing tumors (Zhu et al. 2018). As for nivolumab monotherapy, Morita et al. reported (Morita et al. 2021) that it significantly prolonged OS and PFS in patients with PD-L1 expression rates of 1% or more. Therefore, treatment of advanced HCC with nivolumab and pembrolizumab may be effective in selected patients with some biomarkers of efficacy.

An immune-responsive microenvironment in which lymphocytes infiltrate the tumor is necessary for the immune response to work well against cancer. On the other hand, if the anti-PD-1/PD-L1 antibody is administered as a single agent in a suppressive immune microenvironment without tumor-infiltrating lymphocytes, the therapeutic effect will be limited.

The antibody atezolizumab directly relieves the immunosuppressive state of the tumor by inhibiting PD-L1 expressed on HCC cells. Bevacizumab in the combination of atezolizumab plus bevacizumab, the first-line therapy for unresectable advanced HCC, is thought to enhance the therapeutic effect of atezolizumab by suppressing VEGF, thereby changing the immunosuppressive microenvironment to immunoreactive.

6 Immune Cycle and Immune Checkpoint Molecules in Liver Cancer

Atezolizumab is an antibody that targets PD-L1, an immune checkpoint molecule, and is used in combination with the anti-VEGF antibody bevacizumab for the treatment of HCC (Finn et al. 2020a). This anti-PD-L1 antibody can enhance the effects of Step 3, T-cell priming, and Step 7, cancer cytotoxicity, of the cancer immune cycle. Moreover, anti-VEGF antibodies inhibit the growth of cancer neovascularization, which may weaken T-cell trafficking and infiltration into cancer in Steps 4 and 5.

6.1 Hot Tumor and Cold Tumor

Recently, cancer has been classified into two types based on the state of immune activation. A tumor with abundant T cells inside the lesion and immunologically activated is called a "hot tumor" type, while a tumor with few T cells inside and immunologically inactive is called a "cold tumor" (Galon and Bruni 2019). Sia et al. classified the hepatocarcinoma immune microenvironment, based on the expression patterns of inflammation-related genes, such as IFN and T-cell-related genes, into an immune class with inflammatory response and non-immune class with no inflammatory response (Sia et al. 2017). The immune class was further divided into active immune class (about 15%), with an adaptive T-cell response, and exhausted immune class (about 10%). Immune class hepatocellular carcinomas showed increased expression of the PD-1/PD-L1 gene in the tumor, and a lower percentage had a β-catenin (*CTNNB1* gene) mutation. Calderaro et al. also investigated multiple samples of HCC by pathological review, immunohistochemistry, gene expression profiling, and sequencing, and reported that hepatocellular carcinomas with a *CTNNB1* mutation had significantly less inflammatory cell infiltration into the tumor than those without the mutation (Calderaro et al. 2017). In a report that investigated the presence or absence of T-cell infiltration in tumors by analyzing their gene expression signatures, only about 25% of HCC cases had T-cell infiltration in the tumor, and about 50% were classified as the "cold tumor" type with no T-cell infiltration in the tumor (Luke et al. 2019).

Thus, a high percentage of HCC cases have tumors that are not immunogenic in nature, and it is assumed that treatment with single-agent ICIs will have limited efficacy. In fact, Harding et al. reported that, among patients with HCC treated with ICIs, those who had mutations in the Wnt/β-catenin pathway (*CTNNB1* or *AXIN1* genes) detected in liver tumor biopsy samples showed lower disease control, while those who did not have these mutations had a response of stable disease or better in about 70% of cases (Harding et al. 2019).

An important issue is which biomarkers to use to screen for such tumors. Currently, the Wnt/β-catenin pathway may play an important role in the hepatocarcinoma immune microenvironment, and the detection of abnormalities in this pathway is an important candidate biomarker for predicting the therapeutic effects of ICIs (Morita et al. 2021).

6.2 Converting Cold Tumors into Immunogenic Hot Tumors

In terms of therapeutic approaches, a combination of ICIs with other therapies, to convert cold tumors into immunogenic hot tumors, also is being investigated.

Anti-VEGF antibodies not only inhibit cancer angiogenesis but also inhibit T-cell trafficking and infiltration into cancer by normalizing vascular structures and reducing vascular permeability, thereby reducing increased stromal pressure in tumor tissue. Because VEGF is also involved in the activation of bone marrow-derived immunosuppressive cells (MDSCs), regulatory T cells, and tumor-associated macrophages that suppress immunity, administration of VEGF inhibitors reportedly enhances the effects of ICIs (Huang et al. 2013).

In general, macrophages are classified into two types, M1 and M2, based on their roles (Orecchioni et al. 2019): M1 macrophages are differentiated from inflammatory monocytes through secreted inflammatory cytokines such as tumor necrosis factor (TNF)-α and IFN-γ, or antigen presentation, and promote innate immune responses such as defense against pathogens and parasites. On the other hand, M2 macrophages are induced by Th2 cytokines such as IL-4, IL-10, and IL-13, and play important roles in immunosuppression, tissue remodeling, and angiogenesis (Orecchioni et al. 2019). In the tumor microenvironment, the polarization state of macrophages is heterogeneous, but in malignant tumors, tumor-associated macrophages often exhibit an M2-like phenotype and act in a tumor-promoting manner (Noy and Pollard 2014). Thus, reprogramming tumor-associated macrophages into M1-like macrophages may be effective in promoting tumor regression, as shown below. Treatments such as administration of some agents and radiation therapy (RT) are candidates for this purpose (Genard et al. 2017).

Macrophage scavenger receptor type A/II (MSR-A), also known as CD204, is a receptor involved in the basic functions of macrophages and is a target for depleting tumor-infiltrating macrophages. An immunotoxin combining anti-CD204 antibody and saporin toxin, a plant enzyme with protein synthesis-inhibitory activity, eliminated tumor-infiltrating macrophages and suppressed the progression of ovarian cancer in mice (Bak et al. 2007).

An immune checkpoint receptor, T-cell immunoglobulin, and mucin domain-containing molecule-3 (TIM-3) play an important role in the activation of tumor-infiltrating macrophages. Anti-TIM-3 antibody suppressed tumorigenesis in a kidney cancer model (Komohara et al. 2015). Macrophage colony-stimulating factor (M-CSF, also known as CSF-1) is required for macrophage differentiation, and CSF-1 receptor inhibitors such as Ki20227 and BLZ945 have been shown to inhibit tumorigenesis (Kubota et al. 2009; Pyonteck et al. 2013).

In the process of cancer invasion into the stroma, proteases play an important role in degrading stromal proteins. It has been reported that cathepsin, matrix metalloproteinase (MMP)2, and MMP9 produced by tumor-infiltrating macrophages promote cancer cell invasion in pancreatic cancer (Gocheva et al. 2010), and colorectal cancer models (Kitamura et al. 2007). The TKI, dasatinib, targeted MMP9-positive macrophages and reduced the amount of MMP9 in the tumor microenvironment of a xenograft model of a human colorectal adenocarcinoma cell line (Liang et al. 2010).

Zoledronic acid, a bisphosphonate, has been found to have not only an inhibitory effect on bone-related events such as osteoporosis but also a possible tumor-suppressive effect (Aft et al. 2010). Zoledronic acid depleted tumor-infiltrating M2-like macrophages expressing MMP9 and inhibited the differentiation of myeloid cells into tumor-infiltrating macrophages in a HER2+ mouse model of breast carcinogenesis (Melani et al. 2007). In two human HCC mouse models, zoledronic acid in combination with sorafenib inhibited tumor progression, angiogenesis, and metastasis (Zhang et al. 2010). Clodronate, a liposome encapsulating another bisphosphonate, is specifically taken up by macrophages and causes macrophage depletion (Rogers et al. 1996). Clodronate is metabolized to an analog of ATP and is cytotoxic to macrophages (Lehenkari et al. 2002).

Trabectedin is a therapeutic agent for soft tissue sarcoma that inhibits nucleotide excision repair mechanisms by binding to the sub-groove of DNA, thereby inducing the death (apoptosis) of cancer cells. Trabectedin also exerts cytotoxic effects and antitumor activity against tumor-infiltrating macrophages (Germano et al. 2013).

Shigella has a needlelike secretion apparatus that pierces host cells to secrete a variety of effectors and induce macrophage cell death. Galmbacher et al. developed an attenuated strain of *Shigella flexneri* and infected breast cancer-bearing mice. Then, the macrophages were depleted, and the tumors of the tumor-bearing mice shrank (Galmbacher et al. 2010).

Komohara et al. assessed macrophage differentiation into the M2 phenotype by quantifying CD163 expression upon tumor cell supernatant stimulation, and IL-10 secretion upon lipopolysaccharide stimulation, and screened for compounds that inhibit macrophage differentiation into the M2 phenotype. They found that a triterpenoid compound, corosolic acid, significantly inhibited the differentiation of macrophages into the M2 phenotype. The compound significantly inhibited subcutaneous tumorigenesis and lung metastasis in a mouse sarcoma model by suppressing the pro-tumor activity of macrophages and the immunosuppressive activity of bone marrow-derived suppressor cells (Fujiwara et al. 2011; Horlad et al. 2013).

7 The Tumor Microenvironment and Macroenvironment

As seen in the case of cachexia and paraneoplastic syndrome, the tumor and its microenvironment interact with several organs of the host. Al-Zoughbi and Hoefler proposed the concept of a "tumor macroenvironment" to define the interactions between the tumor cells and tumor microenvironment with other organs and body systems (Al-Zoughbi et al. 2014; Al-Zoughbi and Hoefler 2020). In fact, aberrant tumor blood vessels drawn into the tumor microenvironment supply oxygen and nutrients to cancer cells, while newly formed blood and lymphatic vessels serve as pathways for the release of angiogenic and tumor growth factors, secreted by the tumor and tumor microenvironment into the surrounding macroenvironment, and for the metastasis of cancer cells to distant organs. Furthermore, cachexia in patients who are near the end of life, a condition in which both adipose tissue and skeletal muscle are depleted, is not caused by mere anorexia but by certain cytokines, secreted by tumor cells and tissue, which trigger systemic inflammatory responses and metabolic abnormalities. On the other hand, systemic metabolic diseases such as obesity and diabetes are known to affect tumor development. In other cases, it was reported recently that cancer cells secrete factors and extracellular vesicles that form pre-metastatic niches at distant locations (Peinado et al. 2017). Thus, an incomplete network of newly formed blood vessels around the tumor cells allows the release of soluble factors and extracellular vesicles from the tumor microenvironment into the systemic circulatory system. This leads to abnormal endocrine effects and inflammation, developing cancer-related systemic syndromes in the tumor macroenvironment through the interaction of the tumor microenvironment with other organs.

7.1 Overnutrition and Chronic Inflammation

Lifestyles such as excessive calorie intake and lack of exercise lead to obesity and diabetes. When adipose tissues are activated by obesity, they secrete transcription factors such as NF-κB and c-Jun N-terminal kinase (JNK), pro-inflammatory cytokines such as TNF-α and MCP-1, and anti-obesity hormones such as leptin. In addition, adipose tissues increase the production of free fatty acids, leading to endoplasmic reticulum stress which causes the development of chronic inflammation, mainly through macrophage infiltration (Sun et al. 2011). Furthermore, such a lifestyle elevates blood glucose levels and leads to persistent hyperinsulinemia, where the expression of insulin receptor substrate 2 (IRS2) is selectively downregulated via macrophage surface insulin receptor kinase, especially in the liver. This is the cause of insulin resistance (Shimomura et al. 2000). In a mouse model of obesity, reduced IRS2 expression suppresses the activity of M2 macrophages. Because M2 macrophages have an anti-inflammatory effect through IL-4, reduced IRS2 expression increases the number of M1 macrophages, which secrete pro-inflammatory cytokines (Kubota et al. 2018). The persistence of such inflammation leads to the continued increase and activation of other immunocompetent cells, such as Th1 cells, CD8+ T cells, and NK cells, and a decrease in regulatory T cells. Such chronic inflammation and abnormalities of the immune system increase the risk of carcinogenesis.

Under conditions of metabolic stress, such as overnutrition, M1 macrophages infiltrate adipose and hepatic tissues and form a crown-like structure (CLS) by surrounding parenchymal cells that have undergone cell death due to excessive lipid accumulation. The CLS contains immunocompetent cells and serves as an origin of chronic inflammation. In addition, macrophages in the CLS highly express a lectin, macrophage-inducible C-type lectin (Mincle), which promotes tissue fibrosis through the activation of cytokines such as TGF-β. Mincle is also involved in fat accumulation in the liver and abnormal glucose metabolism (Tanaka et al. 2014).

7.2 Organ-Specific Metastases and Metastatic Niches

In a paper published in 1978, Schofield proposed the concept of a "stem cell niche," a specialized microenvironment that can sustain hematopoietic stem cells (Schofield 1978). Later, this concept was extended to the idea of "metastatic niches" as microenvironments suitable for the survival and proliferation of disseminated cancer cells, and, in 2005, Kaplan et al. demonstrated for the first time the existence of "pre-metastatic niches," i.e., localized microenvironments that are already formed before cancer cells arrive at the target organ (Kaplan et al. 2005). Pre-metastatic niches are characterized by the presence of pre-metastatic niche regulators (growth factors, cytokines, and extracellular vesicles) produced by the primary tumor. In pre-metastatic niches, the primary tumor secretes metastatic niche regulators (such as growth factors, cytokines, and extracellular vesicles), which act on the target organ prior to cancer metastasis, mobilizing bone marrow-derived cells, and build a microenvironment suitable for cancer metastatic growth. Wong et al. found that the collagen-modifying enzyme, lysyl oxidase-like 2 (LOXL-2), is significantly overexpressed in the tumor tissues and sera of patients with HCC and plays an essential role in modifying extracellular matrix components in the microenvironment and metastatic niches of HCC. Wong et al. also reported that LOXL2 promotes intrahepatic metastasis by increasing the stiffness of the liver and facilitates the development of extrahepatic metastasis by promoting the mobilization of bone marrow-derived cells to metastatic foci (Wong et al. 2014).

On the other hand, the existence of a post-metastatic niche, which is formed after cancer cells arrive at the target organ, has also been reported (Oskarsson et al. 2011). It is thought that, after a small number of cancer cells arrive at the target organ from the primary tumor, they secrete paracrine factors and form a microenvironment suitable for cancer cell survival and proliferation.

7.3 Exosomes as Pre-metastatic Niches

Cancer cell-derived exosomes also play an important role as initiators of pre-metastatic niche formation, and Hoshino et al. investigated their organ-specific distribution using human breast cancer cell lines that metastasize primarily to the lung, bone, and brain. The bodily distribution of fluorescently labeled exosomes was analyzed, and it was confirmed that cancer cell-derived exosomes were distributed to each metastatic site. In addition, a comprehensive proteomic analysis of the proteins from more than 30 cancer cell-derived exosomes with organ specificity revealed that, regardless of the cancer type, exosomes from metastatic lung cancer cells expressed integrin α6β4, a membrane protein involved in cell adhesion, and exosomes from metastatic liver cancer cells expressed integrin αvβ4 (Hoshino et al. 2015). These results suggest that the measurement of integrin levels in exosomes from cancer in patients' blood may be useful for the diagnosis and guide the treatment of metastatic tumors.

Furthermore, to investigate the role of exosomes in the microenvironment of metastatic sites, Hoshino et al. fluorescently labeled exosomes derived from metastatic cancer cells from the lung and liver and administered them to mice. In the lung, the exosomes were selectively taken up by fibroblasts and epithelial cells and, in the liver, by Kupffer cells. In addition, Kupffer cells were treated with exosomes derived from cancer cells with knockdown of integrin β5, and their gene expression was examined; the expression of some of the *S100* genes encoding inflammatory proteins (S100A8, S100P) was increased. Because exosomes form a pre-metastatic niche that is suitable for metastasis, these findings may enable the development of therapeutic agents that inhibit the distribution of exosomes to organs (Hoshino et al. 2015).

8 Conclusion

Cancer cells interact with the surrounding cells, including stromal cells, fibroblasts, immunocompetent cells, and vascular endothelial cells, and form a tumor tissue with a complex microenvironment (Fig. 1). Recent innovations in the techniques of cell biological analysis have made it possible to analyze this complex environment in greater detail than ever before, revealing the diversity of cancer cells and the tumor microenvironment, as well as the incredible offense and defense between cancer cells and the body. Tomizo Yoshida (1903–1973), a Japanese pathologist famous for the world's first induced liver cancer in laboratory animals and the discovery of "Yoshida sarcoma" (Tsuchiya et al. 1994), said, "What happens inside the human body is linked to society." Looking at the environment surrounding the liver disease, although the number of hepatocellular carcinomas caused by infectious diseases such as hepatitis C has been rapidly decreasing in recent years, thanks to the introduction of new drugs, the incidence of lifestyle-related HCC has been increasing rapidly because of the worldwide increase in obesity and diabetes (Yang et al. 2019). The characteristics of the HCC microenvironment may also change with changing lifestyles, as described in the main text. The advent of ICIs has totally changed how advanced HCC is treated, but a new challenge also has emerged: how to treat tumors that are not immunogenic. With a combination of recent technological innovations and a more detailed analysis of metastasis, we should be able to develop novel biomarkers for metastatic cancer and new treatment methods. We would like to close this chapter with the hope that we can overcome liver cancer in the near future by improving diagnosis and treatment methods, through more comprehensive and integrated analysis techniques, and by reducing the incidence of HCC through improvements in our living environment.

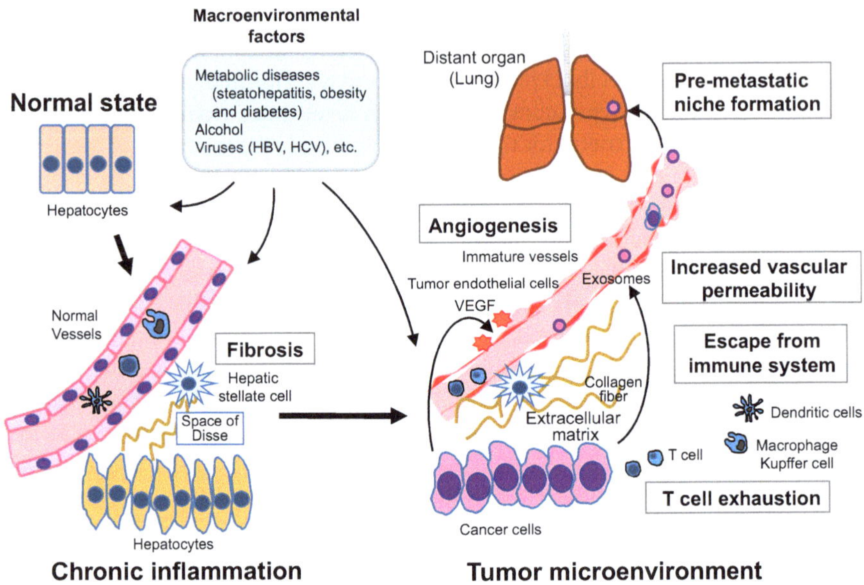

Fig. 1 The components of the microenvironment and the related mechanisms

References

Aft R, Naughton M, Trinkaus K, Watson M, Ylagan L, Chavez-Macgregor M, Zhai J, Kuo S, Shannon W, Diemer K, Herrmann V, Dietz J, Ali A, Ellis M, Weiss P, Eberlein T, Ma C, Fracasso PM, Zoberi I, Taylor M, Gillanders W, Pluard T, Mortimer J, Weilbaecher K (2010) Effect of zoledronic acid on disseminated tumour cells in women with locally advanced breast cancer: an open label, randomised, phase 2 trial. Lancet Oncol 11:421–428

Al-Zoughbi W, Hoefler G (2020) Tumor macroenvironment: an update. Pathobiology 87:58–60

Al-Zoughbi W, Al-Zhoughbi W, Huang J, Paramasivan GS, Till H, Pichler M, Guertl-Lackner B, Hoefler G (2014) Tumor macroenvironment and metabolism. Semin Oncol 41:281–295

Arizumi T, Ueshima K, Minami T, Kono M, Chishina H, Takita M, Kitai S, Inoue T, Yada N, Hagiwara S, Minami Y, Sakurai T, Nishida N, Kudo M (2015) Effectiveness of Sorafenib in patients with Transcatheter arterial chemoembolization (TACE) refractory and intermediate-stage hepatocellular carcinoma. Liver Cancer 4:253–262

Bak SP, Walters JJ, Takeya M, Conejo-Garcia JR, Berwin BL (2007) Scavenger receptor-A-targeted leukocyte depletion inhibits peritoneal ovarian tumor progression. Cancer Res 67: 4783–4789

Bekki Y, Yoshizumi T, Shimoda S, Itoh S, Harimoto N, Ikegami T, Kuno A, Narimatsu H, Shirabe K, Maehara Y (2017) Hepatic stellate cells secreting WFA(+) -M2BP: its role in biological interactions with Kupffer cells. J Gastroenterol Hepatol 32:1387–1393

Blomhoff R, Wake K (1991) Perisinusoidal stellate cells of the liver: important roles in retinol metabolism and fibrosis. FASEB J 5:271–277

Borghaei H, Paz-Ares L, Horn L, Spigel DR, Steins M, Ready NE, Chow LQ, Vokes EE, Felip E, Holgado E, Barlesi F, Kohlhaufl M, Arrieta O, Burgio MA, Fayette J, Lena H, Poddubskaya E, Gerber DE, Gettinger SN, Rudin CM, Rizvi N, Crino L, Blumenschein GR Jr, Antonia SJ,

Dorange C, Harbison CT, Graf Finckenstein F, Brahmer JR (2015) Nivolumab versus docetaxel in advanced nonsquamous non-small-cell lung cancer. N Engl J Med 373:1627–1639
Boucher Y, Jain RK (1992) Microvascular pressure is the principal driving force for interstitial hypertension in solid tumors: implications for vascular collapse. Cancer Res 52:5110–5114
Breedis C, Young G (1954) The blood supply of neoplasms in the liver. Am J Pathol 30:969–977
Burnet FM (1970) The concept of immunological surveillance. Prog Exp Tumor Res 13:1–27
Cairns RA, Harris IS, Mak TW (2011) Regulation of cancer cell metabolism. Nat Rev Cancer 11: 85–95
Calderaro J, Couchy G, Imbeaud S, Amaddeo G, Letouze E, Blanc JF, Laurent C, Hajji Y, Azoulay D, Bioulac-Sage P, Nault JC, Zucman-Rossi J (2017) Histological subtypes of hepatocellular carcinoma are related to gene mutations and molecular tumour classification. J Hepatol 67:727–738
Chattergoon MA, Latanich R, Quinn J, Winter ME, Buckheit RW 3rd, Blankson JN, Pardoll D, Cox AL (2014) HIV and HCV activate the inflammasome in monocytes and macrophages via endosomal toll-like receptors without induction of type 1 interferon. PLoS Pathog 10:e1004082
Chen DS, Mellman I (2013) Oncology meets immunology: the cancer-immunity cycle. Immunity 39:1–10
Chen DS, Mellman I (2017) Elements of cancer immunity and the cancer-immune set point. Nature 541:321–330
Chen DS, Irving BA, Hodi FS (2012) Molecular pathways: next-generation immunotherapy--inhibiting programmed death-ligand 1 and programmed death-1. Clin Cancer Res 18:6580–6587
Colegio OR, Chu NQ, Szabo AL, Chu T, Rhebergen AM, Jairam V, Cyrus N, Brokowski CE, Eisenbarth SC, Phillips GM, Cline GW, Phillips AJ, Medzhitov R (2014) Functional polarization of tumour-associated macrophages by tumour-derived lactic acid. Nature 513:559–563
Cortes-Ciriano I, Lee S, Park WY, Kim TM, Park PJ (2017) A molecular portrait of microsatellite instability across multiple cancers. Nat Commun 8:15180
D'Alessio A, Cammarota A, Prete MG, Pressiani T, Rimassa L (2021) The evolving treatment paradigm of advanced hepatocellular carcinoma: putting all the pieces back together. Curr Opin Oncol 33:386–394
Dong LQ, Peng LH, Ma LJ, Liu DB, Zhang S, Luo SZ, Rao JH, Zhu HW, Yang SX, Xi SJ, Chen M, Xie FF, Li FQ, Li WH, Ye C, Lin LY, Wang YJ, Wang XY, Gao DM, Zhou H, Yang HM, Wang J, Zhu SD, Wang XD, Cao Y, Zhou J, Fan J, Wu K, Gao Q (2020) Heterogeneous immunogenomic features and distinct escape mechanisms in multifocal hepatocellular carcinoma. J Hepatol 72:896–908
Donisi C, Puzzoni M, Ziranu P, Lai E, Mariani S, Saba G, Impera V, Dubois M, Persano M, Migliari M, Pretta A, Liscia N, Astara G, Scartozzi M (2020) Immune checkpoint inhibitors in the treatment of HCC. Front Oncol 10:601240
Doroshow DB, Bhalla S, Beasley MB, Sholl LM, Kerr KM, Gnjatic S, Wistuba II, Rimm DL, Tsao MS, Hirsch FR (2021) PD-L1 as a biomarker of response to immune-checkpoint inhibitors. Nat Rev Clin Oncol 18:345–362
Dunn GP, Bruce AT, Ikeda H, Old LJ, Schreiber RD (2002) Cancer immunoediting: from immunosurveillance to tumor escape. Nat Immunol 3:991–998
Eso Y, Shimizu T, Takeda H, Takai A, Marusawa H (2020) Microsatellite instability and immune checkpoint inhibitors: toward precision medicine against gastrointestinal and hepatobiliary cancers. J Gastroenterol 55:15–26
Ettinger DS, Wood DE, Aisner DL, Akerley W, Bauman JR, Bharat A, Bruno DS, Chang JY, Chirieac LR, D'amico TA, Dilling TJ, Dowell J, Gettinger S, Gubens MA, Hegde A, Hennon M, Lackner RP, Lanuti M, Leal TA, Lin J, Loo BW Jr, Lovly CM, Martins RG, Massarelli E, Morgensztern D, Ng T, Otterson GA, Patel SP, Riely GJ, Schild SE, Shapiro TA, Singh AP, Stevenson J, Tam A, Yanagawa J, Yang SC, Gregory KM, Hughes M (2021) NCCN guidelines insights: non-small cell lung cancer, version 2.2021. J Natl Compr Cancer Netw 19: 254–266

Ferguson TA, Choi J, Green DR (2011) Armed response: how dying cells influence T-cell functions. Immunol Rev 241:77–88

Finn RS, Qin S, Ikeda M, Galle PR, Ducreux M, Kim TY, Kudo M, Breder V, Merle P, Kaseb AO, Li D, Verret W, Xu DZ, Hernandez S, Liu J, Huang C, Mulla S, Wang Y, Lim HY, Zhu AX, Cheng AL, Investigators IM (2020a) Atezolizumab plus Bevacizumab in Unresectable Hepatocellular Carcinoma. N Engl J Med 382:1894–1905

Finn RS, Ryoo BY, Merle P, Kudo M, Bouattour M, Lim HY, Breder V, Edeline J, Chao Y, Ogasawara S, Yau T, Garrido M, Chan SL, Knox J, Daniele B, Ebbinghaus SW, Chen E, Siegel AB, Zhu AX, Cheng AL (2020b) Pembrolizumab as second-line therapy in patients with advanced hepatocellular carcinoma in KEYNOTE-240: a randomized, double-blind, phase III trial. J Clin Oncol 38:193–202

Franciszkiewicz K, Boissonnas A, Boutet M, Combadiere C, Mami-Chouaib F (2012) Role of chemokines and chemokine receptors in shaping the effector phase of the antitumor immune response. Cancer Res 72:6325–6332

Franciszkiewicz K, Le Floc'h A, Boutet M, Vergnon I, Schmitt A, Mami-Chouaib F (2013) CD103 or LFA-1 engagement at the immune synapse between cytotoxic T cells and tumor cells promotes maturation and regulates T-cell effector functions. Cancer Res 73:617–628

Fujiwara Y, Komohara Y, Ikeda T, Takeya M (2011) Corosolic acid inhibits glioblastoma cell proliferation by suppressing the activation of signal transducer and activator of transcription-3 and nuclear factor-kappa B in tumor cells and tumor-associated macrophages. Cancer Sci 102: 206–211

Fukumura D, Kloepper J, Amoozgar Z, Duda DG, Jain RK (2018) Enhancing cancer immunotherapy using antiangiogenics: opportunities and challenges. Nat Rev Clin Oncol 15:325–340

Galmbacher K, Heisig M, Hotz C, Wischhusen J, Galmiche A, Bergmann B, Gentschev I, Goebel W, Rapp UR, Fensterle J (2010) Shigella mediated depletion of macrophages in a murine breast cancer model is associated with tumor regression. PLoS One 5:e9572

Galon J, Bruni D (2019) Approaches to treat immune hot, altered and cold tumours with combination immunotherapies. Nat Rev Drug Discov 18:197–218

Gantumur D, Harimoto N, Muranushi R, Hoshino K, Batbayar C, Hagiwara K, Yamanaka T, Ishii N, Tsukagoshi M, Igarashi T, Watanabe A, Kubo N, Araki K, Yokobori T, Aishima S, Shirabe K (2021) Hepatic stellate cell as a Mac-2-binding protein-producing cell in patients with liver fibrosis. Hepatol Res 51:1058–1063

Genard G, Lucas S, Michiels C (2017) Reprogramming of tumor-associated macrophages with anticancer therapies: radiotherapy versus chemo- and immunotherapies. Front Immunol 8:828

Germano G, Frapolli R, Belgiovine C, Anselmo A, Pesce S, Liguori M, Erba E, Uboldi S, Zucchetti M, Pasqualini F, Nebuloni M, VAN Rooijen N, Mortarini R, Beltrame L, Marchini S, Fuso Nerini I, Sanfilippo R, Casali PG, Pilotti S, Galmarini CM, Anichini A, Mantovani A, D'incalci M, Allavena P (2013) Role of macrophage targeting in the antitumor activity of trabectedin. Cancer Cell 23:249–262

Giraud J, Chalopin D, Blanc JF, Saleh M (2021) Hepatocellular carcinoma immune landscape and the potential of immunotherapies. Front Immunol 12:655697

Global Burden of Disease Cancer, C, Fitzmaurice C, Abate D, Abbasi N, Abbastabar H, Abd-Allah F, Abdel-Rahman O, Abdelalim A, Abdoli A, Abdollahpour I, Abdulle ASM, Abebe ND, Abraha HN, Abu-Raddad LJ, Abualhasan A, Adedeji IA, Advani SM, Afarideh M, Afshari M, Aghaali M, Agius D, Agrawal S, Ahmadi A, Ahmadian E, Ahmadpour E, Ahmed MB, Akbari ME, Akinyemiju T, Al-Aly Z, Alabdulkader AM, Alahdab F, Alam T, Alamene GM, Alemnew BTT, Alene KA, Alinia C, Alipour V, Aljunid SM, Bakeshei FA, Almadi MAH, Almasi-Hashiani A, Alsharif U, Alsowaidi S, Alvis-Guzman-N, Amini E, Amini S, Amoako YA, Anbari Z, Anber NH, Andrei CL, Anjomshoa M, Ansari F, Ansariadi A, Appiah SCY, Arab-Zozani M, Arabloo J, Arefi Z, Aremu O, Areri HA, Artaman A, Asayesh H, Asfaw ET, Ashagre AF, Assadi R, Ataeinia B, Atalay HT, Ataro Z, Atique S, Ausloos M, Avila-Burgos L, Avokpaho E, Awasthi A, Awoke N, Ayala Quintanilla BP, Ayanore MA, Ayele HT, Babaee E, Bacha U, Badawi A, Bagherzadeh M, Bagli E,

Balakrishnan S, Balouchi A, Barnighausen TW, Battista RJ, Behzadifar M, Behzadifar M, Bekele BB, Belay YB, Belayneh YM, Berfield KKS, Berhane A, Bernabe E, Beuran M, Bhakta N, Bhattacharyya K, Biadgo B, Bijani A, Bin Sayeed MS, Birungi C et al (2019) Global, regional, and National Cancer Incidence, mortality, years of life lost, years lived with disability, and disability-adjusted life-years for 29 cancer groups, 1990 to 2017: A systematic analysis for the global Burden of disease study. JAMA Oncol 5:1749–1768

Gocheva V, Wang HW, Gadea BB, Shree T, Hunter KE, Garfall AL, Berman T, Joyce JA (2010) IL-4 induces cathepsin protease activity in tumor-associated macrophages to promote cancer growth and invasion. Genes Dev 24:241–255

Hanahan D, Weinberg RA (2000) The hallmarks of cancer. Cell 100:57–70

Hanahan D, Weinberg RA (2011) Hallmarks of cancer: the next generation. Cell 144:646–674

Hara M, Kono H, Furuya S, Hirayama K, Tsuchiya M, Fujii H (2014) Macrophage colony-stimulating factor plays a pivotal role in chemically induced hepatocellular carcinoma in mice. Hepatol Res 44:798–811

Harding JJ, Nandakumar S, Armenia J, Khalil DN, Albano M, Ly M, Shia J, Hechtman JF, Kundra R, El Dika I, Do RK, Sun Y, Kingham TP, D'angelica MI, Berger MF, Hyman DM, Jarnagin W, Klimstra DS, Janjigian YY, Solit DB, Schultz N, Abou-Alfa GK (2019) Prospective genotyping of hepatocellular carcinoma: clinical implications of next-generation sequencing for matching patients to targeted and immune therapies. Clin Cancer Res 25:2116–2126

Hedegaard DL, Tully DC, Rowe IA, Reynolds GM, Bean DJ, Hu, K., Davis, C., Wilhelm, A., Ogilvie, C. B., Power, K. A., Tarr, A. W., Kelly, D., Allen, T. M., Balfe, P. & Mckeating, J. A. (2017) High resolution sequencing of hepatitis C virus reveals limited intra-hepatic compartmentalization in end-stage liver disease. J Hepatol 66:28–38

Hegde PS, Wallin JJ, Mancao C (2018) Predictive markers of anti-VEGF and emerging role of angiogenesis inhibitors as immunotherapeutics. Semin Cancer Biol 52:117–124

Hernandez-Gea V, Toffanin S, Friedman SL, Llovet JM (2013) Role of the microenvironment in the pathogenesis and treatment of hepatocellular carcinoma. Gastroenterology 144:512–527

Horlad H, Fujiwara Y, Takemura K, Ohnishi K, Ikeda T, Tsukamoto H, Mizuta H, Nishimura Y, Takeya M, Komohara Y (2013) Corosolic acid impairs tumor development and lung metastasis by inhibiting the immunosuppressive activity of myeloid-derived suppressor cells. Mol Nutr Food Res 57:1046–1054

Hoshino A, Costa-Silva B, Shen TL, Rodrigues G, Hashimoto A, Tesic Mark M, Molina H, Kohsaka S, Di Giannatale A, Ceder S, Singh S, Williams C, Soplop N, Uryu K, Pharmer L, King T, Bojmar L, Davies AE, Ararso Y, Zhang T, Zhang H, Hernandez J, Weiss JM, Dumont-Cole VD, Kramer K, Wexler LH, Narendran A, Schwartz GK, Healey JH, Sandstrom P, Labori KJ, Kure EH, Grandgenett PM, Hollingsworth MA, De Sousa M, Kaur S, Jain M, Mallya K, Batra SK, Jarnagin WR, Brady MS, Fodstad O, Muller V, Pantel K, Minn AJ, Bissell MJ, Garcia BA, Kang Y, Rajasekhar VK, Ghajar CM, Matei I, Peinado H, Bromberg J, Lyden D (2015) Tumour exosome integrins determine organotropic metastasis. Nature 527:329–335

Huang Y, Stylianopoulos T, Duda DG, Fukumura D, Jain RK (2013) Benefits of vascular normalization are dose and time dependent--letter. Cancer Res 73:7144–7146

Ichikawa Y, Joshita S, Umemura T, Shobugawa Y, Usami Y, Shibata S, Yamazaki T, Fujimori N, Komatsu M, Matsumoto A, Tanaka E (2017) Serum Wisteria floribunda agglutinin-positive human mac-2 binding protein may predict liver fibrosis and progression to hepatocellular carcinoma in patients with chronic hepatitis B virus infection. Hepatol Res 47:226–233

Iio E, Ocho M, Togayachi A, Nojima M, Kuno A, Ikehara Y, Hasegawa I, Yatsuhashi H, Yamasaki K, Shimada N, Ide T, Shinkai N, Nojiri S, Fujiwara K, Joh T, Mizokami M, Narimatsu H, Tanaka Y (2016) A novel glycobiomarker, Wisteria floribunda agglutinin macrophage colony-stimulating factor receptor, for predicting carcinogenesis of liver cirrhosis. Int J Cancer 138:1462–1471

Izzedine H, Ederhy S, Goldwasser F, Soria JC, Milano G, Cohen A, Khayat D, Spano JP (2009) Management of hypertension in angiogenesis inhibitor-treated patients. Ann Oncol 20:807–815

Jenne CN, Kubes P (2013) Immune surveillance by the liver. Nat Immunol 14:996–1006

Kang YK, Boku N, Satoh T, Ryu MH, Chao Y, Kato K, Chung HC, Chen JS, Muro K, Kang WK, Yeh KH, Yoshikawa T, Oh SC, Bai LY, Tamura T, Lee KW, Hamamoto Y, Kim JG, Chin K, Oh DY, Minashi K, Cho JY, Tsuda M, Chen LT (2017) Nivolumab in patients with advanced gastric or gastro-oesophageal junction cancer refractory to, or intolerant of, at least two previous chemotherapy regimens (ONO-4538-12, ATTRACTION-2): a randomised, double-blind, placebo-controlled, phase 3 trial. Lancet 390:2461–2471

Kaplan RN, Riba RD, Zacharoulis S, Bramley AH, Vincent L, Costa C, Macdonald DD, Jin DK, Shido K, Kerns SA, Zhu Z, Hicklin D, Wu Y, Port JL, Altorki N, Port ER, Ruggero D, Shmelkov SV, Jensen KK, Rafii S, Lyden D (2005) VEGFR1-positive haematopoietic bone marrow progenitors initiate the pre-metastatic niche. Nature 438:820–827

Ke L, Qu H, Nagy JA, Eckelhoefer IA, Masse EM, Dvorak AM, Dvorak HF (1996) Vascular targeting of solid and ascites tumours with antibodies to vascular endothelial growth factor. Eur J Cancer 32A:2467–2473

Kim KJ, Li B, Winer J, Armanini M, Gillett N, Phillips HS, Ferrara N (1993) Inhibition of vascular endothelial growth factor-induced angiogenesis suppresses tumour growth in vivo. Nature 362: 841–844

Kitamura T, Kometani K, Hashida H, Matsunaga A, Miyoshi H, Hosogi H, Aoki M, Oshima M, Hattori M, Takabayashi A, Minato N, Taketo MM (2007) SMAD4-deficient intestinal tumors recruit CCR1+ myeloid cells that promote invasion. Nat Genet 39:467–475

Knolle PA, Thimme R (2014) Hepatic immune regulation and its involvement in viral hepatitis infection. Gastroenterology 146:1193–1207

Kobayashi H, Enomoto A, Woods SL, Burt AD, Takahashi M, Worthley DL (2019) Cancer-associated fibroblasts in gastrointestinal cancer. Nat Rev Gastroenterol Hepatol 16:282–295

Komohara Y, Morita T, Annan DA, Horlad H, Ohnishi K, Yamada S, Nakayama T, Kitada S, Suzu S, Kinoshita I, Dosaka-Akita H, Akashi K, Takeya M, Jinushi M (2015) The coordinated actions of TIM-3 on cancer and myeloid cells in the regulation of Tumorigenicity and clinical prognosis in clear cell renal cell carcinomas. Cancer Immunol Res 3:999–1007

Kubota Y, Takubo K, Shimizu T, Ohno H, Kishi K, Shibuya M, Saya H, Suda T (2009) M-CSF inhibition selectively targets pathological angiogenesis and lymphangiogenesis. J Exp Med 206: 1089–1102

Kubota T, Inoue M, Kubota N, Takamoto I, Mineyama T, Iwayama K, Tokuyama K, Moroi M, Ueki K, Yamauchi T, Kadowaki T (2018) Downregulation of macrophage Irs2 by hyperinsulinemia impairs IL-4-indeuced M2a-subtype macrophage activation in obesity. Nat Commun 9:4863

Kudo M, Finn RS, Qin S, Han KH, Ikeda K, Piscaglia F, Baron A, Park JW, Han G, Jassem J, Blanc JF, Vogel A, Komov D, Evans TRJ, Lopez C, Dutcus C, Guo M, Saito K, Kraljevic S, Tamai T, Ren M, Cheng AL (2018) Lenvatinib versus sorafenib in first-line treatment of patients with unresectable hepatocellular carcinoma: a randomised phase 3 non-inferiority trial. Lancet 391: 1163–1173

Kudo M, Han KH, Ye SL, Zhou J, Huang YH, Lin SM, Wang CK, Ikeda M, Chan SL, Choo SP, Miyayama S, Cheng AL (2020) A changing paradigm for the treatment of intermediate-stage hepatocellular carcinoma: Asia-Pacific primary liver cancer expert consensus statements. Liver Cancer 9:245–260

Kudo M, Galle PR, Brandi G, Kang YK, Yen CJ, Finn RS, Llovet JM, Assenat E, Merle P, Chan SL, Palmer DH, Ikeda M, Yamashita T, Vogel A, Huang YH, Abada PB, Yoshikawa R, Shinozaki K, Wang C, Widau RC, Zhu AX (2021) Effect of ramucirumab on ALBI grade in patients with advanced HCC: results from REACH and REACH-2. JHEP Rep 3:100215

Kuno A, Ikehara Y, Tanaka Y, Ito K, Matsuda A, Sekiya S, Hige S, Sakamoto M, Kage M, Mizokami M, Narimatsu H (2013) A serum "sweet-doughnut" protein facilitates fibrosis evaluation and therapy assessment in patients with viral hepatitis. Sci Rep 3:1065

Kurebayashi Y, Kubota N, Sakamoto M (2021) Immune microenvironment of hepatocellular carcinoma, intrahepatic cholangiocarcinoma and liver metastasis of colorectal adenocarcinoma: relationship with histopathological and molecular classifications. Hepatol Res 51:5–18

Kurkjian C, Kim ES (2012) Risks and benefits with bevacizumab: evidence and clinical implications. Ther Adv Drug Saf 3:59–69

Le DT, Uram JN, Wang H, Bartlett BR, Kemberling H, Eyring AD, Skora AD, Luber BS, Azad NS, Laheru D, Biedrzycki B, Donehower RC, Zaheer A, Fisher GA, Crocenzi TS, Lee JJ, Duffy SM, Goldberg RM, De La Chapelle A, Koshiji M, Bhaijee F, Huebner T, Hruban RH, Wood LD, Cuka N, Pardoll DM, Papadopoulos N, Kinzler KW, Zhou S, Cornish TC, Taube JM, Anders RA, Eshleman JR, Vogelstein B, Diaz LA Jr (2015) PD-1 blockade in tumors with mismatch-repair deficiency. N Engl J Med 372:2509–2520

Le DT, Durham JN, Smith KN, Wang H, Bartlett BR, Aulakh LK, Lu S, Kemberling H, Wilt C, Luber BS, Wong F, Azad NS, Rucki AA, Laheru D, Donehower R, Zaheer A, Fisher GA, Crocenzi TS, Lee JJ, Greten TF, Duffy AG, Ciombor KK, Eyring AD, Lam BH, Joe A, Kang SP, Holdhoff M, Danilova L, Cope L, Meyer C, Zhou S, Goldberg RM, Armstrong DK, Bever KM, Fader AN, Taube J, Housseau F, Spetzler D, Xiao N, Pardoll DM, Papadopoulos N, Kinzler KW, Eshleman JR, Vogelstein B, Anders RA, Diaz LA Jr (2017) Mismatch repair deficiency predicts response of solid tumors to PD-1 blockade. Science 357:409–413

Lee MS, Ryoo BY, Hsu CH, Numata K, Stein S, Verret W, Hack SP, Spahn J, Liu B, Abdullah H, Wang Y, He AR, Lee KH (2020) Atezolizumab with or without bevacizumab in unresectable hepatocellular carcinoma (GO30140): an open-label, multicentre, phase 1b study. Lancet Oncol 21:808–820

Lee JC, Green MD, Huppert LA, Chow C, Pierce RH, Daud AI (2022) The liver-immunity nexus and cancer immunotherapy. Clin Cancer Res 28:5–12

Lehenkari PP, Kellinsalmi M, Näpänkangas JP, Ylitalo KV, Mönkkönen J, Rogers MJ, Azhayev A, Väänänen HK, Hassinen IE (2002) Further insight into mechanism of action of clodronate: inhibition of mitochondrial ADP/ATP translocase by a nonhydrolyzable, adenine-containing metabolite. Mol Pharmacol 61:1255–1262

Liang W, Kujawski M, Wu J, Lu J, Herrmann A, Loera S, Yen Y, Lee F, Yu H, Wen W, Jove R (2010) Antitumor activity of targeting SRC kinases in endothelial and myeloid cell compartments of the tumor microenvironment. Clin Cancer Res 16:924–935

Lin YY, Tan CT, Chen CW, Ou DL, Cheng AL, Hsu C (2018) Immunomodulatory effects of current targeted therapies on hepatocellular carcinoma: implication for the future of immunotherapy. Semin Liver Dis 38:379–388

Lippitz BE (2013) Cytokine patterns in patients with cancer: a systematic review. Lancet Oncol 14: e218–e228

Llovet JM, Ricci S, Mazzaferro V, Hilgard P, Gane E, Blanc JF, de Oliveira AC, Santoro A, Raoul JL, Forner A, Schwartz M, Porta C, Zeuzem S, Bolondi L, Greten TF, Galle PR, Seitz JF, Borbath I, Häussinger D, Giannaris T, Shan M, Moscovici M, Voliotis D, Bruix J (2008) Sorafenib in advanced hepatocellular carcinoma. N Engl J Med 359:378–390

Llovet JM, Villanueva A, Marrero JA, Schwartz M, Meyer T, Galle PR, Lencioni R, Greten TF, Kudo M, Mandrekar SJ, Zhu AX, Finn RS, Roberts LR (2021) Trial design and endpoints in hepatocellular carcinoma: AASLD consensus conference. Hepatology 73(Suppl 1):158–191

Lopes AR, Kellam P, Das A, Dunn C, Kwan A, Turner J, Peppa D, Gilson RJ, Gehring A, Bertoletti A, Maini MK (2008) Bim-mediated deletion of antigen-specific CD8 T cells in patients unable to control HBV infection. J Clin Invest 118:1835–1845

Luke JJ, Bao R, Sweis RF, Spranger S, Gajewski TF (2019) WNT/beta-catenin pathway activation correlates with immune exclusion across human cancers. Clin Cancer Res 25:3074–3083

Mack M, Yanagita M (2015) Origin of myofibroblasts and cellular events triggering fibrosis. Kidney Int 87:297–307

Melani C, Sangaletti S, Barazzetta FM, Werb Z, Colombo MP (2007) Amino-biphosphonate-mediated MMP-9 inhibition breaks the tumor-bone marrow axis responsible for myeloid-derived suppressor cell expansion and macrophage infiltration in tumor stroma. Cancer Res 67:11438–11446

Mellman I, Coukos G, Dranoff G (2011) Cancer immunotherapy comes of age. Nature 480:480–489

Meylan M, Petitprez F, Lacroix L, DI Tommaso L, Roncalli M, Bougoüin A, Laurent A, Amaddeo G, Sommacale D, Regnault H, Derman J, Charpy C, Lafdil F, Pawlotsky JM, Sautès-Fridman C, Fridman WH, Calderaro J (2020) Early hepatic lesions display immature tertiary lymphoid structures and show elevated expression of immune inhibitory and immunosuppressive molecules. Clin Cancer Res 26:4381–4389

Morita M, Nishida N, Sakai K, Aoki T, Chishina H, Takita M, Ida H, Hagiwara S, Minami Y, Ueshima K, Nishio K, Kobayashi Y, Kakimi K, Kudo M (2021) Immunological microenvironment predicts the survival of the patients with hepatocellular carcinoma treated with anti-PD-1 antibody. Liver Cancer 10:380–393

Noe JT, Rendon BE, Geller AE, Conroy LR, Morrissey SM, Young LEA, Bruntz RC, Kim EJ, Wise-Mitchell A, Barbosa De Souza Rizzo M, Relich ER, Baby BV, Johnson LA, Affronti HC, Mcmasters KM, Clem BF, Gentry MS, Yan J, Wellen KE, Sun RC, Mitchell RA (2021) Lactate supports a metabolic-epigenetic link in macrophage polarization. Sci Adv 7:eabi8602

Noy R, Pollard JW (2014) Tumor-associated macrophages: from mechanisms to therapy. Immunity 41:49–61

Ocho M, Togayachi A, Iio E, Kaji H, Kuno A, Sogabe M, Korenaga M, Gotoh M, Tanaka Y, Ikehara Y, Mizokami M, Narimatsu H (2014) Application of a glycoproteomics-based biomarker development method: alteration in glycan structure on colony stimulating factor 1 receptor as a possible glycobiomarker candidate for evaluation of liver cirrhosis. J Proteome Res 13: 1428–1437

Ohashi T, Aoki M, Tomita H, Akazawa T, Sato K, Kuze B, Mizuta K, Hara A, Nagaoka H, Inoue N, Ito Y (2017) M2-like macrophage polarization in high lactic acid-producing head and neck cancer. Cancer Sci 108:1128–1134

Ohga N, Ishikawa S, Maishi N, Akiyama K, Hida Y, Kawamoto T, Sadamoto Y, Osawa T, Yamamoto K, Kondoh M, Ohmura H, Shinohara N, Nonomura K, Shindoh M, Hida K (2012) Heterogeneity of tumor endothelial cells: comparison between tumor endothelial cells isolated from high- and low-metastatic tumors. Am J Pathol 180:1294–1307

Orecchioni M, Ghosheh Y, Pramod AB, Ley K (2019) Macrophage polarization: different gene signatures in M1(LPS+) vs. classically and M2(LPS-) vs. alternatively activated macrophages. Front Immunol 10:1084

Oskarsson T, Acharyya S, Zhang XH, Vanharanta S, Tavazoie SF, Morris PG, Downey RJ, Manova-Todorova K, Brogi E, Massagué J (2011) Breast cancer cells produce tenascin C as a metastatic niche component to colonize the lungs. Nat Med 17:867–874

Paget S (1989) The distribution of secondary growths in cancer of the breast. 1889. Cancer Metastasis Rev 8:98–101

Pan Q, Chanthery Y, Liang WC, Stawicki S, Mak J, Rathore N, Tong RK, Kowalski J, Yee SF, Pacheco G, Ross S, Cheng Z, le Couter J, Plowman G, Peale F, Koch AW, Wu Y, Bagri A, Tessier-Lavigne M, Watts RJ (2007) Blocking neuropilin-1 function has an additive effect with anti-VEGF to inhibit tumor growth. Cancer Cell 11:53–67

Peinado H, Zhang H, Matei IR, Costa-Silva B, Hoshino A, Rodrigues G, Psaila B, Kaplan RN, Bromberg JF, Kang Y, Bissell MJ, Cox TR, Giaccia AJ, Erler JT, Hiratsuka S, Ghajar CM, Lyden D (2017) Pre-metastatic niches: organ-specific homes for metastases. Nat Rev Cancer 17: 302–317

Peng W, Liu C, Xu C, Lou Y, Chen J, Yang Y, Yagita H, Overwijk WW, Lizee G, Radvanyi L, Hwu P (2012) PD-1 blockade enhances T-cell migration to tumors by elevating IFN-gamma inducible chemokines. Cancer Res 72:5209–5218

Pfister D, Núñez NG, Pinyol R, Govaere O, Pinter M, Szydlowska M, Gupta R, Qiu M, Deczkowska A, Weiner AM, Ller F, Sinha A, Friebel E, Engleitner T, Lenggenhager D, Moncsek A, Heide D, Stirm K, Kosla J, Kotsiliti E, Leone V, Dudek M, Yousuf S, Inverso D, Singh I, Teijeiro A, Castet F, Montironi C, Haber PK, Tiniakos D, Bedossa P, Cockell S, Younes R, Vacca M, Marra F, Schattenberg JM, Allison M, Bugianesi E, Ratziu V, Pressiani T, D'alessio A, Personeni N, Rimassa L, Daly AK, Scheiner B, Pomej K, Kirstein MM, Vogel A, Peck-Radosavljevic M, Hucke F, Finkelmeier F, Waidmann O, Trojan J,

Schulze K, Wege H, Koch S, Weinmann A, Bueter M, Rössler F, Siebenh Ner A, De Dosso S, Mallm JP, Umansky V, Jugold M, Luedde T, Schietinger A, Schirmacher P, Emu B, Augustin HG, Billeter AM, Ller-Stich B, Kikuchi H, Duda DGK, Tting F, Waldschmidt DT, Ebert MP, Rahbari N, Mei HE, Schulz AR, Ringelhan M, Malek N, Spahn S, Bitzer M, Ruiz De Galarreta M, Lujambio A, Dufour JF, Marron TU, Kaseb A, Kudo M, Huang YH, Djouder N, Wolter K, Zender L, Marche PN, Decaens T, Pinato DJ, Rad R, Mertens JC, Weber A, Unger K et al (2021) NASH limits anti-tumour surveillance in immunotherapy-treated HCC. Nature 592:450–456

Presta LG, Chen H, O'connor SJ, Chisholm V, Meng YG, Krummen L, Winkler M, Ferrara N (1997) Humanization of an anti-vascular endothelial growth factor monoclonal antibody for the therapy of solid tumors and other disorders. Cancer Res 57:4593–4599

Protzer U, Maini MK, Knolle PA (2012) Living in the liver: hepatic infections. Nat Rev Immunol 12:201–213

Pyonteck SM, Akkari L, Schuhmacher AJ, Bowman RL, Sevenich L, Quail DF, Olson OC, Quick ML, Huse JT, Teijeiro V, Setty M, Leslie CS, Oei Y, Pedraza A, Zhang J, Brennan CW, Sutton JC, Holland EC, Daniel D, Joyce JA (2013) CSF-1R inhibition alters macrophage polarization and blocks glioma progression. Nat Med 19:1264–1272

Rappaport AM (1980) Hepatic blood flow: morphologic aspects and physiologic regulation. Int Rev Physiol 21:1–63

Rey S, Semenza GL (2010) Hypoxia-inducible factor-1-dependent mechanisms of vascularization and vascular remodelling. Cardiovasc Res 86:236–242

Robinson MW, Harmon C, O'farrelly C (2016) Liver immunology and its role in inflammation and homeostasis. Cell Mol Immunol 13:267–276

Rogers MJ, Chilton KM, Coxon FP, Lawry J, Smith MO, Suri S, Russell RG (1996) Bisphosphonates induce apoptosis in mouse macrophage-like cells in vitro by a nitric oxide-independent mechanism. J Bone Miner Res 11:1482–1491

Ruiz de Galarreta M, Bresnahan E, Molina-Sanchez P, Lindblad KE, Maier B, Sia D, Puigvehi M, Miguela V, Casanova-Acebes M, Dhainaut M, Villacorta-Martin C, Singhi AD, Moghe A, Von Felden J, Tal Grinspan L, Wang S, Kamphorst AO, Monga SP, Brown BD, Villanueva A, Llovet JM, Merad M, Lujambio A (2019) Beta-catenin activation promotes immune escape and resistance to anti-PD-1 therapy in hepatocellular carcinoma. Cancer Discov 9:1124–1141

Sahai E, Astsaturov I, Cukierman E, Denardo DG, Egeblad M, Evans RM, Fearon D, Greten FR, Hingorani SR, Hunter T, Hynes RO, Jain RK, Janowitz T, Jorgensen C, Kimmelman AC, Kolonin MG, Maki RG, Powers RS, Puré E, Ramirez DC, Scherz-Shouval R, Sherman MH, Stewart S, Tlsty TD, Tuveson DA, Watt FM, Weaver V, Weeraratna AT, Werb Z (2020) A framework for advancing our understanding of cancer-associated fibroblasts. Nat Rev Cancer 20:174–186

Schofield R (1978) The relationship between the spleen colony-forming cell and the haemopoietic stem cell. Blood Cells 4:7–25

Schreiber RD, Old LJ, Smyth MJ (2011) Cancer immunoediting: integrating immunity's roles in cancer suppression and promotion. Science 331:1565–1570

Schultz GS, Sibbald RG, Falanga V, Ayello EA, Dowsett C, Harding K, Romanelli M, Stacey MC, Teot L, Vanscheidt W (2003) Wound bed preparation: a systematic approach to wound management. Wound Repair Regen 11(Suppl 1):S1–S28

Senger DR, Galli SJ, Dvorak AM, Perruzzi CA, Harvey VS, Dvorak HF (1983) Tumor cells secrete a vascular permeability factor that promotes accumulation of ascites fluid. Science 219:983–985

Shankaran V, Ikeda H, Bruce AT, White JM, Swanson PE, Old LJ, Schreiber RD (2001) IFNgamma and lymphocytes prevent primary tumour development and shape tumour immunogenicity. Nature 410:1107–1111

Sharma A, Seow JJW, Dutertre CA, Pai R, Blériot C, Mishra A, Wong RMM, Singh GSN, Sudhagar S, Khalilnezhad S, Erdal S, Teo HM, Khalilnezhad A, Chakarov S, Lim TKH, Fui ACY, Chieh AKW, Chung CP, Bonney GK, Goh BK, Chan JKY, Chow PKH, Ginhoux F,

Dasgupta R (2020) Onco-fetal reprogramming of endothelial cells drives immunosuppressive macrophages in hepatocellular carcinoma. Cell 183:377–394.e21
Shimomura I, Matsuda M, Hammer RE, Bashmakov Y, Brown MS, Goldstein JL (2000) Decreased IRS-2 and increased SREBP-1c lead to mixed insulin resistance and sensitivity in livers of lipodystrophic and Ob/Ob mice. Mol Cell 6:77–86
Shirabe K, Bekki Y, Gantumur D, Araki K, Ishii N, Kuno A, Narimatsu H, Mizokami M (2018) Mac-2 binding protein glycan isomer (M2BPGi) is a new serum biomarker for assessing liver fibrosis: more than a biomarker of liver fibrosis. J Gastroenterol 53:819–826
Sia D, Jiao Y, Martinez-Quetglas I, Kuchuk O, Villacorta-Martin C, Castro De Moura M, Putra J, Camprecios G, Bassaganyas L, Akers N, Losic B, Waxman S, Thung SN, Mazzaferro V, Esteller M, Friedman SL, Schwartz M, Villanueva A, Llovet JM (2017) Identification of an immune-specific class of hepatocellular carcinoma, based on molecular features. Gastroenterology 153:812–826
Siegel AB, Cohen EI, Ocean A, Lehrer D, Goldenberg A, Knox JJ, Chen H, Clark-Garvey S, Weinberg A, Mandeli J, Christos P, Mazumdar M, Popa E, Brown RS Jr, Rafii S, Schwartz JD (2008) Phase II trial evaluating the clinical and biologic effects of bevacizumab in unresectable hepatocellular carcinoma. J Clin Oncol 26:2992–2998
Sun K, Kusminski CM, Scherer PE (2011) Adipose tissue remodeling and obesity. J Clin Invest 121:2094–2101
Sun Y, Wu L, Zhong Y, Zhou K, Hou Y, Wang Z, Zhang Z, Xie J, Wang C, Chen D, Huang Y, Wei X, Shi Y, Zhao Z, Li Y, Guo Z, Yu Q, Xu L, Volpe G, Qiu S, Zhou J, Ward C, Sun H, Yin Y, Xu X, Wang X, Esteban MA, Yang H, Wang J, Dean M, Zhang Y, Liu S, Yang X, Fan J (2021) Single-cell landscape of the ecosystem in early-relapse hepatocellular carcinoma. Cell 184:404–421.e16
Szabo G, Petrasek J (2015) Inflammasome activation and function in liver disease. Nat Rev Gastroenterol Hepatol 12:387–400
Takayasu K, Muramatsu Y, Furukawa H, Wakao F, Moriyama N, Takayama T, Yamasaki S, Sakamoto M, Hirohashi S (1995) Early hepatocellular carcinoma: appearance at CT during arterial portography and CT arteriography with pathologic correlation. Radiology 194:101–105
Tanaka M, Ikeda K, Suganami T, Komiya C, Ochi K, Shirakawa I, Hamaguchi M, Nishimura S, Manabe I, Matsuda T, Kimura K, Inoue H, Inagaki Y, Aoe S, Yamasaki S, Ogawa Y (2014) Macrophage-inducible C-type lectin underlies obesity-induced adipose tissue fibrosis. Nat Commun 5:4982
Taube JM, Klein A, Brahmer JR, Xu H, Pan X, Kim JH, Chen L, Pardoll DM, Topalian SL, Anders RA (2014) Association of PD-1, PD-1 ligands, and other features of the tumor immune microenvironment with response to anti-PD-1 therapy. Clin Cancer Res 20:5064–5074
The L (2018) Age-related macular degeneration: treatment at what cost? Lancet 392:1090
Tohyama O, Matsui J, Kodama K, Hata-Sugi N, Kimura T, Okamoto K, Minoshima Y, Iwata M, Funahashi Y (2014) Antitumor activity of lenvatinib (e7080): an angiogenesis inhibitor that targets multiple receptor tyrosine kinases in preclinical human thyroid cancer models. J Thyroid Res 2014:638747
Tsuchida T, Friedman SL (2017) Mechanisms of hepatic stellate cell activation. Nat Rev Gastroenterol Hepatol 14:397–411
Tsuchiya H, Tsuchiya Y, Kobayashi T, Kikuchi Y, Hino O (1994) Isolation of genes differentially expressed between the Yoshida sarcoma and long-survival Yoshida sarcoma variants: origin of Yoshida sarcoma revisited. Jpn J Cancer Res 85:1099–1104
Ueshima K, Nishida N, Hagiwara S, Aoki T, Minami T, Chishina H, Takita M, Minami Y, Ida H, Takenaka M, Sakurai T, Watanabe T, Morita M, Ogawa C, Hiraoka A, Johnson P, Kudo M (2019) Impact of baseline ALBI grade on the outcomes of hepatocellular carcinoma patients treated with lenvatinib: a multicenter study. Cancers (Basel) 11:952
Vogelstein B, Kinzler KW (1993) The multistep nature of cancer. Trends Genet 9:138–141
Warburg O (1956) On the origin of cancer cells. Science 123:309–314
Weiss RA (2004) Multistage carcinogenesis. Br J Cancer 91:1981–1982

Wolf MJ, Adili A, Piotrowitz K, Abdullah Z, Boege Y, Stemmer K, Ringelhan M, Simonavicius N, Egger M, Wohlleber D, Lorentzen A, Einer C, Schulz S, Clavel T, Protzer U, Thiele C, Zischka H, Moch H, Tschöp M, Tumanov AV, Haller D, Unger K, Karin M, Kopf M, Knolle P, Weber A, Heikenwalder M (2014) Metabolic activation of intrahepatic CD8+ T cells and NKT cells causes nonalcoholic steatohepatitis and liver cancer via cross-talk with hepatocytes. Cancer Cell 26:549–564

Wong CC, Tse AP, Huang YP, Zhu YT, Chiu DK, Lai RK, Au SL, Kai AK, Lee JM, Wei LL, Tsang FH, Lo RC, Shi J, Zheng YP, Wong CM, Ng IO (2014) Lysyl oxidase-like 2 is critical to tumor microenvironment and metastatic niche formation in hepatocellular carcinoma. Hepatology 60:1645–1658

Yamamoto Y, Matsui J, Matsushima T, Obaishi H, Miyazaki K, Nakamura K, Tohyama O, Semba T, Yamaguchi A, Hoshi SS, Mimura F, Haneda T, Fukuda Y, Kamata JI, Takahashi K, Matsukura M, Wakabayashi T, Asada M, Nomoto KI, Watanabe T, Dezso Z, Yoshimatsu K, Funahashi Y, Tsuruoka A (2014) Lenvatinib, an angiogenesis inhibitor targeting VEGFR/FGFR, shows broad antitumor activity in human tumor xenograft models associated with microvessel density and pericyte coverage. Vasc Cell 6:18

Yang JD, Hainaut P, Gores GJ, Amadou A, Plymoth A, Roberts LR (2019) A global view of hepatocellular carcinoma: trends, risk, prevention and management. Nat Rev Gastroenterol Hepatol 16:589–604

Yarchoan M, Xing D, Luan L, Xu H, Sharma RB, Popovic A, Pawlik TM, Kim AK, Zhu Q, Jaffee EM, Taube JM, Anders RA (2017) Characterization of the immune microenvironment in hepatocellular carcinoma. Clin Cancer Res 23:7333–7339

Zhang W, Zhu XD, Sun HC, Xiong YQ, Zhuang PY, Xu HX, Kong LQ, Wang L, Wu WZ, Tang ZY (2010) Depletion of tumor-associated macrophages enhances the effect of sorafenib in metastatic liver cancer models by antimetastatic and antiangiogenic effects. Clin Cancer Res 16: 3420–3430

Zhao L, Burt AD (2007) The diffuse stellate cell system. J Mol Histol 38:53–64

Zhu AX, Finn RS, Edeline J, Cattan S, Ogasawara S, Palmer D, Verslype C, Zagonel V, Fartoux L, Vogel A, Sarker D, Verset G, Chan SL, Knox J, Daniele B, Webber AL, Ebbinghaus SW, Ma J, Siegel AB, Cheng AL, Kudo M (2018) Pembrolizumab in patients with advanced hepatocellular carcinoma previously treated with sorafenib (KEYNOTE-224): a non-randomised, open-label phase 2 trial. Lancet Oncol 19:940–952

Zhu AX, Kang YK, Yen CJ, Finn RS, Galle PR, Llovet JM, Assenat E, Brandi G, Pracht M, Lim HY, Rau KM, Motomura K, Ohno I, Merle P, Daniele B, Shin DB, Gerken G, Borg C, Hiriart JB, Okusaka T, Morimoto M, Hsu Y, Abada PB, Kudo M (2019) Ramucirumab after sorafenib in patients with advanced hepatocellular carcinoma and increased α-fetoprotein concentrations (REACH-2): a randomised, double-blind, placebo-controlled, phase 3 trial. Lancet Oncol 20: 282–296

Metabolic Alterations of Hepatocellular Cancer Stem Cells

Jan Caira David, Marta Bento Afonso, and Cecília Maria Pereira Rodrigues

Abstract

Hepatocellular carcinoma is the most common type of liver cancer and the third leading cause of cancer-related deaths. Tumour cells exhibit phenotypic and functional heterogeneity because of genetic and environmental factors. Metastatic spread and disease recurrence are believed to be owed to the existence of cancer stem cells (CSCs), a subpopulation of tumour cells capable of self-renewal and production of heterogeneous progeny cells. CSCs do not only exhibit unique metabolic properties, but also possess interesting metabolic plasticity, allowing them to exploit several metabolic and signalling pathways to survive and evolve. Being localised in a niche within the tumour microenvironment (TME), CSCs can manipulate stromal cells to their needs, while their metabolic phenotype is likewise dependent on their surrounding. Recent findings evidentiate TME involvement in stemness promotion and maintenance, suggesting that novel treatments targeting both CSCs and components of the TME, which also show capabilities of metabolic reprogramming, may prove more effective. Since CSCs are intrinsically resistant to conventional chemotherapeutic regimens, targeting the unique CSC metabolism, different from that of normal cells and other cancer cells, can potentially facilitate a selective therapeutic approach. This could then be used in rational drug combinations to improve patient treatment and prognosis.

Keywords

Cancer stem cells · Drug resistance · Hepatocellular carcinoma · Metabolism

J. C. David · M. B. Afonso · C. M. P. Rodrigues (✉)
Research Institute for Medicines (iMed.ULisboa), Faculty of Pharmacy, Universidade de Lisboa, Lisbon, Portugal
e-mail: cmprodrigues@ff.ulisboa.pt

Interdisciplinary Cancer Research, https://doi.org/10.1007/16833_2022_43
Published online: 6 October 2022

1 Introduction

Primary liver cancer ranks as the seventh most common cancer and the third leading cause of cancer-related deaths in 2020. Among all types of liver cancer, hepatocellular carcinoma (HCC) is the most common and accounts for 75–85% of cases (Sung et al. 2021). HCC arises almost exclusively in a background of chronic liver disease characterised by continuous hepatocyte death and compensatory proliferation, whereby chronic infection with hepatitis B virus (HBV) and hepatitis C virus (HCV), excessive alcohol consumption, and ingestion of aflatoxin B1 are the most prominent risk factors (Sung et al. 2021). Besides these, cases of HCC developed in the background of the metabolic disorders non-alcoholic fatty liver disease (NAFLD) and nonalcoholic steatohepatitis (NASH) have been steadily rising in recent years and are expected to keep increasing, in parallel with the rise of obesity (Estes et al. 2018; Ramani et al. 2022; Younossi et al. 2015). Thus, HCC remains a global public health concern, associated with high morbidity and mortality, as well as a significant economic burden (Yang et al. 2019). While hepatic resection, liver transplantation and locoregional therapies have been the principal curative treatments in early and intermediate stage HCC, the first-line treatment for patients with advanced unresectable HCC continues to be Sorafenib, an oral multiple-target tyrosine kinase inhibitor (TKI) (Galle et al. 2018). Despite being the most effective single-drug treatment, compiled evidence from the European Association for the Study of Liver (EASL) showed that systemic chemotherapy is not recommended for HCC treatment, nor as a control regimen for any trial because of its well-known toxic effects. Besides this, another systemic drug called Lenvatinib, an inhibitor of vascular endothelial growth factor receptors 1-3, fibroblast growth factor receptors 1-4, platelet-derived growth factor receptor α, RET and KIT has been highly recommended and recognised as non-inferior to Sorafenib as first-line treatment (Galle et al. 2018). Immune-checkpoint inhibitors are also promising agents for the management of HCC; however, certain aetiologies of HCC could limit anti-tumour surveillance in immunotherapy-treated HCC (Pfister et al. 2021). Noteworthy, tumour cells exhibit phenotypic and functional heterogeneity because of genetic and epigenetic changes, as well as molecular and environmental differences, highlighting the importance of new treatments due to the prevalent Sorafenib resistance (Cabral et al. 2020). The low survival of HCC patients relates to high recurrence rates as well as therapy resistance, both being increasingly thought to result from intra-tumour heterogeneity, stemming from clonal evolution and selection of cancer stem cells (CSCs).

CSCs or tumour-initiating cells represent a small subpopulation of cancer cells in various types of cancers that dictate the whole tumorigenic process, including initiation, metastatic spread and disease recurrence. Like normal stem cells, these malignant cells display stem cell markers, possess the ability to self-renew and produce heterogeneous progeny cells (Meacham and Morrison 2013). These cells have been discovered in many types of solid cancers such as pancreatic, lung, colon, prostate, ovarian breast, brain, melanoma and liver carcinomas (Collins et al. 2005; Fang et al. 2005; Ginestier et al. 2007; Hermann et al. 2007; Kim et al. 2005;

Ma et al. 2007; O'Brien et al. 2007; Singh et al. 2003; Szotek et al. 2006), and may be derived from normal stem cells or differentiated cells. A growing topic in the literature indicates that CSCs are empowered by metabolic plasticity conferring a remarkable capacity of metabolic adaptation to physiopathological stresses, such as decreased energy source, hypoxia and pH of the microenvironment, when compared to the tumour bulk cells (Kaur and Bhattacharyya 2021). Altered glucose ('Warburg effect') and lipid metabolism have been frequently considered cardinal features of CSCs, making them important hallmarks for therapies targeting cancer metabolism. Consequently, a lot of effort is being given into characterising and targeting the metabolic alterations of hepatic CSCs (HCSCs) using state of the art techniques such as time-of-flight mass spectrometry, isobaric tag labelling with mass spectrometry and miRNA-based technologies (Hur et al. 2017; Ko et al. 2013).

Indeed, since CSCs are intrinsically resistant to conventional chemotherapeutic regimens, targeting the unique CSC metabolism, different from that of normal cells and other cancer cells, can potentially facilitate a selective therapeutic approach that could be used in rational drug combinations to improve patient treatment and prognosis.

2 Origins and Biomarkers of Hepatic Cancer Stem Cells

It is well established that tumours are a heterogeneous mass of cells. The two most accepted but also controversial models that explain the heterogeneity and progression of tumours are known as the stochastic or clonal evolution model, and the hierarchy or CSC model (Shackleton et al. 2009). The stochastic model postulates that every cell within a tumour has the same tumour-initiating capacity (Fig. 1). This model relies on the premise that cancer is a disease defined by hyperproliferation and sequential acquisition of oncogenic genetic mutations either caused by intrinsic or extrinsic factors, followed by subsequent clonal expansions. On the other hand, the hierarchical CSC model argues that the heterogeneity and tumour-initiating activity is restricted to a specific population of tumour cells (Fig. 1). Specifically, the model posits that most cancers must have a hierarchical organisation, where similarly to just non-transformed healthy stem cells, CSCs can proliferate asymmetrically by either self-renewing their own population with long-term clone-propagating capacity or differentiate into short-lived progeny with self-limited proliferative capacity, constituting the tumour bulk (Reya et al. 2001).

The hierarchical model was first demonstrated in acute myeloid leukaemia, in which a subset of leukaemia cells expressed stem cell markers and harboured the potential of self-renewal, propagation, and differentiation (Bonnet and Dick 1997). In solid tumours, CSCs were first shown in breast cancer, as they were particularly efficient in establishing tumours upon their isolation from the tumour bulk and transplantation into mice (Al-Hajj et al. 2003). Since then, the existence of CSCs has been shown in various cancer types including HCC (Ma et al. 2007). Although seemingly opposite, the two models might not be entirely exclusive but rather

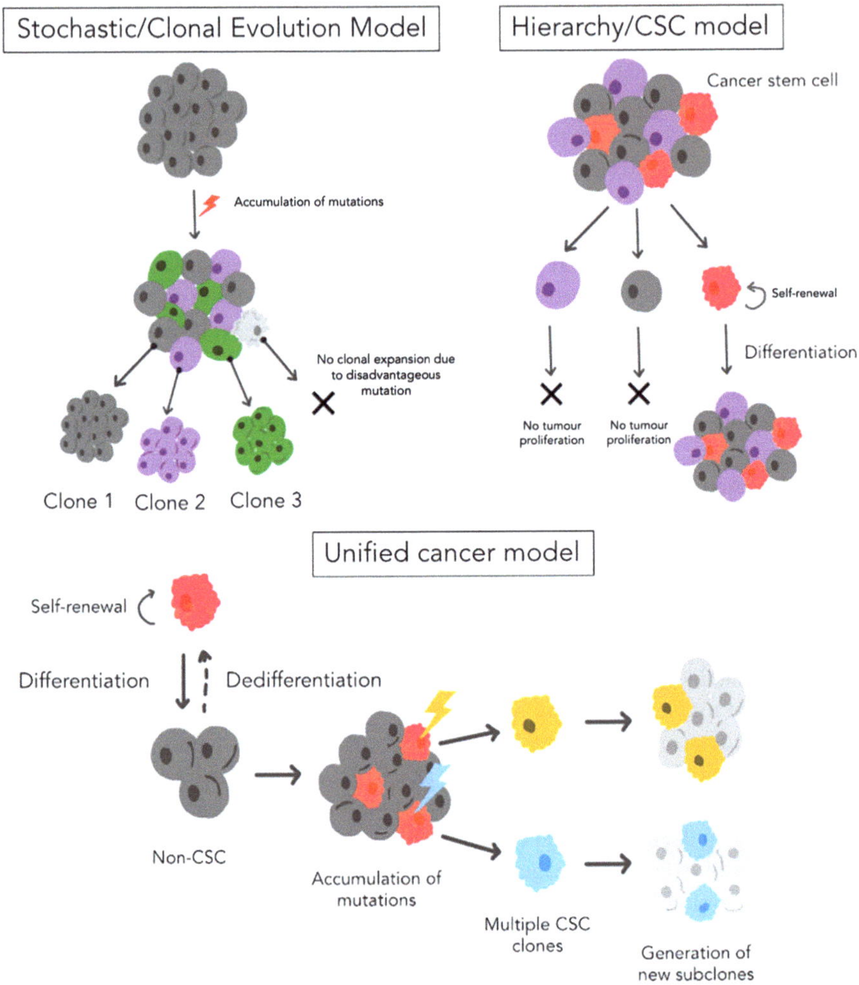

Fig. 1 Schematic representation of the two most accepted models for tumour heterogeneity, and modern unified cancer model. The stochastic or clonal evolution model states that heterogeneity arises within tumours through stochastic genetic and epigenetic mutations resulting in phenotypic and functional differences that can be passed down through clonal expansion. Additionally, mutations can be disadvantageous, resulting in non-proliferative clones. The hierarchy or CSC model postulates that a tumorigenic subpopulation called CSCs exist within the tumour, found at the apex of tumour initiation, and are capable of self-renewal much like normal stem cells. These cells give rise to phenotypically diverse non-tumorigenic cells making up the tumour bulk. Finally, the unified cancer model argues that heterogeneity is not entirely exclusive to one model and is rather coexistent, thereby making CSCs also subject to clonal evolution resulting in multiple CSC clones (yellow and blue) within the tumour, that can give rise to new subclones composing the tumour bulk

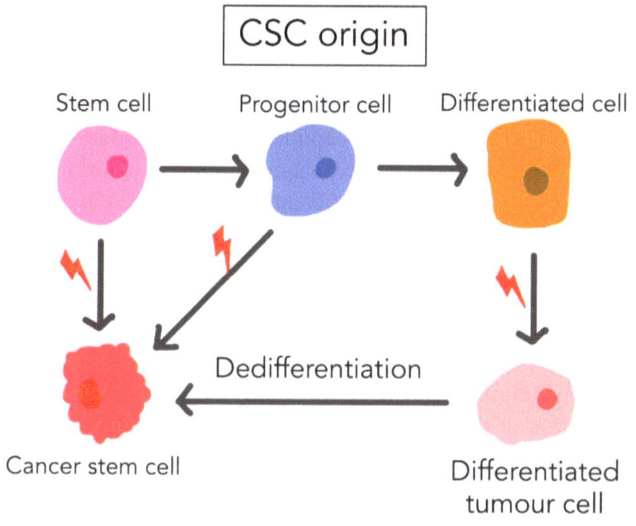

Fig. 2 Schematic representation of CSC origin. CSCs can either originate from normal tissue resident stem cells, progenitor cells, differentiated cells through accumulation of genetic or epigenetic mutations (represented by red lightning), or even non-transformed cancer cells through dedifferentiation programmes

dynamic and concurrent (Wang et al. 2014), thus co-existing multiple CSC clones within the tumour selected by clonal evolution within the stem cell pool.

Tumour initiation through CSCs can either originate from transformed tissue resident stem cells, progenitor cells, differentiated cells and even normal cancer cells (Fig. 2). Liver malignancies develop as a result of an enduring repetitive inflammation/regeneration during chronic liver disease, leading to a continuous cycle of normal stem cell expansion, accumulation of genetic or epigenetic changes and alteration of the microenvironment, which might result in the transformation of hepatic stem cells into HCSCs and cancer development (He et al. 2013; Tang et al. 2008; Wu et al. 2012). Nevertheless, HCSCs may not only originate from transformed normal stem cells, but also from their progeny. The de-differentiation programme where cells acquire a stem-like characteristic was observed using cell tracking, showing that HCC can originate from adult differentiated hepatocytes (Mu et al. 2015). Further, in vitro and in vivo studies proved that any cell type in the mouse hepatic lineage can undergo oncogenic reprogramming into CSCs, by activating different cell type-specific pathways (Holczbauer et al. 2013). Likewise, de-differentiation of this sub-population into differentiated tumour cells does not seem to be unidirectional. A tumour CSC pool can be replenished by reprogramming of tumour bulk cells which has been observed in various cancers, including HCC (Chaffer et al. 2011; Mani et al. 2008). Liu and colleagues observed that the genomic amplification of chromodomain-helicase-DNA-binding protein 1-like (CHD1L), a chromatin remodelling factor that promotes the transcriptional activation of key developmental factors associated with pluripotency and differentiation, such as estrogen-related receptor-beta *(ESRRB)* and transcription factor 4 *(TCF4)*, further facilitates HCC lineage reversion (Liu et al. 2016).

Besides self-renewal and pluripotency, stem cell surface markers are shared between normal stem cells and CSCs. HCSCs are identified and classified using normal liver stem cell markers, such as epithelial cell adhesion molecule (EpCAM) (Schmelzer et al. 2007), leucine-rich repeat-containing G-protein coupled receptor 5 (Lgr5) (Huch et al. 2013), cluster of differentiation (CD)133 (Rountree et al. 2007), CD44 (Schmelzer et al. 2007), CD24 (Qiu et al. 2011) and CD90 (Yang et al. 2008). Thus, single HCSC markers are not specific and are not reliable on their own, calling the need for combination with other markers for stemness/differentiation, and/or with functional methodologies. Several assays based on the functional properties of CSCs include: (1) the aldehyde dehydrogenase (ALDH) activity measurement (Ma et al. 2008); (2) tumour-sphere formation assay which focuses on selection of CSCs and evaluation of their long-term self-renewal capability (Raggi et al. 2014) and (3) limiting dilution analysis that determines the tumour-initiating capacity in vivo in immune-deficient mice (Raggi et al. 2014, 2017a).

CSCs exhibit inherent plasticity and dormancy comparable to that of normal stem cells (Batlle and Clevers 2017). On top of that, CSC phenotypes appear to be heterogenous within each cancer type which may require a more personalised diagnosis and patient treatment. Conventional chemo/radiotherapeutic approaches that target highly proliferative tumour cells prove to be ineffective against CSCs given their quiescent features and slow growing rate in vivo, which allows survival and subsequent cancer relapse and metastasis. CSCs can also undergo epithelial mesenchymal transition (EMT) and transdifferentiation, a process where CSCs differentiate into other cell lineages that further promote tumour proliferation (Shekhani et al. 2013). Overall, it is owed to the dynamic nature of HCSCs that relapse is a common occurrence even after tumour removal, consequently making HCC treatment more challenging.

3 Metabolic Phenotypes

3.1 Source of Energy

3.1.1 The 'Warburg Effect'

The liver plays a highly important role in blood glucose homeostasis by balancing between energy generation (glycolysis) during fasting and energy storage (glycogenesis) during nourishment. Cancer cells have been observed to take advantage of the glycolytic pathway rather than mitochondrial oxidative phosphorylation (OXPHOS), even in the presence of oxygen, a phenomenon called the 'Warburg effect'. During this aerobic glycolysis, cancer cells can increase the rate of glucose uptake and produce energy through ATP production. Despite being less efficient in terms of ATP generation compared to OXPHOS, enhanced glycolysis seems to be preferred during rapid proliferation due to the generation of intermediates that can be used in anabolic reactions. Notably, the activation of the pentose phosphate pathway

(PPP) using glucose-6-phosphate (G6P) can generate NADPH and ribose 5-phosphate (R5P), which are important metabolites involved in tumour cell proliferation (Ge et al. 2020). Besides this, increasing PPP metabolic flux can enhance the level of the antioxidant glutathione (GSH) and suppress reactive oxygen species (ROS) levels. Consequently, cancer cells strategically avoid ROS accumulation while maintaining a more favourable environment for survival.

Particularly, several studies specifically focused on CSCs in various tumour types, such as glioblastoma, lung cancer, osteosarcoma, breast cancer, ovarian cancer, colon cancer and HCC, suggest that CSCs strongly favour glycolysis, compared to other differentiated cancer cells in vitro and in vivo (Ciavardelli et al. 2014; Kai-Yuan Chen et al. 2014; Liao et al. 2014; Song et al. 2015; Zhou et al. 2011). Indeed, it has been shown that $CD133^+$ HCSCs displayed an increased glycolysis rate and capacity, coupled with a decreased oxygen consumption rate (OCR) (Song et al. 2015). Genes involved in glycolysis, such as glucose transporter 1 (*GLUT1*), hexokinase 2 (*HK2*), pyruvate dehydrogenase kinase (*PDK*), and phosphoglycerate mutase 1 (*PGAM1*) were also found to be up-regulated in $CD133^+$ HCSCs, while genes involved in gluconeogenesis, glucose-6-phosphatase catalytic subunit (*G6PC*) and phosphoenolpyruvate carboxykinase (*PEPCK*) were down-regulated, supporting a dependence on aerobic glycolysis and reduction of gluconeogenesis (Song et al. 2015). It has also been proven that HCSCs preferentially utilise glycolysis rather than mitochondrial respiration as shown by augmented basal and maximal OCR in $CD133^+$ cells compared with $CD133^-$ counterparts. Furthermore, the intracellular ATP level was lower in $CD133^+$ HCSCs, reflecting the lower ATP yield from glycolysis comparing with OXPHOS. Knockdown of the glycolytic enzymes, lactate dehydrogenase A (LDHA) and pyruvate dehydrogenase kinase 4 (PDK4) in $CD133^+$ cells has resulted in decreased protein levels of several stemness markers (NANOG, octamer-binding transcription factor 4 (Oct4), and SRY-Box Transcription Factor 4 (Sox4)) and restored Sorafenib sensitivity, suggesting that glycolysis is an important factor in sustaining stemness and drug resistance (Song et al. 2015). Similarly, a global metabolic analysis has demonstrated an increased proliferation of $CD133^+$ HCC cells through enhancing glycolytic metabolism, compared with $CD133^-$ HCC cells (Hur et al. 2017). Likewise, it has been shown that knockdown of NANOG in HCC impairs glycolytic activity as demonstrated by a decreased extracellular acidification rate, indicative of a diminished glycolytic flux (Chen et al. 2016). Moreover, NANOG directly represses OXPHOS and maintains low intracellular ROS levels in HCSCs, a mechanism that is required for the preservation of CSCs properties. In turn, restoration of OXPHOS activity renders HCSCs more susceptible to standard care anti-HCC drugs (Chen et al. 2016). Overall, the metabolic shift to augmented glycolysis contributes to the maintenance of HCSCs resistance to chemotherapeutic drugs.

miRNAs, in turn, participate in the regulation of multiple biological processes, implicated in normal organism homeostasis. In addition, they have also been identified as tumour suppressors or oncogenic drivers in various cancer types (Zhang et al. 2007). Evidence for the role of miRNAs in HCSCs metabolism is scarce but its potential as a treatment for targeted therapy in HCSC eradication is being explored. A study involving HCSCs with low levels of miRNA-192-5p, the second most abundant miRNA in the liver (Gu et al. 2019), showed that these cells are hyperglycolytic, presenting up-regulated glycolytic genes and high glucose consumption (Gu et al. 2020). An increase in extracellular acidification rate that resulted from high concentrations of lactic acid also proved to be capable of increasing HCC malignant features via interacting with the environmental non-tumour cells. On the other hand, miRNA-122 has been found to inhibit glycolysis. This liver-specific miRNA has been known to regulate cellular metabolism including lipid and cholesterol metabolism as well as act as a pivotal tumour suppressor in the liver (Esau et al. 2006). miRNA-122 is frequently decreased in human HCCs and $CD133^+$ CSCs (Song et al. 2015), while its overexpression in $CD133^+$ cells inhibits aerobic glycolysis by directly targeting PDK4, consequently leading to decreased expression of stemness genes and reduced spheroid formation. A miRNA-122 mimic also decreased the spheroid formation efficiency of $CD133^+$ (Song et al. 2015), further supporting the therapeutic exploitation of this miRNA in targeted therapy for metabolic regulation and HCSC eradication.

3.1.2 Mitochondrial Respiration

Healthy differentiated cells largely rely on OXPHOS as the most efficient energy source. As mentioned previously, normal cancer cells mostly opt for glycolysis. As opposed to tumour bulk cells solely exploiting the 'Warburg effect', CSCs appear to rely either on glycolytic or OXPHOS metabolism. Interestingly, despite the well-established Warburg theory, it has been demonstrated that CSCs in various cancer types such as leukaemia and pancreatic cancer display a preference for OXPHOS as the source of energy production, as well as to support self-renewal and drug resistance (Kuntz et al. 2017; Valle et al. 2020). Similarly, in cholangiocarcinoma (CCA), the second most common malignancy arising from the liver, CSCs showed a more efficient respiratory phenotype due to OXPHOS. Targeting complex I using metformin and peroxisome proliferator-activated receptor-γ coactivator-1α (PGC-1α) silencing, whose overexpression is associated with shorter patient survival, impaired sphere formation as well as surface markers related to pluripotency and EMT (Raggi et al. 2021).

Recent studies demonstrated that even if the glycolytic pathway is indeed more robust in HCSCs, pyruvate is transformed into acetyl-CoA and goes into the mitochondria for OXPHOS instead of being converted into lactic acid (Liu et al. 2020). In turn, others showed that HCSCs exhibit increased mitochondrial respiratory capacity, and the regulation of mitochondrial respiration by overexpression of complex I is necessary for stemness properties of HCSCs (Wei et al. 2019). The team also verified that this cell subpopulation had higher OCR and ROS as well as less lactic acid concentration, further indicating a reduced 'Warburg effect'. This metabolic flexibility was shown to be owed to the Sirtuin-1/mitochondrial ribosomal protein S5 (MRPS5) axis. Indeed, MRPS5 knockdown, a protein critical in promoting the function of complex I and whose overexpression is associated with poor patient prognosis, significantly reduced NAD^+ and ROS concentration. SIRT1, a deacetylase that is also overexpressed in HCSCs could reduce acetylation of

MRPS5, therefore reducing its nuclear localisation and expression of glycolysis-related genes. These observations imply that SIRT1/MRPS5 plays a key role as a metabolic switch between mitochondrial aerobic respiration and glycolysis, which may serve as a promising therapeutic target (Wei et al. 2019).

To explain these apparent discrepant findings, it was first hypothesised that CSCs should possess a metabolic phenotype reminiscent of normal tissue hierarchy, where multipotent stem cells are essentially glycolytic, while differentiated cells rely on OXPHOS for ATP production (Folmes et al. 2012). However, a study on ovarian and cervical squamous CSCs demonstrated that cancer metabolism is not limited to aerobic glycolysis (i.e. the 'Warburg effect') but is flexible and context-dependent (Sato et al. 2016). The tricarboxylic (TCA) cycle, a cellular respiration process common to all aerobic organisms that supply OXPHOS, has been found down-regulated in tumour bulk cells (Sancho et al. 2016). In turn, ovarian and cervical squamous CSCs exhibit increased levels of TCA metabolites and seem to obtain energy via OXPHOS by increasing the TCA cycle (Sato et al. 2016). Colorectal CSCs from the HCT-116 cell line also showed significant up-regulation of TCA cycle metabolites such as pyruvate, citric acid, malic acid and succinic acid (Sun and Yang 2019). While the same up-regulation of these metabolites was described in HCSCs compared to non-stem counterparts, there was simultaneously a high concentration of lactic acid (Hur et al. 2017). Since lactic acid results from the Warburg effect and high TCA metabolites point to more active bioenergetics, this discovery might suggest that HCSCs display high metabolic plasticity and maintain homeostasis by switching their energy metabolism pathways between OXPHOS and glycolysis.

Finally, mitochondrial fragmentation has been found to increase in malignant cells and to promote EMT and tumour cell invasion (Gundamaraju et al. 2021). Likewise, reorganisation of mitochondrial structure through mitochondrial fusion and fission is known to be associated with (cancer) stem cell functional outcomes. Recently, mitochondrial fission has been shown as a requirement for liver CSC maintenance (M. Tang et al. 2021). Mitochondrial fission factor (MFF), which is overexpressed in liver CSCs, enhanced stemness and tumour initiating capacity in their non-stem counterparts. The MFF-induced mitochondrial fission promoted asymmetric stem cell division as well as a metabolic shift from OXPHOS to glycolysis, thus decreasing ROS, and prevented ROS-mediated degradation of stem-cell marker OCT4. Curiously, HCSCs treated with Mdivi-1, a commonly used inhibitor of mitochondrial division and a selective inhibitor of dynamin-related protein 1 (DRP1), which is a key protein regulating mitochondrial division, promoted differentiation and weakened stemness properties by reducing OXPHOS activity through inhibition of mitochondrial fission (Liu et al. 2020). Supporting these results, a study in human breast cancer cells showed increased mitochondrial fission as a driver for cancer stemness (Peiris-Pagès et al. 2018).

3.1.3 Fatty Acid Oxidation

It is widely accepted that lipid metabolism rewiring is a hallmark of cancer cells, including alterations in fatty acid transport, de novo lipogenesis, storage as lipid droplets and β-oxidation (Bian et al. 2021). The role of β-oxidation/fatty acid oxidation (FAO) as a source of energy for HCSCs is not consistent among studies. Enhancement of FAO appears to be essential in some CSC-enriched cultures, such as leukaemia CSCs expressing CD36, a fatty acid translocase (Ye et al. 2016). This translocase was also found in oral squamous cell carcinomas, melanoma and breast cancer (Pascual et al. 2017). The overexpression of CD36 in CSCs may induce lipolysis and fuel the FAO, contributing to chemotherapy resistance. The same impact of FAO on HCC stemness has been described in cells expressing the stem cell marker NANOG (Chen et al. 2016). Increased NANOG activity simultaneously reduced the expression of mitochondrial OXPHOS genes and elevated the expression of mitochondrial FAO genes in HCSCs. Targeted expression analysis showed that genes associated with the FAO pathway [acyl-CoA dehydrogenase very long chain (*Acadvl*), enoyl coenzyme A hydratase short chain 1 (*ECHS1*) and acyl-CoA dehydrogenase (*Acads*)] were significantly expressed in this subpopulation of cells, while NANOG knockdown resulted in a down-regulation of FAO-associated genes, emphasising the dependence of HCSCs on NANOG for FAO activation to satisfy the needs for stemness maintenance and ATP generation (Chen et al. 2016).

Conversely, peroxisome proliferator-activated receptor α (PPARα) and PPARγ expression levels on $CD133^+$ HCSCs were decreased in comparison with non-stem counterparts (Hur et al. 2017). PPARα regulates the expression of genes involved in FAO, while PPARγ signalling has been shown to regulate lipogenesis and lipid accumulation in the liver (Lamichane et al. 2018). In agreement, decreased levels of PPARγ and FAO-related proteins, namely carnitine palmitoytransterase-1 (CPT-1) and PGC-1α were found in HCSCs (Bort et al. 2020).

NASH is an advanced form of NAFLD characterised by inflammation and liver damage. Although this metabolic disorder has been one of the fastest growing etiologies for HCC, the exact mechanism of progression to HCC is currently not clear and no effective drugs have been developed yet (Foerster et al. 2022). Of note, upregulation of the lipoprotein lipase (LPL)/fatty acid-binding protein 4 (FABP4)/CPT1 molecular axis occurred specifically during the NASH phase and is essential for HCSC maintenance, acting synergistically with a variety of up-regulated oncogenic signals that drive the hepatocyte-HCSCs transformation during NASH to HCC progression (Yang et al. 2021). The metabolic reprogramming of FAO through this axis may create a favourable microenvironment by providing fatty acids for building blocks, as well as energy through FAO, therefore supporting the tumour-initiating capacity of HCSCs. Indeed, the pharmacological inhibition of LPL/FABP4/CPT1 reduced sphere-forming, proliferation and clonality of CSCs in vitro, while also delaying tumour growth in an experimental animal model of NASH-related liver carcinogenesis. This suggests that targeting this axis may play an important role in NASH to HCC prevention (Yang et al. 2021).

Collectively, the available evidence suggests that the impact of FAO in HCSCs still needs to be tackled and might be context-dependent and/or play differential roles in HCSCs properties.

3.2 Lipid Biosynthesis

As described in the above section, fatty acid metabolism is often reprogrammed in CSCs. Lipids, either through exogenous uptake or endogenous synthesis, have a great impact on maintaining stemness properties of cancer cells (Li et al. 2020). It has been shown that enriched human colon cancer and ovarian CSCs have relatively higher levels of monounsaturated fatty acids (MUFAs) and unsaturated lipids compared to non-stem counterparts (Sun and Yang 2019). Unsaturated lipids are mainly produced by stearoyl-CoA desaturase-1 (SCD1), whereby SCD1 activity has been increased in CSCs (Sun and Yang 2019), including in HCSCs (Qin et al. 2020). Furthermore, the genetic or pharmacological inhibition of PPARα/SCD1 axis reduced the in vitro tumour sphere formation and stem cell markers in cultures enriched in HCSCs (Ma et al. 2019; Qin et al. 2020).

Relatively higher abundances of unsaturated lipids in CSCs are likely related to the formation of lipid droplets, the lipid storage compartment containing variable ratios of neutral lipids, such as cholesteryl esters and triglycerides (Farese and Walther 2009; Mukherjee et al. 2017). As the de novo fatty acid synthesis pathway is upregulated in CSCs, lipid droplets accumulate. Protein Kinase B or AKT regulates neutral lipid metabolism through sterol regulatory element-binding protein 1c (SREBP1c), an important transcription factor for the de novo lipogenesis (Wang et al. 2015). The phosphoinositide 3-kinase (PI3K) and AKT signalling pathway has been observed to be essential for the survival and progression of HCSCs (Kahraman et al. 2019). Mechanistic target of rapamycin (mTOR) inhibited liver CSC enrichment and suppressed the upregulation of interleukin 8 (IL-8) triggered by Sorafenib (Kahraman et al. 2019). Accordingly, combined targeting of AKT and mTOR in $EpCAM^+$ and $CD90^+$ HCC cells showed synergistic results in vitro, while restoration of phosphorylated AKT was observed during conventional HCC therapy, underlining the importance of restored AKT signalling as a resistance mechanism (Moustafa et al. 2022). Fluorescence microscopy analysis revealed that HCSCs had increased and enlarged cytoplasmic lipid droplets, confirming lipid accumulation in those cells (Bort et al. 2020). Long-term treatment with increasing concentrations of Sorafenib selected cancer cells with higher expression of the efflux transporter ATP-dependent translocase ABCB1 and stemness transcription factors OCT4 and NANOG, along with increased levels of SREBP1c (Bort et al. 2020). SREBP1c stimulates the expression of ATP–citrate lyase (ACLY), acetyl-CoA carboxylase (ACC) and fatty acid synthase (FASN), which were found to be significantly expressed in HCSCs (Bort et al. 2020).

NAFLD is a disease driven, at least in part, by overwhelming hepatic fat accumulation. On the other hand, HCSCs display increased lipid biosynthesis, whereby a link between lipogenesis and HCSC maintenance and aggressiveness in NAFLD might be hypothesised. In fact, the palmitoylation of proteins in HepG2 cell line, comprising the covalent attachment of palmitic acid to cysteine and serine and threonine residues of proteins, promoted the activation of CSC-like properties and an increase in sphere formation (Chong et al. 2020). Although yet to be proven whether aberrant palmitoylation can contribute to tumour initiation, its potential implication in CSC activation opens doors for a promising strategy to treat patients with NAFLD (Chong et al. 2020). Altered hepatic lipid metabolism promotes HCC development through inflammation and activation of oncogenes. Growth differentiation factor 11 (GDF11), a member of the transforming growth factor-β (TGF-β) superfamily is an anti-aging factor that can alleviate high-fat diet, induced obesity, hyperglycaemia, insulin resistance and NAFLD (Frohlich et al. 2021). Supplementation of this factor has been shown to promote pro-lipogenic gene expression and lipid accumulation in HCC. Further, evidence shows that de novo lipogenesis was activated through PI3K/AKT pathway, making GDF11 a possible key player in lipid metabolic reprogramming in HCC (Frohlich et al. 2021).

Noteworthy, the Myc proto-oncogene protein family members, one of them being MYCN, play dual roles in regulating normal stem cell-mediated tissue regeneration and CSC-mediated tumorigenesis (Dang 2012). During repeated liver damage and compensatory regeneration, aberrant stabilisation and activation of c-Myc have contributed to the development of liver cancers (Dauch et al. 2016). Wnt/β-catenin signalling is also critical for maintaining stem cell pluripotency. The correlation between the expression levels of MYCN and Wnt/β-catenin signalling markers in human HCC suggested that the role played by MYCN in liver CSCs is probably related to its regulation of Wnt/β-catenin signalling. EpCAM, a known cell surface marker of HCSCs, is a direct transcriptional target of Wnt/β-catenin signalling (Yamashita et al. 2009), and is functionally linked with MYCN in human HCC (Yamashita et al. 2008), whereby MYCN expression is positively correlated with EpCAM in HCSCs (Qin et al. 2020). Up-regulation of MYC was also found in HCSCs derived from the Huh7 cell line, accompanied by higher gene expression of ACC1 and ACC2, important enzymes to fatty acid biosynthesis (Hur et al. 2017).

Overall, HCSCs are associated with lipid metabolism rewiring, where both the amount and type of lipids are strong contributors to their phenotype.

3.3 Iron Metabolism

Iron is a fundamental element utilised by living organisms in various cellular processes such as metabolism, replication and growth. Nonetheless, excess free Fe^{2+} can promote the generation of highly reactive oxidants through Fenton reaction and lead to DNA damage and eventually cell death. Indeed, dysregulation of the iron metabolism, particular high iron concentrations, has been linked to various diseases such as HCC (Jung et al. 2019). As mentioned in the previous sections, CSCs endure extreme stress such as hypoxia, acidosis and oxidative stress. Given that iron is involved in these processes, the disruption of iron homeostasis is a potential target for CSC elimination. Accordingly, accumulating evidence show that increased iron in the liver positively correlates with HCC development and progression (Fargion et al. 2014). Ferritin heavy chain (FTH), a major iron storing nanocage that harbours ferroxidase activity to prevent iron-mediated production of ROS, showed to be upregulated in HCC and is associated with its poor prognosis (Hu et al. 2021). FTH overexpression resulted in accelerated HCC proliferation, migration and resistance to ferroptosis, a distinct regulated cell death process described by accumulation of iron-dependent lipid ROS (Hu et al. 2021). This apparent iron addiction seems to be particularly prevalent in CSCs compared to non-stem tumour cells as reviewed elsewhere (Cosialls et al. 2021).

Transferrin receptor 1 (TFR1) is an important protein during iron uptake and has been demonstrated to be significantly dysregulated in many cancers including HCC (Y. Shen et al. 2018). Indeed, TFR1 is highly expressed in HCC and induces aggressive phenotypes in human derived HCSCs (Xiao et al. 2020). Knockdown of the protein resulted in decreased sphere formation and protection against erastin, a widely used inducer of ferroptosis, suggesting that TFR1 expression has an important role in stemness maintenance through iron accumulation and ferroptosis sensitivity (Xiao et al. 2020). Similarly, upregulation of TFR1 and decreased iron exporter ferroportin (FPN) have also been found in CCA and related to worse patient prognosis (Raggi et al. 2017b). Sphere forming CCA cells showed iron dependency and ROS production, contributing to modulation of stem associated features (Raggi et al. 2017b).

EMT is related to CSC development and was recently discovered to be another iron-related cellular mechanism promoting HCC predisposition/progression (Mehta and Sharp 2020). Core biomarkers related to EMT, such as E-cadherin, N-cadherin and vimentin were upregulated in the presence of increased iron intracellular storage (Mehta and Sharp 2020). Conversely, studies focusing on iron deficiency in HCC resulted in activation of transcription factor SP1 and hypoxia-inducible factor 1 α (HIF1α) leading to overexpression of sphingosine-1-phosphate transporter spinster homolog 2 (SPNS2), a protein that promotes tumour progression by modulation of EMT and invasion (Li et al. 2022; Wang et al. 2022). Thus, further studies are needed to better clarify the link between iron metabolism and EMT, to provide a more reliable treatment for HCC, in which augmented iron levels and metastasis originated from CSCs are common features.

4 Alternative Metabolic Routes in Harsh Conditions

HCSCs are highly adaptable and have a remarkable ability to reprogramme their metabolism to survive adverse conditions such as starvation, acidosis and hypoxia. Reprogramming energy metabolism promotes tumour cells to use alternative carbon sources such as glutamine, considered to be the second source of nutrients after glucose. Glutamine is the most abundant amino acid in cells, and one of the most important nutrients present in the tumour microenvironment. Cancer cells are known to develop an addiction to glutamine, relying on it as an energy source for survival and proliferation, provided that glutamine catabolism results in several amounts of cellular precursors, including glutamate, aspartate, pyruvate, lactate, alanine, and citrate (Li et al. 2015, 2019). Equally important is glutamine role as a precursor for glutathione synthesis to maintain redox balance. In fact, stem-like subpopulations of human non-small cell lung carcinoma and pancreatic cancer cell lines showed to depend heavily on glutamine for the maintenance of the stem cell phenotype by promoting glutathione synthesis, which leads to redox balance mediated by β-catenin (Liao et al. 2017). Its deprivation can lead to decreased CSC dominance, reflected by diminished stem cell markers SOX-2 and ATP-binding cassette superfamily G member 2 (ABCG2) (Liao et al. 2017). Similarly, the stemness phenotype in HCSCs has been associated with high expression of glutaminase 1 (GLS1), which converts glutamine to glutamate (Li et al. 2019). Glutamine metabolism is essential for the activation of the Wnt/β-catenin pathway. Targeting GLS1 attenuated stemness properties in HCC in vitro and in vivo, indicating that GLS1 might be a therapeutic target to eliminate CSCs. Overall, it seems that targeting glutamine metabolism may be a possible strategy to eliminate CSCs and overcome drug resistance.

Hypoxia is another important hallmark in solid cancers and tumour progression (Keith et al. 2012). As the existing literature suggests, the hypoxic microenvironment is a fundamental stem cell niche that promotes the persistence of CSCs in tumours (Lin and Yun 2010). HIFs are key transcription factors that allow cancer cells to survive in hypoxia, composed of the stable HIF-1β subunit and the oxygen-sensitive subunit HIF-1/2α. In HCC, HIF-1α is closely associated with poor prognosis, where high expression is related to venous and lymph node metastasis (Zheng et al. 2013). Moreover, both HIF-1α and HIF-2α can be induced by Sorafenib in HCC cells, contributing to drug resistance (Cui et al. 2017). Induced cancer stem-like enriched spheres from HCC cells had higher HIF-1α mRNA levels and lower ROS (Muramatsu et al. 2013), while the knockdown of HIF-1α or HIF-2α diminished the expression of CD133 in HCSCs (Lai et al. 2016). Echinomycin, an inhibitor of HIF-1α DNA binding activity also greatly reduced the stemness of HCC cells (Muramatsu et al. 2013).

Specific small ubiquitin-like modifier (SUMO) protease 1 (SENP1), in turn, is essential for the stability and activation of HIF-1α and can deSUMOylate HIF-1α, increasing its stability in hypoxia (Cheng et al. 2007). Reports have also shown that SENP1 increases the stabilisation and transcriptional activity of HIF-1α during hypoxia in HCC, enhancing stemness and CSCs subpopulations (Cui et al. 2017) Additionally, SENP1 is a direct target gene of HIFs, feeding a positive feedback loop that contributes to HCC stemness and tumourigenesis (Cui et al. 2017).

Studying the crosstalk between hypoxia conditioned mesenchymal stem cells (MSC) and HCC development showed that Yes-associated protein (YAP) activation in HCC cells activated AKT/mTOR/SREBP1 pathway, which then enhanced lipogenesis and accelerated the growth of HCC cells (Liu et al. 2019). SOX9, a downstream transcription target of YAP was required for hepatocyte fate plasticity by converting less differentiated hepatocytes to biliary epithelial cell-like cells in hepatocarcinogenesis (Ding and Sancho-Bru 2022). These findings were recently compared with patient liver tumours and results showed that YAP activity is highly correlated with poorly differentiated HCC, which translates to poor survival (Ding and Sancho-Bru 2022).

The gain of mesenchymal properties is critical in the process of cancer invasion and metastasis. Expression of ETS domain-containing protein Elk-3 (ELK3), an oncogenic transcription factor that regulates HIF-1α expression by modulating

HIF-1α stability (Lee et al. 2017), was found to be up-regulated in CD133⁺/CD44⁺ HCC cells and is suggested to be associated with cell migration and invasion. At the same time, silencing the expression of *ELK3* in HCSCs resulted in a down-regulation of their metastatic potential by modulating HIF-1α expression (Lee et al. 2017). The relationship between HIF-1α and CD133 expression to HCC stemness has been further proven by showing that the knockdown of HIF-1α expression led to a reduction of mRNA expression levels of the stem-cell markers NANOG and EpCAM (Lai et al. 2016). Additionally, lipopolysaccharide (LPS) has been associated with stimulation of nuclear factor κB (NF-κB)/HIF-1α signalling, enabling the maintenance of HCC stemness. The fact that patients with chronic liver diseases exhibit elevated levels of LPS in the liver and systemic circulation due to gut dysbiosis, increased intestinal permeability and subsequent bacterial translocation, makes this discovery an important stepping stone in revealing mechanisms governing HCSC persistence.

Altogether, these studies further highlight the metabolic plasticity of HCSCs, showcasing their capabilities in exploiting various signalling pathways to survive and maintain high tumorigenicity.

5 The Metabolic Interplay Between HCSC and Tumour Microenvironment

Similar to normal stem cells, CSCs are believed to be localised in niches, which are part of the TME. The TME functions through a complex signalling network between its two components: stromal cells and the extracellular matrix (ECM). The first includes tumour cells, stromal fibroblasts, endothelial cells and immune cells including macrophages and lymphocytes; while the other component consists of non-cellular components such as collagen, fibronectin and laminin (Baghban et al. 2020).

CSCs can escape recognition by innate and immune immunity and shape the TME into an immunosuppressive, pro-tumorigenic landscape. In turn, functional properties of tumour-infiltrating immune cells impact on the phenotype and differentiation of tumour cells. Although the specific underlying mechanisms remain to be fully understood, extensive evidence has demonstrated the contribution of inflammatory signals secreted by tumour-associated macrophages (TAMs) to CSC maintenance (Allavena et al. 2021). Indeed, IL-6 was found to be highly produced by TAMs co-cultured with CD44⁺ HCSCs and promoted sphere expansion via signal transducer and activator of transcription 3 (STAT3) signalling as well as formation of tumour xenografts in mice (Wan et al. 2014). Importantly, the functions of immune cells in the TME are closely related to abnormal lipid metabolism, much like CSCs. In fact, TAMs were shown to rely on increased lipid uptake through surface scavenger receptor CD36 and FAO for energy generation instead of glycolysis, which promoted inflammation and ROS production, correlated to polarisation and protumoral functions both in vivo and in vitro (Su et al. 2020).

Cancer associated fibroblasts (CAFs) have been well observed in various cancers as an indicator of poor prognosis through supporting tumour growth, ECM remodeling, angiogenesis and tumour-promoting inflammation (Monteran and Erez 2019). Recent evidence shows that they also promote formation of tumour spheres and stemness maintenance (Chan et al. 2019), and possess metabolic adaptability driven by TGF-β, platelet-derived growth factor (PDGF), HIF-1α and ROS-mediated caveolin 1, triggering a switch from OXPHOS to aerobic glycolysis (Kalluri 2016). Moreover, CAFs are associated with creating an EMT favourable microenvironment through TGF-β secretion. Incidentally, inhibitors of TGF-β signalling stunt HCC progression and showed improved clinical outcome (Giannelli et al. 2014). This metabolic reprogramming would be interesting to explore in the context of HCSCs, given their proneness to EMT and metabolic plasticity.

Recently, voltage-gated calcium channel $\alpha 2\delta 1$subunit$^+$ HCC cells were shown to initiate extracellular matrix remodelling by specifically secreting lysyl oxidase (LOX), which leads to the crosslinking of collagen (Zhao et al. 2022). In consequence, a stiff CSC niche was formed, which promoted the formation of tumour-initiating cells with a stiff mechanical trait and was subsequently required for stemness maintenance through the integrin subunit alpha 7 (ITGA7)/focal adhesion kinase (FAK)/ mitogen-activated protein kinase (ERK)1/2 signalling pathway (Zhao et al. 2022). Another study using a 3D tumour model of some HCC cell lines showed that stiffness and hypoxia resulted in doxorubicin and sorafenib resistance, mainly through altered drug metabolism of the cytochrome P450 3A4 (CYP3A4) enzyme (Özkan et al. 2021). Altogether, it is evident that extracellular matrix remodeling has a role in CSC maintenance and contributes to drug resistance.

The flexibility and adaptability of CSCs to the TME is a big hurdle to overcoming conventional therapy resistance and metastasis. Nonetheless, a lot of effort is being given into understanding the underlying mechanisms between the two and its applicability in HCC treatment.

6 Conclusions

Metabolic studies on CSCs are a growing field dedicated to the discovery of potential selective treatments for aggressive cancers such as HCC (Table 1). Sorafenib continues to be considered the standard of care for patients with advanced unresectable HCC (Galle et al. 2018). While it is currently an effective first-line therapy, Sorafenib resistance is becoming a common occurrence (Tang et al. 2020). The role of HCSCs in replenishing the tumour pool and causing chemoresistance is evident, which makes CSC eradication of utmost importance.

Although glycolysis and OXPHOS seem to be the primary source of energy generation in HCSCs, FAO also remains as a possible candidate for ATP source despite the controversies. In any case, drugs that can inhibit these pathways are important in HCSCs eradication. A pre-clinical study on metformin, an anti-diabetic drug that inhibits mitochondrial complex-I, has been shown to specifically target pancreatic CSCs in patient-derived xenograft mice through OXPHOS inhibition,

Table 1 Summary of reported metabolic phenotypes of HCSCs with respective potential therapeutic target

Observation	Model	Potential therapeutic target	Reference
↑Glycolysis	*(Human)* PLC/PRF/5	LDHA, PDK4, miRNA-122	Song et al. (2015)
	(Human) Huh1, Huh7, Hep3B, SK-Hep1, HLE, HLF, 293T	miRNA-192-5p	Gu et al. (2019)
↑OXPHOS	*(Human)* HCCLM3	Glycolysis (using 2-deoxy-D-glucose)	Liu et al. (2020)
	(Human) PLC/PRF/5, Huh7, HEK293T	SIRT1/MRPS5 axis	Wei et al. (2019)
↑Glycolysis; ↑OXPHOS; ↓FAO	*(Human)* Huh7	MYC	Hur et al. (2017)
↑Glycolysis; ↑FAO	*(Human)* HEK293T, Huh7	NANOG	Chen et al. (2016)
↑Lipogenesis	*(Human)* JCRB0405	SCD1, MYCN	Qin et al. (2020)
	(Human) Huh7, Hep3b	PPARα-SCD1	Ma et al. (2019)
	(Human) HepG2, Huh7	SREBP1c, AMPK, NF-kB, ADLH1A1	Bort et al. (2020)
	(Human) 7402, Hep3b	YAP	Liu et al. (2019)
↑Glutaminolysis	*(Human)* HCCLM3, MHCC97H, HepG2, LO2, Huh7, Hep3B, SMMC-7721, HepG2/2.2.15, PLC/PRF/5, MHCC97L, BEL7402, BEL7404 *(Mouse)* Hepa1–6, H22	GLS1	Li et al. (2019)
Hypoxia-induced	*(Human)* Huh7, PLC/PRF/5, MHCC-97L, MHCC-97H, BEL-7402, SMMC-7721, MIHA	SENP1/HIF-1α	Cui et al. (2017)
	(Human) Huh7 and Hep3B, SKHep1, HepG2, PLC/PRF/5	ELK3	Lee et al. (2017)
	(Human) Huh7	LPS, HIF-1/2α, NF-kB/HIF-1α	Lai et al. (2016)

which translates to energy crisis and eventually, apoptosis (Sancho et al. 2015). Subjecting HCSCs to metformin treatment has been shown to result in diminished expression of complex I at the protein level, leading to decreased efficiency of CSC colony formation (Wei et al. 2019). Moreover, combination with Sorafenib completely suppressed the colonies, suggesting that a combined treatment is more effective. Interestingly, metformin also reportedly inhibits proliferation and induces apoptosis of cancer cells as a result of decreased levels of glycolytic enzymes, including HK2 in HCC (DeWaal et al. 2018) and pyruvate kinase M2 (PKM2) in breast cancer (Silvestri et al. 2015).

In a pre-clinical study on HepG2 xenograft mice, an anti-malarial drug, Atovaquone, targeting OXPHOS by inhibition of mitochondrial complex III has been found to exhibit anti-tumour effects in vivo by inducing DNA damage with no obvious evidence of toxicity (Gao et al. 2018). By lowering OCR, tumour hypoxia is alleviated resulting in delayed tumour growth. The specific action of complex III has never been studied in HCSCs but the fact that it can inhibit OXPHOS sparks some interest. Besides glycolysis, OXPHOS and FAO, glutamine is also being used by CSCs as an alternative carbon source. Still, in reality, targeting glutamine metabolism is a challenging task. Systemic approaches to cause glutamine deprivation are difficult since glutamine can be synthesised de novo by non-cancerous tissues (Curthoys and Watford 1995). A potential target for glutamine metabolism is GLS1, whose pharmacological or genetic knockdown leads to suppression of HCSC cell markers in vitro (Li et al. 2019).

Lipid biosynthesis being dysregulated in HCSCs make SREBPs important targets to inhibit de novo lipogenesis. A metabolic analysis showed that HCSCs display decreased AMP-activated protein kinase (AMPK) protein levels, while overexpression of AMPK through plasmid transfection effectively inhibited SREBP1c (Bort et al. 2020). Moreover, HCSCs have high levels of unsaturated lipids. Inhibiting the activities of SCD1, NF-κB, and ALDH1A1 in HCSCs might provide useful insights, since the inhibition of these targets in enriched human colon cancer cell line HCT-116-enriched CSCs lead to reduced abundance of unsaturated lipids and hindered spheroid formation (Sun and Yang 2019). Additionally, although CSCs display increased antioxidant capacity, drugs aiming at selectively targeting HCSCs could be designed by exploiting HCSCs iron metabolic rewiring-associated vulnerabilities, namely through ferroptosis-inducing approaches. Interestingly, a recent study put forward four iron metabolism related genes, ribonucleotide reductase regulatory subunit M2 (*RRM2*), formimidoyltransferase-cyclodeaminase (*FTCD*), ATPase H+ transporting V1 subunit C1 (*ATP6V1C1*) and *CYP2C9* as prognostic predictors for HCC patients (H. Shen et al. 2021). CYP2C9 and STAT3 activation were shown to sensitise HCSCs to anti-cancer treatments specially in advanced stages (Myung et al. 2012), further highlighting the importance of this potential strategy.

Overall, the elucidation of the metabolic phenotype of HCSCs and its implications in HCC development and progression are key for the discovery of novel effective therapeutic approaches for HCC treatment and relapse prevention. Fortunately, the potential of CSC metabolism is being increasingly recognised.

Acknowledgments The authors thank all members of the Cellular Function and Therapeutic Targeting lab at iMed.ULisboa for insightful discussions.

Authors' Contributions Conceptualization, MBA; writing – original draft preparation, JCD; writing – review and editing, CMPR and MBA; artwork – JCD; funding acquisition, CMPR. All authors have read and agreed to the published version of the manuscript.

Conflicts of Interest The authors declare no conflict of interest.

Funding Main funding is provided by Fundação para a Ciência e a Tecnologia (PTDC/MED-FAR/3492/2021) and La Caixa Foundation (LCF/PR/HR21/52410028) to CMPR.

References

Al-Hajj M, Wicha MS, Benito-Hernandez A, Morrison SJ, Clarke MF (2003) Prospective identification of tumorigenic breast cancer cells. Proc Natl Acad Sci 100(7):3983–3988. https://doi.org/10.1073/pnas.0530291100

Allavena P, Digifico E, Belgiovine C (2021) Macrophages and cancer stem cells: a malevolent alliance. Mol Med 27(1):121. https://doi.org/10.1186/s10020-021-00383-3

Baghban R, Roshangar L, Jahanban-Esfahlan R, Seidi K, Ebrahimi-Kalan A, Jaymand M, Kolahian S, Javaheri T, Zare P (2020) Tumor microenvironment complexity and therapeutic implications at a glance. Cell Commun Signal 18(1):59. https://doi.org/10.1186/s12964-020-0530-4

Batlle E, Clevers H (2017) Cancer stem cells revisited. Nat Med 23(10):1124–1134. https://doi.org/10.1038/nm.4409

Bian X, Liu R, Meng Y, Xing D, Xu D, Lu Z (2021) Lipid metabolism and cancer. J Exp Med 218(1). https://doi.org/10.1084/jem.20201606

Bonnet D, Dick JE (1997) Human acute myeloid leukemia is organized as a hierarchy that originates from a primitive hematopoietic cell. Nat Med 3(7):730–737. https://doi.org/10.1038/nm0797-730

Bort A, Sánchez BG, de Miguel I, Mateos-Gómez PA, Diaz-Laviada I (2020) Dysregulated lipid metabolism in hepatocellular carcinoma cancer stem cells. Mol Biol Rep 47(4):2635–2647. https://doi.org/10.1007/s11033-020-05352-3

Cabral LKD, Tiribelli C, Sukowati CHC (2020) Sorafenib resistance in hepatocellular carcinoma: the relevance of genetic heterogeneity. Cancers 12(6):1576. https://doi.org/10.3390/cancers12061576

Chaffer CL, Brueckmann I, Scheel C, Kaestli AJ, Wiggins PA, Rodrigues LO, Brooks M, Reinhardt F, Su Y, Polyak K, Arendt LM, Kuperwasser C, Bierie B, Weinberg RA (2011) Normal and neoplastic nonstem cells can spontaneously convert to a stem-like state. Proc Natl Acad Sci 108(19):7950–7955. https://doi.org/10.1073/pnas.1102454108

Chan T-S, Shaked Y, Tsai KK (2019) Targeting the interplay between cancer fibroblasts, mesenchymal stem cells, and cancer stem cells in desmoplastic cancers. Front Oncol 9. https://doi.org/10.3389/fonc.2019.00688

Chen C-L, Uthaya Kumar DB, Punj V, Xu J, Sher L, Tahara SM, Hess S, Machida K (2016) NANOG metabolically reprograms tumor-initiating stem-like cells through tumorigenic changes in oxidative phosphorylation and fatty acid metabolism. Cell Metab 23(1):206–219. https://doi.org/10.1016/j.cmet.2015.12.004

Cheng J, Kang X, Zhang S, Yeh ETH (2007) SUMO-specific protease 1 is essential for stabilization of HIF1α during hypoxia. Cell 131(3):584–595. https://doi.org/10.1016/j.cell.2007.08.045

Chong L-W, Tsai C-L, Yang K-C, Liao C-C, Hsu Y-C (2020) Targeting protein palmitoylation decreases palmitate-induced sphere formation of human liver cancer cells. Mol Med Rep 22(2):939–947. https://doi.org/10.3892/mmr.2020.11172

Ciavardelli D, Rossi C, Barcaroli D, Volpe S, Consalvo A, Zucchelli M, de Cola A, Scavo E, Carollo R, D'Agostino D, Forlì F, D'Aguanno S, Todaro M, Stassi G, di Ilio C, de Laurenzi V, Urbani A (2014) Breast cancer stem cells rely on fermentative glycolysis and are sensitive to 2-deoxyglucose treatment. Cell Death Dis 5(7):e1336–e1336. https://doi.org/10.1038/cddis.2014.285

Collins AT, Berry PA, Hyde C, Stower MJ, Maitland NJ (2005) Prospective identification of tumorigenic prostate cancer stem cells. Cancer Res 65(23):10946–10951. https://doi.org/10.1158/0008-5472.CAN-05-2018

Cosialls E, el Hage R, dos Santos L, Gong C, Mehrpour M, Hamaï A (2021) Ferroptosis: cancer stem cells rely on iron until "to die for" it. Cell 10(11):2981. https://doi.org/10.3390/cells10112981

Cui C-P, Wong CC-L, Kai AK-L, Ho DW-H, Lau EY-T, Tsui Y-M, Chan L-K, Cheung T-T, Chok KS-H, Chan ACY, Lo RC-L, Lee JM-F, Lee TK-W, Ng IOL (2017) SENP1 promotes hypoxia-induced cancer stemness by HIF-1α deSUMOylation and SENP1/HIF-1α positive feedback loop. Gut 66(12):2149–2159. https://doi.org/10.1136/gutjnl-2016-313264

Curthoys NP, Watford M (1995) Regulation of glutaminase activity and glutamine metabolism. Annu Rev Nutr 15(1):133–159. https://doi.org/10.1146/annurev.nu.15.070195.001025

Dang CV (2012) MYC on the path to cancer. Cell 149(1):22–35. https://doi.org/10.1016/j.cell.2012.03.003

Dauch D, Rudalska R, Cossa G, Nault J-C, Kang T-W, Wuestefeld T, Hohmeyer A, Imbeaud S, Yevsa T, Hoenicke L, Pantsar T, Bozko P, Malek NP, Longerich T, Laufer S, Poso A, Zucman-Rossi J, Eilers M, Zender L (2016) A MYC–aurora kinase a protein complex represents an actionable drug target in p53-altered liver cancer. Nat Med 22(7):744–753. https://doi.org/10.1038/nm.4107

DeWaal D, Nogueira V, Terry AR, Patra KC, Jeon S-M, Guzman G, Au J, Long CP, Antoniewicz MR, Hay N (2018) Hexokinase-2 depletion inhibits glycolysis and induces oxidative phosphorylation in hepatocellular carcinoma and sensitizes to metformin. Nat Commun 9(1):446. https://doi.org/10.1038/s41467-017-02733-4

Ding W-X, Sancho-Bru P (2022) SOX9 acts downstream of YAP to decide liver cell fate and tumor types. J Hepatol 76(3):503–505. https://doi.org/10.1016/j.jhep.2021.12.008

Esau C, Davis S, Murray SF, Yu XX, Pandey SK, Pear M, Watts L, Booten SL, Graham M, McKay R, Subramaniam A, Propp S, Lollo BA, Freier S, Bennett CF, Bhanot S, Monia BP (2006) miR-122 regulation of lipid metabolism revealed by in vivo antisense targeting. Cell Metab 3(2):87–98. https://doi.org/10.1016/j.cmet.2006.01.005

Estes C, Razavi H, Loomba R, Younossi Z, Sanyal AJ (2018) Modeling the epidemic of nonalcoholic fatty liver disease demonstrates an exponential increase in burden of disease. Hepatology 67(1):123–133. https://doi.org/10.1002/hep.29466

Fang D, Nguyen TK, Leishear K, Finko R, Kulp AN, Hotz S, Van Belle PA, Xu X, Elder DE, Herlyn M (2005, Oct 15) A tumorigenic subpopulation with stem cell properties in melanomas. Cancer Res 65(20):9328–9337. https://doi.org/10.1158/0008-5472.CAN-05-1343

Farese RV, Walther TC (2009) Lipid droplets finally get a little R-E-S-P-E-C-T. Cell 139(5):855–860. https://doi.org/10.1016/j.cell.2009.11.005

Fargion S, Valenti L, Fracanzani AL (2014) Role of iron in hepatocellular carcinoma. Clin Liver Dis 3(5):108–110. https://doi.org/10.1002/cld.350

Foerster F, Gairing SJ, Müller L, Galle PR (2022) NAFLD-driven HCC: safety and efficacy of current and emerging treatment options. J Hepatol 76(2):446–457. https://doi.org/10.1016/j.jhep.2021.09.007

Folmes CDL, Dzeja PP, Nelson TJ, Terzic A (2012) Metabolic plasticity in stem cell homeostasis and differentiation. Cell Stem Cell 11(5):596–606. https://doi.org/10.1016/j.stem.2012.10.002

Frohlich J, Mazza T, Sobolewski C, Foti M, Vinciguerra M (2021) GDF11 rapidly increases lipid accumulation in liver cancer cells through ALK5-dependent signaling. Biochim Biophys Acta Mol Cell Biol Lipids 1866(6):158920. https://doi.org/10.1016/j.bbalip.2021.158920

Galle PR, Forner A, Llovet JM, Mazzaferro V, Piscaglia F, Raoul J-L, Schirmacher P, Vilgrain V (2018) EASL clinical practice guidelines: management of hepatocellular carcinoma. J Hepatol 69(1):182–236. https://doi.org/10.1016/j.jhep.2018.03.019

Gao X, Liu X, Shan W, Liu Q, Wang C, Zheng J, Yao H, Tang R, Zheng J (2018) Anti-malarial atovaquone exhibits anti-tumor effects by inducing DNA damage in hepatocellular carcinoma. Am J Cancer Res 8(9):1697–1711

Ge T, Yang J, Zhou S, Wang Y, Li Y, Tong X (2020) The role of the pentose phosphate pathway in diabetes and cancer. Front Endocrinol 11. https://doi.org/10.3389/fendo.2020.00365

Giannelli G, Villa E, Lahn M (2014) Transforming growth factor-β as a therapeutic target in hepatocellular carcinoma. Cancer Res 74(7):1890–1894. https://doi.org/10.1158/0008-5472.CAN-14-0243

Ginestier C, Hur MH, Charafe-Jauffret E, Monville F, Dutcher J, Brown M, Jacquemier J, Viens P, Kleer CG, Liu S, Schott A, Hayes D, Birnbaum D, Wicha MS, Dontu G (2007) ALDH1 is a marker of normal and malignant human mammary stem cells and a predictor of poor clinical outcome. Cell Stem Cell 1(5):555–567. https://doi.org/10.1016/j.stem.2007.08.014

Gu Y, Wei X, Sun Y, Gao H, Zheng X, Wong LL, Jin L, Liu N, Hernandez B, Peplowska K, Zhao X, Zhan Q-M, Feng X-H, Tang Z-Y, Ji J (2019) miR-192-5p silencing by genetic aberrations is a key event in hepatocellular carcinomas with cancer stem cell features. Cancer Res 79(5):941–953. https://doi.org/10.1158/0008-5472.CAN-18-1675

Gu Y, Ji F, Liu N, Zhao Y, Wei X, Hu S, Jia W, Wang XW, Budhu A, Ji J, Zhao B, Roessler S, Zheng X, Ji J (2020) Loss of miR-192-5p initiates a hyperglycolysis and stemness positive feedback in hepatocellular carcinoma. J Exp Clin Cancer Res 39(1):268. https://doi.org/10.1186/s13046-020-01785-7

Gundamaraju R, Lu W, Manikam R (2021) Revisiting mitochondria scored cancer progression and metastasis. Cancers 13(3):432. https://doi.org/10.3390/cancers13030432

He G, Dhar D, Nakagawa H, Font-Burgada J, Ogata H, Jiang Y, Shalapour S, Seki E, Yost SE, Jepsen K, Frazer KA, Harismendy O, Hatziapostolou M, Iliopoulos D, Suetsugu A, Hoffman RM, Tateishi R, Koike K, Karin M (2013) Identification of liver cancer progenitors whose malignant progression depends on autocrine IL-6 signaling. Cell 155(2):384–396. https://doi.org/10.1016/j.cell.2013.09.031

Hermann PC, Huber SL, Herrler T, Aicher A, Ellwart JW, Guba M, Bruns CJ, Heeschen C (2007) Distinct populations of cancer stem cells determine tumor growth and metastatic activity in human pancreatic cancer. Cell Stem Cell 1(3):313–323. https://doi.org/10.1016/j.stem.2007.06.002

Holczbauer Á, Factor VM, Andersen JB, Marquardt JU, Kleiner DE, Raggi C, Kitade M, Seo D, Akita H, Durkin ME, Thorgeirsson SS (2013) Modeling pathogenesis of primary liver cancer in lineage-specific mouse cell types. Gastroenterology 145(1):221–231. https://doi.org/10.1053/j.gastro.2013.03.013

Hu W, Zhou C, Jing Q, Li Y, Yang J, Yang C, Wang L, Hu J, Li H, Wang H, Yuan C, Zhou Y, Ren X, Tong X, Du J, Wang Y (2021) FTH promotes the proliferation and renders the HCC cells specifically resist to ferroptosis by maintaining iron homeostasis. Cancer Cell Int 21(1):709. https://doi.org/10.1186/s12935-021-02420-x

Huch M, Dorrell C, Boj SF, van Es JH, Li VSW, van de Wetering M, Sato T, Hamer K, Sasaki N, Finegold MJ, Haft A, Vries RG, Grompe M, Clevers H (2013) In vitro expansion of single Lgr5+ liver stem cells induced by Wnt-driven regeneration. Nature 494(7436):247–250. https://doi.org/10.1038/nature11826

Hur W, Ryu JY, Kim HU, Hong SW, Lee EB, Lee SY, Yoon SK (2017) Systems approach to characterize the metabolism of liver cancer stem cells expressing CD133. Sci Rep 7(1):45557. https://doi.org/10.1038/srep45557

Jung M, Mertens C, Tomat E, Brüne B (2019) Iron as a central player and promising target in cancer progression. Int J Mol Sci 20(2):273. https://doi.org/10.3390/ijms20020273

Kahraman DC, Kahraman T, Cetin Atalay R (2019) Targeting PI3K/Akt/mTOR pathway identifies differential expression and functional role of IL-8 in liver cancer stem cell enrichment. Mol Cancer Ther. https://doi.org/10.1158/1535-7163.MCT-19-0004

Kai-Yuan Chen, Xiaojing Liu, Pengcheng Bu, Chieh-Sheng Lin, Rakhilin N, Locasale JW, Xiling Shen (2014) A metabolic signature of colon cancer initiating cells. 2014 36th Annual International Conference of the IEEE Engineering in Medicine and Biology Society, 4759–4762. https://doi.org/10.1109/EMBC.2014.6944688

Kalluri R (2016) The biology and function of fibroblasts in cancer. Nat Rev Cancer 16(9):582–598. https://doi.org/10.1038/nrc.2016.73

Kaur J, Bhattacharyya S (2021) Cancer stem cells: metabolic characterization for targeted cancer therapy. Front Oncol 11. https://doi.org/10.3389/fonc.2021.756888

Keith B, Johnson RS, Simon MC (2012) HIF1α and HIF2α: sibling rivalry in hypoxic tumour growth and progression. Nat Rev Cancer 12(1):9–22. https://doi.org/10.1038/nrc3183

Kim CFB, Jackson EL, Woolfenden AE, Lawrence S, Babar I, Vogel S, Crowley D, Bronson RT, Jacks T (2005) Identification of bronchioalveolar stem cells in normal lung and lung cancer. Cell 121(6):823–835. https://doi.org/10.1016/j.cell.2005.03.032

Ko C-H, Cheng C-F, Lai C-P, Tzu T-H, Chiu C-W, Lin M-W, Wu S-Y, Sun C-Y, Tseng H-W, Wang C-C, Kuo Z-K, Wang L-M, Chen S-F (2013) Differential proteomic analysis of cancer stem cell properties in hepatocellular carcinomas by isobaric tag labeling and mass spectrometry. J Proteome Res 12(8):3573–3585. https://doi.org/10.1021/pr4004294

Kuntz EM, Baquero P, Michie AM, Dunn K, Tardito S, Holyoake TL, Helgason GV, Gottlieb E (2017) Targeting mitochondrial oxidative phosphorylation eradicates therapy-resistant chronic myeloid leukemia stem cells. Nat Med 23(10):1234–1240. https://doi.org/10.1038/nm.4399

Lai F-B, Liu W-T, Jing Y-Y, Yu G-F, Han Z-P, Yang X, Zeng J-X, Zhang H-J, Shi R-Y, Li X-Y, Pan X-R, Li R, Zhao Q-D, Wu M-C, Zhang P, Liu J-F, Wei L-X (2016) Lipopolysaccharide supports maintaining the stemness of CD133+ hepatoma cells through activation of the NF-κB/HIF-1α pathway. Cancer Lett 378(2):131–141. https://doi.org/10.1016/j.canlet.2016.05.014

Lamichane S, Dahal Lamichane B, Kwon S-M (2018) Pivotal roles of peroxisome proliferator-activated receptors (PPARs) and their signal cascade for cellular and whole-body energy homeostasis. Int J Mol Sci 19(4):949. https://doi.org/10.3390/ijms19040949

Lee JH, Hur W, Hong SW, Kim J-H, Kim SM, Lee EB, Yoon SK (2017) ELK3 promotes the migration and invasion of liver cancer stem cells by targeting HIF-1α. Oncol Rep 37(2): 813–822. https://doi.org/10.3892/or.2016.5293

Li C, Zhang G, Zhao L, Ma Z, Chen H (2015) Metabolic reprogramming in cancer cells: glycolysis, glutaminolysis, and Bcl-2 proteins as novel therapeutic targets for cancer. World J Surg Oncol 14(1):15. https://doi.org/10.1186/s12957-016-0769-9

Li B, Cao Y, Meng G, Qian L, Xu T, Yan C, Luo O, Wang S, Wei J, Ding Y, Yu D (2019) Targeting glutaminase 1 attenuates stemness properties in hepatocellular carcinoma by increasing reactive oxygen species and suppressing Wnt/beta-catenin pathway. EBioMedicine 39:239–254. https://doi.org/10.1016/j.ebiom.2018.11.063

Li H, Feng Z, He M-L (2020) Lipid metabolism alteration contributes to and maintains the properties of cancer stem cells. Theranostics 10(16):7053–7069. https://doi.org/10.7150/thno.41388

Li M, Tang Y, Wang D, Zhai X, Shen H, Zhong C, Yao M, Jin A, Zhou Z, Zhou S, Fan J, Ling C, Ling C (2022) Sphingosine-1-phosphate transporter spinster homolog 2 is essential for iron-regulated metastasis of hepatocellular carcinoma. Mol Ther 30(2):703–713. https://doi.org/10.1016/j.ymthe.2021.09.012

Liao J, Qian F, Tchabo N, Mhawech-Fauceglia P, Beck A, Qian Z, Wang X, Huss WJ, Lele SB, Morrison CD, Odunsi K (2014) Ovarian cancer spheroid cells with stem cell-like properties contribute to tumor generation, metastasis and chemotherapy resistance through hypoxia-resistant metabolism. PLoS One 9(1):e84941. https://doi.org/10.1371/journal.pone.0084941

Liao J, Liu P-P, Hou G, Shao J, Yang J, Liu K, Lu W, Wen S, Hu Y, Huang P (2017) Regulation of stem-like cancer cells by glutamine through β-catenin pathway mediated by redox signaling. Mol Cancer 16(1):51. https://doi.org/10.1186/s12943-017-0623-x

Lin Q, Yun Z (2010) Impact of the hypoxic tumor microenvironment on the regulation of cancer stem cell characteristics. Cancer Biol Ther 9(12):949–956. https://doi.org/10.4161/cbt.9.12.12347

Liu M, Chen L, Ma N-F, Chow RKK, Li Y, Song Y, Chan THM, Fang S, Yang X, Xi S, Jiang L, Li Y, Zeng T-T, Li Y, Yuan Y-F, Guan X-Y (2016) CHD1L promotes lineage reversion of hepatocellular carcinoma through opening chromatin for key developmental transcription factors. Hepatology 63(5):1544–1559. https://doi.org/10.1002/hep.28437

Liu Y, Ren H, Zhou Y, Shang L, Zhang Y, Yang F, Shi X (2019) The hypoxia conditioned mesenchymal stem cells promote hepatocellular carcinoma progression through YAP mediated lipogenesis reprogramming. J Exp Clin Cancer Res 38(1):228. https://doi.org/10.1186/s13046-019-1219-7

Liu G, Luo Q, Li H, Liu Q, Ju Y, Song G (2020) Increased oxidative phosphorylation is required for stemness maintenance in liver cancer stem cells from hepatocellular carcinoma cell line HCCLM3 cells. Int J Mol Sci 21(15):5276. https://doi.org/10.3390/ijms21155276

Ma S, Chan K, Hu L, Lee TK, Wo JY, Ng IO, Zheng B, Guan X (2007) Identification and characterization of tumorigenic liver cancer stem/progenitor cells. Gastroenterology 132(7): 2542–2556. https://doi.org/10.1053/j.gastro.2007.04.025

Ma S, Chan KW, Lee TK-W, Tang KH, Wo JY-H, Zheng B-J, Guan X-Y (2008) Aldehyde dehydrogenase discriminates the CD133 liver cancer stem cell populations. Mol Cancer Res 6(7):1146–1153. https://doi.org/10.1158/1541-7786.MCR-08-0035

Ma X-L, Sun Y-F, Wang B-L, Shen M-N, Zhou Y, Chen J-W, Hu B, Gong Z-J, Zhang X, Cao Y, Pan B, Zhou J, Fan J, Guo W, Yang X-R (2019) Sphere-forming culture enriches liver cancer stem cells and reveals Stearoyl-CoA desaturase 1 as a potential therapeutic target. BMC Cancer 19(1):760. https://doi.org/10.1186/s12885-019-5963-z

Mani SA, Guo W, Liao M-J, Eaton EN, Ayyanan A, Zhou AY, Brooks M, Reinhard F, Zhang CC, Shipitsin M, Campbell LL, Polyak K, Brisken C, Yang J, Weinberg RA (2008) The epithelial-mesenchymal transition generates cells with properties of stem cells. Cell 133(4):704–715. https://doi.org/10.1016/j.cell.2008.03.027

Meacham CE, Morrison SJ (2013) Tumour heterogeneity and cancer cell plasticity. Nature 501(7467):328–337. https://doi.org/10.1038/nature12624

Mehta KJ, Sharp PA (2020) Iron elevates mesenchymal and metastatic biomarkers in HepG2 cells. Sci Rep 10(1):21926. https://doi.org/10.1038/s41598-020-78348-5

Monteran L, Erez N (2019) The dark side of fibroblasts: cancer-associated fibroblasts as mediators of immunosuppression in the tumor microenvironment. Front Immunol 10. https://doi.org/10.3389/fimmu.2019.01835

Moustafa M, Dähling K-K, Günther A, Riebandt L, Smit DJ, Riecken K, Schröder C, Zhuang R, Krech T, Kriegs M, Fehse B, Izbicki JR, Fischer L, Nashan B, Li J, Jücker M (2022) Combined targeting of AKT and mTOR inhibits tumor formation of EpCAM+ and CD90+ human hepatocellular carcinoma cells in an orthotopic mouse model. Cancers 14(8):1882. https://doi.org/10.3390/cancers14081882

Mu X, Español-Suñer R, Mederacke I, Affò S, Manco R, Sempoux C, Lemaigre FP, Adili A, Yuan D, Weber A, Unger K, Heikenwälder M, Leclercq IA, Schwabe RF (2015) Hepatocellular carcinoma originates from hepatocytes and not from the progenitor/biliary compartment. J Clin Investig 125(10):3891–3903. https://doi.org/10.1172/JCI77995

Mukherjee A, Kenny HA, Lengyel E (2017) Unsaturated fatty acids maintain cancer cell stemness. Cell Stem Cell 20(3):291–292. https://doi.org/10.1016/j.stem.2017.02.008

Muramatsu S, Tanaka S, Mogushi K, Adikrisna R, Aihara A, Ban D, Ochiai T, Irie T, Kudo A, Nakamura N, Nakayama K, Tanaka H, Yamaoka S, Arii S (2013) Visualization of stem cell features in human hepatocellular carcinoma reveals *in vivo* significance of tumor-host interaction and clinical course. Hepatology 58(1):218–228. https://doi.org/10.1002/hep.26345

Myung SJ, Yoon J-H, Yu SJ (2012) STAT3 & Cytochrome P450 2C9: a novel signaling pathway in liver cancer stem cells. Biomed Pharmacother 66(8):612–616. https://doi.org/10.1016/j.biopha.2012.08.011

O'Brien CA, Pollett A, Gallinger S, Dick JE (2007) A human colon cancer cell capable of initiating tumour growth in immunodeficient mice. Nature 445(7123):106–110. https://doi.org/10.1038/nature05372

Özkan A, Stolley DL, Cressman ENK, McMillin M, DeMorrow S, Yankeelov TE, Rylander MN (2021) Tumor microenvironment alters chemoresistance of hepatocellular carcinoma through CYP3A4 metabolic activity. Front Oncol 11. https://doi.org/10.3389/fonc.2021.662135

Pascual G, Avgustinova A, Mejetta S, Martín M, Castellanos A, Attolini CS-O, Berenguer A, Prats N, Toll A, Hueto JA, Bescós C, di Croce L, Benitah SA (2017) Targeting metastasis-initiating cells through the fatty acid receptor CD36. Nature 541(7635):41–45. https://doi.org/10.1038/nature20791

Peiris-Pagès M, Bonuccelli G, Sotgia F, Lisanti MP (2018) Mitochondrial fission as a driver of stemness in tumor cells: mDIVI1 inhibits mitochondrial function, cell migration and cancer stem cell (CSC) signalling. Oncotarget 9(17):13254–13275. https://doi.org/10.18632/oncotarget.24285

Pfister D, Núñez NG, Pinyol R, Govaere O, Pinter M, Szydlowska M, Gupta R, Qiu M, Deczkowska A, Weiner A, Müller F, Sinha A, Friebel E, Engleitner T, Lenggenhager D, Moncsek A, Heide D, Stirm K, Kosla J et al (2021) NASH limits anti-tumour surveillance in immunotherapy-treated HCC. Nature 592(7854):450–456. https://doi.org/10.1038/s41586-021-03362-0

Qin X-Y, Su T, Yu W, Kojima S (2020) Lipid desaturation-associated endoplasmic reticulum stress regulates MYCN gene expression in hepatocellular carcinoma cells. Cell Death Dis 11(1):66. https://doi.org/10.1038/s41419-020-2257-y

Qiu Q, Hernandez JC, Dean AM, Rao PH, Darlington GJ (2011) CD24-positive cells from normal adult mouse liver are hepatocyte progenitor cells. Stem Cells Dev 20(12):2177–2188. https://doi.org/10.1089/scd.2010.0352

Raggi C, Factor VM, Seo D, Holczbauer A, Gillen MC, Marquardt JU, Andersen JB, Durkin M, Thorgeirsson SS (2014) Epigenetic reprogramming modulates malignant properties of human liver cancer. Hepatology 59(6):2251–2262. https://doi.org/10.1002/hep.27026

Raggi C, Correnti M, Sica A, Andersen JB, Cardinale V, Alvaro D, Chiorino G, Forti E, Glaser S, Alpini G, Destro A, Sozio F, di Tommaso L, Roncalli M, Banales JM, Coulouarn C, Bujanda L, Torzilli G, Invernizzi P (2017a) Cholangiocarcinoma stem-like subset shapes tumor-initiating niche by educating associated macrophages. J Hepatol 66(1):102–115. https://doi.org/10.1016/j.jhep.2016.08.012

Raggi C, Gammella E, Correnti M, Buratti P, Forti E, Andersen JB, Alpini G, Glaser S, Alvaro D, Invernizzi P, Cairo G, Recalcati S (2017b) Dysregulation of iron metabolism in cholangiocarcinoma stem-like cells. Sci Rep 7(1):17667. https://doi.org/10.1038/s41598-017-17804-1

Raggi C, Taddei ML, Sacco E, Navari N, Correnti M, Piombanti B, Pastore M, Campani C, Pranzini E, Iorio J, Lori G, Lottini T, Peano C, Cibella J, Lewinska M, Andersen JB, di Tommaso L, Viganò L, di Maira G et al (2021) Mitochondrial oxidative metabolism contributes to a cancer stem cell phenotype in cholangiocarcinoma. J Hepatol 74(6):1373–1385. https://doi.org/10.1016/j.jhep.2020.12.031

Ramani K, Robinson AE, Berlind J, Fan W, Abeynayake A, Binek A, Barbier-Torres L, Noureddin M, Nissen NN, Yildirim Z, Erbay E, Mato JM, van Eyk JE, Lu SC (2022) S-adenosylmethionine inhibits la ribonucleoprotein domain family member 1 in murine liver and human liver cancer cells. Hepatology 75(2):280–296. https://doi.org/10.1002/hep.32130

Reya T, Morrison SJ, Clarke MF, Weissman IL (2001) Stem cells, cancer, and cancer stem cells. Nature 414(6859):105–111. https://doi.org/10.1038/35102167

Rountree CB, Barsky L, Ge S, Zhu J, Senadheera S, Crooks GM (2007) A CD133-expressing murine liver oval cell population with bilineage potential. Stem Cells 25(10):2419–2429. https://doi.org/10.1634/stemcells.2007-0176

Sancho P, Burgos-Ramos E, Tavera A, Bou Kheir T, Jagust P, Schoenhals M, Barneda D, Sellers K, Campos-Olivas R, Graña O, Viera CR, Yuneva M, Sainz B, Heeschen C (2015) MYC/PGC-1α balance determines the metabolic phenotype and plasticity of pancreatic cancer stem cells. Cell Metab 22(4):590–605. https://doi.org/10.1016/j.cmet.2015.08.015

Sancho P, Barneda D, Heeschen C (2016) Hallmarks of cancer stem cell metabolism. Br J Cancer 114(12):1305–1312. https://doi.org/10.1038/bjc.2016.152

Sato M, Kawana K, Adachi K, Fujimoto A, Yoshida M, Nakamura H, Nishida H, Inoue T, Taguchi A, Takahashi J, Eguchi S, Yamashita A, Tomio K, Wada-Hiraike O, Oda K, Nagamatsu T, Osuga Y, Fujii T (2016) Spheroid cancer stem cells display reprogrammed metabolism and obtain energy by actively running the tricarboxylic acid (TCA) cycle. Oncotarget 7(22):33297–33305. https://doi.org/10.18632/oncotarget.8947

Schmelzer E, Zhang L, Bruce A, Wauthier E, Ludlow J, Yao H, Moss N, Melhem A, McClelland R, Turner W, Kulik M, Sherwood S, Tallheden T, Cheng N, Furth ME, Reid LM (2007) Human hepatic stem cells from fetal and postnatal donors. J Exp Med 204(8):1973–1987. https://doi.org/10.1084/jem.20061603

Shackleton M, Quintana E, Fearon ER, Morrison SJ (2009) Heterogeneity in cancer: cancer stem cells versus clonal evolution. Cell 138(5):822–829. https://doi.org/10.1016/j.cell.2009.08.017

Shekhani MT, Jayanthy A-S, Maddodi N, Setaluri V (2013) Cancer stem cells and tumor transdifferentiation: implications for novel therapeutic strategies. Am J Stem Cells 2(1):52–61

Shen Y, Li X, Dong D, Zhang B, Xue Y, Shang P (2018) Transferrin receptor 1 in cancer: a new sight for cancer therapy. Am J Cancer Res 8(6):916–931

Shen H, Wu H, Sun F, Qi J, Zhu Q (2021) A novel four-gene of iron metabolism-related and methylated for prognosis prediction of hepatocellular carcinoma. Bioengineered 12(1):240–251. https://doi.org/10.1080/21655979.2020.1866303

Silvestri A, Palumbo F, Rasi I, Posca D, Pavlidou T, Paoluzi S, Castagnoli L, Cesareni G (2015) Metformin induces apoptosis and downregulates pyruvate kinase M2 in breast cancer cells only when grown in nutrient-poor conditions. PLoS One 10(8):e0136250. https://doi.org/10.1371/journal.pone.0136250

Singh SK, Clarke ID, Terasaki M, Bonn VE, Hawkins C, Squire J, Dirks PB (2003) Identification of a cancer stem cell in human brain tumors. Cancer Res 63(18):5821–5828

Song K, Kwon H, Han C, Zhang J, Dash S, Lim K, Wu T (2015) Active glycolytic metabolism in CD133(+) hepatocellular cancer stem cells: regulation by MIR-122. Oncotarget 6(38): 40822–40835. https://doi.org/10.18632/oncotarget.5812

Su P, Wang Q, Bi E, Ma X, Liu L, Yang M, Qian J, Yi Q (2020) Enhanced lipid accumulation and metabolism are required for the differentiation and activation of tumor-associated macrophages. Cancer Res 80(7):1438–1450. https://doi.org/10.1158/0008-5472.CAN-19-2994

Sun M, Yang Z (2019) Metabolomic studies of live single cancer stem cells using mass spectrometry. Anal Chem 91(3):2384–2391. https://doi.org/10.1021/acs.analchem.8b05166

Sung H, Ferlay J, Siegel RL, Laversanne M, Soerjomataram I, Jemal A, Bray F (2021) Global cancer statistics 2020: GLOBOCAN estimates of incidence and mortality worldwide for 36 cancers in 185 countries. CA Cancer J Clin 71(3):209–249. https://doi.org/10.3322/caac.21660

Szotek PP, Pieretti-Vanmarcke R, Masiakos PT, Dinulescu DM, Connolly D, Foster R, Dombkowski D, Preffer F, MacLaughlin DT, Donahoe PK (2006) Ovarian cancer side population defines cells with stem cell-like characteristics and Mullerian Inhibiting Substance responsiveness. Proc Natl Acad Sci 103(30):11154–11159. https://doi.org/10.1073/pnas.0603672103

Tang Y, Kitisin K, Jogunoori W, Li C, Deng C-X, Mueller SC, Ressom HW, Rashid A, He AR, Mendelson JS, Jessup JM, Shetty K, Zasloff M, Mishra B, Reddy EP, Johnson L, Mishra L (2008) Progenitor/stem cells give rise to liver cancer due to aberrant TGF-β and IL-6 signaling. Proc Natl Acad Sci 105(7):2445–2450. https://doi.org/10.1073/pnas.0705395105

Tang W, Chen Z, Zhang W, Cheng Y, Zhang B, Wu F, Wang Q, Wang S, Rong D, Reiter FP, de Toni EN, Wang X (2020) The mechanisms of sorafenib resistance in hepatocellular carcinoma: theoretical basis and therapeutic aspects. Signal Transduct Target Ther 5(1):87. https://doi.org/10.1038/s41392-020-0187-x

Tang M, Yang M, Wu G, Mo S, Wu X, Zhang S, Yu R, Hu Y, Xu Y, Li Z, Liao X, Li J, Song L (2021) Epigenetic induction of mitochondrial fission is required for maintenance of liver cancer–initiating cells. Cancer Res 81(14):3835–3848. https://doi.org/10.1158/0008-5472.CAN-21-0436

Valle S, Alcalá S, Martin-Hijano L, Cabezas-Sáinz P, Navarro D, Muñoz ER, Yuste L, Tiwary K, Walter K, Ruiz-Cañas L, Alonso-Nocelo M, Rubiolo JA, González-Arnay E, Heeschen C, Garcia-Bermejo L, Hermann PC, Sánchez L, Sancho P, Fernández-Moreno MÁ, Sainz B (2020) Exploiting oxidative phosphorylation to promote the stem and immunoevasive properties of pancreatic cancer stem cells. Nat Commun 11(1):5265. https://doi.org/10.1038/s41467-020-18954-z

Wan S, Zhao E, Kryczek I, Vatan L, Sadovskaya A, Ludema G, Simeone DM, Zou W, Welling TH (2014) Tumor-associated macrophages produce interleukin 6 and signal via STAT3 to promote expansion of human hepatocellular carcinoma stem cells. Gastroenterology 147(6):1393–1404. https://doi.org/10.1053/j.gastro.2014.08.039

Wang W, Quan Y, Fu Q, Liu Y, Liang Y, Wu J, Yang G, Luo C, Ouyang Q, Wang Y (2014) Dynamics between cancer cell subpopulations reveals a model coordinating with both hierarchical and stochastic concepts. PLoS One 9(1):e84654. https://doi.org/10.1371/journal.pone.0084654

Wang Y, Viscarra J, Kim S-J, Sul HS (2015) Transcriptional regulation of hepatic lipogenesis. Nat Rev Mol Cell Biol 16(11):678–689. https://doi.org/10.1038/nrm4074

Wang D, Li M, Shen H, Yang J, Gao Z, Tang Y (2022) Iron deficiency increases phosphorylation of SP1 to upregulate SPNS2 expression in hepatocellular carcinoma. Biol Trace Elem Res. https://doi.org/10.1007/s12011-022-03296-2

Wei Z, Jia J, Heng G, Xu H, Shan J, Wang G, Liu C, Xia J, Zhou H, Wu M, Yang Z, Wang M, Xiong Z, Huang H, Liu L, Qian C (2019) Sirtuin-1/mitochondrial ribosomal protein S5 axis enhances the metabolic flexibility of liver cancer stem cells. Hepatology 70(4):1197–1213. https://doi.org/10.1002/hep.30622

Wu K, Ding J, Chen C, Sun W, Ning B-F, Wen W, Huang L, Han T, Yang W, Wang C, Li Z, Wu M-C, Feng G-S, Xie W-F, Wang H-Y (2012) Hepatic transforming growth factor beta gives rise to tumor-initiating cells and promotes liver cancer development. Hepatology 56(6):2255–2267. https://doi.org/10.1002/hep.26007

Xiao C, Fu X, Wang Y, Liu H, Jiang Y, Zhao Z, You F (2020) Transferrin receptor regulates malignancies and the stemness of hepatocellular carcinoma-derived cancer stem-like cells by affecting iron accumulation. PLoS One 15(12):e0243812. https://doi.org/10.1371/journal.pone.0243812

Yamashita T, Forgues M, Wang W, Kim JW, Ye Q, Jia H, Budhu A, Zanetti KA, Chen Y, Qin L-X, Tang Z-Y, Wang XW (2008) EpCAM and α-fetoprotein expression defines novel prognostic subtypes of hepatocellular carcinoma. Cancer Res 68(5):1451–1461. https://doi.org/10.1158/0008-5472.CAN-07-6013

Yamashita T, Ji J, Budhu A, Forgues M, Yang W, Wang H, Jia H, Ye Q, Qin L, Wauthier E, Reid LM, Minato H, Honda M, Kaneko S, Tang Z, Wang XW (2009) EpCAM-positive hepatocellular carcinoma cells are tumor-initiating cells with stem/progenitor cell features. Gastroenterology 136(3):1012–1024.e4. https://doi.org/10.1053/j.gastro.2008.12.004

Yang ZF, Ho DW, Ng MN, Lau CK, Yu WC, Ngai P, Chu PWK, Lam CT, Poon RTP, Fan ST (2008) Significance of CD90+ cancer stem cells in human liver cancer. Cancer Cell 13(2):153–166. https://doi.org/10.1016/j.ccr.2008.01.013

Yang JD, Hainaut P, Gores GJ, Amadou A, Plymoth A, Roberts LR (2019) A global view of hepatocellular carcinoma: trends, risk, prevention and management. Nat Rev Gastroenterol Hepatol 16(10):589–604. https://doi.org/10.1038/s41575-019-0186-y

Yang H, Deng Q, Ni T, Liu Y, Lu L, Dai H, Wang H, Yang W (2021) Targeted inhibition of LPL/FABP4/CPT1 fatty acid metabolic axis can effectively prevent the progression of nonalcoholic steatohepatitis to liver cancer. Int J Biol Sci 17(15):4207–4222. https://doi.org/10.7150/ijbs.64714

Ye H, Adane B, Khan N, Sullivan T, Minhajuddin M, Gasparetto M, Stevens B, Pei S, Balys M, Ashton JM, Klemm DJ, Woolthuis CM, Stranahan AW, Park CY, Jordan CT (2016) Leukemic stem cells evade chemotherapy by metabolic adaptation to an adipose tissue niche. Cell Stem Cell 19(1):23–37. https://doi.org/10.1016/j.stem.2016.06.001

Younossi ZM, Otgonsuren M, Henry L, Venkatesan C, Mishra A, Erario M, Hunt S (2015) Association of nonalcoholic fatty liver disease (NAFLD) with hepatocellular carcinoma (HCC) in the United States from 2004 to 2009. Hepatology 62(6):1723–1730. https://doi.org/10.1002/hep.28123

Zhang B, Pan X, Cobb GP, Anderson TA (2007) microRNAs as oncogenes and tumor suppressors. Dev Biol 302(1):1–12. https://doi.org/10.1016/j.ydbio.2006.08.028

Zhao W, Lv M, Yang X, Zhou J, Xing B, Zhang Z (2022) Liver tumor-initiating cells initiate the formation of a stiff cancer stem cell microenvironment niche by secreting LOX. Carcinogenesis. https://doi.org/10.1093/carcin/bgac035

Zheng S-S, Chen X-H, Yin X, Zhang B-H (2013) Prognostic significance of HIF-1α expression in hepatocellular carcinoma: a meta-analysis. PLoS One 8(6):e65753. https://doi.org/10.1371/journal.pone.0065753

Zhou Y, Zhou Y, Shingu T, Feng L, Chen Z, Ogasawara M, Keating MJ, Kondo S, Huang P (2011) Metabolic alterations in highly tumorigenic glioblastoma cells. J Biol Chem 286(37): 32843–32853. https://doi.org/10.1074/jbc.M111.260935

The Role of Senescence in NASH-Related HCC

Lampros Chrysavgis, Grigorios Papadopoulos, and Antonios Chatzigeorgiou

Abstract

Nonalcoholic fatty liver disease (NAFLD) has emerged as the predominant cause of chronic liver injury worldwide, mainly due to the epidemic spread of obesity, type II diabetes mellitus (T2DM), and aging. Although NAFLD is considered a benign disease, it can progress to nonalcoholic steatohepatitis (NASH) characterized by the additional emergence of inflammation and ballooning or even to liver fibrosis and hepatocellular carcinoma (HCC). Cellular senescence, a process whereby diploid differentiated cells lose their ability to proliferate and are driven to permanent cell cycle arrest, has generated a lot of interest among researchers due to its involvement in several chronic metabolic and inflammatory diseases, as well as carcinogenesis. In this chapter, we aim to summarize the current knowledge on the impact of senescence in NAFLD and to provide more insight into the implication of senescence in NASH-related HCC.

Evidence from murine and human studies highlighted the implication of senescence in the development of NAFLD and its progression to NASH. Importantly, the current evidence suggests a bidirectional role of senescence in NASH-related HCC since the former exerts both pro- and anti-tumorigenic actions, mediated mostly via the senescence-associated secretory phenotype (SASP).

L. Chrysavgis · G. Papadopoulos
Department of Physiology, Medical School of National and Kapodistrian University of Athens, Athens, Greece

A. Chatzigeorgiou (✉)
Department of Physiology, Medical School of National and Kapodistrian University of Athens, Athens, Greece

Institute for Clinical Chemistry and Laboratory Medicine, University Hospital and Faculty of Medicine Carl Gustav Carus of TU Dresden, Dresden, Germany
e-mail: achatzig@med.uoa.gr

Interdisciplinary Cancer Research, https://doi.org/10.1007/16833_2022_87
Published online: 23 December 2022

Senescence is a beneficial process for the prevention of tumorigenesis. However, intriguingly, the accumulation of senescent cells facilitates the onset of NASH-related HCC. Bearing this in mind, novel senolytic drugs may be a therapeutic option by delaying the progression of NAFLD and in turn the development of NASH-related HCC.

Keywords

Cellular senescence · DNA damage · Hepatocellular carcinoma · Immunosurveillance · Nonalcoholic fatty liver disease · p53 restoration

1 Introduction

Nonalcoholic fatty liver disease (NAFLD) has emerged as the predominant cause of chronic liver dysfunction affecting approximately one-fourth of the global population (Younossi et al. 2016; Katsarou et al. 2020), while its incidence is estimated to be increased along with the elevated prevalence of obesity, type II diabetes mellitus (T2DM), and aging (Zhang et al. 2021). NAFLD is defined as the presence of lipid deposition in more than 5% of the hepatic parenchyma in the absence of excess alcohol consumption (<30g/day for men and <20g/day for women) and secondary causes of chronic liver injury, such as viral hepatitis and administration of steatogenic drugs (i.e., tamoxifen, corticosteroids, etc.) (European Association for the Study of the et al. 2016). It encompasses a wide spectrum of disease severity stages ranging from the simple accumulation of fat in hepatocytes, defined as steatosis, to the more aggressive form of the disease, known as nonalcoholic steatohepatitis (NASH) (Nati et al. 2016; Huang et al. 2021). The latter is featured by hepatocytes' ballooning and inflammation with or without fibrosis and can progress to liver cirrhosis or even to hepatocellular carcinoma (HCC) (Forner et al. 2018). Importantly NAFLD is a multisystemic disease rather than a chronic disease affecting the hepatic parenchyma per se. For instance, NAFLD is considered an independent risk factor for cardiovascular disease (CVD) (Targher et al. 2016), while concurrently, CVD is the leading cause of mortality among patients with NAFLD (Ekstedt et al. 2006). In addition, NAFLD has been associated with endothelial dysfunction (Ogresta et al. 2022), lower skeletal muscle mass, sarcopenia, and frailty (Karakousis et al. 2022). In the last decades, mostly due to the Western lifestyle, the global incidences of obesity and T2DM, well-established risk factors of NAFLD, are increasing in epidemic proportion, especially in the industrialized world. Thus, the need for liver transplantation due to NAFLD progression and more importantly NASH-related HCC are projected to emerge as a major public health issue in the near future.

2 NAFLD and Its Progression to NASH

Despite that NAFLD is considered a benign disease, the progression from NAFLD to NASH along with liver fibrosis and liver cirrhosis is the turning point for NAFLD-related morbidity and mortality. NAFLD to NASH progression can be triggered by numerous molecular mechanisms including insulin resistance, lipotoxicity, oxidative stress, inflammation, tissue damage, and aging as well as genetic predisposition. The upcoming section will summarize the most important mechanisms contributing to NASH pathogenesis and progression.

First, the increased levels of free fatty acids (FFAs) are central in the progression of simple steatosis to NASH. FFAs are derived either by the lipolysis of triglycerides in the adipose tissue and delivered via the blood circulation into the liver or by de novo lipogenesis, a process during which hepatocytes convert the excess glucose and fructose into FFAs (Donnelly et al. 2005; Legaki et al. 2022). The elevated levels of FFA lead to augmented mitochondrial and peroxisomal β-oxidation, formation of lipotoxic molecules such as ceramides (Luukkonen et al. 2016), cholesterol (Ioannou 2016), phospholipids, and diacylglycerols (Perry et al. 2014), which in turn provoke oxidative and endoplasmic reticulum (ER) stress as well as activation of the inflammasome. The latter constitutes a major link between the initial metabolic dysfunction during steatosis and the subsequent hepatocyte damage and scar tissue formation since its presence has been associated with hepatocellular expression of inflammatory cytokines namely interleukin (IL)-1β and IL-18, which promote caspase-1-mediated apoptosis of hepatocytes (Henao-Mejia et al. 2012; Szabo and Petrasek 2015). Apart from the inflammasome-related apoptosis, the presence of ER stress in hepatocytes facilitates the production of reactive oxygen/nitrogen species (ROS/RNS) that leads to insufficient mitochondrial respiratory chain activity and can trigger apoptotic death pathways, such as Jun-(N)-terminal kinase (JNK) and receptor-interacting protein 1 (RIP1) and RIP3, which further promote liver inflammation and fibrosis development and advancement (Gautheron et al. 2014; Afonso et al. 2015). Concurrently, the oxidative stress in hepatocytes promotes an increase in calcium release from ER that leads to mitochondrial permeabilization, disrupted ER function, liver cell injury, and ultimately progression to NASH (Fu et al. 2011). Of importance, insulin resistance is also a common trigger for NAFLD development, as well as for its progression to NASH. More precisely, the impaired insulin signaling is associated with an increased influx of the FFA to the liver as an alternative source of energy, thus providing an excess of substrates for hepatic lipogenesis and lipid deposition and an increased lipotoxic environment (Bugianesi et al. 2005; Loomba et al. 2012). Along this line, intensification of insulin resistance can occur through crosstalk between non-hepatic tissues, such as the adipose tissue, with the liver. For instance, the obesity-driven release of dipeptidyl peptidase 4 (DPP4) by hepatocytes can stimulate pro-inflammatory signaling and overexpression of Mcp1 and IL-6 in the visceral adipose tissue (VAT)-residing macrophages, promoting their accumulation and inducing metabolic inflammation and impaired insulin sensitivity in the adipose tissue as outlined in an obese rodent model (Ghorpade et al. 2018).

Additionally, the gut–liver axis can be impacted by dysbiosis, a condition referring to the dysregulation of the gut microbiome's composition, which can occur by adopting harmful dietary habits (Friedman et al. 2018; Marra and Svegliati-Baroni 2018). Dysregulated gut microbiota can provoke increased intestinal permeability and promote the progression of NAFLD by enabling the entrance of bacterial products and damaging factors in the liver. Specifically, gut-microbiota-derived lipopolysaccharides (LPS) that enter the hepatic tissue stimulate the activation of innate immune cells, thus perpetuating the maintenance of an inflammatory microenvironment in the liver (Leung et al. 2016; Ferro et al. 2020). In parallel, the dysregulated gut-microbiome can produce secondary messengers, like lipid species that simulate signaling molecules which activate predominantly G-protein coupled receptor (GPCR) signaling cataracts, leading to disruption of normal hepatic metabolism of glucose and lipids in the context of NAFLD (Cohen et al. 2017; Friedman et al. 2018; Ferro et al. 2020).

Moreover, besides hepatocytes' injury and ballooning along with inflammation, NASH can also be accompanied by hepatic fibrosis. Fibrosis is induced by the chronic inflammatory responses to liver injury, which activate hepatic stellate cells (HSCs) to produce and deposit extracellular matrix proteins into the hepatic parenchyma, thereby provoking the formation of scar tissue (Tacke and Weiskirchen 2021). The latter causes progressive loss of liver function, which renders the basis for the emergence of cirrhosis and eventually HCC. Apart from their transition to myofibroblasts (MFBs) during NASH and the subsequent deposition of fibrotic connective tissue, HSCs can also crosstalk with several immune cell types in the liver. These cell types include Kupffer cells (Cao et al. 2022), as well as dendritic cells, T-cells, and natural killer (NK) and natural killer-T (NKT) cells; interactions with these immune populations modulate HSC's properties including their capacity to recognize factors secreted by hepatocytes, usually described as pathogen-associated molecular patterns (PAMPs) and damage-associated molecular patterns (DAMPs) (Tacke and Weiskirchen 2021). Cellular and molecular responses to PAMPs and DAMPs can aggravate the inflammatory *milieu* in the hepatic parenchyma during NASH.

Of note, the genetic predisposition of NAFLD patients seems to be significantly implicated in the progression of the disease to its more advanced phenotype (Trepo and Valenti 2020). In fact, hepatic fat accumulation strongly depends on genetics, and it is considered to be a hereditable feature to a significant extent. Indeed, it is estimated that the genetic background accounts for 25% to 75% for the emergence of steatosis among NAFLD patients, with an individual's nationality and family history being of major importance (Guerrero et al. 2009; Eslam et al. 2018; Trepo and Valenti 2020). Recent genome-wide association studies (GWASs) have given prominence to specific genes associated with lipid metabolism and lipid droplet dynamics in NAFLD, as well as with inflammation and fibrosis development (Trepo and Valenti 2020). Specific variants of these genes are thought to be major determinants of NAFLD and concern either loss-of-function or gain-of-function effects on the respective proteins. For instance, the patatin-like phospholipase-containing domain 3 (PNPLA3) gene encodes for a regulator of lipid droplet lipolysis in adipose tissue that can acquire the pathogenic mutation I148M, which causes impaired insulin sensitivity, aberrant fat accumulation, and increased vulnerability to NASH, whereas it has been identified as a predictive factor of the occurrence of HCC (Romeo et al. 2008; Huang et al. 2010; Grimaudo et al. 2020). This variant has also been correlated with increased mortality among NAFLD patients (Grimaudo et al. 2020). In addition, the dysfunctional variant P446L is thought to be responsible for the gain-of-function state of glucokinase regulatory protein (GCKR), which is involved in the regulation of both glucose and triglyceride plasma levels. In fact, therapeutic attempts targeting these pathogenic variants of PNPLA3 and GCKR have shown substantial therapeutic responses in humans (Trepo and Valenti 2020). Similarly, the pathogenic variant E167K of transmembrane 6 superfamily member 2 (TM6SF2) favors hepatic steatosis and fibrosis by restricting lipoprotein-mediated lipid export and increasing the risk for liver damage (Chambers et al. 2011; Speliotes et al. 2011). Along this line, membrane-bound *O*-acyltransferase domain-containing 7 (MBOAT7) is a modifier of lipids, inhibitory mutations of which have been related to steatosis under obese conditions (Helsley et al. 2019). Finally, the loss-of-function variants of 17-beta hydroxysteroid dehydrogenase 13 (HSD17B13), which encodes the corresponding hepatic lipid protein (hydroxysteroid 17-beta dehydrogenase 13), are associated with decreased risk of progression from steatosis to NASH as evaluated by histopathological analysis of 2391 human liver samples (Abul-Husn et al. 2018). Moreover, it has been recently identified that some HSD17B13 variants might protect against HCC development (Abul-Husn et al. 2018; Ma et al. 2019; Wang et al. 2020).

3 Mechanisms and Pathophysiology of NASH-Related HCC

HCC is the principal histological type of liver cancer and the second most common cause of cancer-related mortality globally (Torre et al. 2015). Although most HCC cases are attributed to chronic viral hepatitis or alcoholic cirrhosis, NAFLD patients are at increased risk of developing HCC. A recent meta-analysis revealed that the prevalence of HCC among NASH patients was almost double (22.5% vs 13.6%) as compared to their non-NASH-related HCC counterparts (Stine et al. 2018). Moreover, bearing in mind the high incidence and prevalence of obesity and T2DM, the impact of NASH on the whole HCC burden is projected to be increased in the coming decades (Chrysavgis et al. 2022). Importantly, although liver fibrosis constitutes the greatest risk parameter that affects the development of HCC (Anstee et al. 2019), the presence of cirrhosis, namely a condition of extended irreversible fibrosis, is not a prerequisite for the onset of NAFLD-related HCC. NAFLD along with hepatitis B virus infection can lead to HCC even in a non-cirrhotic liver. For instance, two independent studies revealed that approximately half of NAFLD-related HCC arose in the absence of cirrhosis (Sanyal et al. 2010; Piscaglia 2016). Apart from liver cirrhosis, demographic, genetic, metabolic, and behavioral parameters are well-recognized risk factors for the development of NASH-related HCC. More precisely, older age (>65 years), Hispanic ethnicity, and the male sex are significantly related to increased risk of HCC among NAFLD patients (Kanwal et al. 2018). Furthermore, overweight and obese patients had 17% and 89% elevated relative risk for HCC, respectively, as compared to normal-weight patients, as outlined in a large meta-analysis investigating liver cancer cases (Larsson and Wolk 2007). Along this line, T2DM is an independent risk factor for HCC since it was associated with a 3-4-fold increased risk for HCC, as identified in several human studies (Kawamura et al. 2012; Turati et al. 2013; Yang et al. 2020). Importantly, Dyson et al. showed that the prevalence of obesity and T2DM, as metabolic risk factors in HCC patients, was growing during a ten-year follow-up (2000–2010), while they highlighted that the metabolic dysfunction was the only identified risk parameter for HCC in one-third of the patient population (Dyson et al. 2014).

Beyond the aforementioned risk factors of NASH-related HCC development, at the molecular level, specific pathogenetic pathways are shown to be implicated in the emergence of NASH-related HCC. The persistent impairment of insulin signaling, which causes a lack of fatty acid homeostasis in the course of NAFLD and NASH and the excessive insulin-like growth factor 1 (IGF-1) signaling occurring during NAFLD, can activate cancer-related pathways, including the mitogen-activated protein kinase (MAPK) and the phosphoinositide tri-phosphate kinase/protein kinase B (PI3K/AKT). The latter are major regulators of the expression of cancer-related genes, namely c-fos and c-jun, which are able to alter cell growth, proliferation, and apoptosis (Kudo et al. 2011; Janku et al. 2014). In addition, the Wnt/b-catenin signaling axis is also a well-studied pro-tumorigenic mechanism in the pathogenesis of HCC that is activated by the MAPK signaling pathway (Chettouh et al. 2015). Besides, during NAFLD, the increased FFA oxidation and peroxidation of lipid species gradually lead to mitochondrial dysfunction and ER stress, respectively, thus constantly increasing the oxidant "load" in the liver and activating the family of JNK signaling pathways. In turn, JNK signaling cascade enables caspase-induced hepatocellular apoptosis and necroptosis, favoring inflammation and subsequently fibrosis advancement and NASH-related tumorigenesis (Fu et al. 2011; Gautheron et al. 2014; Afonso et al. 2015; Bozaykut et al. 2016).

Of note, several studies have highlighted the implication of a plethora of micro-RNAs (miRNAs) in the development of NASH-related HCC (Schulze et al. 2015). Particularly, increased expression of miR-19a and miR-21 has been correlated with NASH-HCC via the activation of the tumorigenic signaling pathway PI3K/AKT/mTOR (de Conti et al. 2017), while decreased expression of miR-122 and miR-34a seem to affect lipid metabolism by provoking exacerbation of fat accumulation and transcriptional activation of the tumour suppressor gene p53, a key gene for the onset of tumorigenesis (Cheung and Sanyal 2010; Tornesello et al. 2013; Takaki et al. 2014). In addition, Wang et al. demonstrated that the expression of the small nucleolar RNA host gene 20 (SNHG20) was significantly upregulated in the NAFLD-related HCC tissue and in liver-derived KCs (Wang et al. 2019). As SNHG20 acts as a regulator of M1/M2 polarization in KCs, SNHG20 overexpression provoked the tumour-promoting M2-like phenotype, a process which was associated with signal transducer and activator of transcription (STAT)-6 pathway activation and thereafter facilitated dietary and chemically induced hepatic tumorigenesis in mice (Wang et al. 2019). On the contrary, silencing of SNHG20 in the mouse cell line of macrophages RAW264.7 drove the shift of KCs towards M1 polarization, a tumour suppressor phenotype, thus delaying the progression of NAFLD to HCC (Wang et al. 2019).

Of importance, obesity, a common feature of NAFLD, is associated with increased lipid storage in the liver with subsequent increase of de novo lipogenesis, overproduction and accumulation of ROS, and elevated ER stress (Vinciguerra et al. 2009), triggering a vicious cycle of hepatic inflammation in NAFLD patients, thus predisposing among others to NASH and HCC. Moreover, since treatment with leptin of HepG2 and Huh7 cell lines in vitro induced increased proliferation and greater invasion potential of HCC cells via the activation of JAK/STAT, AKT, and extracellular signal-regulated kinase (ERK) signaling pathways, the high leptin levels during obesity may further aggravate the onset and the progression of NASH-related HCC (Saxena et al. 2007).

Beyond the well-established molecular pathways involved in the progression of NAFLD to NASH and HCC, recent evidence has pinpointed the critical and relatively new-identified link between cellular senescence and NAFLD and, more importantly, the association between senescence and NASH-related HCC. In the upcoming lines of the current chapter, we aim to provide a comprehensive summary of the role of cellular senescence in NAFLD and its progression to NASH to thereafter discuss in-depth the role of senescence in NASH-related HCC.

4 Senescence in NAFLD and NASH

Senescence represents a cellular state mainly characterized by irreversible cell cycle arrest, perturbation of cellular structure, and a characteristic secretory phenotype widely known as the senescence-associated secretory phenotype (SASP) (Lopez-Otin et al. 2013; Aravinthan 2015). It can be induced by exogenous or endogenous stimuli that affect the genomic structure and can cause chromatin and telomere damage (Coppe et al. 2010; Papatheodoridi et al. 2020). Those stimuli encompass natural aging, intra-tissular accumulation of ROS/RNS, oncogene expression, and chemical inducers such as in the case of chemotherapy (Regulski 2017; Papatheodoridi et al. 2020). There are two distinct types of cellular senescence; replicative senescence, which relies on the distortion of telomeres' length and structure and is predominantly associated with aging, and stress-induced senescence, which represents the exogenous, premature, and non-telomere-associated cause of DNA damage (Aravinthan 2015; Di Micco et al. 2021). Both of those types are capable of initiating the DNA damage response (DDR) pathway, which can result either in the genomic repair of the cell or the inhibition of the cell cycle (Jackson and Bartek 2009).

The mechanistic aspects of senescence have been greatly elucidated in recent years. Specifically, some of the most important mechanisms regarding senescence-induced cell cycle arrest are the ataxia-telangiectasia mutated (ATM) kinase-dependent activation of p53 and p21 (Bakkenist and Kastan 2003; Tornesello et al. 2013; Georgakilas et al. 2017), which are major oncogenes regulating cell cycle, along with the p16-dependent regulation of retinoblastoma protein (Rb)-E2F transcription factor interaction that finally hinders cell cycle progression (Rubin et al. 2005; Papatheodoridi et al. 2020).

Several molecular features that the cells adapt during senescence can be used as biomarkers of this cellular state. Apart from cells' morphological changes such as cell and nuclei enlargement and flattening, the presence of senescence-associated β-galactosidase and senescence-associated heterochromatin foci (SAHF) are considered major senescence biomarkers. The latter consists of heterochromatin and a group of proteins that contribute to senescence by mitigating the expression of genes associated with cell proliferation. Along this line, the SASP, a combination of inflammatory cytokines, chemokines, ROS species, proteases, and other mediators, serves as an important molecular signature that provokes vicious cycles of senescence-driven immune activation (Itahana et al. 2007; Acosta et al. 2008, 2013; Zhang et al. 2012). Besides, the permanent expression of SASP creates a pro-inflammatory environment regionally, which can favor the expansion of senescence toward other cells. This phenomenon has been observed in the context of NAFLD and NASH (Zhang et al. 2012; Papatheodoridi et al. 2020) as well as in other age-associated chronic pathologies (Rossi et al. 2007; Wei and Ji 2018), including neurological disorders, but also metabolic dysregulation and malignancies (Fig. 1).

Senescent hepatocytes can be normally detected in the liver at any point and at low percentages of about 5%, but their frequency can be largely increased during different stages of hepatic disease (Paradis et al. 2001; Wiemann et al. 2002). Obesity, which is associated with high fat diet (HFD) and is a cardinal cause of NAFLD, has been correlated in several studies with the induction of senescence in the context of NAFLD and NASH. For example, p16- and p21-induced hepatocellular senescence has been demonstrated in studies incorporating obese diets in rodent models. Specifically, Zhang et al. demonstrated, at the chromatin, transcriptional and protein levels that obesity-prone rats that displayed NAFLD had increased expression of the cyclin-dependent kinase inhibitors p16 and p21, which led to less Rb phosphorylation (Zhang et al. 2012). Moreover, increased adiposity has been observed to exist in parallel with a greater number of senescent hepatocytes and elevated SASP components in hepatocytes, namely IL-6, MCP-1, and SA-β-Gal, which in cooperation with an increased hepatic influx of FFA and ROS production restrain lipid catabolic processes in NAFLD, such as fatty-acid β-oxidation (Skurk et al. 2007; Wree et al. 2014; Gustafson et al. 2019; Moustakas II et al. 2021). Besides, obesity has been connected to increased hepatocyte expression of the γH2AX histone family member, which is phosphorylated by ATM, and underlines the presence of damaged genomic regions that are detected during senescence (Nakajima et al. 2010; Aravinthan 2015; Ping et al. 2017; Papatheodoridi et al. 2020). The upregulation of histone γH2AX is correlated with increased levels of p21 and p16 and cell cycle arrest in NAFLD animal models and humans suffering from NAFLD (Aravinthan 2015; Ogrodnik et al. 2019). Furthermore, Daugherity et al. showed that inactivation of the ATM-C/EBPα-PPARg axis blocked senescence-associated progression to NASH in obese rodents (Daugherity et al. 2012). In addition, the methylation status of genomic DNA and histone modifications are

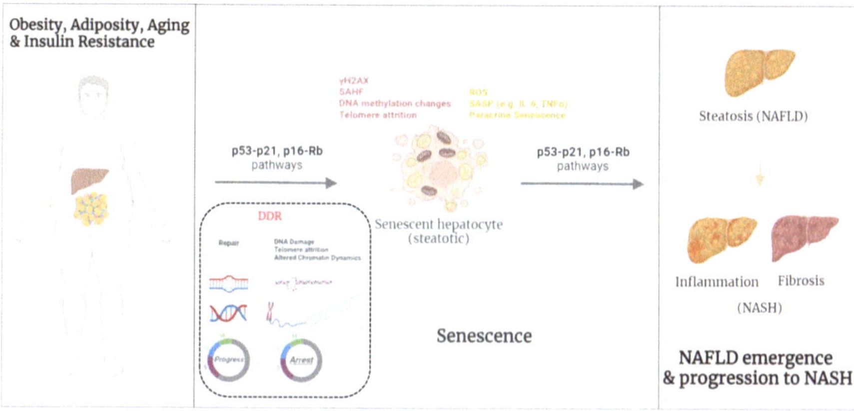

Fig. 1 Role of senescence in NAFLD and NASH
Senescence can be induced by several stress stimuli, such as aging, reactive oxygen species, mitochondrial oxidative stress, and anticancer treatments, which can lead to changes in DNA structure and activate a DNA damage response (DDR) by the affected cells. Obesity and the associated features of the metabolic syndrome, apart from being well-recognized causal factors of nonalcoholic fatty liver disease (NAFLD), can concomitantly induce hepatocellular senescence via the p53-induced activation of p21 and p16-driven regulation of retinoblastoma protein (Rb). Therefore, hepatocytes acquire a senescent phenotype that encompasses specific genomic features, such as altered DNA methylation patterns, SAHF, the presence of histone γH2AX, and shortened telomeres. The latter in combination with SASP features and ROS production can favor inflammation and fibrosis in the liver, while further facilitate the paracrine expansion of senescence toward other hepatic and non-hepatic cell populations, thus perpetuating the progression of NAFLD toward NASH.
Abbreviations: *DDR* DNA damage response, *NAFLD* nonalcoholic fatty liver disease, *NASH* nonalcoholic steatohepatitis, *Rb* retinoblastoma protein, *ROS* reactive oxygen species, *SAHF* senescence-associated heterochromatin foci, *SASP* senescence-associated secretory phenotype

considered solid indicators of aging as well as of hepatic disease stage related to senescence, as shown in several studies (Horvath 2013; Nishida et al. 2016; Hotta et al. 2018; Loomba et al. 2018). For instance, Tryndyak et al. demonstrated that the induction of NAFLD using choline- and folate-deficient diet into distinct mouse strains, which display different susceptibility to steatosis (A/J and WSB/EiJ mice), resulted in different DNA methylation patterns of genes related to NAFLD, which in turn modulated the respective gene expression levels (Tryndyak et al. 2016).

Apart from hepatocytes, the process of senescence also affects other cell types of the hepatic parenchyma, influencing the pathophysiological framework of NAFLD. For instance, senescent HSCs seem to act as a double-edged sword for NAFLD and its transition to NASH. Senescence of HSCs results in reduced deposition of fibrotic components and connective tissue in the liver, thereby limiting fibrosis advancement (Krizhanovsky et al. 2008). However, their senescence can also exacerbate inflammation in the hepatic microenvironment through SASP. Indeed, Schnabl et al. showed that replicative senescence of activated human HSCs limits their fibrogenic potential but exacerbates their surrounding inflammatory microenvironment

(Schnabl et al. 2003). Accordingly, in a rodent NASH model, steatosis, inflammation, and fibrosis were found to be reduced when inducing senescence in HSCs via IGF-1, and this was attributed to the action of the senescence-related p53 protein (Nishizawa et al. 2016). Finally, the senescence of cholangiocytes, KCs, and liver sinusoidal endothelial cells is secondary but of importance for NAFLD and NASH. These cell types can also acquire SASP during NAFLD, affecting the biliary function, the levels of pro-inflammatory cytokines such as IL-6, and the sinusoidal function/fenestration, respectively (Harada et al. 2001; Sasaki et al. 2010; Lee et al. 2017). Considering the importance of senescence in NAFLD to NASH transition, it is currently of high scientific interest whether it also plays a role in the further progression of the disease into HCC. In the next section, we aim to provide more insight into the role of cellular senescence in NASH-related HCC and discuss the current state-of-the-art on senescence-related therapeutics.

5 The Involvement of Senescence in NASH-Associated HCC

NASH-associated HCC is a chronic process with a gradual transition from the state of fatty liver to the state of NASH, liver cirrhosis, and HCC. Liver cirrhosis is considered the most significant risk factor for HCC development (Gines et al. 2021). During NASH and cirrhosis, the chronic inflammatory process that takes place in the liver is accompanied by repeated cycles of hepatocytes' death and subsequent liver regeneration (Pellicoro et al. 2014). In turn, the gradual replacement of the healthy liver tissue with fibrotic scar and regenerative nodules, along with constant cell growth, likely predisposes tumor development (Pellicoro et al. 2014).

Noteworthy, in patients with NAFLD, HCC can arise even in the absence of cirrhosis, and a lot of non-cirrhotic-related HCCs are associated with NAFLD/NASH (Gawrieh et al. 2019). Insulin resistance, increased lipid peroxidation, and the activation of the immune system are well-recognized pathogenetic mechanisms of HCC development in the non-cirrhotic liver. Importantly, in recent years, cellular senescence has generated a lot of interest among researchers due to its potential involvement as a bidirectional factor acting in favor of HCC development but also against it (Fig. 2).

Under normotropic conditions, the secretion of SASP mediators by senescent cells leads to immune cell recruitment and subsequent clearance of senescent cells (Campisi and d'Adda di Fagagna 2007; Aravinthan 2015). Similarly, under conditions of competent immune surveillance, the genetic alterations happening in pre-malignant or even malignant cells are likely perceived as DNA damage phenomena, driving these cells to acquire a senescent phenotype, and making them susceptible to immune clearance. Indeed, oncogenic stimuli leading to DNA damage drive the ATM-mediated phosphorylation of the tumor suppressor protein p-53, which in turn stimulates the expression of p21 (Bakkenist and Kastan 2003; Georgakilas et al. 2017). The activation of p21 leads to the deactivation of the Rb protein, promoting its binding to the E2F transcription factor, thus halting the repeated cell division and tumorigenesis (Bakkenist and Kastan 2003; Georgakilas

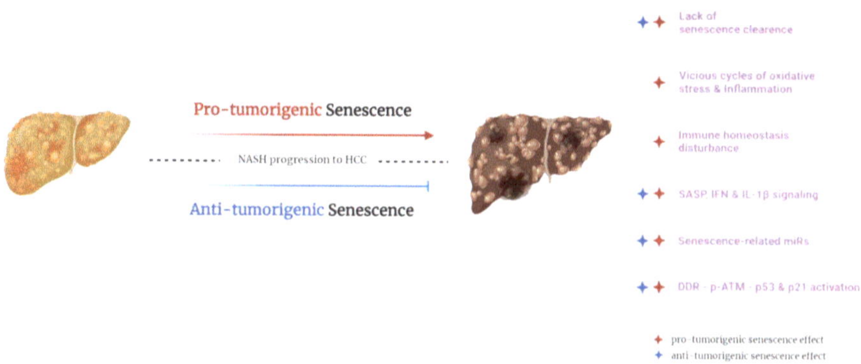

Fig. 2 Implication of senescence in the emergence of NASH-related HCC
Mechanisms implicated in the hepatocellular senescence have been observed to exert pro-tumorigenic and anti-tumorigenic effects. The impaired immunosurveillance results in accumulation of senescent cells which acquiring SASP-associated features can aggravate the pro-tumorigenic inflammatory hepatic microenviroment during NASH. Concurrently, the sufficient cellular senescence of pre-malignant cells substantially impedes their ability to promote onset of HCC. Therefore, senescence is a bidirectional factor that affects the development and progression of HCC.
Abbreviations: *ATM* ataxia-telangiectasia mutated, *DDR* DNA damage response, *HCC* hepatocellular carcinoma, *IFN* interferon, *IL-1β* interleukin-1β, *NASH* nonalcoholic steatohepatitis, *SASP* senescence-associated secretory phenotype

et al. 2017). More specifically, Xue et al. demonstrated that the restoration of p53 signaling in a liver-specific p53-deficient mouse could result in tumour regression via induction of cellular senescence (Xue et al. 2007). Upon p53 reactivation, the murine HCC cells, since then expressing the oncogenic ras or Akt, were led to senescence as evaluated by the accumulation of SA-β-Gal activity and the increased expression of the senescent markers p16INK4a, DcR2, and p15INK4b. Importantly, p53 reactivation was accompanied by a progressive inflammatory response and increased recruitment of innate immune cells such as macrophages, neutrophils, and NKT cells. Their accumulation was initially observed at the peri-tumoral regions and subsequently throughout the tumor, leading to the elimination of malignant cells and immunological clearance of the HCC (Xue et al. 2007). More recently, Wu et al. proposed that microRNA-21 may act as a link between the tumor suppressor p53 and NAFLD-related HCC (Wu et al. 2016). They showed that the lipid accumulation-mediated overexpression of microRNA-21 in NAFLD acts antagonistically to p53 expression since the transcriptional factor of p53, namely HBP1, is also the target of micro-RNA-21 (Wu et al. 2016). Besides, miRNA-21 knockdown prevented G1/S transition and HCC proliferation since the tumorigenic-related pathways of G2/mitotic-specific cyclin-B1, G1/S-specific cyclin-D1, and Sterol regulatory element-binding protein 1 were inhibited (Wu et al. 2016). Of note, a mutation or even a loss of heterozygosity of p53 has been observed in almost 90% of HCC patients (Thorgeirsson and Grisham 2002).

Other studies have focused on whether oncogenic stimuli could provoke a senescent phenotype and through this tumor suppression (Kang et al. 2011). For instance, Kang et al. used transposable elements expressing the oncogenic Nras ($Nras^{G12V}$) into pre-malignant murine hepatocytes in vivo and that provoked their senescence 12 days after the delivery of transposon constructs (Kang et al. 2011). Furthermore, the immunological clearance of the senescent hepatocytes was mediated by the liver-infiltrating accumulation of neutrophils, monocytes, NKT cells, macrophages, and CD8+ and CD4+ cells in senescent murine livers (Kang et al. 2011). Noteworthy, by utilizing C.B-17 SCID and C.B-17 SCID/beige mice as well as their wild type counterparts, the authors showed that the clearance and the overall immunological surveillance of the pre-malignant senescent hepatocytes are governed by antigen-specific CD4+ T cells (Kang et al. 2011). Those findings were also confirmed in a more recent study, in which a hydrodynamic injection of a transposon that co-expressed NRasV12 and luciferase into hepatocytes of wild-type mice induced cellular senescence, SASP, and subsequently elimination of the senescent cells by innate immune cells (Dou et al. 2017). Beyond that, the authors also highlighted that the presence of DNA in the cytoplasm acted as a trigger of cellular senescence (Dou et al. 2017). More specifically, the presence of cytoplasmic chromatin activated the Stimulator of Interferon Genes (STING) pathway via the formation of cyclic GMP-AMP (cGAMP), leading to activation of two pro-inflammatory downstream pathways, namely type I interferon and nuclear factor kappa light chain enhancer of activated B cells (NF-κB) signaling (Dou et al. 2017). That activation facilitated the immune cell accumulation and thereafter the elimination of the senescent cells (Dou et al. 2017). These data implied that during the progression of NASH to HCC, genomic DNA likely escapes the nuclei and is released into the cytoplasm, thus triggering senescence of the pre-malignant cells and therefore preventing tumor development. Despite that those findings were further supported by Gluck et al. (2017), showing that the cGAS-STING pathway promoted senescence in a paracrine manner, we should note that during that process augmented expression of SASP factors was present, creating an inflammatory environment that may in long term promote HCC. Indeed, although the progression of NAFLD to NASH and HCC is characterized by increased senescence, when senescent cells either bypass their cell cycle arrest or are not cleared by the immunological response, it can give rise to HCC.

Although cellular senescence is considered a fundamental tumor suppressing mechanism of organisms, a growing body of evidence suggests that under certain conditions senescence may act bidirectionally, promoting the onset and progression of carcinogenesis. In general, under special circumstances, senescent cells can potently bypass the senescence barrier by activating various pre-tumorigenic pathways such as transformation/transcription domain-associated protein/lysine acetyltransferase 5 (TRAP/KAT5) complex, topoisomerase II alpha, kinesin family member 14 and 23, and micro-RNAs namely mir-139 and mir-4800, thus facilitating HCC proliferation (Coppe et al. 2010; Lee and Schmitt 2019). In the setting of NASH-associated HCC, a link between obesity, a major feature of NAFLD, and HCC development was identified. Interestingly, that relation was not mediated by the increased body weight but by the alteration of the gut microbiota under obese conditions. More precisely, Yoshimoto et al. used p21-p-luc mice, which were fed a HFD and treated with the carcinogen 7,12-dimethylbenz(a)anthracene (DMBA) and demonstrated that obesity increased deoxycholic acid (DCA) production, a secondary bile acid product. This increase provoked DNA damage via the production of ROS and promoted SASP in HSCs through enterohepatic circulation, thus facilitating carcinogenesis (Yoshimoto et al. 2013). Indeed, the depletion of senescent HSCs or elimination of IL-1β, a SASP promoter in mice, prevented HCC development (Yoshimoto et al. 2013). Along this line, the administration of vancomycin eliminated gut microbiota in obese mice and subsequently attenuated the senescence of HSCs and the development of HCC (Yoshimoto et al. 2013). Consistently, another study in mice further supported the aforementioned findings, demonstrating that obesity-induced production of the Gram-positive gut microbe component lipoteichoic acid (LTA) leads to HSCs' senescence (Loo et al. 2017). Specifically, in a Toll-like receptor 2-dependent manner, LTA in cooperation with DCA induced the expression of SASP factors in murine primary HSCs, namely IL-1β, growth-regulated oncogene α, and IL-6 as well as cyclooxygenase-2 (COX2) (Loo et al. 2017). Importantly, the latter drives the production of prostaglandin E2, which suppresses antitumor immunosurveillance, and along with the abovementioned hepatic translocation of LTA and DCA that induced senescence, further promoted HCC onset and progression (Loo et al. 2017). Importantly, Li et al. highlighted an additional pathway by which HSCs' senescence facilitated HCC progression by investigating both human RNA-sequencing data from the Cancer Genome Atlas and liver tumors from $Trp53^{flox/flox}$ and DEN-treated mice (Li et al. 2020). Specifically, they demonstrated that ablation of the fructose 1,6-biphosphatase-1 (FBP-1) enzyme, the rate-limiting enzyme of gluconeogenesis, in DEN-treated mice led to metabolic dysregulation, increased SASP expression, and by that, senescence of HSCs and ultimately HCC progression (Li et al. 2020). By performing secretome profiling and analyzing 1,944 distinct proteins, they revealed that the expression of high mobility group protein B1 (HMGB1) has been associated with chronic liver injury and HSC activation acts as a link during FBP-1 deficiency and the subsequent SASP-mediated HSC activation (Li et al. 2020). The expression levels of FBP-1 were significantly decreased in liver tumors of murine NAFLD-associated HCC as compared to the corresponding expression at the NASH stage, indicating that FBP-1 expression is diminished during HCC progression (Li et al. 2020). Importantly, myeloid-derived suppressor cells and NK-cells were reduced in FBP-1-deficient livers, indicating a compromised surveillance by the immune system, thus facilitating the accumulation of senescent HSCs (Li et al. 2020). Of interest, in human NAFLD, an increased expression of the senescent markers $p21^{Waf1/Cip1}$, γ-H2AX, and IL-6 was identified in tumoral and nontumoral fibroblasts of NASH-associated HCC as compared to other non-NASH-associated HCC, indicating the importance of SASP factors in the development of NASH-related HCC (Loo et al. 2017). Along this line, Avarithan et al. showed that senescence of hepatocytes, as evaluated by the increased hepatocyte-specific p21 expression and enlarged nuclear area of the hepatocytes, was associated with an adverse liver-related outcome, including the development of HCC (Aravinthan et al. 2013). However, it should be noted that in the latter study, the patients with NASH-related HCC compared to their non-NASH-HCC counterparts were older and had an increased prevalence of T2DM and other comorbidities, and consequently, those findings should be interpreted cautiously.

More recently, Yoshioka et al. utilized obese melanocortin 4 receptor-deficient mice, which were fed a Western diet along with a chemical carcinogen and treated with tofogliflozin, a sodium-glucose co-transporter inhibitor (Yoshioka et al. 2021). They showed that even though there was no significant difference in the mRNA expression levels of SASP-related genes, namely p21, C-X-C Motif Chemokine Ligand 1 (Cxcl1), and Matrix metalloproteinase 12/13 (Mmp12/13) in tumoral regions of treated compared to untreated mice, tofogliflozin prevented further HCC progression (Yoshioka et al. 2021). That finding was mainly attributed to the attenuated mRNA expression of SASP factors, namely p21, Cxcl1, and Mmp12/13 in surrounding stromal cells in nontumoral regions of treated mice, indicating that alleviation of SASP *milieu* could prevent the progression of liver tumors (Yoshioka et al. 2021).

Finally, the increased levels of ceramides, species which are accumulated in the livers of NAFLD patients and of obese individuals in general, might be bidirectionally implicated as a tumor-suppressing and tumor-promoting factor in senescence-driven NAFLD-HCC. Importantly, it has been identified that hepatic levels of ceramides are significantly elevated in multiple mouse models of steatosis such as mice fed a methionine choline-deficient diet (Mas et al. 2009), genetically ob/ob mice (Yang et al. 2009), and methionine adenosyl transferase 1A knockout mice (Alonso et al. 2017). In parallel, ceramides have been shown to provoke cellular senescence in in vitro models by regulating a plethora of senescent factors, namely p21 (Lee et al. 2000), p38, telomerase activity (Wooten-Blanks et al. 2007), Rb dephosphorylation, and p53-tumour growth arrest (Castro et al. 2008), thus likely promoting senescence-associated HCC during NAFLD/NASH as well. In contrast, a study by Liu et al. demonstrated that treatment of tumor-bearing mice with LCL521, a lysosomotropic inhibitor of acid ceramidase, resulted in decreased numbers of myeloid derived suppressor cells (MDSC) in vivo, which favour tumor development. Utilizing an in vitro model of MDSC-like cells, the authors showed that LCL521 induces autophagy and in turn cell death, by targeting lysosomes to activate cathepsin B and cathepsin D (Liu et al. 2016).

6 Conclusions

Even though recent findings have not clarified yet whether senescence is a cause or a consequence of NAFLD, robust evidence has highlighted the detrimental association between NAFLD and senescence. Of note, the association between senescence and NASH-related HCC is more complicated. Senescence exerts both anti-tumorigenic and pro-tumorigenic effects since it seems to prevent the onset of malignancy via induction of cell cycle arrest, though likely promoting the progression of tumour mainly via the formation of SASP-associated inflammatory environment.

Cellular senescence is profoundly associated with aging, T2DM, and obesity, all well-established risk factors for NAFLD and NASH. Concurrently, despite the vast scientific effort globally, no approved medication for NASH exists so far, and the current management of those patients mostly focuses on lifestyle interventions such as weight loss and increased physical exercise. Meanwhile, senolytic drugs have already been developed and their efficacy is under assessment in rodent studies and novel clinical trials. The combination of dasatinib, a tyrosine kinase inhibitor, and quercetin, an antioxidant agent, has shown promising results in in vitro and in vivo models of NAFLD (Ogrodnik et al. 2017). More precisely, the aforementioned drug combination eliminated the hepatic senescent cells in INK-ATTAC mice and reduced age-associated liver fat deposition, alleviating steatosis (Ogrodnik et al. 2017). Nevertheless, no human clinical trials of dasatinib plus quercetin in NAFLD patients have so far been performed, while that combination has been only assessed in two phase-1 pilot studies for the treatment of idiopathic pulmonary fibrosis (Justice et al. 2019) and chronic diabetic kidney disease (Hickson et al. 2019), respectively. Concerning the latter study, senolytic therapy led to a significant reduction of p-16^{INK4A} and p-21^{CIP} expressing cells in adipose tissue along with decreased SA-β-gal activity and accumulation of CD68+ macrophages (Hickson et al. 2019). Besides, SASP factors including MMP-9 and 12, IL-1a, and IL-6 were also decreased following senolytic therapy (Hickson et al. 2019). Of importance, no serious adverse effects, such as hospitalization or death, were reported during the aforementioned human pilot studies (Hickson et al. 2019; Justice et al. 2019) while the non-severe effects were primarily mild to moderate respiratory symptoms, skin irritation, and gastrointestinal discomfort. This pinpoints that the risk of using dasatinib/quercetin is quite mild in relation to its clinical benefit.

Furthermore, besides direct elimination of senescent cells, targeting senescence could be performed by inhibiting SASP by engaging agents, defined as senomorphics. As far as now, metformin (Moiseeva et al. 2013), ruxolitinib, rapamycin (Singh et al. 2016), and glucocorticoids (Laberge et al. 2012) have been used as senomorphic medication. Specifically, under Ras-induced tumorigenic conditions, treatment with metformin of human IMR90 fibroblasts in vitro inhibited the expression of SASP-related genes CXCL5, IL-1b, IL-6, and IL-8. Moreover, while conditioned medium from senescent IMR90 fibroblasts promoted the growth of PC3 prostate cancer cells, pre-treatment of senescent cells with metformin significantly repressed that effect (Moiseeva et al. 2013). Consistently, corticosterone significantly suppressed senescence-associated secretion of IL-6 in a dose-dependent manner in HCA2 fibroblast cultures and in an in vitro model of human IMR-90 fibroblasts, both of which had previously been induced to senescence by ionizing radiation or lentiviral-mediated overexpression of oncogenic MAP kinase 6 or RAS (Laberge et al. 2012). Lastly, the effect of rapamycin in SASP components was demonstrated in a human pilot study. More precisely, elderly patients who underwent cardiac rehabilitation were administered rapamycin daily for 12 weeks.

At the end of this period, the adipose tissue mRNA expression of IL-8, TNF-α, and p16 was significantly decreased whereas SA-β-GAL activity also showed a similar trend (Singh et al. 2016). Based on the abovementioned studies, senomorphics mostly target the signaling pathways associated with the regulation of the SASP such as JAK-STAT (Xu et al. 2015), rapamycin complex 2 (mTORC1), and the IκB kinase-nuclear factor (IKK-NF-κB) and via them likely alleviate the SASP environment (Khosla et al. 2020).

Although senolytics or senomorphics seem an appealing therapeutic opportunity for many senescence-related diseases, including NAFLD/NASH, their potential clinical applicability has some serious limitations. First, cellular senescence is a normal process for diploid organisms that eliminates damaged cells or cells escaping the cell cycle, thus preventing rampant cell proliferation and tumorigenesis. Therefore, senolytics may mitigate the progression of NAFLD to NASH but also may increase the risk of NAFLD-associated HCC. Indeed, although induction of HSC senescence can lead to reduced production of extracellular matrix components and ameliorated fibrosis (Krizhanovsky et al. 2008), concurrently, as above outlined, senescent HSCs play a central role in the onset of NAFLD-related HCC (Yoshimoto et al. 2013; Loo et al. 2017; Li et al. 2020). Moreover, despite that senolytic drugs have shown a favorable safety profile in the current pilot trials (Hickson et al. 2019; Justice et al. 2019), large clinical studies with increased sample size are needed to determine the potent adverse effects of these drugs. Of importance, since senescence is a chronic process, the administration period of senolytics is an open question to be addressed. Few data derived by human and rodent studies suggest that even an intermittent administration of senolytics might be effective in reducing the burden of senescent cells, accumulation of which is a determinant factor for the onset of NASH-associated HCC. Therefore, such short-administration strategies might be preferable since the risk of adverse effects is significantly reduced as compared to a potent continuous treatment (Hickson et al. 2019). Importantly, the senomorphics that target specific SASP components may require a continuous pattern of administration in order to alleviate the SASP *milieu*. Nevertheless, the therapeutic potential of senolytics and senomorphics remains elusive, and further work is required to determine the benefits and risks of these drugs in several age-related diseases in clinical practice.

Compliance with Ethical Standards

Disclosure of Potential Conflicts of Interest The authors have nothing to disclose related to the manuscript.

Funding Information No financial support.

Data Availability Statement Data sharing is not applicable to this article as no datasets were generated or analyzed during the current study.

References

Abul-Husn NS, Cheng X, Li AH, Xin Y, Schurmann C, Stevis P, Liu Y, Kozlitina J, Stender S, Wood GC, Stepanchick AN, Still MD, McCarthy S, O'Dushlaine C, Packer JS, Balasubramanian S, Gosalia N, Esopi D, Kim SY, Mukherjee S, Lopez AE, Fuller ED, Penn J, Chu X, Luo JZ, Mirshahi UL, Carey DJ, Still CD, Feldman MD, Small A, Damrauer SM, Rader DJ, Zambrowicz B, Olson W, Murphy AJ, Borecki IB, Shuldiner AR, Reid JG, Overton JD, Yancopoulos GD, Hobbs HH, Cohen JC, Gottesman O, Teslovich TM, Baras A, Mirshahi T, Gromada J, Dewey FE (2018) A protein-truncating HSD17B13 variant and protection from chronic liver disease. N Engl J Med 378(12):1096–1106

Acosta JC, O'Loghlen A, Banito A, Guijarro MV, Augert A, Raguz S, Fumagalli M, Da Costa M, Brown C, Popov N, Takatsu Y, Melamed J, d'Adda di Fagagna F, Bernard D, Hernando E, Gil J (2008) Chemokine signaling via the CXCR2 receptor reinforces senescence. Cell 133(6): 1006–1018

Acosta JC, Banito A, Wuestefeld T, Georgilis A, Janich P, Morton JP, Athineos D, Kang TW, Lasitschka F, Andrulis M, Pascual G, Morris KJ, Khan S, Jin H, Dharmalingam G, Snijders AP, Carroll T, Capper D, Pritchard C, Inman GJ, Longerich T, Sansom OJ, Benitah SA, Zender L, Gil J (2013) A complex secretory program orchestrated by the inflammasome controls paracrine senescence. Nat Cell Biol 15(8):978–990

Afonso MB, Rodrigues PM, Carvalho T, Caridade M, Borralho P, Cortez-Pinto H, Castro RE, Rodrigues CM (2015) Necroptosis is a key pathogenic event in human and experimental murine models of non-alcoholic steatohepatitis. Clin Sci (Lond) 129(8):721–739

Alonso C, Fernandez-Ramos D, Varela-Rey M, Martinez-Arranz I, Navasa N, Van Liempd SM, Lavin Trueba JL, Mayo R, Ilisso CP, de Juan VG, Iruarrizaga-Lejarreta M, DelaCruz-Villar L, Minchole I, Robinson A, Crespo J, Martin-Duce A, Romero-Gomez M, Sann H, Platon J, Van Eyk J, Aspichueta P, Noureddin M, Falcon-Perez JM, Anguita J, Aransay AM, Martinez-Chantar ML, Lu SC, Mato JM (2017) Metabolomic identification of subtypes of nonalcoholic steatohepatitis. Gastroenterology 152(6):1449–1461 e1447

Anstee QM, Reeves HL, Kotsiliti E, Govaere O, Heikenwalder M (2019) From NASH to HCC: current concepts and future challenges. Nat Rev Gastroenterol Hepatol 16(7):411–428

Aravinthan A (2015) Cellular senescence: a hitchhiker's guide. Hum Cell 28(2):51–64

Aravinthan A, Scarpini C, Tachtatzis P, Verma S, Penrhyn-Lowe S, Harvey R, Davies SE, Allison M, Coleman N, Alexander G (2013) Hepatocyte senescence predicts progression in non-alcohol-related fatty liver disease. J Hepatol 58(3):549–556

Bakkenist CJ, Kastan MB (2003) DNA damage activates ATM through intermolecular autophosphorylation and dimer dissociation. Nature 421(6922):499–506

Bozaykut P, Sahin A, Karademir B, Ozer NK (2016) Endoplasmic reticulum stress related molecular mechanisms in nonalcoholic steatohepatitis. Mech Ageing Dev 157:17–29

Bugianesi E, McCullough AJ, Marchesini G (2005) Insulin resistance: a metabolic pathway to chronic liver disease. Hepatology 42(5):987–1000

Campisi J, d'Adda di Fagagna F (2007) Cellular senescence: when bad things happen to good cells. Nat Rev Mol Cell Biol 8(9):729–740

Cao Y, Mai W, Li R, Deng S, Li L, Zhou Y, Qin Q, Zhang Y, Zhou X, Han M, Liang P, Yan Y, Hao Y, Xie W, Yan J, Zhu L (2022) Macrophages evoke autophagy of hepatic stellate cells to promote liver fibrosis in NAFLD mice via the PGE2/EP4 pathway. Cell Mol Life Sci 79(6):303

Castro ME, Ferrer I, Cascon A, Guijarro MV, Lleonart M, Ramon y Cajal S, Leal JF, Robledo M, Carnero A (2008) PPP1CA contributes to the senescence program induced by oncogenic Ras. Carcinogenesis 29(3):491–499

Chambers JC, Zhang W, Sehmi J, Li X, Wass MN, Van der Harst P, Holm H, Sanna S, Kavousi M, Baumeister SE, Coin LJ, Deng G, Gieger C, Heard-Costa NL, Hottenga JJ, Kuhnel B, Kumar V, Lagou V, Liang L, Luan J, Vidal PM, Leach IM, O'Reilly PF, Peden JF, Rahmioglu N, Soininen P, Speliotes EK, Yuan X, Thorleifsson G, Alizadeh BZ, Atwood LD, Borecki IB, Brown MJ, Charoen P, Cucca F, Das D, de Geus EJ, Dixon AL, Doring A, Ehret G, Eyjolfsson

GI, Farrall M, Forouhi NG, Friedrich N, Goessling W, Gudbjartsson DF, Harris TB, Hartikainen AL, Heath S, Hirschfield GM, Hofman A, Homuth G, Hypponen E, Janssen HL, Johnson T, Kangas AJ, Kema IP, Kuhn JP, Lai S, Lathrop M, Lerch MM, Li Y, Liang TJ, Lin JP, Loos RJ, Martin NG, Moffatt MF, Montgomery GW, Munroe PB, Musunuru K, Nakamura Y, O'Donnell CJ, Olafsson I, Penninx BW, Pouta A, Prins BP, Prokopenko I, Puls R, Ruokonen A, Savolainen MJ, Schlessinger D, Schouten JN, Seedorf U, Sen-Chowdhry S, Siminovitch KA, Smit JH, Spector TD, Tan W, Teslovich TM, Tukiainen T, Uitterlinden AG, Van der Klauw MM, Vasan RS, Wallace C, Wallaschofski H, Wichmann HE, Willemsen G, Wurtz P, Xu C, Yerges-Armstrong LM, Alcohol Genome-wide Association C, Diabetes Genetics R, Meta-analyses S, Genetic Investigation of Anthropometric Traits C, Global Lipids Genetics C, Genetics of Liver Disease C, International Consortium for Blood P, Meta-analyses of, C. Insulin-Related Traits G, Abecasis GR, Ahmadi KR, Boomsma DI, Caulfield M, Cookson WO, van Duijn CM, Froguel P, Matsuda K, McCarthy MI, Meisinger C, Mooser V, Pietilainen KH, Schumann G, Snieder H, Sternberg MJ, Stolk RP, Thomas HC, Thorsteinsdottir U, Uda M, Waeber G, Wareham NJ, Waterworth DM, Watkins H, Whitfield JB, Witteman JC, Wolffenbuttel BH, Fox CS, Ala-Korpela M, Stefansson K, Vollenweider P, Volzke H, Schadt EE, Scott J, Jarvelin MR, Elliott P, Kooner JS (2011) Genome-wide association study identifies loci influencing concentrations of liver enzymes in plasma. Nat Genet 43(11):1131–1138

Chettouh H, Lequoy M, Fartoux L, Vigouroux C, Desbois-Mouthon C (2015) Hyperinsulinaemia and insulin signalling in the pathogenesis and the clinical course of hepatocellular carcinoma. Liver Int 35(10):2203–2217

Cheung O, Sanyal AJ (2010) Role of microRNAs in non-alcoholic steatohepatitis. Curr Pharm Des 16(17):1952–1957

Chrysavgis L, Giannakodimos I, Diamantopoulou P, Cholongitas E (2022) Non-alcoholic fatty liver disease and hepatocellular carcinoma: clinical challenges of an intriguing link. World J Gastroenterol 28(3):310–331

Cohen LJ, Esterhazy D, Kim SH, Lemetre C, Aguilar RR, Gordon EA, Pickard AJ, Cross JR, Emiliano AB, Han SM, Chu J, Vila-Farres X, Kaplitt J, Rogoz A, Calle PY, Hunter C, Bitok JK, Brady SF (2017) Commensal bacteria make GPCR ligands that mimic human signalling molecules. Nature 549(7670):48–53

Coppe JP, Desprez PY, Krtolica A, Campisi J (2010) The senescence-associated secretory phenotype: the dark side of tumor suppression. Annu Rev Pathol 5:99–118

Daugherity EK, Balmus G, Al Saei A, Moore ES, Abi Abdallah D, Rogers AB, Weiss RS, Maurer KJ (2012) The DNA damage checkpoint protein ATM promotes hepatocellular apoptosis and fibrosis in a mouse model of non-alcoholic fatty liver disease. Cell Cycle 11(10):1918–1928

de Conti A, Ortega JF, Tryndyak V, Dreval K, Moreno FS, Rusyn I, Beland FA, Pogribny IP (2017) MicroRNA deregulation in nonalcoholic steatohepatitis-associated liver carcinogenesis. Oncotarget 8(51):88517–88528

Di Micco R, Krizhanovsky V, Baker D, d'Adda di Fagagna F (2021) Cellular senescence in ageing: from mechanisms to therapeutic opportunities. Nat Rev Mol Cell Biol 22(2):75–95

Donnelly KL, Smith CI, Schwarzenberg SJ, Jessurun J, Boldt MD, Parks EJ (2005) Sources of fatty acids stored in liver and secreted via lipoproteins in patients with nonalcoholic fatty liver disease. J Clin Invest 115(5):1343–1351

Dou Z, Ghosh K, Vizioli MG, Zhu J, Sen P, Wangensteen KJ, Simithy J, Lan Y, Lin Y, Zhou Z, Capell BC, Xu C, Xu M, Kieckhaefer JE, Jiang T, Shoshkes-Carmel M, Tanim K, Barber GN, Seykora JT, Millar SE, Kaestner KH, Garcia BA, Adams PD, Berger SL (2017) Cytoplasmic chromatin triggers inflammation in senescence and cancer. Nature 550(7676):402–406

Dyson J, Jaques B, Chattopadyhay D, Lochan R, Graham J, Das D, Aslam T, Patanwala I, Gaggar S, Cole M, Sumpter K, Stewart S, Rose J, Hudson M, Manas D, Reeves HL (2014) Hepatocellular cancer: the impact of obesity, type 2 diabetes and a multidisciplinary team. J Hepatol 60(1):110–117

Ekstedt M, Franzen LE, Mathiesen UL, Thorelius L, Holmqvist M, Bodemar G, Kechagias S (2006) Long-term follow-up of patients with NAFLD and elevated liver enzymes. Hepatology 44(4): 865–873

Eslam M, Valenti L, Romeo S (2018) Genetics and epigenetics of NAFLD and NASH: Clinical impact. J Hepatol 68(2):268–279

European Association for the Study of the L, European Association for the Study of D, European Association for the Study of O (2016) EASL-EASD-EASO clinical practice guidelines for the management of non-alcoholic fatty liver disease. J Hepatol 64:1388–1402

Ferro D, Baratta F, Pastori D, Cocomello N, Colantoni A, Angelico F, Del Ben M (2020) New insights into the pathogenesis of non-alcoholic fatty liver disease: gut-derived lipopolysaccharides and oxidative stress. Nutrients 12(9)

Forner A, Reig M, Bruix J (2018) Hepatocellular carcinoma. Lancet 391(10127):1301–1314

Friedman SL, Neuschwander-Tetri BA, Rinella M, Sanyal AJ (2018) Mechanisms of NAFLD development and therapeutic strategies. Nat Med 24(7):908–922

Fu S, Yang L, Li P, Hofmann O, Dicker L, Hide W, Lin X, Watkins SM, Ivanov AR, Hotamisligil GS (2011) Aberrant lipid metabolism disrupts calcium homeostasis causing liver endoplasmic reticulum stress in obesity. Nature 473(7348):528–531

Gautheron J, Vucur M, Reisinger F, Cardenas DV, Roderburg C, Koppe C, Kreggenwinkel K, Schneider AT, Bartneck M, Neumann UP, Canbay A, Reeves HL, Luedde M, Tacke F, Trautwein C, Heikenwalder M, Luedde T (2014) A positive feedback loop between RIP3 and JNK controls non-alcoholic steatohepatitis. EMBO Mol Med 6(8):1062–1074

Gawrieh S, Dakhoul L, Miller E, Scanga A, deLemos A, Kettler C, Burney H, Liu H, Abu-Sbeih H, Chalasani N, Wattacheril J (2019) Characteristics, aetiologies and trends of hepatocellular carcinoma in patients without cirrhosis: a United States multicentre study. Aliment Pharmacol Ther 50(7):809–821

Georgakilas AG, Martin OA, Bonner WM (2017) p21: a two-faced genome guardian. Trends Mol Med 23(4):310–319

Ghorpade DS, Ozcan L, Zheng Z, Nicoloro SM, Shen Y, Chen E, Bluher M, Czech MP, Tabas I (2018) Hepatocyte-secreted DPP4 in obesity promotes adipose inflammation and insulin resistance. Nature 555(7698):673–677

Gines P, Krag A, Abraldes JG, Sola E, Fabrellas N, Kamath PS (2021) Liver cirrhosis. Lancet 398(10308):1359–1376

Gluck S, Guey B, Gulen MF, Wolter K, Kang TW, Schmacke NA, Bridgeman A, Rehwinkel J, Zender L, Ablasser A (2017) Innate immune sensing of cytosolic chromatin fragments through cGAS promotes senescence. Nat Cell Biol 19(9):1061–1070

Grimaudo S, Pipitone RM, Pennisi G, Celsa C, Camma C, Di Marco V, Barcellona MR, Boemi R, Enea M, Giannetti A, Spatola F, Marchesini G, Craxi A, Petta S (2020) Association between PNPLA3 rs738409 C>G variant and liver-related outcomes in patients with nonalcoholic fatty liver disease. Clin Gastroenterol Hepatol 18(4):935–944 e933

Guerrero R, Vega GL, Grundy SM, Browning JD (2009) Ethnic differences in hepatic steatosis: an insulin resistance paradox? Hepatology 49(3):791–801

Gustafson B, Nerstedt A, Smith U (2019) Reduced subcutaneous adipogenesis in human hypertrophic obesity is linked to senescent precursor cells. Nat Commun 10(1):2757

Harada K, Furubo S, Ozaki S, Hiramatsu K, Sudo Y, Nakanuma Y (2001) Increased expression of WAF1 in intrahepatic bile ducts in primary biliary cirrhosis relates to apoptosis. J Hepatol 34(4): 500–506

Helsley RN, Varadharajan V, Brown AL, Gromovsky AD, Schugar RC, Ramachandiran I, Fung K, Kabbany MN, Banerjee R, Neumann CK, Finney C, Pathak P, Orabi D, Osborn LJ, Massey W, Zhang R, Kadam A, Sansbury BE, Pan C, Sacks J, Lee RG, Crooke RM, Graham MJ, Lemieux ME, Gogonea V, Kirwan JP, Allende DS, Civelek M, Fox PL, Rudel LL, Lusis AJ, Spite M, Brown JM (2019) Obesity-linked suppression of membrane-bound O-acyltransferase 7 (MBOAT7) drives non-alcoholic fatty liver disease. Elife 8

Henao-Mejia J, Elinav E, Jin C, Hao L, Mehal WZ, Strowig T, Thaiss CA, Kau AL, Eisenbarth SC, Jurczak MJ, Camporez JP, Shulman GI, Gordon JI, Hoffman HM, Flavell RA (2012) Inflammasome-mediated dysbiosis regulates progression of NAFLD and obesity. Nature 482(7384):179–185

Hickson LJ, Langhi Prata LGP, Bobart SA, Evans TK, Giorgadze N, Hashmi SK, Herrmann SM, Jensen MD, Jia Q, Jordan KL, Kellogg TA, Khosla S, Koerber DM, Lagnado AB, Lawson DK, LeBrasseur NK, Lerman LO, McDonald KM, McKenzie TJ, Passos JF, Pignolo RJ, Pirtskhalava T, Saadiq IM, Schaefer KK, Textor SC, Victorelli SG, Volkman TL, Xue A, Wentworth MA, Wissler Gerdes EO, Zhu Y, Tchkonia T, Kirkland JL (2019) Senolytics decrease senescent cells in humans: preliminary report from a clinical trial of Dasatinib plus Quercetin in individuals with diabetic kidney disease. EBioMedicine 47:446–456

Horvath S (2013) DNA methylation age of human tissues and cell types. Genome Biol 14(10):R115

Hotta K, Kitamoto A, Kitamoto T, Ogawa Y, Honda Y, Kessoku T, Yoneda M, Imajo K, Tomeno W, Saito S, Nakajima A (2018) Identification of differentially methylated region (DMR) networks associated with progression of nonalcoholic fatty liver disease. Sci Rep 8(1):13567

Huang Y, He S, Li JZ, Seo YK, Osborne TF, Cohen JC, Hobbs HH (2010) A feed-forward loop amplifies nutritional regulation of PNPLA3. Proc Natl Acad Sci U S A 107(17):7892–7897

Huang DQ, El-Serag HB, Loomba R (2021) Global epidemiology of NAFLD-related HCC: trends, predictions, risk factors and prevention. Nat Rev Gastroenterol Hepatol 18(4):223–238

Ioannou GN (2016) The role of cholesterol in the pathogenesis of NASH. Trends Endocrinol Metab 27(2):84–95

Itahana K, Campisi J, Dimri GP (2007) Methods to detect biomarkers of cellular senescence: the senescence-associated beta-galactosidase assay. Methods Mol Biol 371:21–31

Jackson SP, Bartek J (2009) The DNA-damage response in human biology and disease. Nature 461(7267):1071–1078

Janku F, Kaseb AO, Tsimberidou AM, Wolff RA, Kurzrock R (2014) Identification of novel therapeutic targets in the PI3K/AKT/mTOR pathway in hepatocellular carcinoma using targeted next generation sequencing. Oncotarget 5(10):3012–3022

Justice JN, Nambiar AM, Tchkonia T, LeBrasseur NK, Pascual R, Hashmi SK, Prata L, Masternak MM, Kritchevsky SB, Musi N, Kirkland JL (2019) Senolytics in idiopathic pulmonary fibrosis: Results from a first-in-human, open-label, pilot study. EBioMedicine 40:554–563

Kang TW, Yevsa T, Woller N, Hoenicke L, Wuestefeld T, Dauch D, Hohmeyer A, Gereke M, Rudalska R, Potapova A, Iken M, Vucur M, Weiss S, Heikenwalder M, Khan S, Gil J, Bruder D, Manns M, Schirmacher P, Tacke F, Ott M, Luedde T, Longerich T, Kubicka S, Zender L (2011) Senescence surveillance of pre-malignant hepatocytes limits liver cancer development. Nature 479(7374):547–551

Kanwal F, Kramer JR, Mapakshi S, Natarajan Y, Chayanupatkul M, Richardson PA, Li L, Desiderio R, Thrift AP, Asch SM, Chu J, El-Serag HB (2018) Risk of hepatocellular cancer in patients with non-alcoholic fatty liver disease. Gastroenterology 155(6):1828–1837 e1822

Karakousis ND, Chrysavgis L, Chatzigeorgiou A, Papatheodoridis G, Cholongitas E (2022) Frailty in metabolic syndrome, focusing on nonalcoholic fatty liver disease. Ann Gastroenterol 35(3): 234–242

Katsarou A, Moustakas I II, Pyrina P, Lembessis MK, Chatzigeorgiou A (2020) Metabolic inflammation as an instigator of fibrosis during non-alcoholic fatty liver disease. World J Gastroenterol 26(17):1993–2011

Kawamura Y, Arase Y, Ikeda K, Seko Y, Imai N, Hosaka T, Kobayashi M, Saitoh S, Sezaki H, Akuta N, Suzuki F, Suzuki Y, Ohmoto Y, Amakawa K, Tsuji H, Kumada H (2012) Large-scale long-term follow-up study of Japanese patients with non-alcoholic Fatty liver disease for the onset of hepatocellular carcinoma. Am J Gastroenterol 107(2):253–261

Khosla S, Farr JN, Tchkonia T, Kirkland JL (2020) The role of cellular senescence in ageing and endocrine disease. Nat Rev Endocrinol 16(5):263–275

Krizhanovsky V, Yon M, Dickins RA, Hearn S, Simon J, Miething C, Yee H, Zender L, Lowe SW (2008) Senescence of activated stellate cells limits liver fibrosis. Cell 134(4):657–667
Kudo Y, Tanaka Y, Tateishi K, Yamamoto K, Yamamoto S, Mohri D, Isomura Y, Seto M, Nakagawa H, Asaoka Y, Tada M, Ohta M, Ijichi H, Hirata Y, Otsuka M, Ikenoue T, Maeda S, Shiina S, Yoshida H, Nakajima O, Kanai F, Omata M, Koike K (2011) Altered composition of fatty acids exacerbates hepatotumorigenesis during activation of the phosphatidylinositol 3-kinase pathway. J Hepatol 55(6):1400–1408
Laberge RM, Zhou L, Sarantos MR, Rodier F, Freund A, de Keizer PL, Liu S, Demaria M, Cong YS, Kapahi P, Desprez PY, Hughes RE, Campisi J (2012) Glucocorticoids suppress selected components of the senescence-associated secretory phenotype. Aging Cell 11(4):569–578
Larsson SC, Wolk A (2007) Overweight, obesity and risk of liver cancer: a meta-analysis of cohort studies. Br J Cancer 97(7):1005–1008
Lee S, Schmitt CA (2019) The dynamic nature of senescence in cancer. Nat Cell Biol 21(1):94–101
Lee JY, Bielawska AE, Obeid LM (2000) Regulation of cyclin-dependent kinase 2 activity by ceramide. Exp Cell Res 261(2):303–311
Lee JS, Yoo JE, Kim H, Rhee H, Koh MJ, Nahm JH, Choi JS, Lee KH, Park YN (2017) Tumor stroma with senescence-associated secretory phenotype in steatohepatitic hepatocellular carcinoma. PLoS One 12(3):e0171922
Legaki AI, Moustakas II, Sikorska M, Papadopoulos G, Velliou RI, Chatzigeorgiou A (2022) Hepatocyte mitochondrial dynamics and bioenergetics in obesity-related non-alcoholic fatty liver disease. Rep, Curr Obes
Leung C, Rivera L, Furness JB, Angus PW (2016) The role of the gut microbiota in NAFLD. Nat Rev Gastroenterol Hepatol 13(7):412–425
Li F, Huangyang P, Burrows M, Guo K, Riscal R, Godfrey J, Lee KE, Lin N, Lee P, Blair IA, Keith B, Li B, Simon MC (2020) FBP1 loss disrupts liver metabolism and promotes tumorigenesis through a hepatic stellate cell senescence secretome. Nat Cell Biol 22(6):728–739
Liu F, Li X, Lu C, Bai A, Bielawski J, Bielawska A, Marshall B, Schoenlein PV, Lebedyeva IO, Liu K (2016) Ceramide activates lysosomal cathepsin B and cathepsin D to attenuate autophagy and induces ER stress to suppress myeloid-derived suppressor cells. Oncotarget 7(51):83907–83925
Loo TM, Kamachi F, Watanabe Y, Yoshimoto S, Kanda H, Arai Y, Nakajima-Takagi Y, Iwama A, Koga T, Sugimoto Y, Ozawa T, Nakamura M, Kumagai M, Watashi K, Taketo MM, Aoki T, Narumiya S, Oshima M, Arita M, Hara E, Ohtani N (2017) Gut microbiota promotes obesity-associated liver cancer through PGE2-mediated suppression of antitumor immunity. Cancer Discov 7(5):522–538
Loomba R, Abraham M, Unalp A, Wilson L, Lavine J, Doo E, Bass NM, N. Nonalcoholic Steatohepatitis Clinical Research (2012) Association between diabetes, family history of diabetes, and risk of nonalcoholic steatohepatitis and fibrosis. Hepatology 56(3):943–951
Loomba R, Gindin Y, Jiang Z, Lawitz E, Caldwell S, Djedjos CS, Xu R, Chung C, Myers RP, Subramanian GM, Goodman Z, Charlton M, Afdhal NH, Diehl AM (2018) DNA methylation signatures reflect aging in patients with nonalcoholic steatohepatitis. JCI Insight 3(2)
Lopez-Otin C, Blasco MA, Partridge L, Serrano M, Kroemer G (2013) The hallmarks of aging. Cell 153(6):1194–1217
Luukkonen PK, Zhou Y, Sadevirta S, Leivonen M, Arola J, Oresic M, Hyotylainen T, Yki-Jarvinen H (2016) Hepatic ceramides dissociate steatosis and insulin resistance in patients with non-alcoholic fatty liver disease. J Hepatol 64(5):1167–1175
Ma Y, Belyaeva OV, Brown PM, Fujita K, Valles K, Karki S, de Boer YS, Koh C, Chen Y, Du X, Handelman SK, Chen V, Speliotes EK, Nestlerode C, Thomas E, Kleiner DE, Zmuda JM, Sanyal AJ, Kedishvili NY, Liang TJ, Rotman Y (2019) 17-beta hydroxysteroid dehydrogenase 13 Is a hepatic retinol dehydrogenase associated with histological features of nonalcoholic fatty liver disease. Hepatology 69(4):1504–1519
Marra F, Svegliati-Baroni G (2018) Lipotoxicity and the gut-liver axis in NASH pathogenesis. J Hepatol 68(2):280–295
Mas E, Danjoux M, Garcia V, Carpentier S, Segui B, Levade T (2009) IL-6 deficiency attenuates murine diet-induced non-alcoholic steatohepatitis. PLoS One 4(11):e7929

Moiseeva O, Deschenes-Simard X, St-Germain E, Igelmann S, Huot G, Cadar AE, Bourdeau V, Pollak MN, Ferbeyre G (2013) Metformin inhibits the senescence-associated secretory phenotype by interfering with IKK/NF-kappaB activation. Aging Cell 12(3):489–498

Moustakas A II, Katsarou AI, Legaki I, Pyrina K, Ntostoglou AM, Papatheodoridi B, Gercken IS, Pateras VG, Gorgoulis M, Koutsilieris TC, Chatzigeorgiou A (2021) Hepatic senescence accompanies the development of NAFLD in non-aged mice independently of obesity. Int J Mol Sci 22(7)

Nakajima T, Nakashima T, Okada Y, Jo M, Nishikawa T, Mitsumoto Y, Katagishi T, Kimura H, Itoh Y, Kagawa K, Yoshikawa T (2010) Nuclear size measurement is a simple method for the assessment of hepatocellular aging in non-alcoholic fatty liver disease: comparison with telomere-specific quantitative FISH and p21 immunohistochemistry. Pathol Int 60(3):175–183

Nati M, Haddad D, Birkenfeld AL, Koch CA, Chavakis T, Chatzigeorgiou A (2016) The role of immune cells in metabolism-related liver inflammation and development of non-alcoholic steatohepatitis (NASH). Rev Endocr Metab Disord 17(1):29–39

Nishida N, Yada N, Hagiwara S, Sakurai T, Kitano M, Kudo M (2016) Unique features associated with hepatic oxidative DNA damage and DNA methylation in non-alcoholic fatty liver disease. J Gastroenterol Hepatol 31(9):1646–1653

Nishizawa H, Iguchi G, Fukuoka H, Takahashi M, Suda K, Bando H, Matsumoto R, Yoshida K, Odake Y, Ogawa W, Takahashi Y (2016) IGF-I induces senescence of hepatic stellate cells and limits fibrosis in a p53-dependent manner. Sci Rep 6:34605

Ogresta D, Mrzljak A, Cigrovski Berkovic M, Bilic-Curcic I, Stojsavljevic-Shapeski S, Virovic-Jukic L (2022) Coagulation and endothelial dysfunction associated with NAFLD: current status and therapeutic implications. J Clin Transl Hepatol 10(2):339–355

Ogrodnik M, Miwa S, Tchkonia T, Tiniakos D, Wilson CL, Lahat A, Day CP, Burt A, Palmer A, Anstee QM, Grellscheid SN, Hoeijmakers JHJ, Barnhoorn S, Mann DA, Bird TG, Vermeij WP, Kirkland JL, Passos JF, von Zglinicki T, Jurk D (2017) Cellular senescence drives age-dependent hepatic steatosis. Nat Commun 8:15691

Ogrodnik M, Salmonowicz H, Gladyshev VN (2019) Integrating cellular senescence with the concept of damage accumulation in aging: relevance for clearance of senescent cells. Aging Cell 18(1):e12841

Papatheodoridi AM, Chrysavgis L, Koutsilieris M, Chatzigeorgiou A (2020) The role of senescence in the development of nonalcoholic fatty liver disease and progression to nonalcoholic steatohepatitis. Hepatology 71(1):363–374

Paradis V, Youssef N, Dargere D, Ba N, Bonvoust F, Deschatrette J, Bedossa P (2001) Replicative senescence in normal liver, chronic hepatitis C, and hepatocellular carcinomas. Hum Pathol 32(3):327–332

Pellicoro A, Ramachandran P, Iredale JP, Fallowfield JA (2014) Liver fibrosis and repair: immune regulation of wound healing in a solid organ. Nat Rev Immunol 14(3):181–194

Perry RJ, Samuel VT, Petersen KF, Shulman GI (2014) The role of hepatic lipids in hepatic insulin resistance and type 2 diabetes. Nature 510(7503):84–91

Ping F, Li ZY, Lv K, Zhou MC, Dong YX, Sun Q, Li YX (2017) Deoxyribonucleic acid telomere length shortening can predict the incidence of non-alcoholic fatty liver disease in patients with type 2 diabetes mellitus. J Diabetes Investig 8(2):174–180

Piscaglia F, Svegliati-Baroni G, Barchetti A, Pecorelli A, Marinelli S, Tiribelli C, Bellentani S, H.-N. I. S. Group (2016) Clinical patterns of hepatocellular carcinoma in nonalcoholic fatty liver disease: a multicenter prospective study. Hepatology 63(3):827–838

Regulski MJ (2017) Cellular senescence: what, why, and how. Wounds 29(6):168–174

Romeo S, Kozlitina J, Xing C, Pertsemlidis A, Cox D, Pennacchio LA, Boerwinkle E, Cohen JC, Hobbs HH (2008) Genetic variation in PNPLA3 confers susceptibility to nonalcoholic fatty liver disease. Nat Genet 40(12):1461–1465

Rossi DJ, Bryder D, Seita J, Nussenzweig A, Hoeijmakers J, Weissman IL (2007) Deficiencies in DNA damage repair limit the function of haematopoietic stem cells with age. Nature 447(7145): 725–729

Rubin SM, Gall AL, Zheng N, Pavletich NP (2005) Structure of the Rb C-terminal domain bound to E2F1-DP1: a mechanism for phosphorylation-induced E2F release. Cell 123(6):1093–1106
Sanyal A, Poklepovic A, Moyneur E, Barghout V (2010) Population-based risk factors and resource utilization for HCC: US perspective. Curr Med Res Opin 26(9):2183–2191
Sasaki M, Ikeda H, Yamaguchi J, Miyakoshi M, Sato Y, Nakanuma Y (2010) Bile ductular cells undergoing cellular senescence increase in chronic liver diseases along with fibrous progression. Am J Clin Pathol 133(2):212–223
Saxena NK, Sharma D, Ding X, Lin S, Marra F, Merlin D, Anania FA (2007) Concomitant activation of the JAK/STAT, PI3K/AKT, and ERK signaling is involved in leptin-mediated promotion of invasion and migration of hepatocellular carcinoma cells. Cancer Res 67(6): 2497–2507
Schnabl B, Purbeck CA, Choi YH, Hagedorn CH, Brenner D (2003) Replicative senescence of activated human hepatic stellate cells is accompanied by a pronounced inflammatory but less fibrogenic phenotype. Hepatology 37(3):653–664
Schulze K, Imbeaud S, Letouze E, Alexandrov LB, Calderaro J, Rebouissou S, Couchy G, Meiller C, Shinde J, Soysouvanh F, Calatayud AL, Pinyol R, Pelletier L, Balabaud C, Laurent A, Blanc JF, Mazzaferro V, Calvo F, Villanueva A, Nault JC, Bioulac-Sage P, Stratton MR, Llovet JM, Zucman-Rossi J (2015) Exome sequencing of hepatocellular carcinomas identifies new mutational signatures and potential therapeutic targets. Nat Genet 47(5):505–511
Singh M, Jensen MD, Lerman A, Kushwaha S, Rihal CS, Gersh BJ, Behfar A, Tchkonia T, Thomas RJ, Lennon RJ, Keenan LR, Moore AG, Kirkland JL (2016) Effect of low-dose rapamycin on senescence markers and physical functioning in older adults with coronary artery disease: results of a pilot study. J Frailty Aging 5(4):204–207
Skurk T, Alberti-Huber C, Herder C, Hauner H (2007) Relationship between adipocyte size and adipokine expression and secretion. J Clin Endocrinol Metab 92(3):1023–1033
Speliotes EK, Yerges-Armstrong LM, Wu J, Hernaez R, Kim LJ, Palmer CD, Gudnason V, Eiriksdottir G, Garcia ME, Launer LJ, Nalls MA, Clark JM, Mitchell BD, Shuldiner AR, Butler JL, Tomas M, Hoffmann U, Hwang SJ, Massaro JM, O'Donnell CJ, Sahani DV, Salomaa V, Schadt EE, Schwartz SM, Siscovick DS, Nash CRN, Consortium G, Investigators M, Voight BF, Carr JJ, Feitosa MF, Harris TB, Fox CS, Smith AV, Kao WH, Hirschhorn JN, Borecki IB, G. Consortium (2011) Genome-wide association analysis identifies variants associated with nonalcoholic fatty liver disease that have distinct effects on metabolic traits. PLoS Genet 7(3): e1001324
Stine JG, Wentworth BJ, Zimmet A, Rinella ME, Loomba R, Caldwell SH, Argo CK (2018) Systematic review with meta-analysis: risk of hepatocellular carcinoma in non-alcoholic steatohepatitis without cirrhosis compared to other liver diseases. Aliment Pharmacol Ther 48(7):696–703
Szabo G, Petrasek J (2015) Inflammasome activation and function in liver disease. Nat Rev Gastroenterol Hepatol 12(7):387–400
Tacke F, Weiskirchen R (2021) Non-alcoholic fatty liver disease (NAFLD)/non-alcoholic steatohepatitis (NASH)-related liver fibrosis: mechanisms, treatment and prevention. Ann Transl Med 9(8):729
Takaki Y, Saito Y, Takasugi A, Toshimitsu K, Yamada S, Muramatsu T, Kimura M, Sugiyama K, Suzuki H, Arai E, Ojima H, Kanai Y, Saito H (2014) Silencing of microRNA-122 is an early event during hepatocarcinogenesis from non-alcoholic steatohepatitis. Cancer Sci 105(10): 1254–1260
Targher G, Byrne CD, Lonardo A, Zoppini G, Barbui C (2016) Non-alcoholic fatty liver disease and risk of incident cardiovascular disease: A meta-analysis. J Hepatol 65(3):589–600
Thorgeirsson SS, Grisham JW (2002) Molecular pathogenesis of human hepatocellular carcinoma. Nat Genet 31(4):339–346
Tornesello ML, Buonaguro L, Tatangelo F, Botti G, Izzo F, Buonaguro FM (2013) Mutations in TP53, CTNNB1 and PIK3CA genes in hepatocellular carcinoma associated with hepatitis B and hepatitis C virus infections. Genomics 102(2):74–83

Torre LA, Bray F, Siegel RL, Ferlay J, Lortet-Tieulent J, Jemal A (2015) Global cancer statistics, 2012. CA Cancer J Clin 65(2):87–108

Trepo E, Valenti L (2020) Update on NAFLD genetics: from new variants to the clinic. J Hepatol 72(6):1196–1209

Tryndyak VP, Han T, Fuscoe JC, Ross SA, Beland FA, Pogribny IP (2016) Status of hepatic DNA methylome predetermines and modulates the severity of non-alcoholic fatty liver injury in mice. BMC Genomics 17:298

Turati F, Talamini R, Pelucchi C, Polesel J, Franceschi S, Crispo A, Izzo F, La Vecchia C, Boffetta P, Montella M (2013) Metabolic syndrome and hepatocellular carcinoma risk. Br J Cancer 108(1):222–228

Vinciguerra M, Carrozzino F, Peyrou M, Carlone S, Montesano R, Benelli R, Foti M (2009) Unsaturated fatty acids promote hepatoma proliferation and progression through downregulation of the tumor suppressor PTEN. J Hepatol 50(6):1132–1141

Wang B, Li X, Hu W, Zhou Y, Din Y (2019) Silencing of lncRNA SNHG20 delays the progression of nonalcoholic fatty liver disease to hepatocellular carcinoma via regulating liver Kupffer cells polarization. IUBMB Life 71(12):1952–1961

Wang P, Wu CX, Li Y, Shen N (2020) HSD17B13 rs72613567 protects against liver diseases and histological progression of nonalcoholic fatty liver disease: a systematic review and meta-analysis. Eur Rev Med Pharmacol Sci 24(17):8997–9007

Wei W, Ji S (2018) Cellular senescence: molecular mechanisms and pathogenicity. J Cell Physiol 233(12):9121–9135

Wiemann SU, Satyanarayana A, Tsahuridu M, Tillmann HL, Zender L, Klempnauer J, Flemming P, Franco S, Blasco MA, Manns MP, Rudolph KL (2002) Hepatocyte telomere shortening and senescence are general markers of human liver cirrhosis. FASEB J 16(9):935–942

Wooten-Blanks LG, Song P, Senkal CE, Ogretmen B (2007) Mechanisms of ceramide-mediated repression of the human telomerase reverse transcriptase promoter via deacetylation of Sp3 by histone deacetylase 1. FASEB J 21(12):3386–3397

Wree A, Schlattjan M, Bechmann LP, Claudel T, Sowa JP, Stojakovic T, Scharnagl H, Kofeler H, Baba HA, Gerken G, Feldstein AE, Trauner M, Canbay A (2014) Adipocyte cell size, free fatty acids and apolipoproteins are associated with non-alcoholic liver injury progression in severely obese patients. Metabolism 63(12):1542–1552

Wu H, Ng R, Chen X, Steer CJ, Song G (2016) MicroRNA-21 is a potential link between non-alcoholic fatty liver disease and hepatocellular carcinoma via modulation of the HBP1-p53-Srebp1c pathway. Gut 65(11):1850–1860

Xu M, Tchkonia T, Ding H, Ogrodnik M, Lubbers ER, Pirtskhalava T, White TA, Johnson KO, Stout MB, Mezera V, Giorgadze N, Jensen MD, LeBrasseur NK, Kirkland JL (2015) JAK inhibition alleviates the cellular senescence-associated secretory phenotype and frailty in old age. Proc Natl Acad Sci U S A 112(46):E6301–E6310

Xue W, Zender L, Miething C, Dickins RA, Hernando E, Krizhanovsky V, Cordon-Cardo C, Lowe SW (2007) Senescence and tumour clearance is triggered by p53 restoration in murine liver carcinomas. Nature 445(7128):656–660

Yang G, Badeanlou L, Bielawski J, Roberts AJ, Hannun YA, Samad F (2009) Central role of ceramide biosynthesis in body weight regulation, energy metabolism, and the metabolic syndrome. Am J Physiol Endocrinol Metab 297(1):E211–E224

Yang JD, Ahmed F, Mara KC, Addissie BD, Allen AM, Gores GJ, Roberts LR (2020) Diabetes is associated with increased risk of hepatocellular carcinoma in patients with cirrhosis from nonalcoholic fatty liver disease. Hepatology 71(3):907–916

Yoshimoto S, Loo TM, Atarashi K, Kanda H, Sato S, Oyadomari S, Iwakura Y, Oshima K, Morita H, Hattori M, Honda K, Ishikawa Y, Hara E, Ohtani N (2013) Obesity-induced gut microbial metabolite promotes liver cancer through senescence secretome. Nature 499(7456): 97–101

Yoshioka N, Tanaka M, Ochi K, Watanabe A, Ono K, Sawada M, Ogi T, Itoh M, Ito A, Shiraki Y, Enomoto A, Ishigami M, Fujishiro M, Ogawa Y, Suganami T (2021) The sodium-glucose cotransporter-2 inhibitor Tofogliflozin prevents the progression of nonalcoholic steatohepatitis-associated liver tumors in a novel murine model. Biomed Pharmacother 140:111738

Younossi ZM, Koenig AB, Abdelatif D, Fazel Y, Henry L, Wymer M (2016) Global epidemiology of nonalcoholic fatty liver disease-Meta-analytic assessment of prevalence, incidence, and outcomes. Hepatology 64(1):73–84

Zhang X, Zhou D, Strakovsky R, Zhang Y, Pan YX (2012) Hepatic cellular senescence pathway genes are induced through histone modifications in a diet-induced obese rat model. Am J Physiol Gastrointest Liver Physiol 302(5):G558–G564

Zhang X, Wu M, Liu Z, Yuan H, Wu X, Shi T, Chen X, Zhang T (2021) Increasing prevalence of NAFLD/NASH among children, adolescents and young adults from 1990 to 2017: a population-based observational study. BMJ Open 11(5):e042843

Regulators of Genetic Risk for the Progression of Non-alcoholic Fatty Liver Disease to Hepatocellular Carcinoma: Reconstruction of Transcriptional Network and Signature-Based Metabolic Profiling

Dora Lucía Vallejo-Ardila, Marco A. De Bastiani, and Diego A. Salazar

Abstract

The spectrum of non-alcoholic fatty liver disease (NAFLD) extends from mild hepatic steatosis to non-alcoholic steatohepatitis (NASH), and by a combination of multiple factors can progress to hepatocellular carcinoma (HCC). However, the biological mechanism underlying the progression to HCC had never been explored systematically, using publicly available data and integrating independent transcriptomic and metabolomic studies.

Herein, we reconstructed two regulatory networks centered on transcriptional factors and their putative targets and analyzed whether these regulons were enriched for deregulated genes across the three different phenotypes – non-alcoholic fatty liver or steatosis, NASH, and HCC – against two reference network backgrounds. One network was normal liver (*TN1*), and the other was cirrhotic liver (*TN2*). Then, we used Recon 2.2 to model the human metabolic effect of gene products regulated by the intersected group of master regulators – *AEBP1*,

Supplementary Information The online version contains supplementary material available at [https://doi.org/10.1007/16833_2022_40].

D. L. Vallejo-Ardila (✉)
Colombian Research Group for Clinical and Molecular Research in Cancer – ONCOLGroup, Bogota, Colombia
e-mail: dlucamed@gmail.com

M. A. De Bastiani
Department of Biochemistry, Universidade Federal do Rio Grande do Sul (UFRGS), Porto Alegre, Rio Grande do Sul, Brazil
e-mail: marco.bastiani@ufrgs.br

D. A. Salazar
Industrial Engineering Department, University of los Andes, Bogota, Colombia

Center for Optimization and Applied Probability, University of los Andes, Bogota, Colombia
e-mail: dazb21@gmail.com

Interdisciplinary Cancer Research, https://doi.org/10.1007/16833_2022_40
Published online: 1 December 2022

AR, and *NR1I3*. Additionally, we performed a parsimonious flux balance analysis coupled with KEGG enrichment analysis to identify the differences in metabolic flux between the *NAFLD-related HCC model* and *normal*.

We found that only *AR*, *AEBP1*, and *NR1I3* are the intersected group of master regulators that may constitute a genetic risk for NAFLD progression to HCC and that these findings had a high stringency threshold (*p-value* = 0.001) with consensus across both inference networks – *TN1* and *TN2*. Of additional importance, we explored the number of transcriptional targets they affected and how closely they were connected within the transcriptional network. Finally, we implemented a signature-based metabolic reconstruction of NAFLD-related HCC in Recon 2.2. We found that differential mode of action of *AR*, *AEBP1*, and *NR1I3* altered several metabolic reactions involved in fatty acid metabolism; pyruvate metabolism; beta-alanine metabolism; valine, leucine, and isoleucine degradation; citrate cycle and nucleotide metabolism; and fatty acid degradation.

Our integrative systems biology approach provides a framework to understand the role of *AR*, *AEBP1*, and *NR1I3* shared regulatory circuits governing the progression of NAFLD-to-HCC and to identify susceptibility factors, potential biochemical biomarkers, and therapeutic targets. In addition, the pipeline of analyses is transferable to other disease settings.

Keywords

Hepatocellular carcinoma · Non-alcoholic fatty liver disease · Signature-based metabolic profiling · Systems biology · Transcriptional regulatory units

Abbreviations

ABAT	4-aminobutyrate aminotransferase/(S)-3-amino-2-methylpropionate transaminase
ACAA1	Acetyl-CoA acyltransferase 1
AEBP1	Adipocyte enhancer-binding protein 1
AP-1	Activator protein-1
AR	Androgen receptor
ARACNe	Algorithm for the Reconstruction of Accurate Cellular Networks
AU	Approximately unbiased
BCAAs	Branched-chain amino acids
BP	Bootstrap probability
CAD	Carbamoyltransferase/dihydroorotase
COBRA	COnstraint-Based Reconstruction and Analysis
CPT1A	Carnitine palmitoyltransferase 1
CPT2	Carnitine O-palmitoyltransferase 2
DLD	Dihydrolipoamide dehydrogenase
DPI	Data processing inequality

DPYS	Dihydropyrimidinase
ES	Enrichment scores
ES	Expression signatures
FFA	Free fatty acids
GSEA	Gene set enrichment analysis
GYS	Glycogen synthase
H3K4Me1 and H3K4Me2	Monomethyl and dimethyl Lys4 of histone H3
HADH	3-hydroxyacyl-CoA dehydrogenase
HCC	Hepatocellular carcinoma
HIF-1α	Hypoxia-inducible factor 1-alpha
HMGCS	Hydroxymethylglutaryl-CoA synthase
HSC	Hepatic stellate cells
Ile	Isoleucine
JNK	c-Jun N-terminal kinase
KEGG	Kyoto Encyclopedia of Genes and Genomes
Leu	Leucine
LPS	Lipopolysaccharide
LSD1	Lysine-specific demethylase 1
MI	Mutual information
MoA	Mode of action
MR	Master regulator
MS	Metabolic syndrome
NAD+	Aldehyde dehydrogenase
NAFL	Non-alcoholic fatty liver
NAFLD	Non-alcoholic fatty liver disease
NASH	Non-alcoholic steatohepatitis
NR1I3	Nuclear receptor subfamily 1 group I member 3
NF-κB	Nuclear factor kappa beta
PNPLA3	Patatin-like phospholipase domain-containing protein 3
Nrf2	Nuclear factor erythroid 2-related factor 2
pFBA	Parsimonious flux balance analysis
PPAR-γ	Peroxisome proliferator-activated receptor-γ
ROS	Reactive oxygen species
RTN	Reconstruction of transcriptional network and analysis of master regulators
SREs	Sterol-regulatory element
SREBPs	Sterol-regulatory element-binding proteins
TFs	Transcriptional factors
Val	Valine

1 Introduction

Non-alcoholic fatty liver disease (NAFLD) is now one of the most prevalent liver pathologies worldwide and is considered to be the hepatic manifestation of metabolic syndrome (MS) (Godoy-Matos et al. 2020; Pouwels et al. 2022). In addition, common clinical factors including obesity, insulin resistance, hyperlipidemia (Godoy-Matos et al. 2020), and hypertension bear an increased risk of developing NAFLD-related hepatocellular carcinoma (HCC) (Kanwal et al. 2020). Moreover, the incidence of NAFLD-related HCC is rapidly increasing, to the point that it is the fastest-growing etiology of HCC (Estes et al. 2018) in the United States (McGlynn et al. 2021), Europe (Dyson et al. 2014), South Korea (Cho et al. 2011), and Southeast Asia (Liew et al. 2019) over the past two decades (Huang et al. 2021). However, there is limited high-level evidence to demonstrate the value of HCC surveillance among the NAFLD population because the basis of this association is epidemiological and inferred from the association between MS and long-term diabetes type 2 as susceptibility factors to developing HCC (Dyson et al. 2014; Streba et al. 2015; McGlynn et al. 2021).

NAFLD diagnosis relies on grading and histological assessment using liver biopsy, which is limited not only by the risk involved in a highly invasive procedure to collect a small amount of tissue for examination relative to the size of the liver but also due to the pathologist variability in the scoring interpretation (Vuppalanchi et al. 2009). Noninvasive diagnostic assessments aim to reduce the need for liver biopsy; nevertheless, an individual patient data meta-analysis of studies evaluating liver stiffness measurement by vibration controlled transient elastography against liver histology has shown a moderate accuracy in detecting NAFLD in a high-risk population (Mózes et al. 2022). Although genetic variability in patatin-like phospholipase domain-containing protein 3 (*PNPLA3*) confers high cross-ethnicity risk for NAFLD and NASH development (Romeo et al. 2008; Sookoian and Pirola 2011; Xu et al. 2015), its value to predict NAFLD-related HCC remains uncertain. NAFLD genetic heterogeneity poses a significant challenge in identifying new biomarkers for the early recognition of non-cirrhotic NAFLD patients who will develop HCC (Grgurevic et al. 2021).

The spectrum of NAFLD ranges from mild hepatic steatosis to non-alcoholic steatohepatitis (NASH) with significant necroinflammation (Weiß et al. 2014). In most cases, NAFLD-related HCCs arise from the cirrhotic liver (70–80%). However, 20–30% of the patients develop HCC without cirrhosis (Kanwal et al. 2018), which further adds to the complexity and multifactorial nature of NAFLD progression to HCC (Friedman et al. 2018). In addition, hepatocarcinogenesis may appear in earlier stages of NAFLD (Takaki et al. 2014). Indeed, experimental models of NAFLD have also confirmed an increased incidence of HCC with aging in the absence of overt hepatic fibrosis (Baffy et al. 2012).

The molecular basis and genetic risk for progression to HCC remain a current topic of investigation (Nahon et al. 2020). The nuclear factor kappa beta (NF-κB) and c-Jun N-terminal kinase (JNK)/activator protein-1 (AP-1), p53, hypoxia-inducible factor 1-alpha (HIF-1α), peroxisome proliferator-activated receptor-γ

(PPAR-γ), β-catenin/Wnt, and nuclear factor erythroid 2-related factor *2* (Nrf2) play a crucial role in the progression of NAFLD to HCC (Stickel and Hellerbrand 2010; Waidmann et al. 2020). Among the regulators involved in the inflammation–fibrosis–cancer process, NF-kB, AP-1, and PPARs can be directly modulated by persistent redox imbalance, promoting chronic inflammation by inducing the transcription of cytokines and enhancing fibrosis deposition (Wang et al. 2014). Thus, the tumor immune microenvironment is reprogrammed toward an immune-suppressed response promoted by Th2-type and myeloid-derived suppressor cells (Dallio et al. 2021). Yet, these biological processes are governed by transcriptional factors (TFs) modulating overall gene expression and not by specific driver genes that exhibit differences in expression downstream (MacNeil and Walhout 2011).

Over 40 years ago, Susumu Ohno's first proposed the use of a "master regulator" to define a "gene that occupies the very top of a regulatory hierarchy" (Ohno 1979). For instance, sterol-regulatory element-binding protein (SREBP) is a transcriptional regulator of lipid synthetic genes for cholesterol and fatty acid synthesis, which also have been identified as highly activated in early response to hepatic steatosis and correlated with poor prognosis in HCC (Eberle et al. 2004). The biological characteristics of master regulators (MRs) such as these include that they are membrane-bound proteins and N-terminal active portions able to enter the nucleus to activate their putative target genes; afterward, they bind and activate not only sterol-regulatory elements (SREs) containing promoters, but also E-boxes, which makes SREBPs able to regulate a widespread range of lipid genes (Shimano 2001).

The biological mechanism behind NAFLD progression to HCC had never been explored in a systematic, unbiased manner exploiting publicly available data from independent transcriptomic studies. This network-based approach aims to unravel the gene expression signatures (ES), which are potential biomarkers for disease progression and targets for drug therapy because they match the strongest ES shared by NAFLD, NASH, and HCC. First, we inferred two liver transcriptional networks (TN) using data from several large-scale microarray studies and exploited these models to explore NAFLD, NASH, and HCC regulatory units. This fact challenges us to understand which and how key transcriptional factors regulate putative genes in the pathological phenotypes compared to normal liver and cirrhotic liver tissues. In a second step, we integrate transcriptomics and proteomics data using a well-validated model of human genome-scale metabolic reconstruction (Swainston et al. 2016), inputting the effect of gene products regulated by MRs in NALFD-related HCC to assess metabolic reactions and biological processes they affect.

2 Methods

2.1 Microarray Data Acquisition

NAFL microarray datasets were obtained from the Gene Expression Omnibus database (GEOdatabase) (Barrett et al. 2013), under the accession numbers GSE48452 and GSE33814, which were originally contributed by Jochen et al. in

2013 (Ahrens et al. 2013) and Starmann et al. 2012 (Starmann et al. 2012), respectively. *NASH* microarray datasets were acquired from the GEOdatabase under the accession numbers GSE48452, GSE33814, GSE24807, and GSE37031, which were originally contributed by Jochen et al. in 2013 (Ahrens et al. 2013), Starmann et al. 2012 (Starmann et al. 2012), Baker et al. 2010 (Liu et al. 2011), and López-Vicario et al. 2013 (Lopez-Vicario et al. 2014), respectively. *HCC* microarray datasets were obtained from GEOdatabase under the accession number GSE45436, contributed initially by Wang et al. in 2014 (Wang et al. 2013). *Normal Liver* microarray dataset was obtained from GEOdatabase under the accession number GSE14520, which was originally contributed by Wang et al. in 2010 (Roessler et al. 2010; Roessler et al. 2012). *Cirrhotic Liver* microarray dataset was obtained from GEOdatabase under the accession number GSE10142, which was originally contributed by Hoshida et al. in 2008 (Hoshida et al. 2008; Villanueva et al. 2011). The selected datasets fulfilled our criteria for inclusion by clustering the expected phenotypes over the available ones (Fig. 1a). *Mus musculus* microarray dataset was obtained from GEOdatabase under the accession number GSE83596 (Barrett et al. 2013), which was originally contributed by de Conti A et al. (2016) (Table 1). Of additional importance, we selected the GEO datasets based on whether the dataset was available at the time of collection (2015–2020) after evaluating the sample size, gene coverage, and disease coverage.

2.2 Hierarchical Clustering Analysis

Hierarchical clustering was computed using the R package *pvclust* (Suzuki and Shimodaira 2006). Besides computing hierarchical clustering based on usual distance and agglomeration algorithms, *pvclust* also assesses how strong the data support each cluster by computing multiscale bootstrap resampling. The metrics for cluster strength evaluation provided are approximately unbiased (AU) and bootstrap probability (BP). The higher these metrics, the more confidently we can reject the hypothesis that the cluster does not exist with a significance level of 0.05. We used the Euclidean distance and average agglomeration methods for hierarchical clustering and 10,000 bootstraps multiscale resample.

2.3 Transcriptional Network Inference Derived from Gene Expression Profiles

We created a genome-wide TN centered on TFs and a group of putative target genes (regulons) using normal liver (TN1) gene expression data and cirrhotic liver (TN2) expression data with the RTN (reconstruction of transcriptional network and analysis of master regulators) R-package (http://bioconductor.org/packages/RTN/) (Gentleman et al. 2004; Castro et al. 2016), which is designated to reconstruct and analyze transcriptional networks based on the mutual information (MI) ARACNe (Algorithm for the Reconstruction of Accurate Cellular Networks)

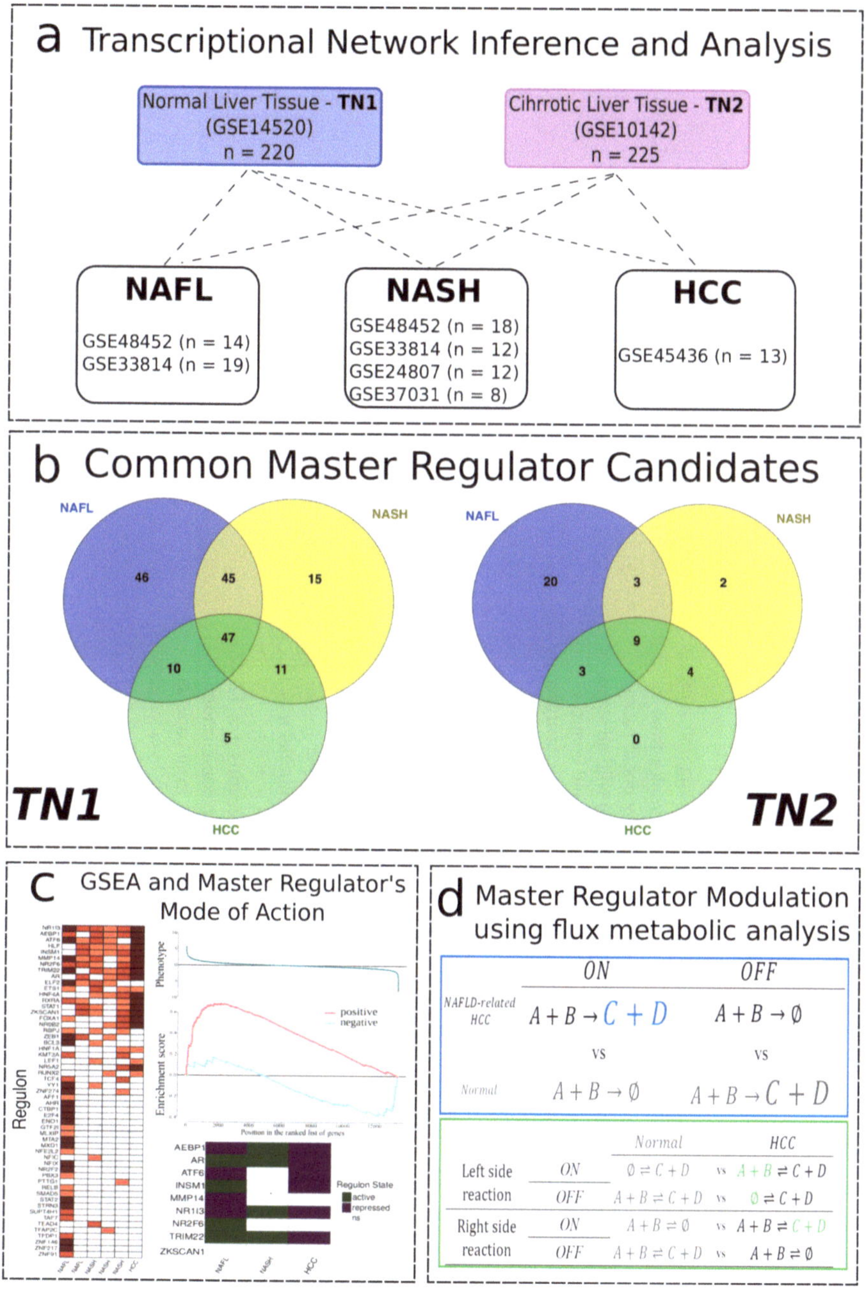

Fig. 1 Analyses pipeline summary. (**a**) Transcriptional inference and analysis of publicly available microarray expression data resulted in two reference networks (TN1 and TN2), upon which different pathological phenotypes were evaluated. (**b**) Venn diagram showing the number of MR candidates in each disease phenotype and their intersection for each reference network. (**c**) One-tailed GSEA (left) was used to obtain the MR candidates, and two-tailed GSEA was used to assess the state of activation of the MR candidates (results using TN1). (**d**) Parsimonious flux balance analysis using COBRAPy

Table 1 Microarray data acquisition

Dataset name	GEO_ID	Summary	Overall design/N° samples	Platforms	Reference
Non-alcoholic fatty liver (NAFL) or steatosis	GSE48452	NAFLD-specific expression and methylation differences were seen for nine genes coding for key enzymes in intermediate metabolism (including PC, ACLY, PLCG1) and insulin/ insulin-like signaling (including IGF1 and IGFBP2 PRKCE)	Liver samples from morbidly obese patients ($N = 45$) with all stages of NAFLD and controls ($N = 18$) were analyzed by array-based DNA methylation and mRNA expression profiling. The samples were grouped into control = 14, healthy, obese = 27, steatosis = 14, and NASH = 18	Affymetrix Human Gene 1.1 ST Array	Ahrens et al. (2013)
	GSE33814	Global gene expression analysis was applied to unravel differentially expressed genes between steatohepatitis compared to steatosis and control samples. Statistical and pathway analysis led to the selection of 46 candidate genes	44 human liver tissue samples (normal $n = 13$; steatosis $n = 19$; steatohepatitis $n = 12$, 8/2 cirrhotic/non-cirrhotic) obtained from patients undergoing liver surgery for HCC, other malignancy/metastatic diseases, benign tumors of the liver and organ dedicated to transplantation	Illumina HumanWG-6 v3.0 expression beadchip	Starmann et al. (2012)

Non-alcoholic steatohepatitis (NASH)	GSE48452	NAFLD-specific expression and methylation differences were seen for nine genes coding for key enzymes in intermediate metabolism (including PC, ACLY, PLCG1) and insulin/ insulin-like signaling (including IGF1 and IGFBP2 PRKCE)	Liver samples from morbidly obese patients ($N = 45$) with all stages of NAFLD and controls ($N = 18$) were analyzed by array-based DNA methylation and mRNA expression profiling. The samples were grouped into control = 14, healthy, obese = 27, steatosis = 14, and NASH = 18	Affymetrix Human Gene 1.1 ST Array	Ahrens et al. (2013)
	GSE33814	Global gene expression analysis was applied to unravel differentially expressed genes between steatohepatitis compared to steatosis and control samples. Statistical and pathway analysis led to the selection of 46 candidate genes	44 human liver tissue samples (normal $n = 13$; steatosis $n = 19$; steatohepatitis $n = 12$, 8/2 cirrhotic/non-cirrhotic) obtained from patients undergoing liver surgery for HCC, other malignancy/metastatic diseases, benign tumors of the liver and organ dedicated to transplantation	Illumina HumanWG-6 v3.0 expression beadchip	Starmann et al. (2012)

(continued)

Table 1 (continued)

Dataset name	GEO_ID	Summary	Overall design/N° samples	Platforms	Reference
	GSE24807	This study investigated whether hemoglobin is expressed in hepatocytes and how it is related to oxidative stress in NASH patients. Analysis of microarray gene expression data has revealed a significant increase in the expression of hemoglobin alpha (HBA1) and beta (HBB) in liver biopsies from NASH patients	Twelve biopsy-diagnosed NASH patients were included in this study. Total RNA from 5 different subjects was purchased from ADMET for control groups. These subjects are free from liver disease	GE Healthcare/Amersham Biosciences CodeLink Human Whole Genome Bioarray	Liu et al. (2011) Feng et al. (2015)
	GSE37031	This study integrated gene expression profiling of liver biopsies of NASH patients with translational studies in a mouse model of steatohepatitis and pharmacological interventions in isolated hepatocytes to identify a novel mechanism implicated in the pathogenesis of NASH	The complete database comprised the expression measurements of 18,185 genes for liver sample groups: 8 non-alcoholic steatohepatitis (NASH) and 7 control samples	Affymetrix Human Genome U133 Plus 2.0 Array [Brainarray Version 13, HGU133Plus2_Hs_ENTREZG]	López-Vicario et al. (2014)

Hepatocellular carcinoma (HCC)	GSE45436	Gene expression profiles of human hepatocellular carcinoma	134 HCC samples	Affymetrix Human Genome U133 Plus 2.0 Array	Wang et al. (2013)
Normal liver tissue	GSE14520	Affymetrix microarray profiling to analyze gene expression patterns in healthy donor liver and tumor and paired non-tumor tissue of HCC patients	220 non-tumor liver samples	Affymetrix GeneChip HG-U133A 2.0 arrays	Roessler et al. (2010) Roessler et al. (2012)
Cirrhotic liver tissue	GSE10142	Gene expression signature correlating with survival is present in liver tissue adjacent to the tumor in patients with hepatocellular carcinoma	225 non-tumor liver tissues surgically resected from patients with HCC	Human 6k Transcriptionally Informative Gene Panel for DASL	Hoshida et al. (2008) Zhao et al. (2015) Wang et al. (2019) Sun et al. (2017) Lu et al. (2015) Chen et al. (2019) Chen et al. (2021) Wang et al. (2021) Li et al. (2022)
Mus musculus	GSE83596	It examined the transcriptomic profiles in the livers of STAMTM mice at different stages of liver carcinogenesis, including steatosis (6-week time interval), NASH (8 weeks), and fibrotic stage (12 weeks), and in full-fledged HCC (20 weeks)	Transcriptomic profile in the liver of STAM mice at 6 weeks (steatosis; $n = 3$), 8 weeks (steatohepatitis; $n = 3$), 12 weeks (fibrosis; $n = 4$), and 20 weeks (HCC-stage tumor tissue, $n = 4$) weeks. Age-matched control samples were also analyzed	SurePrint G3 Mouse Gene Expression v2, 8x60K; Agilent Technologies, Santa Clara, CA	Not published https://www.ncbi.nlm.nih.gov/geo/query/acc.cgi?acc=GSE83596

method. The transcription factors selected in the reference networks are all the genes in the normal datasets that have been annotated as "transcription factor activity, sequence-specific DNA binding" (GO:0003700) by Gene Ontology. Briefly, the regulatory structure of the network is derived by mapping significant associations between known TFs and all potential targets, and interactions that were below the minimum mutual information threshold were eliminated by permutation analysis. Unstable interactions are additionally removed by bootstrap analysis using 1000 bootstrap samples to create the consensus bootstrap network. In a final step, the data processing inequality (DPI) algorithm was applied with a tolerance equal to 0.0 to eliminate interactions likely to be mediated by another TF (Steuer et al. 2002; Margolin et al. 2006a, b; Mutwil et al. 2011).

Thus, the TN1 and TN2 transcriptional networks composed a TF-target pair's most significant MI estimated values. The *p-value* cutoff parameter used to compute each consensus TN was 0.001.

All computational analyses were performed in R statistical environment. In addition, network figures were constructed in the RedeR graphical platform to explore biological networks (Castro et al. 2012), and other plots were created using *ggplot2* (Wickham 2016) (Fig. 1b).

2.4 Master Regulators and Gene Set Enrichment Analysis

Gene set enrichment analysis (GSEA) is a rank-based scoring test that assesses whether a given list of genes is enriched between two phenotypes (Subramanian et al. 2005). In our analyses, each regulon (from TN1 or TN2) is treated as the list of genes. The log fold change (logFC) metric used to obtain the ranked phenotypes required for the GSEA was computed using the Bioconductor package *limma* (Ritchie et al. 2015) by contrasting normal liver tissue versus NAFL, NASH, or HCC, and Mus musculus from the previously selected datasets (Table 1). The GSEA was performed using 1000 permutations and the FDR adjustment method. Significantly enriched regulatory units were considered master regulator candidates given the pathological phenotype (Fig. 1c).

2.5 Two-Tailed GSEA and Master Regulator's Mode of Action

The two-tailed GSEA was also performed using the RTN package with 1000 permutations, as previously described elsewhere (Castro et al. 2016). Concisely, the regulatory units were split into positive and negative targets. Next, the association of each subgroup was assessed by GSEA statistics in each ranked phenotype, resulting in independent enrichment scores (ES). Additionally, a differential enrichment was performed among subgroups (Es_A-Es_B). Thus, a high positive differential score implies that the regulon is induced in the phenotype, while a high negative differential score indicates that the regulon is repressed in the phenotype (NAFL, GSE48452; NASH, GSE24807; HCC, GSE45436; and Mus musculus, GSE83596) (Fig. 1c).

2.6 Flux Metabolic Analysis Using Recon v2.2

We used the genome-scale metabolic model Recon v2.2, which contains all known metabolic reactions and genes involved in human metabolism. We downloaded it from BioModels using the identifier MODEL1603150001 (Swainston et al. 2016). To read, analyze, and modify this model, we used COBRAPy, an interface for reconstructing and analyzing genome-scale models implemented in Python (Ebrahim et al. 2013).

We evaluated the effect of *AR* (androgen receptor), *AEBP1* (adipocyte enhancer-binding protein 1), and *NR1I3* (nuclear receptor subfamily 1 group I member 3) on human metabolism. First, we identified the enzyme-encoding genes regulated by these transcription factors. We performed this process for the identified genes using GSE10142 (normal tissue liver) and GSE14520 (cirrhotic tissue liver) as the background. Second, we established the mechanism of action on these genes, either positive or negative regulation. Third, we mapped regulated genes into the gene metadata of each reaction in the Recon v2.2. We assumed that positively regulated genes are associated with positive enzyme activity, while negative regulation would have the opposite effect.

In Recon v2.2, two types of reactions are defined by their thermodynamic reversibility – reversible or irreversible. We set the flux to positive values for positively regulated reactions, while we set the flux to zero for negative regulation in both reversible and irreversible reactions. Subsequently, we performed a parsimonious flux balance analysis (pFBA) (Jenior et al. 2020) using COnstraint-Based Reconstruction and Analysis (COBRA) for Python (Ebrahim et al. 2013). In general, pFBA optimizes the distribution of fluxes to meet an objective function while minimizing the total sum of fluxes (Orth et al. 2010; Toroghi et al. 2016). paw uses a stoichiometric matrix (S), with $m \times n$ dimensions, representing the contribution of each metabolite (m) per reaction (n). The flux through these reactions is represented by the vector υ, which has constraints known as upper ($\upsilon_{\max}$) and lower ($\upsilon_{\min}$) bounds which limit the space of a reaction in the model. The model is set in a steady state when the equation $Sv = 0$ is satisfied (Orth et al. 2010). In pFBA, the fluxes in a steady state are calculated by solving two optimization problems. The first maximizes the objective function ($c^{\mathrm{T}}\upsilon$) and the second minimizes the sum of fluxes (υ). In this function, c corresponds to the weights associated with each flux (i.e., the contribution of a reaction to the objective function):

$$\max c^{\mathrm{T}}\upsilon$$

$$\min \sum |\upsilon|$$

$$\text{subject to } S\upsilon = 0$$

$$\upsilon_{\min} \leq \upsilon \leq \upsilon_{\max}.$$

We performed pFBA for two models; the first model, *NALFD-related HCC*, contains the reactions regulated by *AEPB1*, *AR*, and *NR1I3*. For the second model, we did not modify any reaction or flux; we named this model *Normal.* We compared the flux distribution obtained for the *NALFD-related HCC* versus the *Normal* model to interpret the results, means by which we explored how the regulation of the transcription factors impacts the direction of the reactions' fluxes in reversible and irreversible reactions. In general, the irreversible reactions could have been turned *on* or *off*, or their flux could have been higher or lower than the fluxes obtained by the *Normal* model, while for the reversible reactions, the flux could also be turned *on* or *off* or change the directionality of the reaction – from *right* to *left* or vice versa (Fig. 1d).

2.7 Enrichment Analysis for Genes and Enzymes Altered NALFD-Related HCC

We performed two types of analysis over the reactions altered in *NALFD-related HCC.* Firstly, we identified the genes associated with each scenario presented, and then we performed a KEGG (Kyoto Encyclopedia of Genes and Genomes) enrichment analysis (Qiu 2013) using clusterProfiler tool version 4.0 (Wu et al. 2021). In addition, for a specific set of genes, we used the Network of Cancer Genes to catalog cancer processes (Repana et al. 2019). Secondly, we identified the reactions altered and mapped in the KEGG pathway (Qiu 2013) and used clusterProfiler v4.0 (Wu et al. 2021).

The analysis pipeline, resampling procedures, and methods used to reconstruct the transcriptional networks and metabolic flux analysis are summarized in Fig. 1.

3 Results

3.1 Transcription Factors Identified as Shared Master Regulators in NAFLD, NASH, and HCC Expression Profiles by Array

We used hierarchical clustering of the expression signatures (ES) to test if our regulon meta-signatures would be able to differentiate the pathological events during NAFLD progression of both human datasets (Supplementary Fig. 2) and Mus musculus datasets (Supplementary Fig. 3). Next, we ranked the most significant common regulons shared by NAFLD, NASH, and HCC accordingly to the three following parameters: (a) the regulon size, which represents the number of targets in a given regulon; (b) the node degree, which represents the number of connections or edges that the node has to other nodes; and (c) the edge width, which corresponds to the number of collective targets between any two regulons assessed by Jaccard coefficient (Fig. 2 and Supplementary Fig. 1).

We inferred the interaction between annotated TFs and all potential targets from each dataset, normal liver tissue (TN1) and cirrhotic liver tissue (TN2), to derive two different TN for reference (Fig. 2). TN1 represents 220-expression patterns in healthy donor liver and non-tumor liver tissue samples (Fig. 2a), whereas TN2 is derived from 225 cirrhotic liver samples adjacent to HCC (Fig. 2b). We used both

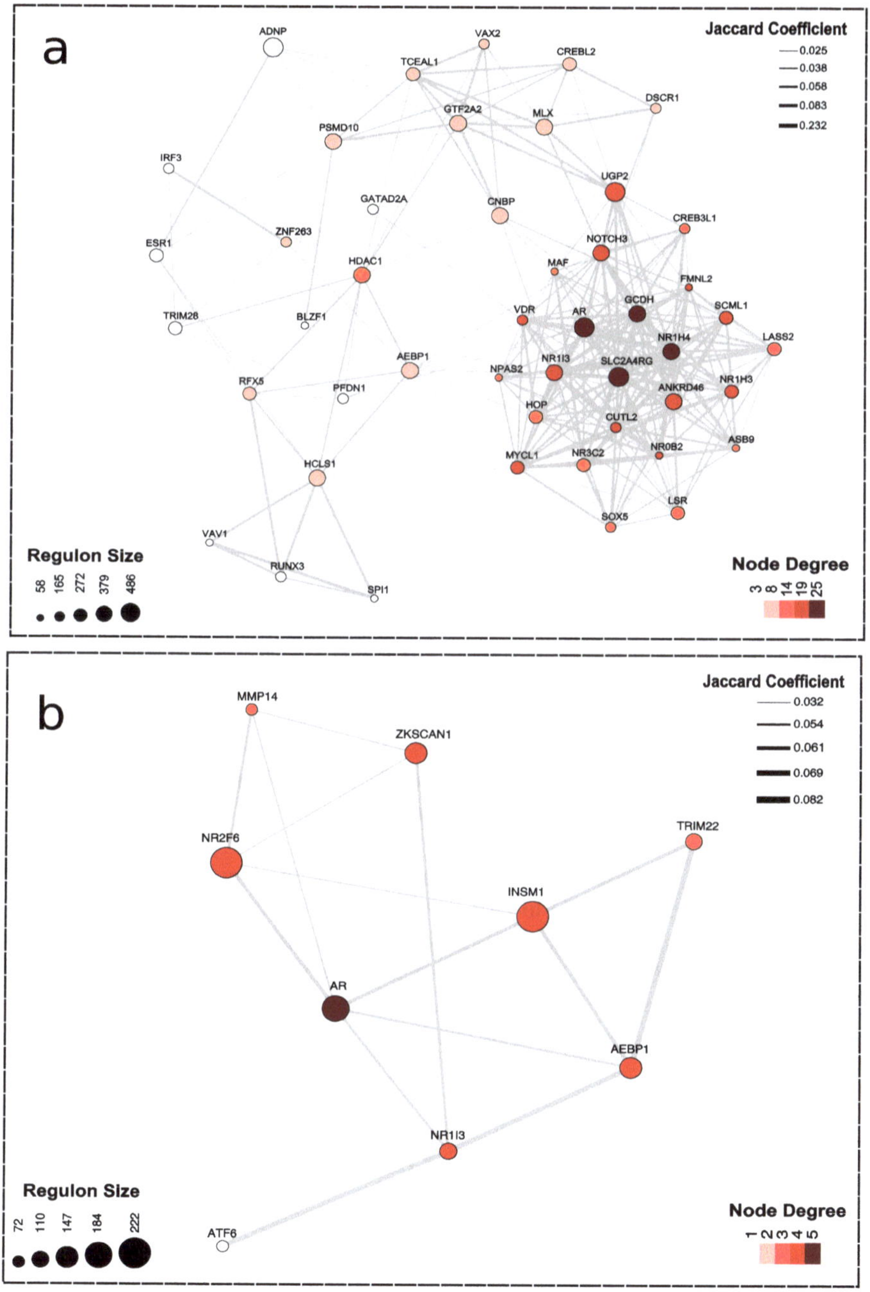

Fig. 2 Regulon association maps of common master regulator candidates for (**a**) TN1 and (**b**) TN 2. Nodes represent the regulatory units of master regulator (MR), node sizes map the number of genes for each regulon, and node colors map the connectivity (node degree) of each MR. Regulon size represents the number of targets inferred for that transcription factor. Node degree is the number of connections with other MR a regulatory unit has. Edge widths map the proportion of overlapping genes between any two regulatory units indexed by the Jaccard coefficient. Jaccard coefficient summarizes the number of shared targets two regulatory units have

TN to contrast and identify differences in the regulons in the disease progression pattern of NAFLD-to-HCC. Additionally, we determined that several TFs appeared only in TN1 but not TN2. Notably, datasets used to infer TN1 and TN2 had a different coverage. Thus, the extent of overlap between them is expected to vary significantly because of annotation issues and biological reasons. This may also explain the difference in the number of shared TFs further identified in the NAFLD-related HCC expression profile (Supplementary Fig. 1).

According to the Jaccard coefficient for each TN identified inside TN1, we determined that NR1H4 shared many targets with *GCDH*, *NR1I3*, *SLC2A4RG*, *CUTL2*, *NOTCH3*, *ANKRD46*, *HOP*, and *LASS2*. Also, *GCDH* shared a large number of shared targets with *SLC2A4RG* and *NOTCH3*. Additionally, we found that *SLC2A4RG* is a further common target with *ANKRD46*. Lastly, we discovered that *NR1I3* shared targets with *HOP*, *LASS2*, and *NR0B2* (Fig. 2a). Inside TN2, AEBP1 shared common targets with TRIM22, NR1I3, INSM1, AR, and MMP14. In addition, we discovered that NR1I3 shared common targets with *ATF6* and *AR*. Also, *AR* shared many common targets with *INSM1*, *NR2F6*, and *MMP4* (Fig. 2b). Among the shared regulatory units of NAFL, NASH, and HCC, we found 47-gene expression signatures significantly enriched in TN1, while only 9-gene expression signatures in TN2 were found (Fig. 3a). Lastly, we determined that *AEBP1*, *AR*, and *NR1I3* were the intersected group of TFs between TN1 and TN2 at the highest stringency threshold (*p-value* cutoff 0.005) (Fig. 3b). One-tailed GSEA of significant consensus MR using TN1 is shown in Fig. 3c.

3.2 Consensus Master Regulators Between Human and Mouse Expression Profiles by Array

We used TN1 to contrast and identify similarities in the regulons and the disease progression pattern of NAFLD-to-HCC between Mus musculus dataset (GSE83596) and human datasets (NAFL, GSE48452; NASH, GSE24807; HCC, GSE45436). We discovered that *AEBP1*, *AR*, and *NR1I3* were also MR at the highest stringency threshold (*p-value* cutoff 0.005) in *Mus musculus* across the four phenotypes associated with liver carcinogenesis, including steatosis, steatohepatitis, fibrosis, and HCC (Fig. 4).

3.3 NAFLD-Related HCC Master Regulator's Mode of Action

In Fig. 5, we illustrate the regulon mode of action (MoA) in both transcriptional networks for the selected datasets of NAFL (GSE48452), NASH (GSE24807), and HCC (GSE45436) (Fig. 5a). We inferred that the MoA of the regulons *AEBP1* was in the active state in NASH but was in the repressed state in NAFL and HCC in TN1 and TN2. *NR1I3* had a similar MoA in the three different phenotypes in TN2. In contrast, *AR* was in the active state in NAFL and NASH (TN2) but repressed in HCC in both TN1 and TN2 (Fig. 5b). Two-tailed GSEA using TN1 for each consensus MR is shown in Fig. 5c.

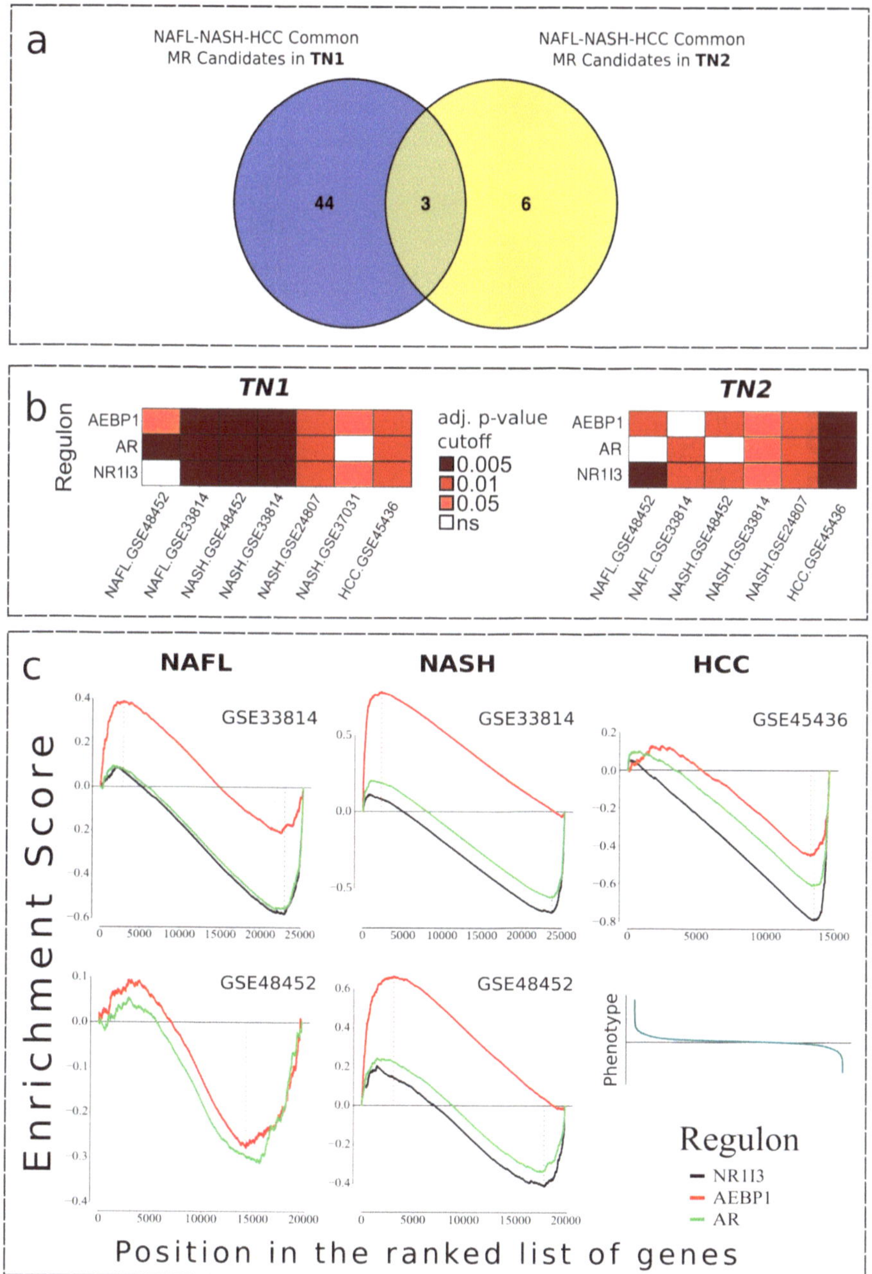

Fig. 3 Consensus master regulators in NALF, NASH, and HCC. (**a**) Venn diagram of common master regulator (MR) candidates obtained from the GSEA analyses of the three disease phenotypes using TN1 and TN2. (**b**) Tile plot showing the consensus MR, defined as the intersected candidates between the two reference networks. (**c**) One-tailed GSEA of significant consensus MR using TN1

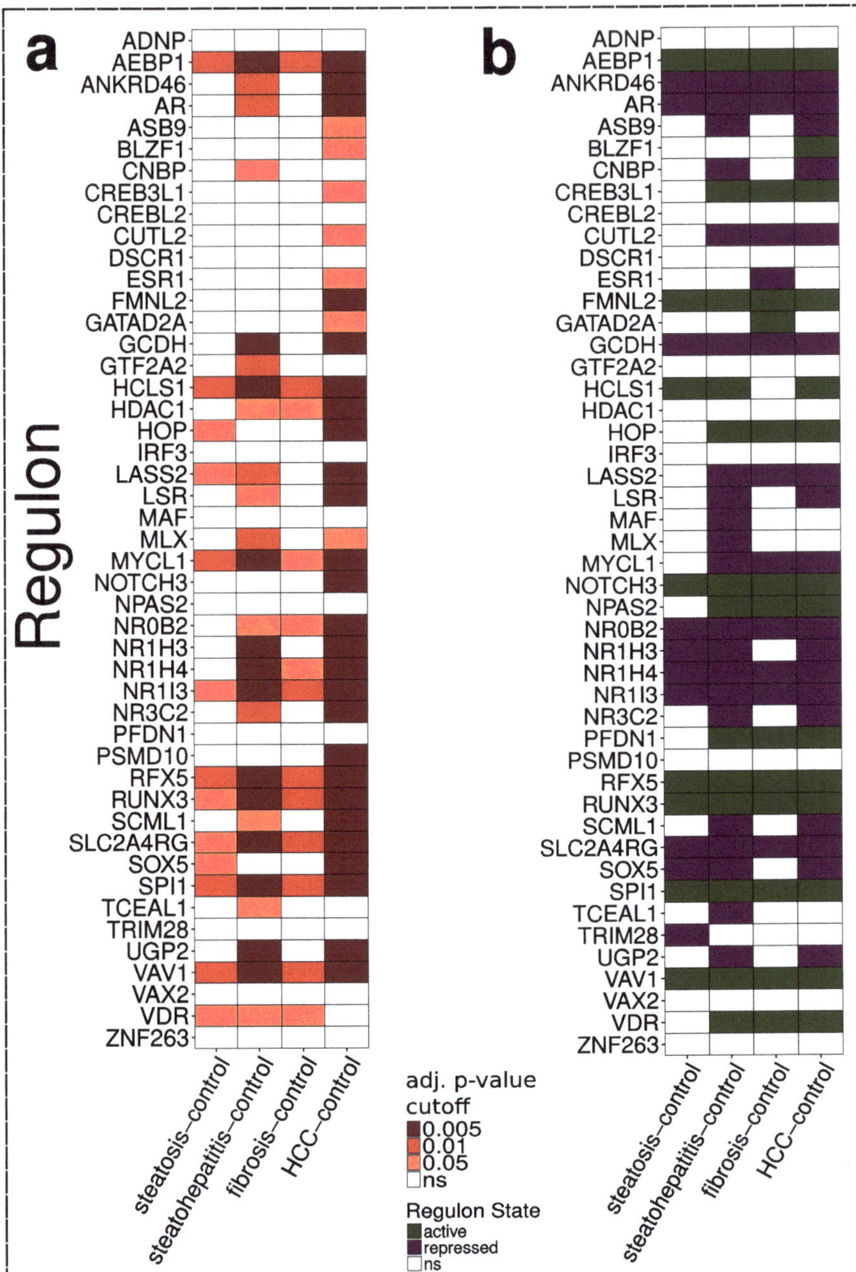

Fig. 4 GSEA analyses using TN1 using the 47 common master regulator candidates on the mouse expression dataset. (**a**) One-tailed GSEA mapping FDR adjusted *p*-values. (**b**) Two-tailed GSEA showing the inferred mode of action of each transcription factor

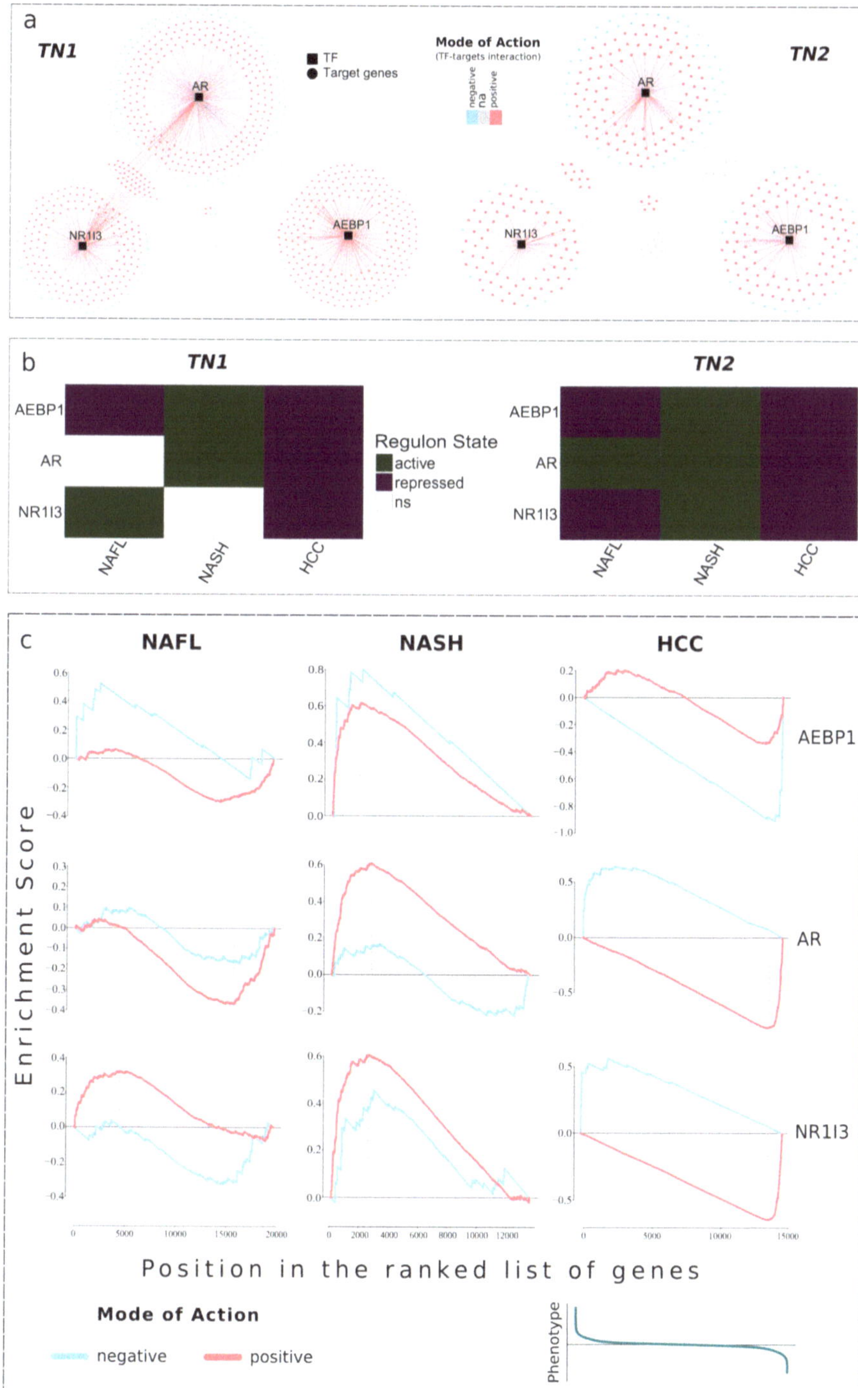

Fig. 5 Consensus master regulators mode of action (MoA). (**a**) Inferred mode of each consensus master regulators (MRs) on its targets in TN1 and TN2. (**b**) State of activation using TN1 and TN2 of each consensus MR using two-tailed GSEA. (**c**) Two-tailed GSEA using TN1 for each consensus MR

We explored the regulon MoA in TN1 for *Mus musculus* (GSE83596) and discovered that the MoA of the regulons *AEBP1*, *FMNL2*, *NOTCH3*, *RFX5*, *RUNX3*, *SPI1*, and *VAV1* was in the active state in all the phenotypes associated with liver carcinogenesis in mice. Alternately, *ANKRD46*, *AR*, *GCDH*, *NR0B2*, *NR1H4*, *NR1I3*, and *SLC2A4RG* displayed the same MoA, being in a repressed state in all stages of progression from NAFLD-to-HCC (Fig. 4).

3.4 Metabolic Flux Differences Between NAFLD-Related HCC and Normal Models

We found that the flux values were the same independently of the gene set background used to adjust the Recon v2.2 model, GSE10142 (normal tissue liver) or GSE14520 (cirrhotic tissue liver) (Supplementary Table 1). Intriguingly, the *NALFD-related HCC* model used almost three times more reactions than the *Normal* model (941 vs. 326). In addition, there was an alteration in 743 reversible reactions and 231 irreversible reactions. As described in the methodology, we identified the behavior of the reactions between the two models. We determined that 134 and 34 irreversible reactions were turned *on* and *off*, respectively, whereas 541 and 41 reversible reactions were turned *on* and *off*, respectively (Supplementary Table 2). In both types of metabolic reactions, it is also possible that in *NALFD-related HCC* and *Normal* model, the flux of some reactions is not equal to zero; however, these types of scenarios were not analyzed in this study.

We implemented a high stringency threshold (*p-value* = 0.01) to explore the differences in the irreversible (Fig. 6a) and reversible (Fig. 6b) reactions in *NALFD-related HCC* compared to *Normal* metabolic flux. The irreversible reactions in which *NALFD-related HCC* (*on* state) had higher flux than *Normal* involved fatty acid metabolism, pyruvate metabolism, beta-alanine metabolism, valine, leucine, and isoleucine degradation, and fatty acid degradation. In contrast, the irreversible reactions that showed a lower flux than *Normal* (*off* state) include citrate cycle and nucleotide metabolism (Fig. 6).

For the reversible reactions, *NALFD-related HCC* (*on* state) had a higher flux to the right than *Normal* in the PPAR signaling pathway and fatty acid metabolism, while nitrogen metabolism, pyrimidine metabolism, and nucleotide metabolism changed the directionality of the reaction to the *left* (Fig. 6b). Likewise, nitrogen metabolism also showed that some reversible reactions had a lower flux than *Normal* (*off* state) and were shifted to the right (Fig. 6b).

PPARα signaling metabolic flux, particularly CD36/FAT (CD36, fatty acid translocase), hydroxymethylglutaryl-CoA synthase (HMGCS), and stearoyl-CoA desaturase (Delta-9 desaturase) and acyl-CoA 6-desaturase (Delta-6 desaturase), had significantly higher flux to the right in the *NAFLD-related HCC* model (Fig. 7a). We also found that free fatty acid (FFA) degradation had significantly higher flux in the *NAFLD-related HCC* model, and those irreversible reactions shifted to the right involving carnitine palmitoyltransferase 1 (CPT1A), liver isoform, and carnitine O-palmitoyltransferase 2 (CPT2) (Fig. 7b). Among other

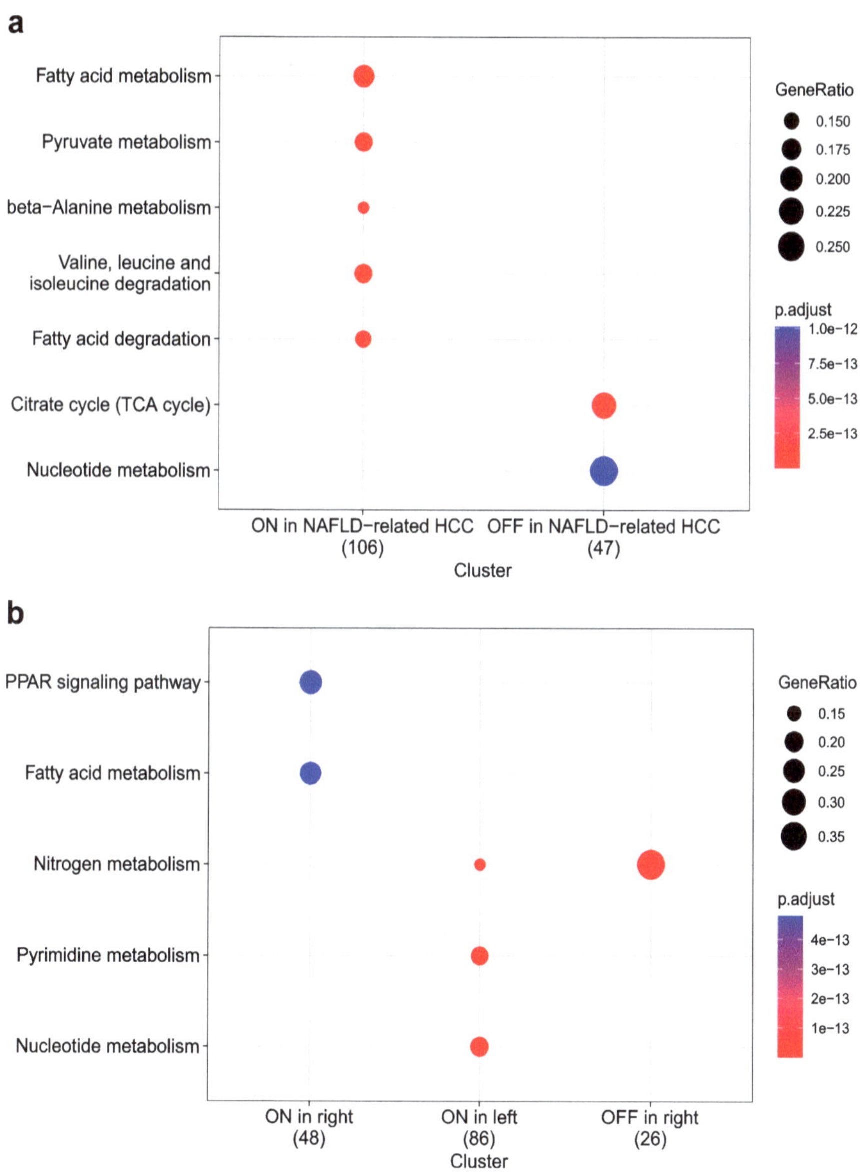

Fig. 6 *NALFD-related HCC* model reconstruction using Reconv2.2. (**a**) Irreversible and (**b**) reversible reactions modified compared to the *Normal* model using parsimonious flux balance analysis using COBRA for Python

enzymes/reactions presented in the Supplementary Table 3. Similarly, CPT1A flux was altered in the glucagon signaling pathway and at the fatty acid oxidation entrance point (Fig. 7c). Also, the reversible rate-limiting reaction that regulates

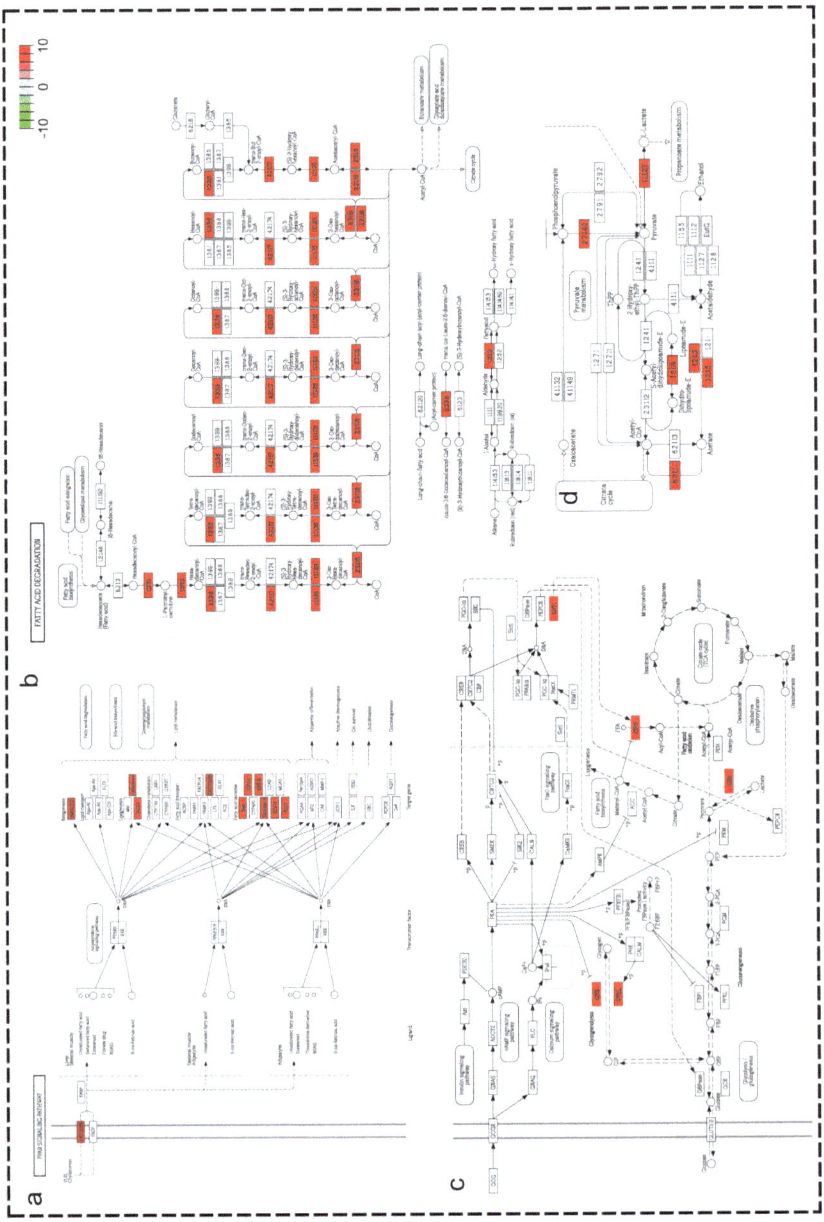

Fig. 7 KEGG enrichment analysis of *NALFD-related HCC* model showing differential metabolic flux in **(a)** PPAR signaling pathway, **(b)** fatty acid degradation, **(c)** insulin signaling pathway, and **(d)** citrate cycle

glycogenolysis had a higher flux exchange between glycogen synthase (GYS) and glycogen phosphorylase (PYG) (Fig. 7c). The metabolic flux analysis of glycolysis/gluconeogenesis was also altered. We found that both reversible and irreversible reactions involving L-lactate dehydrogenase, dihydrolipoamide dehydrogenase, aldehyde dehydrogenase (NAD(P)+), aldehyde dehydrogenase (NAD+), pyruvate kinase, and acetyl-CoA synthetase had higher flux. Furthermore, those reactions were shifted to the right in the citrate cycle if reversible (Fig. 7d).

Within pyrimidine metabolism, we found that carbamoyl-phosphate synthase/aspartate carbamoyltransferase/dihydroorotase (CAD) had higher flux and was shifted to the left, modifying alanine, aspartate, and glutamine metabolism (Fig. 8a). Likewise, other reactions/enzymes feed into the β-alanine metabolism (Fig. 8b), and valine (Val), leucine (Leu), and isoleucine (Ile) degradation (Fig. 8c), which are described in Supplementary Table 3. Interestingly, several enzymes involved in irreversible reactions of the biosynthesis of branched-chain amino acids (BCAAs) had higher flux in the *NAFLD-related HCC* model than in the *Normal* model, altering the activity of dihydrolipoamide dehydrogenase (DLD), enoyl-CoA hydratase (paaF), and 3-hydroxyacyl-CoA dehydrogenase (HADH) among others. Finally, we identified that only irreversible reactions were *on* affecting dihydropyrimidinase (DPYS), beta-ureidopropionase (UPB1), aldehyde dehydrogenase (NAD+), aldehyde dehydrogenase (NAD(P)+), and 4-aminobutyrate aminotransferase/(S)-3-amino-2-methylpropionate transaminase (ABAT) in β-alanine metabolism in the *NAFLD-related HCC* model (Fig. 8b). Last of all, these metabolic processes convey into the pyruvate metabolism (Fig. 8d).

The specific enzymes altered in each of the metabolic pathways mentioned above are summarized in Supplementary Table 3.

4 Discussion

The nature of the progression of NAFLD-to-HCC is multifactorial, as proposed in the "multiple-hit" hypothesis. Buzzetti's theory emphasizes the relevance of genetic and epigenetics factors acting synergistically with lipid accumulation, oxidative stress, gut microbiota, inflammation, and DNA damage to explain the underlying mechanism of the disease progression (Buzzetti et al. 2016; Dallio et al. 2021). However, there is limited understanding of the susceptibility risk of individuals that further develop NAFLD-related HCC (Huang et al. 2021). Therefore, the biological foundation of our study relies on tracing all genes/enzymes that can be considered as potential biomarkers or therapeutic targets for intervention at the earliest transition from steatosis to NASH instead of the late stage of HCC.

In this study, we first used a network-based approach with the intent to shed light on the molecular mechanism that underlies the genetic risk of NAFLD-related HCC. Our results had a high stringency threshold and accordance across both inference networks – TN1 and TN2 – and showed that only *AR*, *AEBP1*, and *NR1I3* were consensus MRs. Second, we integrated signature-based transcriptomic scale data and thermodynamic parameter values into quantitative constraints on metabolic

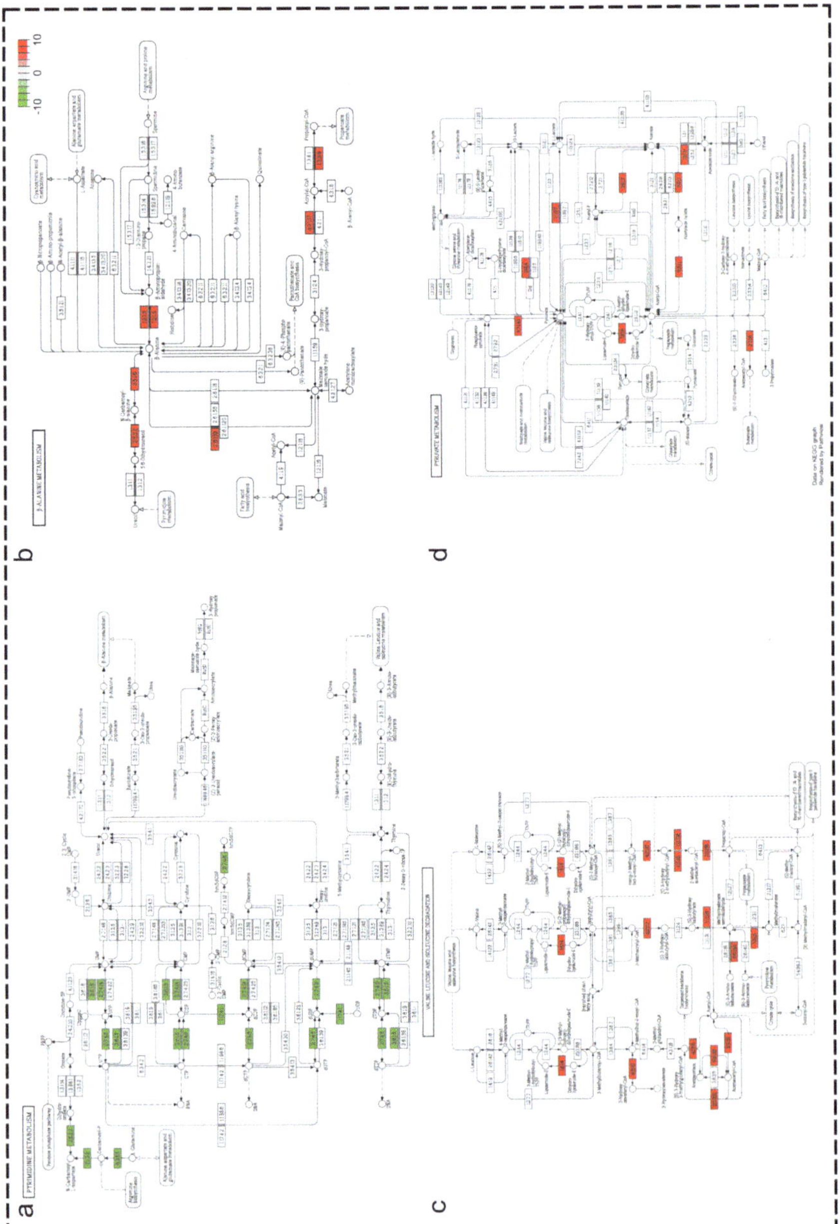

Fig. 8 KEGG enrichment analysis of *NALFD-related HCC* model showing differential metabolic flux in (**a**) pyrimidine metabolism; (**b**) β-alanine metabolism; (**c**) valine, leucine, and isoleucine degradation; and (**d**) pyruvate metabolism

fluxes, enabling accurate metabolic predictions in the *NAFLD-related HCC model*. Our bioinformatics modeling pipeline is, to the best of our knowledge, the first study to integrate multiple omics to examine whether risk-associated genes share regulatory mechanisms in the progression of NALFD-to-HCC.

We found that *AR* regulates approximately 400 putative target genes across 3 different phenotypes and is one of the most connected regulons within both TN. Furthermore, *AR* mediates the development of steatosis and cancer progression by binding to genomic androgen response elements (Bolton et al. 2007), which influence the transcription of androgen-responsive genes (ARGs) (Ma et al. 2014). Accordingly, we inferred that the *AR* MoA was in its active state in the NAFL and NASH phenotypes. In addition to this, we found that *AR* MoA shares many targets with *AEBP1* and *NR1I3*. This intersection of regulons may also reflect the involvement of *AR* and *AEBP1* in a sex-specific susceptibility to diet-induced obesity by the androgen- and estrogen-signaling pathways, respectively (Zhang et al. 2005). Indeed, sexually dimorphic differences in the responses to changes in hepatic androgen–AR signaling were similar to humans in the hepatic AR-deficient (H-AR$^{-/-}$) mice (Lin et al. 2008). Male H-AR$^{-/y}$ had an increased body weight and insulin resistance, resulting from decreased fatty acid β-oxidation, enhanced de novo lipid synthesis arising from decreased PPARα, and increased SREBP1c (Lin et al. 2008). However, female H-AR$^{-/-}$ mice fed with a high-fat diet did not have such a phenotype. There is limited and controversial evidence in the value of serum total testosterone to predict the association with NAFLD (Kim et al. 2012; Mody et al. 2015). Sexually dimorphic differences in the MoA of *AR* need further investigation in NAFLD-related HCC to evaluate other potential metabolites used as biomarkers.

AR is repressed in HCC, which is congruent with the strategy to therapeutic target it in early-stage HCC with sorafenib and agents that enhance the functional expression of *AR* to prevent the progression of late-stage HCC (Ma et al. 2014). However, its activity and value as a biomarker in the prognosis of HCC remain inconclusive because the level of *AR* expression is qualitatively different among histological grades (Acosta-Lopez et al. 2020).

Our analysis predicted that PPARα signaling metabolic flux, particularly CD36/FAT, had significantly higher flux to the right in the *NAFLD-related HCC* model. These findings align with previous studies that CD36 fatty acid translocase drives hepatosteatosis onset and is a potential biomarker for NAFLD diagnosis and patient stratification (Rada et al. 2020). Moreover, a clinical study found that elevated circulating levels of a soluble form of CD36 correlate positively with the histological grade of hepatic steatosis (García-Monzón et al. 2014). Likewise, the PPARα signaling pathway is a crucial regulator of acylcarnitine metabolism (van Vlies et al. 1767). We found that a higher metabolic flux impacts the degradation of BCAAs, which could be explained by the increased lipid processing that produces acylcarnitine products in NAFLD (Schooneman et al. 2013). In addition, BCAA catabolism mediates insulin sensitivity response due to the accumulation of intermediary metabolites such as acylcarnitines (Bishop et al. 2022). BCAA metabolomic profile confirmed the increases in the levels of leucine (127% of normal), isoleucine

(139%), and valine (147%) while transitioning from steatosis to NAFLD (Lake et al. 2015); however, its value as a biomarker in NAFLD progression to HCC needs to be assessed.

NAFLD progression results from FFA-induced lipotoxicity caused by increased uptake and impaired degradation of FFA (López-Vicario et al. 2014; Pardo et al. 2015). We also found that FFA degradation had significantly higher flux in the *NAFLD-related HCC* model, and those irreversible reactions shifted to the right involving CPT1A/2. A previous study discovered that *AR* might directly regulate CPT1A and exert its function by HMG-CoA (Lund et al. 1991).

Over the past few years, the association between increased gut permeability and lipopolysaccharide (LPS) levels with NASH has been reported (Gabele et al. 2011; Bajaj et al. 2012; Ferolla et al. 2014). In fact, several studies support the hypothesis that bacterial translocation is related to increased fat intake in the liver (Wiest et al. 2014). LPS is known to induce the expression of *AEBP1* during monocyte differentiation. *AEBP1*-mediated foam cell formation inside macrophages is an LPS-induced downregulation of its cholesterol efflux mediators (Majdalawieh and Ro 2009). We have described that *AEBP1* is in its active state in NASH, supported by recent evidence demonstrating that *AEBP1* overexpression is higher in activated hepatic stellate cells (HSC), which induced hepatic fibrosis (Gerhard et al. 2019). The same study described that *AEBP1* blocks the PPAR-γ-dependent inhibition of activated HSC. The metabolic reconstruction of the *NAFLD-related HCC* model indicated a significantly higher flux shifted to the right of PPAR-γ-dependent signaling of thiolase (also called acetyl-CoA acyltransferase 1, ACAA1) in the adipocyte. A different form, "adipogenic hepatic steatosis," resulting from PPAR-γ1 overexpression, is associated with adipogenic gene expression in which ACAA1 was found upregulated by sixfold (Yu et al. 2003). Another systems biology approach also predicted that *AEBP1* activation in the NASH state is associated with the production and maintenance of the extracellular matrix (Lou et al. 2017) and regulates myofibroblast transition (Steensels et al. 2020). Similarly, a preclinical study discovered that *AEBP1* expression increases in NASH compared to simple steatosis in the livers of $ApoE^{+/-}$ mice (Teratani et al. 2018). In our inference network, *AEBP1* is in the repressed state in NAFLD and HCC, which needs to be further investigated.

NR1I3/CAR activation protects against hepatic steatosis by regulating lipid metabolism, suppressing lipogenesis and gluconeogenesis, and inducing fatty acid β-oxidation (Dong et al. 2009; Režen et al. 2009; Giriş et al. 2014; Yang et al. 2020). In addition, *NR1I3* is a critical regulator of xenobiotic and endobiotic metabolism (Huang et al. 2004; Li and Wang 2010). In our analysis, *NR1I3* was found in a repressed state in steatosis, which would support the view that is triggering the NAFLD-related HCC. Furthermore, our metabolic flux analysis of glycolysis/gluconeogenesis found that both reversible reactions involving NAD(P)+ and NAD+ had higher flux. Those reactions were shifted to the right in the citrate cycle. This change in redox metabolism might be caused by the FFA overload diverted to both FA degradation and citrate cycle, from which $FADH_2$/NADH is generated, respectively. At last, both end products transfer electrons to the respiratory chain couple to the ATP synthesis, and the excess causes ROS inducing the inflammatory response in NAFLD progression to HCC (Serviddio et al. 2008; Longo et al. 2021). The direct regulatory activity of *NR1I3/CAR* associated with these enzymes needs further research.

Pharmacological stimulation of the *CAR* agonist 1,4-bis-(2-(3,5-dichloropyridyloxy)) benzene, 3,3′,5,5′-tetrachloro-1,4-bis (pyridyloxy) benzene (TCPOBOP) was reported to prevent hepatic steatosis in high-fat-fed mice and obese mice (Gao et al. 2009). However, persistent *NR1I3/CAR* activation TCPOBOP alone induced HCC development (Mattu et al. 2018). Another study also suggested that persistent *NR1I3* activation promoted liver injury and the development of HCC (Choi et al. 2013), and a splicing isoform of gene *NR1I3* was found to be upregulated in most HCC samples (Zhang et al. 2016). Altogether, the previous findings differ from our result of the *NR1I3*-repressed state in HCC. Nevertheless, *NR1I3* with the splicing event is a different isoform to the one included in our study.

Systems biology analyses of transcriptional regulatory networks shed light on how transcription factors can induce collective responses leading to a system-level gene expression deregulation in disease (Margolin et al. 2006a, b; Carro et al. 2010; Piao et al. 2012; Castro et al. 2016; Pfaffenseller et al. 2016). In the present study, we implemented a well-validated approach to reconstruct the NAFLD-related HCC transcriptional regulatory networks, which have been used to examine whether risk-associated genes share regulatory mechanisms in breast cancer (Castro et al. 2016), Parkinson's disease (Vargas et al. 2021), Huntington's disease, and early-onset Alzheimer's (Carello-Collar et al. 2021) and major psychiatric disorders (Bristot et al. 2020). Moreover, other studies have successfully integrated this transcriptomic analysis into a transcription factor-centered drug repositioning approach (De Bastiani et al. 2018; Vargas et al. 2018; De Bastiani and Klamt 2019).

We extended the capability of this strategy to compare transcriptomic similarities and differences between mouse models and human pathophysiology, similar to previous studies (De Bastiani et al. 2022). We examined the transcriptomic profiles in the livers of STAM mice at different stages of liver carcinogenesis, including steatosis (6-week time interval), NASH (8 weeks), fibrotic stage (12 weeks), and full-fledged HCC (20 weeks). The detailed results of this microarray analysis are not published yet. Moreover, we found that the MoA of the regulon AEBP*1* was in the active state across all the phenotypes, whereas *AR* and *NR1I3* were in the repressed state in the STAM model of liver carcinogenesis (GSE83596). This model resembles the human HCC development associated with progression from simple steatosis to NASH, fibrosis, and HCC. Nevertheless, we found differences in the MoA of the three MRs in the NAFLD-related HCC between humans and the STAM model; we considered that this model recapitulates genetic features of progression from NAFLD-to-HCC.

We recognize that batch effect control is necessary when combining multiple microarray experiments. This was not the case in our study, though. Each dataset was treated individually (either the data for normal inference or the data of case-control studies), and the data we obtained from Gene Expression Omnibus was already normalized. The main goal of our approach using several studies is to observe consistency between the several independent and individually tested datasets. Also, we realize the methodological shortcomings of using microarray gene expression platforms instead of RNA-Seq to obtain a more comprehensive dynamic range of differential express genes and specific insights into the molecular mechanism of NAFLD-related HCC (Rao et al. 2019). However, a limited number of publicly available RNA-Seq studies satisfied the quality criteria to carry on the reconstruction of the transcriptional regulatory network, including tissue samples of the three phenotypes. Therefore, we incorporated the signature-transcriptomic profile to explore metabolic flux distribution differences using a wide-used model, Recon v2.2 (Salazar et al. 2016; Lewis et al. 2021). Lastly, our analysis contributes to improving our understanding of how upstream TFs' early activation of *AR* and repression of *AEBP1* and *NR1I3* impact the metabolic reactions involved in fatty acid metabolism; pyruvate metabolism; beta-alanine metabolism; valine, leucine, and isoleucine degradation; citrate cycle and nucleotide metabolism; and fatty acid degradation. Finally, these model-identified targets need to be experimentally validated as potential biomarkers and therapeutic targets.

5 Conclusions

In summary, these findings suggest that the progression pattern of NAFLD-to-HCC may be initiated by the early activation of *AR* and repression of *AEBP1* and *NR1I3*, which may act as "major drivers" by either activating or repressing target genes downstream for NAFLD, NASH, and HCC. Furthermore, we identified well-known and new signaling pathways and metabolic reactions altered in NAFLD-related HCC. Interestingly, we found that *AR* and *AEBP1* sex-specific susceptibility to diet-induced obesity needs to be further investigated in our model, and CPT1A persistent metabolic flux overload could be a therapeutic target to prevent the development of HCC. Ultimately, as with any biological network, we believe that these networks encompass many dynamical associations that might be coordinated by common transcriptional amplification cascades that need to be extensively studied and experimentally validated. In addition, the pipeline of analyses is transferable to other disease settings.

Acknowledgments The authors wish to thank Mauro A. Castro for his valuable insights into the computational procedures of the manuscript. DL Vallejo-Ardila is supported by the Department of Science, Technology, and Innovation (COLCIENCIAS-Colombia) (PhD overseas 679 program). Salazar DA is supported by the Department of Science, Technology, and Innovation (COLCIENCIAS-Colombia) (PhD overseas 785 program).

Author Contributions DVA and MAB selected the databases and performed transcriptomics analysis. DVA, MAB, and DS analyzed and interpreted the data and conceived figures and tables. DVA, MAB, and DS wrote the manuscript. DVA conceived and supervised the project. DVA, MAB, and DS edited and revised the manuscript for intellectual content.

Conflict of Interest The authors declare that they have no competing interests.

Ethical Approval This article does not contain any studies with animals performed by any of the authors. Microarray datasets were obtained from the Gene Expression Omnibus database (GEOdatabase).

Financial Disclosure The authors received no specific funding for this work. MAB received support from CNPq [150512/2022-8]. DL Vallejo-Ardila is supported by the Department of Science, Technology, and Innovation (COLCIENCIAS-Colombia) (PhD overseas 679 program). Salazar DA is supported by the Department of Science, Technology, and Innovation (COLCIENCIAS-Colombia) (PhD overseas 785 program).

Supporting Information Supplementary Figs. 1, 2, and 3.
Supplementary Tables 1, 2, and 3.

Data Availability Source data for the main Figs. 1, 2, 3, 4, 5, 6, 7, and 8 are provided in the paper. Additional data is provided in the supplementary data file. Metabolic reconstruction code is available at https://github.com/dasalazarb/NALFD-related_HCC_Recon_2.2. Other data that support the findings are available from the corresponding author (DA) upon reasonable request.

References

Acosta-Lopez S, Diaz-Bethencourt D, Concepción-Massip T, Martin-Fernandez de Basoa MC, Plata-Bello A, Gonzalez-Rodriguez A, Perez-Hernandez F, Plata-Bello J (2020) The androgen receptor expression and its activity have different relationships with prognosis in hepatocellular carcinoma. Sci Rep 10:22046

Ahrens M, Ammerpohl O, von Schönfels W, Kolarova J, Bens S, Itzel T, Teufel A, Herrmann A, Brosch M, Hinrichsen H, Erhart W, Egberts J, Sipos B, Schreiber S, Häsler R, Stickel F, Becker T, Krawczak M, Röcken C, Siebert R, Schafmayer C, Hampe J (2013) DNA methylation analysis in nonalcoholic fatty liver disease suggests distinct disease-specific and remodeling signatures after bariatric surgery. Cell Metab 18:296–302

Baffy G, Brunt EM, Caldwell SH (2012) Hepatocellular carcinoma in non-alcoholic fatty liver disease: an emerging menace. J Hepatol 56:1384–1391

Bajaj JS, Hylemon PB, Younossi Z (2012) The intestinal microbiota and liver disease. Am J Gastroenterol Suppl 1:9–14

Barrett T, Wilhite SE, Ledoux P, Evangelista C, Kim IF, Tomashevsky M, Marshall KA, Phillippy KH, Sherman PM, Holko M, Yefanov A, Lee H, Zhang N, Robertson CL, Serova N, Davis S, Soboleva A (2013) NCBI GEO: archive for functional genomics data sets—update. Nucleic Acids Res 41:D991–D995

Bishop CA, Machate T, Henning T, Henkel J, Püschel G, Weber D, Grune T, Klaus S, Weitkunat K (2022) Detrimental effects of branched-chain amino acids in glucose tolerance can be attributed to valine induced glucotoxicity in skeletal muscle. Nutr Diabetes 12:20

Bolton EC, So AY, Chaivorapol C, Haqq CM, Li H, Yamamoto KR (2007) Cell- and gene-specific regulation of primary target genes by the androgen receptor. Genes Dev 21:2005–2017

Bristot G, De Bastiani MA, Pfaffenseller B, Kapczinski F, Kauer-Sant'Anna M (2020) Gene regulatory network of dorsolateral prefrontal cortex: a master regulator analysis of major psychiatric disorders. Mol Neurobiol 57:1305–1316

Buzzetti E, Pinzani M, Tsochatzis EA (2016) The multiple-hit pathogenesis of non-alcoholic fatty liver disease (NAFLD). Metabolism 65:1038–1048

Carello-Collar G, De Bastiani MA, Zimmer ER (2021) Shared transcriptomic signature between Huntington's disease and early-onset Alzheimer's disease. Alzheimers Dement 17:e053759

Carro MS, Lim WK, Alvarez MJ, Bollo RJ, Zhao X, Snyder EY, Sulman EP, Anne SL, Doetsch F, Colman H, Lasorella A, Aldape K, Califano A, Iavarone A (2010) The transcriptional network for mesenchymal transformation of brain tumours. Nature 463:318–325

Castro MA, Wang X, Fletcher MN, Meyer KB, Markowetz F (2012) RedeR: R/Bioconductor package for representing modular structures, nested networks and multiple levels of hierarchical associations. Genome Biol 13:1–11

Castro MA, de Santiago I, Campbell TM, Vaughn C, Hickey TE, Ross E, Tilley WD, Markowetz F, Ponder BA, Meyer KB (2016) Regulators of genetic risk of breast cancer identified by integrative network analysis. Nat Genet 48:12–21

Chen S, Fang H, Li J, Shi J, Zhang J, Wen P, Wang Z, Yang H, Cao S, Zhang H, Tang H, Guo W, Zhang S (2019) Microarray analysis for expression profiles of lncRNAs and circRNAs in rat liver after brain-dead donor liver transplantation. Biomed Res Int 2019:5604843

Chen SL, Zhu ZX, Yang X, Liu LL, He YF, Yang MM, Guan XY, Wang X, Yun JP (2021) Cleavage and polyadenylation specific factor 1 promotes tumor progression via alternative polyadenylation and splicing in hepatocellular carcinoma. Front Cell Dev Biol 9:616835

Cho EJ, Kwack MS, Jang ES, You SJ, Lee JH, Kim YJ, Yoon JH, Lee HS (2011) Relative etiological role of prior hepatitis B virus infection and nonalcoholic fatty liver disease in the development of non-B non-C hepatocellular carcinoma in a hepatitis B-endemic area. Digestion 84(Suppl 1):17–22

Choi EJ, Jang YJ, Cha EY, Shin JG, Lee SS (2013) Identification and characterization of novel alternative splice variants of human constitutive androstane receptor in liver samples of Koreans and Caucasians. Drug Metab Dispos 41:888–896

Dallio M, Sangineto M, Romeo M, Villani R, Romano AD, Loguercio C, Serviddio G, Federico A (2021) Immunity as cornerstone of non-alcoholic fatty liver disease: the contribution of oxidative stress in the disease progression. Int J Mol Sci 22:436

De Bastiani MA, Klamt F (2019) Integrated transcriptomics reveals master regulators of lung adenocarcinoma and novel repositioning of drug candidates. Cancer Med 8:6717–6729

De Bastiani MA, Pfaffenseller B, Klamt F (2018) Master regulators connectivity map: a transcription factors-centered approach to drug repositioning. Front Pharmacol 9:697–697

De Bastiani MA, Bellaver B, Carello-Collar G, Zimmermann M, Kunach P, Lima-Filho RAS, Forner S, Martini AC, Pascoal TA, Lourenco MV, Rosa-Neto P, Zimmer ER (2022) Transcriptomic similarities and differences between mouse models and human Alzheimer's disease. bioRxiv. https://doi.org/10.1101/2021.2006.2009.447404

Dong B, Saha PK, Huang W, Chen W, Abu-Elheiga LA, Wakil SJ, Stevens RD, Ilkayeva O, Newgard CB, Chan L, Moore DD (2009) Activation of nuclear receptor CAR ameliorates diabetes and fatty liver disease. Proc Natl Acad Sci U S A 106:18831–18836

Dyson J, Jaques B, Chattopadyhay D, Lochan R, Graham J, Das D, Aslam T, Patanwala I, Gaggar S, Cole M, Sumpter K, Stewart S, Rose J, Hudson M, Manas D, Reeves HL (2014) Hepatocellular cancer: the impact of obesity, type 2 diabetes and a multidisciplinary team. J Hepatol 60:110–117

Eberle D, Hegarty B, Bossard P, Ferre P, Foufelle F (2004) SREBP transcription factors: master regulators of lipid homeostasis. Biochimie 86:839–848

Ebrahim A, Lerman JA, Palsson BO, Hyduke DR (2013) COBRApy: COnstraints-based reconstruction and analysis for python. BMC Syst Biol 7:74

Estes C, Razavi H, Loomba R, Younossi Z, Sanyal AJ (2018) Modeling the epidemic of nonalcoholic fatty liver disease demonstrates an exponential increase in burden of disease. Hepatology 67:123–133

Feng Q, Baker SS, Liu W, Arbizu RA, Aljomah G, Khatib M, Nugent CA, Baker RD, Forte TM, Hu Y, Zhu L (2015) Increased apolipoprotein A5 expression in human and rat non-alcoholic fatty livers. Pathology 47:341–348

Ferolla SM, Armiliato GNA, Couto CA, Ferrari TCA (2014) The role of intestinal bacteria overgrowth in obesity-related nonalcoholic fatty liver disease. Nutrients 6:5583–5599

Friedman SL, Neuschwander-Tetri BA, Rinella M, Sanyal AJ (2018) Mechanisms of NAFLD development and therapeutic strategies. Nat Med 24:908–922

Gabele E, Dostert K, Hofmann C, Wiest R, Scholmerich J, Hellerbrand C, Obermeier F (2011) DSS induced colitis increases portal LPS levels and enhances hepatic inflammation and fibrogenesis in experimental NASH. J Hepatol 55:1391–1399

Gao J, He J, Zhai Y, Wada T, Xie W (2009) The constitutive androstane receptor is an anti-obesity nuclear receptor that improves insulin sensitivity. J Biol Chem 284:25984–25992

García-Monzón C, Lo Iacono O, Crespo J, Romero-Gómez M, García-Samaniego J, Fernández-Bermejo M, Domínguez-Díez A, Rodríguez de Cía J, Sáez A, Porrero JL, Vargas-Castrillón J, Chávez-Jiménez E, Soto-Fernández S, Díaz A, Gallego-Durán R, Madejón A, Miquilena-Colina ME (2014) Increased soluble CD36 is linked to advanced steatosis in nonalcoholic fatty liver disease. Eur J Clin Investig 44:65–73

Gentleman RC, Carey VJ, Bates DM, Bolstad B, Dettling M, Dudoit S, Ellis B, Gautier L, Ge Y, Gentry J, Hornik K, Hothorn T, Huber W, Iacus S, Irizarry R, Leisch F, Li C, Maechler M, Rossini AJ, Sawitzki G, Smith C, Smyth G, Tierney L, Yang JY, Zhang J (2004) Bioconductor: open software development for computational biology and bioinformatics. Genome Biol 5:R80

Gerhard GS, Hanson A, Wilhelmsen D, Piras IS, Still CD, Chu X, Petrick AT, DiStefano JK (2019) AEBP1 expression increases with severity of fibrosis in NASH and is regulated by glucose, palmitate, and miR-372-3p. PLoS One 14:e0219764–e0219764

Giriş M, Doğru-Abbasoğlu S, Kumral A, Olgaç V, Koçak-Toker N, Uysal M (2014) Effect of carnosine alone or combined with α-tocopherol on hepatic steatosis and oxidative stress in fructose-induced insulin-resistant rats. J Physiol Biochem 70:385–395

Godoy-Matos AF, Silva Júnior WS, Valerio CM (2020) NAFLD as a continuum: from obesity to metabolic syndrome and diabetes. Diabetol Metab Syndr 12:60

Grgurevic I, Bozin T, Mikus M, Kukla M, O'Beirne J (2021) Hepatocellular carcinoma in non-alcoholic fatty liver disease: from epidemiology to diagnostic approach. Cancers 13:5844

Hoshida Y, Villanueva A, Kobayashi M, Peix J, Chiang DY, Camargo A, Gupta S, Moore J, Wrobel MJ, Lerner J, Reich M, Chan JA, Glickman JN, Ikeda K, Hashimoto M, Watanabe G, Daidone MG, Roayaie S, Schwartz M, Thung S, Salvesen HB, Gabriel S, Mazzaferro V, Bruix J, Friedman SL, Kumada H, Llovet JM, Golub TR (2008) Gene expression in fixed tissues and outcome in hepatocellular carcinoma. N Engl J Med 359:1995–2004

Huang W, Zhang J, Wei P, Schrader WT, Moore DD (2004) Meclizine is an agonist ligand for mouse constitutive androstane receptor (CAR) and an inverse agonist for human CAR. Mol Endocrinol 18:2402–2408

Huang DQ, El-Serag HB, Loomba R (2021) Global epidemiology of NAFLD-related HCC: trends, predictions, risk factors and prevention. Nat Rev Gastroenterol Hepatol 18:223–238

Jenior ML, Moutinho TJ Jr, Dougherty BV, Papin JA (2020) Transcriptome-guided parsimonious flux analysis improves predictions with metabolic networks in complex environments. PLoS Comput Biol 16:e1007099

Kanwal F, Kramer JR, Mapakshi S, Natarajan Y, Chayanupatkul M, Richardson PA, Li L, Desiderio R, Thrift AP, Asch SM, Chu J, El-Serag HB (2018) Risk of hepatocellular cancer in patients with non-alcoholic fatty liver disease. Gastroenterology 155:1828–1837.e1822

Kanwal F, Kramer JR, Li L, Dai J, Natarajan Y, Yu X, Asch SM, El-Serag HB (2020) Effect of metabolic traits on the risk of cirrhosis and hepatocellular cancer in nonalcoholic fatty liver disease. Hepatology 71:808–819

Kim S, Kwon H, Park JH, Cho B, Kim D, Oh SW, Lee CM, Choi HC (2012) A low level of serum total testosterone is independently associated with nonalcoholic fatty liver disease. BMC Gastroenterol 12:69

Lake AD, Novak P, Shipkova P, Aranibar N, Robertson DG, Reily MD, Lehman-McKeeman LD, Vaillancourt RR, Cherrington NJ (2015) Branched chain amino acid metabolism profiles in progressive human nonalcoholic fatty liver disease. Amino Acids 47:603–615

Lewis JE, Forshaw TE, Boothman DA, Furdui CM, Kemp ML (2021) Personalized genome-scale metabolic models identify targets of redox metabolism in radiation-resistant tumors. Cell Syst 12:68–81.e11

Li H, Wang H (2010) Activation of xenobiotic receptors: driving into the nucleus. Expert Opin Drug Metab Toxicol 6:409–426

Li Z, Kwon SM, Li D, Li L, Peng X, Zhang J, Sueyoshi T, Raufman JP, Negishi M, Wang XW, Wang H (2022) Human constitutive androstane receptor represses liver cancer development and hepatoma cell proliferation by inhibiting erythropoietin signaling. J Biol Chem 298:101885

Liew ZH, Goh GB, Hao Y, Chang PE, Tan CK (2019) Comparison of hepatocellular carcinoma in patients with cryptogenic versus hepatitis B etiology: a study of 1079 cases over 3 decades. Dig Dis Sci 64:585–590

Lin H-Y, Yu I-C, Wang R-S, Chen Y-T, Liu N-C, Altuwaijri S, Hsu C-L, Ma W-L, Jokinen J, Sparks JD, Yeh S, Chang C (2008) Increased hepatic steatosis and insulin resistance in mice lacking hepatic androgen receptor. Hepatology 47:1924–1935

Liu W, Baker SS, Baker RD, Nowak NJ, Zhu L (2011) Upregulation of hemoglobin expression by oxidative stress in hepatocytes and its implication in nonalcoholic steatohepatitis. PLoS One 6: e24363

Longo M, Paolini E, Meroni M, Dongiovanni P (2021) Remodeling of mitochondrial plasticity: the key switch from NAFLD/NASH to HCC. Int J Mol Sci 22:4173

López-Vicario C, González-Périz A, Rius B, Morán-Salvador E, García-Alonso V, Lozano JJ, Bataller R, Cofán M, Kang JX, Arroyo V, Clària J, Titos E (2014) Molecular interplay between Δ5/Δ6 desaturases and long-chain fatty acids in the pathogenesis of non-alcoholic steatohepatitis. Gut 63:344–355

Lou Y, Chen YD, Sun FR, Shi JP, Song Y, Yang J (2017) Potential regulators driving the transition in nonalcoholic fatty liver disease: a stage-based view. Cell Physiol Biochem 41:239–251

Lu Y, Xu W, Ji J, Feng D, Sourbier C, Yang Y, Qu J, Zeng Z, Wang C, Chang X, Chen Y, Mishra A, Xu M, Lee MJ, Lee S, Trepel J, Linehan WM, Wang X, Yang Y, Neckers L (2015) Alternative splicing of the cell fate determinant Numb in hepatocellular carcinoma. Hepatology 62:1122–1131

Lund J, Zaphiropoulos P, Mode A, Warner M, Gustafsson J (1991) Hormonal regulation of cytochrome P-450 gene expression. Adv Pharmacol 22:325–354

Ma W-L, Lai H-C, Yeh S, Cai X, Chang C (2014) Androgen receptor roles in hepatocellular carcinoma, cirrhosis, and hepatitis. Endocr Relat Cancer 21:R165–R182

MacNeil LT, Walhout AJM (2011) Gene regulatory networks and the role of robustness and stochasticity in the control of gene expression. Genome Res 21:645–657

Majdalawieh A, Ro HS (2009) LPS-induced suppression of macrophage cholesterol efflux is mediated by adipocyte enhancer-binding protein 1. Int J Biochem Cell Biol 41:1518–1525

Margolin AA, Nemenman I, Basso K, Wiggins C, Stolovitzky G, Favera RD, Califano A (2006a) ARACNE: an algorithm for the reconstruction of gene regulatory networks in a mammalian cellular context. BMC Bioinform 7:S7–S7

Margolin AA, Wang K, Lim WK, Kustagi M, Nemenman I, Califano A (2006b) Reverse engineering cellular networks. Nat Protoc 1:662–671

Mattu S, Saliba C, Sulas P, Zavattari P, Perra A, Kowalik MA, Monga SP, Columbano A (2018) High frequency of β-catenin mutations in mouse hepatocellular carcinomas induced by a nongenotoxic constitutive androstane receptor agonist. Am J Pathol 188:2497–2507

McGlynn KA, Petrick JL, El-Serag HB (2021) Epidemiology of hepatocellular carcinoma. Hepatology 73:4–13

Mody A, White D, Kanwal F, Garcia JM (2015) Relevance of low testosterone to non-alcoholic fatty liver disease. Cardiovasc Endocrinol 4:83–89

Mózes FE, Lee JA, Selvaraj EA, Jayaswal ANA, Trauner M, Boursier J, Fournier C, Staufer K, Stauber RE, Bugianesi E, Younes R, Gaia S, Lupşor-Platon M, Petta S, Shima T, Okanoue T, Mahadeva S, Chan W-K, Eddowes PJ, Hirschfield GM, Newsome PN, Wong VW-S, de Ledinghen V, Fan J, Shen F, Cobbold JF, Sumida Y, Okajima A, Schattenberg JM, Labenz C, Kim W, Lee MS, Wiegand J, Karlas T, Yılmaz Y, Aithal GP, Palaniyappan N, Cassinotto C, Aggarwal S, Garg H, Ooi GJ, Nakajima A, Yoneda M, Ziol M, Barget N, Geier A, Tuthill T, Brosnan MJ, Anstee QM, Neubauer S, Harrison SA, Bossuyt PM, Pavlides M (2022) Diagnostic accuracy of non-invasive tests for advanced fibrosis in patients with NAFLD: an individual patient data meta-analysis. Gut 71:1006–1019

Mutwil M, Klie S, Tohge T, Giorgi FM, Wilkins O, Campbell MM, Fernie AR, Usadel B, Nikoloski Z, Persson S (2011) PlaNet: combined sequence and expression comparisons across plant networks derived from seven species. Plant Cell 23:895–910

Nahon P, Allaire M, Nault J-C, Paradis V (2020) Characterizing the mechanism behind the progression of NAFLD to hepatocellular carcinoma. Hepat Oncol 7:HEP36–HEP36

Ohno S (1979) The number of genes in the mammalian genome and the need for master regulatory genes. Major sex-determining genes. Springer Berlin Heidelberg, Berlin, Heidelberg, pp 17–21

Orth JD, Thiele I, Palsson BØ (2010) What is flux balance analysis? Nat Biotechnol 28:245–248

Pardo V, González-Rodríguez Á, Muntané J, Kozma SC, Valverde ÁM (2015) Role of hepatocyte S6K1 in palmitic acid-induced endoplasmic reticulum stress, lipotoxicity, insulin resistance and in oleic acid-induced protection. Food Chem Toxicol 80:298–309

Pfaffenseller B, da Silva Magalhaes PV, De Bastiani MA, Castro MA, Gallitano AL, Kapczinski F, Klamt F (2016) Differential expression of transcriptional regulatory units in the prefrontal cortex of patients with bipolar disorder: potential role of early growth response gene 3. Transl Psychiatry 6:e805

Piao G, Saito S, Sun Y, Liu Z-P, Wang Y, Han X, Wu J, Zhou H, Chen L, Horimoto K (2012) A computational procedure for identifying master regulator candidates: a case study on diabetes progression in Goto-Kakizaki rats. BMC Syst Biol 6:S2–S2

Pouwels S, Sakran N, Graham Y, Leal A, Pintar T, Yang W, Kassir R, Singhal R, Mahawar K, Ramnarain D (2022) Non-alcoholic fatty liver disease (NAFLD): a review of pathophysiology, clinical management and effects of weight loss. BMC Endocr Disord 22:63

Qiu Y-Q (2013) KEGG pathway database. In: Dubitzky W, Wolkenhauer O, Cho K-H, Yokota H (eds) Encyclopedia of systems biology. Springer New York, New York, pp 1068–1069

Rada P, González-Rodríguez Á, García-Monzón C, Valverde ÁM (2020) Understanding lipotoxicity in NAFLD pathogenesis: is CD36 a key driver? Cell Death Dis 11:802

Rao MS, Van Vleet TR, Ciurlionis R, Buck WR, Mittelstadt SW, Blomme EAG, Liguori MJ (2019) Comparison of RNA-Seq and microarray gene expression platforms for the toxicogenomic evaluation of liver from short-term rat toxicity studies. Front Genet 9:636

Repana D, Nulsen J, Dressler L, Bortolomeazzi M, Venkata SK, Tourna A, Yakovleva A, Palmieri T, Ciccarelli FD (2019) The Network of Cancer Genes (NCG): a comprehensive catalogue of known and candidate cancer genes from cancer sequencing screens. Genome Biol 20:1

Režen T, Tamasi V, Lövgren-Sandblom A, Björkhem I, Meyer UA, Rozman D (2009) Effect of CAR activation on selected metabolic pathways in normal and hyperlipidemic mouse livers. BMC Genomics 10:1–15

Ritchie ME, Phipson B, Wu D, Hu Y, Law CW, Shi W, Smyth GK (2015) Limma powers differential expression analyses for RNA-sequencing and microarray studies. Nucleic Acids Res 43:e47

Roessler S, Jia HL, Budhu A, Forgues M, Ye QH, Lee JS, Thorgeirsson SS, Sun Z, Tang ZY, Qin LX, Wang XW (2010) A unique metastasis gene signature enables prediction of tumor relapse in early-stage hepatocellular carcinoma patients. Cancer Res 70:10202–10212

Roessler S, Long EL, Budhu A, Chen Y, Zhao X, Ji J, Walker R, Jia HL, Ye QH, Qin LX, Tang ZY, He P, Hunter KW, Thorgeirsson SS, Meltzer PS, Wang XW (2012) Integrative genomic identification of genes on 8p associated with hepatocellular carcinoma progression and patient survival. Gastroenterology 142:957–966.e912

Romeo S, Kozlitina J, Xing C, Pertsemlidis A, Cox D, Pennacchio LA, Boerwinkle E, Cohen JC, Hobbs HH (2008) Genetic variation in PNPLA3 confers susceptibility to nonalcoholic fatty liver disease. Nat Genet 40:1461–1465

Salazar DA, Rodríguez-López A, Herreño A, Barbosa H, Herrera J, Ardila A, Barreto GE, González J, Alméciga-Díaz CJ (2016) Systems biology study of mucopolysaccharidosis using a human metabolic reconstruction network. Mol Genet Metab 117:129–139

Schooneman MG, Vaz FM, Houten SM, Soeters MR (2013) Acylcarnitines: reflecting or inflicting insulin resistance? Diabetes 62:1–8

Serviddio G, Bellanti F, Tamborra R, Rollo T, Capitanio N, Romano AD, Sastre J, Vendemiale G, Altomare E (2008) Uncoupling protein-2 (UCP2) induces mitochondrial proton leak and increases susceptibility of non-alcoholic steatohepatitis (NASH) liver to ischaemia-reperfusion injury. Gut 57:957–965

Shimano H (2001) Sterol regulatory element-binding proteins (SREBPs): transcriptional regulators of lipid synthetic genes. Prog Lipid Res 40:439–452

Sookoian S, Pirola CJ (2011) Meta-analysis of the influence of I148M variant of patatin-like phospholipase domain containing 3 gene (PNPLA3) on the susceptibility and histological severity of nonalcoholic fatty liver disease. Hepatology 53:1883–1894

Starmann J, Fälth M, Spindelböck W, Lanz KL, Lackner C, Zatloukal K, Trauner M, Sültmann H (2012) Gene expression profiling unravels cancer-related hepatic molecular signatures in steatohepatitis but not in steatosis. PLoS One 7:e46584

Steensels S, Qiao J, Ersoy BA (2020) Transcriptional regulation in non-alcoholic fatty liver disease. Meta 10:283

Steuer R, Kurths J, Daub CO, Weise J, Selbig J (2002) The mutual information: detecting and evaluating dependencies between variables. Bioinformatics 18:S231–S240

Stickel F, Hellerbrand C (2010) Non-alcoholic fatty liver disease as a risk factor for hepatocellular carcinoma: mechanisms and implications. Gut 59:1303–1307

Streba LAM, Vere CC, Rogoveanu I, Streba CT (2015) Nonalcoholic fatty liver disease, metabolic risk factors, and hepatocellular carcinoma: an open question. World J Gastroenterol 21:4103–4110

Subramanian A, Tamayo P, Mootha VK, Mukherjee S, Ebert BL, Gillette MA, Paulovich A, Pomeroy SL, Golub TR, Lander ES, Mesirov JP (2005) Gene set enrichment analysis: a knowledge-based approach for interpreting genome-wide expression profiles. Proc Natl Acad Sci 102:15545–15550

Sun Y, Ji F, Kumar MR, Zheng X, Xiao Y, Liu N, Shi J, Wong L, Forgues M, Qin LX, Tang ZY, Zhao X, Wang XW, Ji J (2017) Transcriptome integration analysis in hepatocellular carcinoma reveals discordant intronic miRNA-host gene pairs in expression. Int J Biol Sci 13:1438–1449

Suzuki R, Shimodaira H (2006) Pvclust: an R package for assessing the uncertainty in hierarchical clustering. Bioinformatics 22:1540–1542

Swainston N, Smallbone K, Hefzi H, Dobson PD, Brewer J, Hanscho M, Zielinski DC, Ang KS, Gardiner NJ, Gutierrez JM, Kyriakopoulos S, Lakshmanan M, Li S, Liu JK, Martínez VS, Orellana CA, Quek LE, Thomas A, Zanghellini J, Borth N, Lee DY, Nielsen LK, Kell DB, Lewis NE, Mendes P (2016) Recon 2.2: from reconstruction to model of human metabolism. Metabolomics 12:109

Takaki Y, Saito Y, Takasugi A, Toshimitsu K, Yamada S, Muramatsu T, Kimura M, Sugiyama K, Suzuki H, Arai E, Ojima H, Kanai Y, Saito H (2014) Silencing of microRNA-122 is an early event during hepatocarcinogenesis from non-alcoholic steatohepatitis. Cancer Sci 105:1254–1260

Teratani T, Tomita K, Suzuki T, Furuhashi H, Irie R, Nishikawa M, Yamamoto J, Hibi T, Miura S, Minamino T, Oike Y, Hokari R, Kanai T (2018) Aortic carboxypeptidase-like protein, a WNT ligand, exacerbates nonalcoholic steatohepatitis. J Clin Invest 128:1581–1596

Toroghi MK, Cluett WR, Mahadevan R (2016) A multi-scale model of the whole human body based on dynamic parsimonious flux balance analysis. IFAC-PapersOnLine 49:937–942

van Vlies N, Ferdinandusse S, Wanders R, Vaz F (1767) PPARalfa-activation results in enhanced carnitine biosynthesis and OCTN2 expression. Biochim Biophys Acta 11:2007

Vargas DM, De Bastiani MA, Zimmer ER, Klamt F (2018) Alzheimer's disease master regulators analysis: search for potential molecular targets and drug repositioning candidates. Alzheimers Res Ther 10:59

Vargas DM, De Bastiani MA, Parsons RB, Klamt F (2021) Parkinson's disease master regulators on substantia nigra and frontal cortex and their use for drug repositioning. Mol Neurobiol 58:1517–1534

Villanueva A, Hoshida Y, Battiston C, Tovar V, Sia D, Alsinet C, Cornella H, Liberzon A, Kobayashi M, Kumada H, Thung SN, Bruix J, Newell P, April C, Fan JB, Roayaie S, Mazzaferro V, Schwartz ME, Llovet JM (2011) Combining clinical, pathology, and gene expression data to predict recurrence of hepatocellular carcinoma. Gastroenterology 140:1501–1512.e1502

Vuppalanchi R, Unalp A, Van Natta ML, Cummings OW, Sandrasegaran KE, Hameed T, Tonascia J, Chalasani N (2009) Effects of liver biopsy sample length and number of readings on sampling variability in nonalcoholic fatty liver disease. Clin Gastroenterol Hepatol 7:481–486

Waidmann O, Pleli T, Weigert A, Imelmann E, Kakoschky B, Schmithals C, Döring C, Frank M, Longerich T, Köberle V, Hansmann M-L, Brüne B, Zeuzem S, Piiper A, Dikic I (2020) Tax1BP1 limits hepatic inflammation and reduces experimental hepatocarcinogenesis. Sci Rep 10:16264

Wang HW, Hsieh TH, Huang SY, Chau GY, Tung CY, Su CW, Wu JC (2013) Forfeited hepatogenesis program and increased embryonic stem cell traits in young hepatocellular carcinoma (HCC) comparing to elderly HCC. BMC Genomics 14:736

Wang X-A, Zhang R, She Z-G, Zhang X-F, Jiang D-S, Wang T, Gao L, Deng W, Zhang S-M, Zhu L-H, Guo S, Chen K, Zhang X-D, Liu D-P, Li H (2014) Interferon regulatory factor 3 constrains IKKβ/NF-κB signaling to alleviate hepatic steatosis and insulin resistance. Hepatology 59:870–885

Wang Y, Gao B, Tan PY, Handoko YA, Sekar K, Deivasigamani A, Seshachalam VP, OuYang HY, Shi M, Xie C, Goh BKP, Ooi LL, Man Hui K (2019) Genome-wide CRISPR knockout screens identify NCAPG as an essential oncogene for hepatocellular carcinoma tumor growth. FASEB J 33:8759–8770

Wang C, Liao Y, He W, Zhang H, Zuo D, Liu W, Yang Z, Qiu J, Yuan Y, Li K, Zhang Y, Wang Y, Shi Y, Qiu Y, Gao S, Yuan Y, Li B (2021) Elafin promotes tumour metastasis and attenuates the anti-metastatic effects of erlotinib via binding to EGFR in hepatocellular carcinoma. J Exp Clin Cancer Res 40:113

Weiß J, Rau M, Geier A (2014) Non-alcoholic fatty liver disease: epidemiology, clinical course, investigation, and treatment. Dtsch Arztebl Int 111:447–452

Wickham H (2016) Elegant graphics for data analysis. Springer International Publishing, Houston

Wiest R, Lawson M, Geuking M (2014) Pathological bacterial translocation in liver cirrhosis. J Hepatol 60:197–209

Wu T, Hu E, Xu S, Chen M, Guo P, Dai Z, Feng T, Zhou L, Tang W, Zhan L, Fu X, Liu S, Bo X, Yu G (2021) clusterProfiler 4.0: a universal enrichment tool for interpreting omics data. Innovation (Camb) 2:100141

Xu R, Tao A, Zhang S, Deng Y, Chen G (2015) Association between patatin-like phospholipase domain containing 3 gene (PNPLA3) polymorphisms and nonalcoholic fatty liver disease: a HuGE review and meta-analysis. Sci Rep 5:9284

Yang X, Gonzalez FJ, Huang M, Bi H (2020) Nuclear receptors and non-alcoholic fatty liver disease: an update. Liver Res 4:88–93

Yu S, Matsusue K, Kashireddy P, Cao W-Q, Yeldandi V, Yeldandi AV, Rao MS, Gonzalez FJ, Reddy JK (2003) Adipocyte-specific gene expression and adipogenic steatosis in the mouse liver due to peroxisome proliferator-activated receptor γ1 (PPARγ1) overexpression*. J Biol Chem 278:498–505

Zhang L, Reidy SP, Nicholson TE, Lee HJ, Majdalawieh A, Webber C, Stewart BR, Dolphin P, Ro HS (2005) The role of AEBP1 in sex-specific diet-induced obesity. Mol Med 11:39–47

Zhang L, Liu X, Zhang X, Chen R (2016) Identification of important long non-coding RNAs and highly recurrent aberrant alternative splicing events in hepatocellular carcinoma through integrative analysis of multiple RNA-Seq datasets. Mol Gen Genomics 291:1035–1051

Zhao X, Parpart S, Takai A, Roessler S, Budhu A, Yu Z, Blank M, Zhang YE, Jia HL, Ye QH, Qin LX, Tang ZY, Thorgeirsson SS, Wang XW (2015) Integrative genomics identifies YY1AP1 as an oncogenic driver in EpCAM(+) AFP(+) hepatocellular carcinoma. Oncogene 34:5095–5104

Tumor Microenvironment and Immunotherapy in Advanced Biliary Tract Cancers

Mustafa Korkmaz and Mehmet Artaç

Abstract

Intrahepatic, perihepatic, and hilar bile duct cancers are defined as cholangiocarcinoma (CCA). Cholangiocarcinomas constitute approximately 3% of all gastrointestinal system malignancies. The main treatment in the nonmetastatic stage is surgery. In metastatic or unresectable disease, the main treatment is systemic therapy, and there is no effective treatment response. Five-year survival in these patients is around 4.1%. With gemcitabine plus cisplatin, which is the standard first-line treatment in metastatic biliary tract cancer (BTC), the median overall survival (mOS) is 11.7 months. Again, with FOLFOX chemotherapy, which is the standard treatment in the second line, the mOS is 5.6 months. Therefore, the search for treatments with more effective response rates and survival results in BTCs continues rapidly. In recent years, with the publication of the cancer genome atlas, the tumor microenvironment has become more understandable. Along with these developments, it has been reported that the most important molecular mechanisms involved in the pathogenesis of CCA are Wnt/β-catenin and nuclear factor kappa B signaling pathways, and these mechanisms play a role in the formation and progression of CCA. The tumor microenvironment is a complex environment that influences the proliferative and invasive properties of epithelial cancer cells. Cancer cells, normal cells, mesenchymal stem cells, fibroblasts, pericytes, adipocytes, follicular dendritic cells, cancer-associated fibroblasts, mesenchymal stem cells, tumor-associated macrophages, regulatory T cells, dendritic cells, cytokines, growth factors, and

M. Korkmaz (✉)
Department of Medical Oncology, Tokat State Hospital, Tokat, Turkey
e-mail: dr.musstafa@gmail.com

M. Artaç
Department of Medical Oncology, Necmettin Erbakan University School of Medicine, Konya, Turkey

Interdisciplinary Cancer Research, https://doi.org/10.1007/16833_2022_41
Published online: 6 October 2022

extracellular matrix constituent components. After this information, immunotherapy with immune checkpoint inhibitors has been proven to be effective in solid cancers such as malign melanoma, metastatic renal cell cancer, and non-small cell lung cancer, and the investigation of the effectiveness of these treatments in other cancer types has accelerated. First, second-subsequent line monotherapy immunotherapy studies were conducted and all are phases 1 and 2, with objective response rates (ORR) reported between 5% and 35%. Afterward, especially in the first line, combination immunotherapy studies, mostly phase 2, were conducted, and the ORR was reported to be around 75% in these studies. These results suggest that first-line immunotherapy combination practices will become the standard of care in the near future.

Keywords

Biliary tract cancers · Biomarker · Cholangiocarcinoma · Immune checkpoint inhibitors · Immunotherapy · Tumor microenvironment

1 Introduction

Cholangiocarcinoma (CCA) is among biliary tract cancers (BTCs) arising in the intrahepatic, perihilar, or distal (extrahepatic) bile tree. In the past, extrahepatic bile duct malignancies were classified as cancers of the gallbladder, extrahepatic ducts, and ampulla of Vater, whereas intrahepatic tumors were classified as primary liver cancers (Banales et al. 2016). Extrahepatic bile ducts, which are divided into perihilar and distal segments, are located proximal to the cystic duct. Perihilar cancers constitute 60–70% of extrahepatic CCAs and are divided into four distinct types according to the involvement pattern of the hepatic ducts by the Bismuth-Corlette classification. Tumors involving the hepatic duct bifurcation (hilar CCA) are collectively called Klatskin tumors. Tumors arising from the small intrahepatic ducts or the large intrahepatic ducts proximal to the bifurcation of the right and left hepatic ducts are termed as intrahepatic CCA and account for approximately 5–10% of all CCAs (Lee and Chun 2018; Vauthey and Blumgart 1994; Bismuth et al. 1992). The CCA index varies by its continents. The incidence is higher in Asia than that in Europe and the USA. While the incidence is low in high-income countries (0.35–2 cases per 100,000 population per year), the incidence is up to 40-fold higher in endemic areas of Thailand and China (Banales et al. 2016; Valle et al. 2021; Globocan 2021). There is a set of identified risk factors for CCA. Of these, the main risk factors in the USA and Europe are primary sclerosing cholangitis (PSC) and fibropolycystic liver disease. Other risk factors include chronic intrahepatic stone disease, chronic liver disease (cirrhosis and hepatitis B-C), parasitic infections (clonorchis and opisthorchis), toxins (thorotrast), iron overload, inflammatory bowel disease (IBD), diabetes, obesity, alcohol, smoking, and host genetic polymorphisms (Tyson and El-Serag 2011; Tran et al. 2019). Surgery is the only potential curative treatment for BTC, but approximately 60–85% of patients are diagnosed at the unresectable or metastatic stage, with a 5-year survival rate of approximately 4.1%. In addition, even if curative surgery is performed, the risk of recurrence is significantly higher, with a 5-year survival rate of approximately 33.3% (Jansen et al. 2020). For unresectable tumors, local treatments including radiofrequency ablation (RFA), transarterial chemoembolization (TACE), and transarterial radioembolization (TARE) can be performed when appropriate (Satiya et al. 2020). The main goal of treatment for metastatic disease is palliation. Gemcitabine-cisplatin combination, which has been shown to be effective in the ABC-02 study, is the standard first-line treatment for biliary tract cancers. In this study, gemcitabine-cisplatin combination and single-agent gemcitabine were compared. During a median follow-up of 8.2 months, the overall survival was 11.7 months in the combination arm and 8.1 months in the monotherapy arm. The median progression-free survival was 8 months in the combination arm and 5 months in the monotherapy arm (Valle et al. 2010). The treatment recommended in the current guidelines on the second line is the FOLFOX chemotherapy regimen. In the ABC-06 study, second-line FOLFOX and activated symptom control (ACS) arms were compared in patients who progressed after first-line gemcitabine-cisplatin combination. Better overall survival rates were achieved in the FOLFOX arm at 6 months (51% versus 36%) and 12 months (26% versus 11%). The median overall survival was 6.2 months in the FOLFOX arm and 5.3 months in the ACS arm (Lamarca et al. 2021). There are no standard guidelines for recommendations after second-line chemotherapy. Because of the modest survival results with cytotoxic chemotherapies in metastatic BTCs, new treatments have emerged. In recent years, it has resulted in a better understanding of the molecular biology of tumors and the detection of targetable mutations. Targeted therapies have been developed for fibroblast growth factor receptor-2 (FGFR2) fusions and isocitrate dehydrogenase-1 (IDH-1) mutations in even duct cancers (Rizvi and Gores 2017). Fibroblast growth factor receptor-2 gene changes are involved in the pathogenesis of cholangiocarcinoma and seen in approximately 14% (Ross et al. 2014). Pemigatinib and infigratinib are selective FGFR 1–3 inhibitors, which increased the objective response rates and durable response in patients with previously treated unresectable locally advanced or metastatic cholangiocarcinoma with FGFR2 fusion or other rearrangement, after which they were FDA approved (Abou-Alfa et al. 2020; Javle et al. 2021; FDA 2020; TRUSELTIQ 2021). The frequency of IDH-1 mutation, another targetable mutation in patients with cholangiocarcinoma, is 25% (Borger et al. 2011). In a placebo-controlled phase III ClarIDHy trial study in patients with cholangiocarcinoma in whom IDH-1 mutation was detected, ivosidenib showed a significant PFS benefit and trend toward better overall survival, and received FDA approval with these results (Zhu et al. 2021). Treatment development studies are ongoing for B-Raf proto-oncogene V600E mutations, neurotrophic tyrosine receptor kinase fusions (NTRK), human epidermal growth factor receptor (HER)-2 amplifications and mutations, and other less commonly detected mutations. Larotrectinib and entrectinib for NTRK fusions, dabrafenib plus trametinib for BRAF V600E mutation, trastuzumab plus pertuzumab, and neratinib for HER-2 amplification or overexpression are recommended (Lamarca et al. 2022).

After demonstrating the effectiveness of immunotherapy in malignant melanoma, lung cancer, and renal cell cancer (RCC), studies have accelerated to demonstrate the effectiveness of immunotherapy in other cancers as well, and studies are continuing to make it a part of the treatment in almost all cancers. In this section, we will extensively review the tumor microenvironment and immunotherapy in BTC.

2 Molecular Mechanisms and Tumor Microenvironment

2.1 Molecular Mechanisms

With increasing studies in recent years, progress has been made in identifying the molecular mechanisms involved in the pathogenesis of CCA. Wnt/β-catenin and nuclear factor kappa B (NF-κB) signaling pathways are thought to be the most important molecular pathways in the formation and progression of CCA. WNT protein inhibits the activity of the complex consisting of disheveled, axin, adenomatous polyposis coli tumor suppressor protein and glycogen synthase kinase, which is activated by binding to its receptors on the cell membrane. This inhibition suppresses β-catenin phosphorylation. When unphosphorylated β-catenin enters the nucleus, it can associate with T-cell factor (TCF)/lymphoid enhancer-binding factor (LEF) transcription factors. Its conjugation leads to the expression of oncogenes involved in CCA tumorigenesis, proliferation, and drug resistance (Boulter et al. 2015; Chen et al. 2016; Zhang et al. 2013; Shen et al. 2013). It is inactivated when bound to the NF-κB inhibitor IκB, but activated when IκBs are phosphorylated or degraded. Activated NF-κB enters the nucleus and binds to DNA, inducing target genes and thereby modulating the growth and development of CCA (Dai et al. 2013). Another pathway involved in cholangiocarcinoma progression is the Notch signaling pathway. After binding of the notch ligands and receptors, the notch intracellular domain (NICD) enters the nucleus after a series of events, forming a complex with the transcription factor CSL (a kind of DNA-binding protein), which activates the expression of the CCA-inducing gene. Activation of notch signaling along with inactivation of the tumor suppressor gene p53 induces CCA tumorigenesis, growth, aggressiveness, and malignant transformation (Razumilava and Gores 2013; Guest et al. 2016; El Khatib et al. 2013). Overexpression of epidermal growth factor receptors (EGFRs) on the cell membrane may be effective prognostic factors for intrahepatic CCA. Downstream pathways include Ras2/Raf2/mitogen-activated protein kinase (MAPK) signaling pathway, phosphatidyl inositol 3-kinase/threonine kinase (PI3K/AKT) signaling pathway, and transducers and activator of transcription (STAT) signaling pathway (Yang et al. 2014; Zhang et al. 2015).

2.2 Tumor Microenvironment

The tumor microenvironment (TME) is a complex environment that influences the proliferative and invasive properties of epithelial cancer cells. The CCA itself and the reactions it causes create this dynamic environment. The TME, which provides support to tumor cells in malignant transformation, consists of cellular and noncellular components (Wang et al. 2017). The cellular component consists of cancer cells, normal cells, mesenchymal stem cells, fibroblasts, pericytes, adipocytes, follicular dendritic cells, cancer-associated fibroblasts (CAFs), mesenchymal stem cells, tumor-associated macrophages (TAMs) or type 2 macrophages, regulatory T cells, dendritic cells, and tumor-infiltrating cells. The noncellular component consists of cytokines, growth factors, soluble factors such as DNA and small RNAs, and the extracellular matrix (ECM) (Balkwill et al. 2012; El-Daly et al. 2020). The interactions of these components with each other form a complex communication network regulated by an ECM and lead to migration, immunosuppression, and induction of angiogenesis, resulting in progression and metastases of CCA (Fabris et al. 2019, 2020).

2.3 Noncellular Components of Tumor Microenvironment

Tumor-stromal cell interaction increases the production of chemokines, growth factors, cytokines, and matrix metalloproteases (MMPs) that facilitate tumor growth and angiogenesis (Patel et al. 2017). In order to invade, cancer cells degrade the basement membrane (BM) and ECM collagen, which contain many proteins such as fibronectin and laminin, with MMP. Matrix metalloproteases are mainly involved in the degradation process of the matrix, promoting cellular migration, regulation of growth factors/cytokines, apoptosis, and neovascularization (Azevedo Martins et al. 2020). Cytokines such as interleukins (IL), tumor necrosis factor (TNF) family, interferons (IFN), and transforming growth factor-β (TGF-β) protein family in noncellular TME may directly or indirectly affect tumor formation (Patel et al. 2017). It shows that various cytokines such as IL-1, IL-6, and IL-8 released from cancer and tumor-infiltrating cells provide support for invasion, proliferation, and epithelial-mesenchymal transition (EMT) formation for the development of TME (Tsai et al. 2014). Epithelial-mesenchymal transition in cancer has been associated with various tumor functions, including tumor initiation, invasion, metastasis, and resistance to therapy (Pastushenko and Blanpain 2019). In addition, IL-33 strongly induces Th2 cells producing IL-4/5/6/10/13 cytokines in TME while suppressing Th1 cells from producing cytokines such as IL-12 and IFN-γ. IL-23, which is responsible for the formation of inflammation, causes the release of TGF-β, which is responsible for the suppression of the immune system and the responses against the immune system. In addition, IL-10 and IL-23 play a role in promoting suppression of the immune system by reducing the penetration of CD4+ and CD8+ T cells into tumor tissues. Extracellular matrix is of great importance in tissue morphogenesis, differentiation, and homeostasis (Frantz et al. 2010). Cancer cells have the ability to penetrate the BM by penetrating the surrounding tissues. The membrane of the ECM provides support to the connective tissue surrounding the epithelial cells and also acts as an important barrier to epithelial cell migration. Cancer cells apply a mechanical force to break through the collagen barrier, which is the most important component of the ECM and determines the primary functional properties of the matrix. Proliferation of epithelial cancer cells increases mechanical stress across the BM and causes membrane rupture, allowing cells to escape from the microenvironment. In the case of such disruptions or accumulations in collagen, changes occur in the homeostasis of the ECM. Increased accumulation of matrix proteins supports tumor progression by interfering with cell adhesion and cell polarity, as well as enhancing growth factor signaling (Walker et al. 2018).

2.4 Cellular Components of Tumor Microenvironment

2.4.1 Fibroblasts

Fibroblasts are transformed into CCA-associated fibroblasts in the CCA tumorigenic region and make up a large proportion of CCA stromal cells. CCA-associated fibroblasts produce alpha-straight actin (α-SMA), fibroblast-specific protein-1, fibroblast activation protein, and PDGFR-β (Sha et al. 2018). PDGF-D interacts with its cognate receptor PDGFR-β, interacting with MAPK and PI3K/Akt to regulate cell survival, cell growth, cell differentiation, cell invasion, and angiogenesis. In these pathways, migration of fibroblasts is controlled by JNK and Rho GTPases (Wang et al. 2010; Yoshida et al. 2005). Alpha-SMA-positive fibroblasts have been shown to induce proliferation of both non-tumorigenic biliary epithelial cells and CCA cells via both secreted substances and cell-cell contact (Thuwajit 2009). Periostin (PN), overexpressed in CCA-associated fibroblasts, induces tumorigenic properties of epithelial cells through activation of integrin receptors. Periostin has been shown to play a role in cell proliferation, migration, invasion/metastasis, and angiogenesis in many cancers. High PN has been shown to be a poor prognostic factor (Utispan et al. 2012).

2.4.2 Tumor-Associated Macrophages

Tumor-associated macrophages (TAMs), a group of plastic immune cells that infiltrate and activate tumor sites, have potent pro-tumor and immunosuppressive properties. Macrophages are divided into two groups according to their state of activation: classically activated (M1 type) and alternatively activated (M2 type) macrophages. M1 macrophages are activated by lipopolysaccharide (LPS), TNF-α and IFN-γ and show pro-inflammatory and anti-tumor properties and surface markers CD80 and CD86. M2 macrophages are mainly stimulated by IL-4 or IL-13 and facilitate tumor-promoting effects, and the surface markers are CD206 and CD163. Tumor-associated macrophages with monocyte chemotactic protein-1 (MCP-1/CCL2), colony-stimulating factor-1 (CSF-1), vascular endothelial growth factor (VEGF)-A, and cytokines such as IL-1β, IL-10, IL-13, and IL-4 are thought to show the main function of macrophages over the M2 type (Zhou et al. 2021).

2.4.3 Dendritic Cells

Dendritic cells are involved in inducing adaptive immunity through antigen presentation. After a series of interactions between cancer-associated fibroblasts and dendritic cells, the expression of antigen-presenting HLA molecules is decreased, resulting in decreased activation of tumor-infiltrating lymphocytes. TILs include CD8+ cytotoxic T lymphocytes and CD4+ T helper lymphocytes, which function to identify and target cells (Bandola-Simon and Roche 2019). Increased CD4+ and CD8+ T lymphocytes are associated with better survival outcomes. The subgroup called Treg suppresses the immune system to maintain self-tolerance. Increased Treg expression and decreased CD4+ and CD8+ T lymphoids are associated with worse survival outcomes (Jung et al. 2018; Høgdall et al. 2018; Goeppert et al. 2013; Oshikiri et al. 2003).

2.4.4 T Lymphocyte

Within the TME, there are many different T-cell populations infiltrating the tumor site. Because cytotoxic CD8+ memory T cells can kill tumor cells, they are associated with good prognosis and are the primary target of tumor cells to escape the immune system. It is also known that CD8+ T cells are supported by CD4+ T cells characterized by IL-2 and IFN-γ. T regulatory cells (Tregs), characterized by surface expression of the transcription factor FOXP3 and CD25, are generally thought to induce tumor growth. Induced Treg cells exhibit an immunosuppressive function with the production of IL-10, preventing the recognition and clearance of tumor cells by the immune system through TGF-β and cytotoxic T lymphocyte-associated antigen-4 (CTLA-4). In addition, the production of TGF-β, IL-35, and galectin-1 suppresses the cytokine production of T effector cells by inhibiting differentiation, proliferation, and activation of T effector cells. Granzyme A/granzyme B secreted from Treg cells causes apoptosis in dendritic cells and T effector cells both in a perforin-dependent and perforin-independent manner. Treg cells, which are abundant in TME, have been associated with poor prognosis in many cancer types due to their immunosuppressive properties. Many immunosuppressive components such as programmed death ligand 1 (PD-L1) have been detected in TME. Various types of stromal cells, CAFs, TAMs, and myeloid-derived suppressor cells (MDSCs) within the tumor, to suppress the function of cytotoxic T lymphocytes (CTLs) and reduce the viability of tumor cells to cooperate with VEGF, cytokines (TGF-β), and metabolic enzymes released by these cells, prevent the spread of tumor-infiltrating CTLs and damage their cytotoxicity (Wang et al. 2017; Long et al. 2016).

2.4.5 B Lymphocyte

It is known that B lymphocytes positively modulate inflammation and immune responses through antibody production, and promote T-cell activation and proliferation by antigen presentation. However, B10 or B regulatory cells (Bregs), which are subsets of B lymphocytes, show immunosuppressive properties by suppressing autoimmune responses against the tumor by secreting anti-inflammatory cytokines such as IL-10 and expressing inhibitory molecules such as PD-L1 (Sarvaria et al. 2017). In neoplasms, Breg cells are produced in response to signals from the TME. It promotes tumor growth of Breg cells, which interact with cells such as Treg cells, MDSCs, TAMs, CD4+ and CD8+ T lymphocytes, NK cells, and directly with tumor cells (Schwartz et al. 2016).

2.5 Therapeutically Targeting the Tumor Microenvironment

Both programmed death-1 (PD-1) and PD-L1 are upregulated in neoplastic cells, and overexpression is associated with increased invasiveness and poor outcome. Conversely, low PD-L1 expression (in combination with high MHC-I expression) is associated with favorable prognosis. The PD-1/PD-L1 and CTLA-4 are negative regulators of the anti-tumor immune response. PD-1 is a member of the B7/CD28 family found on the surface of T cells. It can reduce the proliferation and survival of T cells in tumors when bound mainly to its own ligand PD-L1 released by tumor cells (Buchbinder and Desai 2016; Fontugne et al. 2017). It has been reported that PD-1/PD-L1 expression is detected at a rate of approximately 35–94% in CCA, and PD-1 or PD-L1 expression in CCA is significantly associated with advanced tumor stage and poor prognosis. CTLA-4, the CD28 homolog, is another immune checkpoint that is usually located in the intracellular compartment of native T cells and can be upregulated at the cell surface by the TME. Binding of CTLA-4 and B7 silences the activity of T cells, resulting in decreased proliferation and survival. Unlike the PD-1 pathway, which functions mainly during the effector phase, CTLA-4 functions during the priming phase of T-cell activation, particularly in peripheral tissues. Therefore, the combination of a PD-1 inhibitor and CTLA-4 inhibitor is one of the most studied combinations, which is expected to give a stronger effect in anti-tumor treatments (Yao and Gong 2021; Egen et al. 2002).

3 Immunotherapy in Advanced Biliary Tract Cancers

In this section, we aimed to evaluate in detail the studies evaluating immune checkpoint inhibitor (ICI) treatments as monotherapy or in combination in patients with advanced BTC.

3.1 Single-Agent Immune Checkpoint Inhibitor Treatment

The first study was a phase 2, NCT02829918 study in 2019, in which the efficacy of nivolumab in the second-line and subsequent therapy in 54 patients with treatment-refractory BTC was investigated. Patients who received at least one line, at most three lines of systemic chemotherapy, received nivolumab 240 mg every 2 weeks for 16 weeks followed by 480 mg every 4 weeks until disease progression or intolerable side effects developed. A total of 54 patients, 27 female and 27 male, were included in the study. The primary sites of tumors were intrahepatic CCA (63%), extrahepatic (11%), and gallbladder (26%). Thirty of the patients included in the study were those who progressed after first-line therapy. Nine patients failed the study. Out of 45 patients whose data were evaluated, partial response (PR) was obtained in 10 (22%) and stable disease (SD) in 17 (37.8%). The disease control rate (DCR) was 60%. All patients who responded were microsatellite stable (MSS). The median follow-up time was 13.34 months, the median PFS was 3.98 months (95% CI:

2.33–5.98), and the median OS was 14.22 months (95% CI: 6.64–NA). PD-L1 expression levels were significantly associated with PFS, but no association with OS was detected. The most common treatment-related adverse event (TRAE) was elevated alkaline phosphatase (ALP) (24.5%). The most common grade 3/4 TRAEs were hyponatremia (6%) and ALP elevation (4%). There were no side effects that required discontinuation of treatment (Kim et al. 2019).

The phase 1 JapicCTI-153098 study evaluated the efficacy of the second-line monotherapy nivolumab after first-line gemcitabine in patients with unresectable or advanced BTC. 30 patients were included in the study. The mOS was 5.2 months, and the mPFS was 1.4 months. Only one had an objective response. The most common TRAEs were decreased appetite (5 patients, 17%), malaise (4 patients, 13%), and pruritus (4 patients, 13%). Grade 3–4 TRAEs were reported in three patients (10%). These were rash, maculopapular rash, and increased amylase. Pleurisy was reported as a serious side effect in one patient (3%) (Ueno et al. 2019).

KEYNOTE-028 and KEYNOTE-158 are two studies carried out together. In the KEYNOTE-028 study, the efficacy of pembrolizumab was investigated in the second-line and subsequent treatment in 24 advanced/metastatic BTC patients. The dose of pembrolizumab was given as 10 mg/kg every 2 weeks. All patients were PD-L1 positive. The median follow-up time was 5.7 months (0.6–55.4), the mOS was 5.7 months (3.1–9.8), and the mPFS was 1.8 months (1.4–3.1). Partial response was observed in three patients and the ORR was 13%. In the KEYNOTE-028 study, 16.7% grade 3 TRAEs were observed; grade 4–5 TRAEs were not observed in any patient.

The KEYNOTE-158 phase 2 study evaluated the efficacy of pembrolizumab in second- and subsequent-line therapy in 104 patients with BTC who progressed after standard first-line therapy. Pembrolizumab was used as a flat dose of 200 mg every 3 weeks. The median follow-up time was 7.5 months (0.6–34.3). The ORR was 5.8% 6/104 (95% CI, 2.1–12.1%); the mOS was 7.4 months (5.5–9.6); the mPFS was 2.0 months (1.9–2.1). There were 61 PD-L1-positive patients, of whom 4 had an objective response and the ORR was 6.6%. There were 34 PD-L1-negative patients with 1 patient having an objective response, with an ORR of 2.9%. In the KEYNOTE-158 study, 13.5% of patients developed grade 3–5 TRAEs, while only 1 patient developed grade 5 renal failure. As a result, it has been shown that pembrolizumab provides a permanent tumor response of 6–13% in advanced BTC; this response is independent of PD-L1 expression and has a manageable side-effect profile (Piha-Paul et al. 2020).

The NCT 01938612 study evaluated the efficacy of durvalumab in the second- and subsequent-line treatment of previously treated patients with advanced BTC. 42 patients were included in the study (59% PD-L1 $\geq$ 1%, 14% PD-L1 $\geq$ 25%). Patients had previously received a median of two lines of treatment. Durvalumab dose was 10 mg/kg q2w. Any grade TRAEs were observed in 64%. Grade $\geq$ 3 TRAEs were observed in 19% of patients, 2 patients discontinued treatment due to toxicity, and no treatment-related death was observed. Partial response was observed in two (%4.8) patients. The 12-week DCR was 16.7%, the median duration of response (DOR) was 9.7 months, the mOS was 8.1 months (95% CI, 5.6–10.1), and the mPFS was 2.0 months (Ioka et al. 2019).

The phase 1 study NCT02699515 evaluated the efficacy and safety of bintrafusp alfa in 30 Asian patients with pretreated BTC. Bintrafusp alfa is a first-in-class bifunctional fusion protein consisting of the extracellular domain of the transforming growth factor-β (TGF-β) RII receptor (a TGF-β "trap") fused to a human IgG1 mAb that blocks PD-L1. Bintrafusp alfa was given at a dose of 1200 mg every other week until progression or unacceptable toxicity in patients who progressed after first-line platinum-based chemotherapy. The primary endpoint was safety and tolerability; the secondary endpoint was best overall response. The most common TRAEs were pyrexia, maculopapular rash, and elevated lipase. Grade ≥3 TRAE developed in ten patients (33.3%). Treatment-related death occurred in three patients, with one due to septic shock and two due to ILD. CR developed in one patient and PR in six patients. The ORR was 23.3% (7/30). The ORR confirmed by PD-L1 expression was 25% and 15.4% in patients with PD-L1 + (≥1%) and PD-L1- tumors, respectively (Yoo et al. 2018).

In the long-term results published later, it was reported that the median follow-up time was 122 weeks, eight patients were alive, and three patients remained in treatment. The safety profile was consistent with the initial results, there were no additional deaths, and 2 new grade ≥3 TRAEs (grade 3 rash and grade 3 keratoacanthoma) were reported. The mOS was 12.7 (95% CI, [6.7–15.8]) months and the mDOR was 18 (range, 2.8–24+) months (Yoo et al. 2020a).

In their most recently published results, patients received a median of 8.9 weeks of treatment, and three patients received treatment for more than 59.7 months. ORR calculated by the independent review committee (IRC) was 20%, the mPFS was 2.5 months, and the mOS was 12.7 months. Clinical activity was independent of PD-L1 expression and MSI-H status (Yoo et al. 2020b).

In the NCT03833661 phase 2 study, the efficacy of bintrafusp alfa as second-line therapy was evaluated in 159 patients, and the ORR was found to be 10% (Schrimpf 2021).

3.2 Combination Immune Checkpoint Inhibitor Treatment

In the CA209-538 phase 2 study, the efficacy of nivolumab-ipilimumab combination in second-line therapy was evaluated in 39 patients with advanced BTC. Nivolumab was given at a dose of 3 mg/kg every 2 weeks up to 96 months until progression or unacceptable toxicity and ipilimumab at a dose of 1 mg/kg every 3 weeks followed by four cycles. Objective response was seen in nine patients (PR, nine patients; CR, zero patient), ORR 23%. DCR is 44% (17 patients). All responding patients received prior chemotherapy and none of them had microsatellite instability. Responses were observed only in intrahepatic cholangiocarcinomas and gallbladder carcinomas. The mPFS was 2.9 (95% CI, 2.2–4.6 months) months, and the mOS is 5.7 (95% CI, 2.7–11.9 months) months. Any grade iTRAE was observed in 49% of patients (19 patients). The most common TRAEs were pruritus and rash, which were seen in 13 patients. Grade 3–4 TRAE developed in 15% of patients. Grade 3–4 TRAEs were observed in six patients. Two patients developed hypophysitis, and other four patients developed hypothyroidism, hepatitis, enterocolitis, and gastritis. In this study, nivolumab-ipilimumab combination was found to be associated with positive outcomes in subsequent therapy in patients with advanced BTC (Klein et al. 2020).

Phase 1 NCT01938612 study is a two-arm study. In this arm, the efficacy of the combination of durvalumab and tremelimumab, an anti-CTLA-4 mAb, was investigated in the second- and subsequent-line treatment in Asian patients with advanced BTC who progressed after previous treatments. The treatment protocol was durvalumab 20 mg/kg + tremelimumab 1 mg/kg every 4 weeks. Sixty-five patients were included in the study. All patients received a median of two series of chemotherapy. PR was achieved in seven patients, the ORR was 10.8%, the 12-week DCR was 32.2%, and the mDOR was 8.5 months. The mOS was 10.1 (95% CI, 6.2–11.4) months. Any grade TRAE was observed in 82% of patients. Grade ≥ 3 adverse events were observed in 23% of patients, 5 patients had to discontinue treatment due to side effects, and 1 patient deceased due to drug-induced liver injury. It has been reported that the treatment combination is tolerable and has promising results for future studies (Ioka et al. 2019).

The efficacy and tolerability of adding immunotherapy monotherapy/ combination to chemotherapy in first-line therapy was investigated in the phase 2 NCT03046862 study. 121 chemo-naive patients were included in this 3-armed study. In the biomarker cohort (BMC), the first arm of 30 patients received 1 cycle of gemcitabine + cisplatin followed by gemcitabine-cisplatin + durvalumab and tremelimumab. Gemcitabine-cisplatin+ durvalumab was used in 3 combo cohorts (3C), the second arms of 45 patients. Gemcitabine-cisplatin + durvalumab + tremelimumab was used in 4 combo cohorts (4C), the third arm of 46 patients. Tissue biopsies were obtained in all cohorts before treatment, after one cycle of treatment, and at progression. Furthermore, blood samples were taken for ctDNA at each cycle. The median follow-up time, ORR, DCR, mDOR, mPFS, and mOS were 28.5 months, 50%, 96.7%, 11 months, 13 months, and 15 months, respectively, in the BMC cohort; 11.3 months, 73.4%, 100%, 9.8 months, 11 months, and 18.1 months, respectively, in the 3C cohort; and 11.9 months, 73.3%, 97.8%, 9.1 months, 11.9 months, and 20.7 months, respectively, in the 4C cohort. The most common TRAEs of any grade were neutropenia (54.5%), nausea (59.5%), and pruritus (55.44%). The most common grade 3/4 AEs were neutropenia (50.4%), anemia (35.5%), and thrombocytopenia (16.5%). Tumor mutation burden (TMB) did not correlate with PFS or OS. ctDNA reductions were more pronounced among responders during early durvalumab+tremelimumab cycles. Pre-treatment PD-L1 expression was not associated with efficacy, but PD-L1 expression after the first gemcitabine-cisplatin cycle trended with improved PFS. This study showed that adding immunotherapy to chemotherapy was associated with better outcomes (Oh et al. 2020).

The TOPAZ-1 (NCT03875235) study, the first results of which were presented at the ASCO 2022 Gastrointestinal Cancer symposium, is the first phase 3 study evaluating the effectiveness of immunotherapy in the first-line treatment in advanced

BTC. A total of 685 patients with unresectable, locally advanced, or metastatic BTC diagnosis who had not received any treatment were included in the study. 341 patients were randomized 1:1 to the durvalumab + gemcitabine-cisplatin arm and 344 patients to the placebo + GemCis arm. Durvalumab (1500 mg every 3 weeks [Q3W]) or placebo + gemcitabine-cisplatin (Gem 1000 mg/m^2 and Cis 25 mg/m^2 on Days 1 and 8 Q3W) for up to 8 cycles, followed by durvalumab (1500 mg Q4W) or placebo until disease progression or unacceptable toxicity. The primary endpoint was OS; the secondary endpoints were PFS, ORR, and safety. It was found that durvalumab added to gemcitabine-cisplatin improved OS significantly more than the placebo arm (median 12.8 versus 11.5 months, HR 0.8, 95% CI 0.66–0.97, $p = 0.021$). PFS was also significantly improved with durvalumab vs placebo (HR, 0.75; 95% CI, 0.64–0.89; $p = 0.001$). The ORR was 26.7% in the durvalumab arm and 18.7% in the placebo arm. Grade 3/4 TRAEs were 62.7% in the durvalumab arm and 64.9% in the placebo arm. Only 13% of patients developed immune-mediated side effects, 2.4% of which were grade 3 or 4. The drug was discontinued due to TRAEs in 8.9% of patients in the durvalumab arm, compared to 11.4% in the placebo arm. In this study, it has been shown that adding durvalumab to gemcitabine-cisplatin in the first-line treatment in patients with advanced BTC provides a significant contribution to OS and PFS and that, while making this contribution, it does not increase the frequency of side effects, and this combination can therefore be a standard first-line treatment (Oh et al. 2022).

In the phase 2 study of JapicCTI-153098, nivolumab (240 mg every 2 weeks) and cisplatin (25 mg/m^2) plus gemcitabine (1000 mg/m^2) chemotherapy were given to 30 chemotherapy-naive patients with unresectable or recurrent BTC. The primary objective was safety and tolerability. The most frequently reported TRAEs were neutropenia and thrombocytopenia. Serious side effects were reported in six patients. The most common one was thrombocytopenia in three patients. The mOS was 15.4 (90% CI 11·8-not estimable) months, and the mPFS 4.2 (90% CI 2.8–5.6) months. The ORR was 36.6% (Ueno et al. 2019).

Another phase 2 study investigated the efficacy of adding nivolumab to gemcitabine-cisplatin in primary care in 32 patients with advanced BTC. The median follow-up time was 12.8 (95% CI 10.8–14.8) months. CR was achieved in five patients and PR in ten patients. The ORR was 55.6%, and the DCR was 92.6%. The mPFS was 6.1 months and the mOS 8.5 months. The most common grade ≥3 TRAEs were thrombocytopenia (56%) and neutropenia (%22). Nivolumab was found to be safe and tolerable in these both studies (Feng et al. 2020).

In a phase 2 study coded NCT03486678, the efficacy of adding camrelizumab to gemcitabine-oxaliplatin in the first-line treatment of patients with advanced BTC was investigated. The results of 37 patients were analyzed. The median follow-up time was 11.8 months. The ORR was 54%, the mPFS 6.1 months, and the mOS 11.8 months. The most frequently observed TRAEs were fatigue (73%) and fever (73%). The most frequently observed grade ≥3 TRAEs were hypokalemia (7 patients, 19%) and fatigue (6 patients, 16%). The ORR was 80% in the group with PD-L1 tumor proportion score (TPS) $>1\%$ and 53.8% in the group with PD-L1 TPS $<1\%$ (Chen et al. 2020).

In the phase 2 NCT03092895 trial, untreated advanced BTC patients received camrelizumab plus FOLFOX4/GEMOX chemotherapy. 92 patients were included in the study. 29 patients were in the camrelizumab-FOLFOX arm and 63 patients in the camrelizumab-GEMOX arm. The ORR was 16.3% (%19 in the GEMOX arm, 10.3% in the FOLFOX arm). The DCR was 75% (95% CI = 64.9–83.4), and the mDOR was 8.7 months (95% CI = 5.1 – not reached). The mPFS was 5.3 months (5.5 months in the GEMOX arm, 3.8 months in the FOLFOX arm), and the mOS was 12.4 months (13.6 months in the GEMOX arm, 12.9 months in the FOLFOX arm). Grade ≥3 TRAEs were 82.8% in the Cam-FOLFOX arm and 68.3% in the cam-GEMOX arm. Six patients discontinued treatment due to TRAEs. Camrelizumab plus FOLFOX4 or GEMOX as first-line therapy was effective and tolerable (Chen et al. 2021).

In the phase 2 study coded NCT03796429, the efficacy of adding the anti-PD-1 agent toripalimab to the gemcitabine-S1 combination in the first-line treatment in patients with advanced BTC was investigated. 39 patients were included in the study. The ORR was 20.6% and the DCR 85.3%. The mPFS was 6.7 months; the mOS data was immature. The most common TRAEs were leukopenia, anemia, and rash. 69.2% of the patients developed grade 3/4 TRAEs, 51.3% of which had leukopenia and 17.9% thrombocytopenia. Eight patients developed grade 3/4 non-hematological side effects (three patients, rash; five patients, infection), six patients developed severe TRAEs (five patients, infection; one patient, mucositis), and one patient deceased due to biliary obstruction. The combination of toripalimab chemotherapy has been found to be safe and effective (Li et al. 2021).

In this study, the efficacy of lenvatinib plus pembrolizumab/nivolumab in the second- and subsequent-line treatment was evaluated in patients with advanced BTC. 56 patients were included in the study (29 patients lenvatinib plus nivolumab, 27 patients lenvatinib plus pembrolizumab). The ORR was 30.4%, the DCR 85.7%, the mPFS 5 months, and the mOS 11 months. Any grade TRAEs developed in 96.4% of patients, but only 19.6% of patients developed grade 3 adverse events. Grade 4/5 side effects did not develop in any of the patients. The most frequently observed TRAEs were fatigue, hypertension, and hypothyroidism. PD-L1-positive tumors were associated with higher ORR and longer PFS times. No correlation was found between TMB, survival, and responses. Lenvatinib plus pembrolizumab/nivolumab has been found to be safe and effective (Lin et al. 2019).

In the study with code NCT02443324, the efficacy of pembrolizumab-ramucirumab treatment in the second- and subsequent-line treatment in patients with advanced BTC was evaluated. Twenty-six patients were included in the study. The ORR was 4%, the mPFS 1.6 months, and the mOS 6.4 months. The ramucirumab-pembrolizumab combination made a limited contribution to this study (Arkenau et al. 2018).

In the NCT03482102 phase 1 study, the efficacy of durvalumab and tremelimumab, along with radiotherapy to a single metastatic site (three fractions of 8 Gy at cycle 2 every other day) in the second- and subsequent-line treatment, was evaluated. 15 patients were included in the study. These patients could not reach radiotherapy. For these, the ORR was 20% (13% PR, 7% CR), and the DCR 27%, whereas the ORR was 25% (PR 17%, CR 8%) and the DCR 33% in those who reached radiotherapy. At time of analysis, the median PFS was 54 days. The duration of response for four patients with DCR was 26, 52, 122, and 254+ days, respectively. Any grade TRAEs were reported in 80% of patients. Grade ≥3 TRAE were reported in nine patients, which included lymphopenia and elevated liver function tests. All patients with disease control had a microsatellite-stable (MSS) tumor (Hong et al. 2020).

In single-agent immunotherapy studies in patients with advanced BTC, the ORR approach ranges between 3% and 22%; in combination treatments, the ORR is approximately >70%. Several clinical trials investigating combinations or monotherapy of immune checkpoint inhibitors are summarized in Table 1.

4 Biomarker

Patients with PD-L1 overexpression, dMMR/MSI-H or TMB-high are more likely to benefit from treatment with ICIs. It is known that ICIs are effective regardless of these biomarkers, especially in intrahepatic CCAs.

4.1 dMMR/MSI-H

It is known that dMMR/MSI-H-positive tumors harbor many more mutations than those of the same type without such mismatch repair defects, and neoepitopes are formed in MSI tumors that promote efficient immune responses. Therefore, patients with more immunogenic MSI tumors are more likely to benefit from ICI therapy (Le et al. 2015). Pembrolizumab was shown to be effective for the first time in colorectal cancer with dMMR. Subsequent studies have shown that ICIs also provide benefits in other solid cancer types with dMMR. In May 2017, the FDA approved pembrolizumab as a tumor-agnostic therapy for the treatment of advanced solid tumors with MSI-H/dMMR, including CCAs, which have progressed following previous therapy and have shown no significant efficacy (Marcus et al. 2019).

The efficacy of pembrolizumab in CCAs with dMMR was obtained in the phase 2 KEYNOTE-158 study evaluating patients with non-colorectal MSI/dMMR tumors. There are 22 BTC patients in this study. Complete response was achieved in two patients and PR in seven patients. The ORR was 40.9%, the median PFS 4.2 months (95%CI: 2.1 – not reached), and the median OS 24.3 months (95%CI: 6.5 – not reached) (Marabelle et al. 2020a).

The dMMR/MSI-H ratio of patients with CCA was found at different rates in various studies. In one study, it was reported as 3% in all CCA and as 5% in gallbladder cancer and extrahepatic CCA in another study and as 10% in intrahepatic CCA and ampullary carcinoma (Le et al. 2017; Silva et al. 2016). When only patients with advanced/metastatic BTC were evaluated, the dMMR rate was reported as 1–1.3% (Le et al. 2017; Middha et al. 2017).

Table 1 Completed clinical trial use of immunotherapy outcomes in biliary tract cancers

Study code	Agents	Targets	Phase	Patients	Setting	ORR (%)	mOS (months)	mPFS (months)
Single-agent immune checkpoint inhibitor treatment								
NCT02829918 (Kim et al. 2019)	Nivolumab	PD-1	2	54	Second line and subsequent	22	14.2	3.68
JapicCTI-153098 (Ueno et al. 2019)	Nivolumab	PD-1	1	30	Second line and subsequent	35	5.2	1.4
KEYNOTE-028 (Piha-Paul et al. 2020)	Pembrolizumab	PD-1	1b	24	Second line and subsequent	13	5.7	1.8
KEYNOTE-158 (Piha-Paul et al. 2020)	Pembrolizumab	PD-1	2	104	Second line and subsequent	5.8	7.4	2
NCT01938612 (Ioka et al. 2019)	Durvalumab	PD-L1	1	42	Second line and subsequent	4.8	8.1	2
NCT02699515 (Yoo et al. 2018; Yoo et al. 2020a, b)	Bintrafusp alpha	PD-L1/TGF-B	1	30	Second line and subsequent	20	12.7	2.5
NCT03833661 (Schrimpf 2021)	Bintrafusp alpha	PD-L1/TGF-B	2	159	Second line and subsequent	10		
Combination immune checkpoint inhibitors treatment								
CA209-538 (Klein et al. 2020)	Nivolumab Ipilimumab	PD-1 CTLA-4	2	39	Second line and subsequent	23	5.7	2.9
NCT01938612 (Ioka et al. 2019)	Durvalumab + tremelimumab	PD-L1 CTLA-4	2	65	Second line and subsequent	10.8	10.1	
NCT03046862 (Oh et al. 2020)	Durvalumab + cisplatin/gemcitabine (biomarker cohort)	PD-L1	2	30	First line	50	15	13
NCT03046862 (Oh et al. 2020)	Durvalumab + cisplatin/gemcitabine	PD-L1	2	45	Firs line	73.3	18.1	11
NCT03046862 (Oh et al. 2020)	Durvalumab + tremelimumab + cisplatin/gemcitabine	PD-L1 CTLA-4	2	46	First line	73.4	20.7	11.9

(continued)

Table 1 (continued)

Study code	Agents	Targets	Phase	Patients	Setting	ORR (%)	mOS (months)	mPFS (months)
TOPAZ-1 (NCT03875235) (Oh et al. 2022)	Durvalumab+ gemcitabine/ cisplatin	PD-L1	3	685	First line	26.7	12.8	
JapicCTI-153098 (Ueno et al. 2019)	Nivolumab + cisplatin/ gemcitabine	PD-1	2	30	First line	36.6	15.4	4.2
NCT03311789 (Feng et al. 2020)	Nivolumab + cisplatin/ gemcitabine	PD-1	2	32	First line	55.6	8.5	6.1
NCT03486678 (Chen et al. 2020)	Camrelizumab + gemcitabine/ oxaliplatin	PD-1	2	37	First line	54	11.8	6.1
NCT03092895 (Chen et al. 2021)	Camrelizumab + gemcitabine/ oxaliplatin or FOLFOX	PD-1	2	92	First line	16.3	12.4	5.3
NCT03796429 (Li et al. 2021)	Toripalimab + gemcitabine-S1	PD-1	2	39	First line	27	16	7
NCT03892577 (Lin et al. 2019)	Pembrolizumab/nivolumab +lenvatinib	PD-1 Multi-tyrosine kinase	1	56	Second line and subsequent	30	11	5
NCT02443324 (Arkenau et al. 2018)	Ramucirumab + pembrolizumab	VEGFR2 PD-1	1	26	Second line and subsequent	4	6.4	1.6
NCT03482102 (Hong et al. 2020)	Durvalumab + tremelimumab + radiotherapy	PD-L1 CTLA-4	1	15	Second line and subsequent	25		

Abbreviations: *CTLA-4* cytotoxic T lymphocyte-associated antigen-4, *mOS* median overall survival, *mPFS* median progression-free survival, *ORR* objective response rate, *PD-1* programmed death-1, *PD-L1* programmed death ligand-1, *TGF-β* transforming growth factor-beta, *VEGFR* vascular endothelial growth factor receptor

Not all patients with MSI respond to ICI treatments. More data are needed to evaluate whether immunotherapy, as monotherapy or in combination, might be more effective in MSI patients with BTC.

4.2 PD-L1

PD-L1 is expressed on tumor cells, macrophages, and lymphocytes, and overexpression of PD-L1 may predict response to ICI therapy (Patel and Kurzrock 2015).

In KEYNOTE-028, patients with PD-L1 $\geq 1\%$ were included. A limited response to pembrolizumab was observed in these patients. Partial response was obtained in three patients. The ORR was 13%, the mPFS 1.8 months, and the mOS 5.7 months. In the KEYNOTE-158 study, there were 61 PD-L1-positive patients, 4 of which had an objective response, and the ORR was 6.6%. There were 34 PD-L1-negative patients and 1 patient with an objective response with an ORR of 2.9% (Piha-Paul et al. 2020).

In another phase II study, tumor PD-L1 expression $>1\%$ was detected in 12 (37%) of 32 patients included. In other patients, PD-L1 was negative or could not be evaluated. As a result, there was no significant difference in PFS among those with PD-L1 expression (Feng et al. 2020).

In the study of Chen et al. investigating the combination of GEMOX plus camrelizumab, better mPFS and mOS were observed in patients with tumor proportion score (TPS) ≥ 1 (Chen et al. 2020).

A variety of factors may explain the discordant results between the concluded studies, including a small number of patients, heterogeneity, lack of standardization in PD-L1 testing, scoring, and analysis.

4.3 Tumor Mutational Burden

In recent years, TMB, defined as the total number of somatic mutations per coding domain of a tumor genome, has been identified as a biomarker that can be used to predict response to immunotherapy (Chan et al. 2019). Biliary tract cancers are generally characterized by low TMB and only 2.9–4.0% of cases have high TMB levels (Mody et al. 2019; Javle et al. 2016; Weinberg et al. 2019; Osipov et al. 2020).

There is limited information on the response of TMB-high CCA to their ICI. One case report reported durable disease control with therapies targeting the PD-1 pathway in a single patient with TMB-high advanced BTC (Gbolahan et al. 2019).

In the KEYNOTE-158 study evaluating the efficacy of pembrolizumab in various malignancies, including CCA, better ORR was reported in patients with higher TMB (Marabelle et al. 2020b). However, none of the patients with CCA included in this study were TMB-high.

In the study evaluating the efficacy of cisplatin/gemcitabine plus nivolumab, no association was observed between TMB and PFS (Feng et al. 2020).

However, based largely on this study, pembrolizumab is now approved for patients with any solid tumor, including CCA, with TMB ≥10 mut/Mb (as defined by FDA-approved concomitant diagnostic testing) after progression on standard regimens.

4.4 Other Types of Immunotherapy

Vaccines and chimeric antigen receptor (CAR)-engineered T-cell therapies were studied in patients with advanced BTC in phase 1/2 studies. Results from a limited number of studies are modest. More studies are needed to evaluate the efficacy of both vaccines and cellular therapies (Aruga et al. 2013; Feng et al. 2017).

Numerous phase 2 and 3 trials investigating the efficacy of ICIs as monotherapy or in combination in the first- and subsequent-line treatments are ongoing.

5 Conclusions

Limited responses are obtained with current therapies in the treatment of BTCs. In order to increase the treatment response and survival time in these patients, the search for new treatment agents is rapidly carried on. After the demonstration of the efficacy of ICIs in RCC, melanoma, and lung cancer, the efficacy of ICIs in all other cancer types is being investigated. Above we summarized a large number of monotherapy ICI, ICI-ICI combination, and ICI-chemotherapy combination studies in patients with BTC. Lower response rates and survival results were obtained in monotherapy immunotherapy studies. However, more satisfactory results were obtained in combination treatments. The results of these studies have not yet been reflected in clinical practice and there is no standardized treatment. With data from completed and ongoing studies, immunotherapy agents are thought to be effective in patients with BTC. At this point, the problem of estimating which patient will benefit from which treatment arises. There are currently no biomarkers that can select patients with BTC who would benefit significantly from treatment with their ICI. Considering that the rate of dMMR is around 1–1.4% in patients with metastatic BTC, we think that it would be more beneficial to determine a new biomarker.

Compliance with Ethical Standards

Competing interests The authors have no conflicts of interest to declare.

Funding There are no financial interests.

References

Abou-Alfa GK, Sahai V, Hollebecque A, Vaccaro G, Melisi D, Al-Rajabi R, Paulson AS, Borad MJ, Gallinson D, Murphy AG, Oh D-Y, Dotan E, Catenacci DV, Cutsem EV, Ji T, Lihou CF, Zhen H, Féliz L, Vogel A (2020) Pemigatinib for previously treated, locally advanced or metastatic cholangiocarcinoma: a multicentre, open-label, phase 2 study. Lancet Oncol 21(5): 671–684. https://doi.org/10.1016/S1470-2045(20)30109-1

Arkenau H-T, Martin-Liberal J, Calvo E, Penel N, Krebs MG, Herbst RS, Walgren RA, Widau RC, Mi G, Jin J, Ferry D, Chau I (2018) Ramucirumab plus pembrolizumab in patients with previously treated advanced or metastatic biliary tract cancer: nonrandomized, open-label, phase I trial (JVDF). The Oncologist [online] 23(12):1407–e136. https://doi.org/10.1634/theoncologist.2018-0044

Aruga A, Takeshita N, Kotera Y, Okuyama R, Matsushita N, Ohta T, Takeda K, Yamamoto M (2013) Long-term vaccination with multiple peptides derived from cancer-testis antigens can maintain a specific T-cell response and achieve disease stability in advanced biliary tract cancer. Clin Cancer Res 19(8):2224–2231. https://doi.org/10.1158/1078-0432.CCR-12-3592

Azevedo Martins JM, Rabelo-Santos SH, do Amaral Westin MC, Zeferino LC (2020) Tumoral and stromal expression of MMP-2, MMP-9, MMP-14, TIMP-1, TIMP-2, and VEGF-A in cervical cancer patient survival: a competing risk analysis. BMC Cancer 20(1). https://doi.org/10.1186/s12885-020-07150-3

Balkwill FR, Capasso M, Hagemann T (2012) The tumor microenvironment at a glance. J Cell Sci 125(23):5591–5596. https://doi.org/10.1242/jcs.116392

Banales JM, Cardinale V, Carpino G, Marzioni M, Andersen JB, Invernizzi P, Lind GE, Folseraas T, Forbes SJ, Fouassier L, Geier A, Calvisi DF, Mertens JC, Trauner M, Benedetti A, Maroni L, Vaquero J, Macias RIR, Raggi C, Perugorria MJ (2016) Cholangiocarcinoma: current knowledge and future perspectives consensus statement from the European Network for the Study of Cholangiocarcinoma (ENS-CCA). Nat Rev Gastroenterol Hepatol 13(5):261–280. https://doi.org/10.1038/nrgastro.2016.51

Bandola-Simon J, Roche PA (2019) Dysfunction of antigen processing and presentation by dendritic cells in cancer. Mol Immunol 113:31–37. https://doi.org/10.1016/j.molimm.2018.03.025

Bismuth H, Nakache R, Diamond T (1992) Management strategies in resection for hilar cholangiocarcinoma. Ann Surg 215(1):31–38. https://doi.org/10.1097/00000658-199201000-00005

Borger DR, Tanabe KK, Fan KC, Lopez HU, Fantin VR, Straley KS, Schenkein DP, Hezel AF, Ancukiewicz M, Liebman HM, Kwak EL, Clark JW, Ryan DP, Deshpande V, Dias-Santagata D, Ellisen LW, Zhu AX, Iafrate AJ (2011) Frequent mutation of isocitrate dehydrogenase (IDH)1 and IDH2 in cholangiocarcinoma identified through broad-based tumor genotyping. Oncologist 17(1):72–79. https://doi.org/10.1634/theoncologist.2011-0386

Boulter L, Guest RV, Kendall TJ, Wilson DH, Wojtacha D, Robson AJ, Ridgway RA, Samuel K, Van Rooijen N, Barry ST, Wigmore SJ, Sansom OJ, Forbes SJ (2015) WNT signaling drives cholangiocarcinoma growth and can be pharmacologically inhibited. J Clin Investig 125(3):1269–1285. https://doi.org/10.1172/jci76452.

Buchbinder EI, Desai A (2016) CTLA-4 and PD-1 pathways. Am J Clin Oncol 39(1):98–106. https://doi.org/10.1097/coc.0000000000000239

Chan TA, Yarchoan M, Jaffee E, Swanton C, Quezada SA, Stenzinger A, Peters S (2019) Development of tumor mutation burden as an immunotherapy biomarker: utility for the oncology clinic. Ann Oncol 30(1):44–56. https://doi.org/10.1093/annonc/mdy495

Chen W, Liang J, Huang L, Cai J, Lei Y, Lai J, Liang L, Zhang K (2016) Characterizing the activation of the Wnt signaling pathway in hilar cholangiocarcinoma using a tissue microarray approach. Eur J Histochem 60(1). https://doi.org/10.4081/ejh.2016.2536

Chen X, Wu X, Wu H, Gu Y, Shao Y, Shao Q, Zhu F, Li X, Qian X, Hu J, Zhao F, Mao W, Sun J, Wang J, Han G, Li C, Xia Y, Seesaha PK, Zhu D, Li H (2020) Camrelizumab plus gemcitabine and oxaliplatin (GEMOX) in patients with advanced biliary tract cancer: a single-arm, open-label, phase II trial. J Immunother Cancer 2:e001240. https://doi.org/10.1136/jitc-2020-001240

Chen X, Qin S, Gu S, Ren Z, Chen Z, Xiong J, Liu Y, Meng Z, Zhang X, Wang L, Zhang X, Zou J (2021) Camrelizumab plus oxaliplatin-based chemotherapy as first-line therapy for advanced biliary tract cancer: A multicenter, phase 2 trial. Int J Cancer 149(11):1944–1954. https://doi.org/10.1002/ijc.33751

Dai J, Wang H, Dong Y, Zhang Y, Wang J (2013) Bile acids affect the growth of human cholangiocarcinoma via NF-kB pathway. Cancer Investig 31(2):111–120. https://doi.org/10.3109/07357907.2012.762781

Egen JG, Kuhns MS, Allison JP (2002) CTLA-4: new insights into its biological function and use in tumor immunotherapy. Nat Immunol 3(7):611–618. https://doi.org/10.1038/ni0702-611

El Khatib M, Bozko P, Palagani V, Malek NP, Wilkens L, Plentz RR (2013) Activation of notch signaling is required for cholangiocarcinoma progression and is enhanced by inactivation of p53 in vivo. PLoS One 8(10):e77433. https://doi.org/10.1371/journal.pone.0077433

El-Daly SM, Bayraktar R, Anfossi S, Calin GA (2020) The interplay between microRNAs and the components of the tumor microenvironment in B-cell malignancies. Int J Mol Sci 21(9):3387. https://doi.org/10.3390/ijms21093387

Fabris L, Perugorria MJ, Mertens J, Björkström NK, Cramer T, Lleo A, Solinas A, Sänger H, Lukacs-Kornek V, Moncsek A, Siebenhüner A, Strazzabosco M (2019) The tumour microenvironment and immune milieu of cholangiocarcinoma. Liver Int 39(S1):63–78. https://doi.org/10.1111/liv.14098

Fabris L, Sato K, Alpini G, Strazzabosco M (2020) The tumor microenvironment in cholangiocarcinoma progression. Hepatology 73(S1):75–85. https://doi.org/10.1002/hep.31410

FDA approves first targeted treatment for patients with cholangiocarcinoma, a cancer of bile ducts. https://www.fda.gov/news-events/press-announcements/fda-approves-first-targeted-treatment-patients-cholangiocarcinoma-cancer-bile-ducts. Accessed 21 Apr 2020

Feng K, Liu Y, Guo Y, Qiu J, Wu Z, Dai H, Yang Q, Wang Y, Han W (2017) Phase I study of chimeric antigen receptor modified T cells in treating HER2-positive advanced biliary tract cancers and pancreatic cancers. Protein Cell 9(10):838–847. https://doi.org/10.1007/s13238-017-0440-4

Feng K, Liu Y, Zhao Y, Yang Q, Dong L, Liu J, Li X, Zhao Z, Mei Q, Han W (2020) Efficacy and biomarker analysis of nivolumab plus gemcitabine and cisplatin in patients with unresectable or metastatic biliary tract cancers: results from a phase II study. J Immunother Cancer 8(1):e000367. https://doi.org/10.1136/jitc-2019-000367

Fontugne J, Augustin J, Pujals A, Compagnon P, Rousseau B, Luciani A, Tournigand C, Cherqui D, Azoulay D, Pawlotsky J-M, Calderaro J (2017) PD-L1 expression in perihilar and intrahepatic cholangiocarcinoma. Oncotarget 8(15):24644–24651. https://doi.org/10.18632/oncotarget.15602

Frantz C, Stewart KM, Weaver VM (2010) The extracellular matrix at a glance. J Cell Sci 123(24):4195–4200. https://doi.org/10.1242/jcs.023820

Gbolahan O, Hashemi-Sadraei N, O'Neil B (2019) Prolonged response to anti–PD-1 antibody therapy in chemotherapy-refractory cholangiocarcinoma with high tumor mutational burden. J Natl Compr Cancer Netw 17(6):644–648. https://doi.org/10.6004/jnccn.2019.7304

Globocan database on global cancer incidence available online at https://gco.iarc.fr/today/home. Accessed 19 May 2021

Goeppert B, Frauenschuh L, Zucknick M, Stenzinger A, Andrulis M, Klauschen F, Joehrens K, Warth A, Renner M, Mehrabi A, Hafezi M, Thelen A, Schirmacher P, Weichert W (2013) Prognostic impact of tumour-infiltrating immune cells on biliary tract cancer. Br J Cancer 109(10):2665–2674. https://doi.org/10.1038/bjc.2013.610

Guest RV, Boulter L, Dwyer BJ, Kendall TJ, Man T-Y, Minnis-Lyons SE, Lu W-Y, Robson AJ, Gonzalez SF, Raven A, Wojtacha D, Morton JP, Komuta M, Roskams T, Wigmore SJ, Sansom OJ, Forbes SJ (2016) Notch3 drives development and progression of cholangiocarcinoma. Proc Natl Acad Sci U S A 113(43):12250–12255. https://doi.org/10.1073/pnas.1600067113

Høgdall D, Lewinska M, Andersen JB (2018) Desmoplastic tumor microenvironment and immunotherapy in cholangiocarcinoma. Trends Cancer 4(3):239–255. https://doi.org/10.1016/j.trecan.2018.01.007

Hong TS, Goyal L, Parikh AR, Yeap BY, Ulysse CA, Drapek LC, Allen JN, Clark JW, Christopher B, Bolton C, Ryan DP, Corcoran RB, Meyerhardt JA, Wo JY-L, Zhu AX (2020) A pilot study of durvalumab/tremelimumab (durva/treme) and radiation (XRT) for metastatic

biliary tract cancer (mBTC): preliminary safety and efficacy. J Clin Oncol 38(4_suppl):547–547. https://doi.org/10.1200/jco.2020.38.4_suppl.547
Ioka T, Ueno M, Oh D-Y, Fujiwara Y, Chen J-S, Doki Y, Mizuno N, Park K, Asagi A, Hayama M, Nii M, Komuro K, Sugimoto M, Vlahovic G, Ikeda M (2019) Evaluation of safety and tolerability of durvalumab (D) with or without tremelimumab (T) in patients (pts) with biliary tract cancer (BTC). J Clin Oncol 37(4_suppl):387–387. https://doi.org/10.1200/jco.2019.37.4_suppl.387
Jansen H, Pape U-F, Utku N (2020) A review of systemic therapy in biliary tract carcinoma. J Gastrointest Oncol 11(4):770–789. https://doi.org/10.21037/jgo-20-203
Javle M, Bekaii-Saab T, Jain A, Wang Y, Kelley RK, Wang K, Kang HC, Catenacci D, Ali S, Krishnan S, Ahn D, Bocobo AG, Zuo M, Kaseb A, Miller V, Stephens PJ, Meric-Bernstam F, Shroff R, Ross J (2016) Biliary cancer: utility of next-generation sequencing for clinical management. Cancer 122(24):3838–3847. https://doi.org/10.1002/cncr.30254
Javle MM, Roychowdhury S, Kelley RK, Sadeghi S, Macarulla T, Waldschmidt DT, Goyal L, Borbath I, El-Khoueiry AB, Yong W-P, Philip PA, Bitzer M, Tanasanvimon S, Li A, Pande A, Shepherd SP, Moran S, Abou-Alfa GK (2021) Final results from a phase II study of infigratinib (BGJ398), an FGFR-selective tyrosine kinase inhibitor, in patients with previously treated advanced cholangiocarcinoma harboring an *FGFR2* gene fusion or rearrangement. J Clin Oncol 39(3_suppl):265–265. https://doi.org/10.1200/jco.2021.39.3_suppl.265
Jung IH, Kim DH, Yoo DK, Baek SY, Jeong SH, Jung DE, Park SW, Chung Y-Y (2018) In vivo study of natural killer (NK) cell cytotoxicity against cholangiocarcinoma in a nude mouse model. In Vivo (Athens, Greece) 32(4):771–781. https://doi.org/10.21873/invivo.11307
Kim RD, Kim DW, Alese OB, Li D, Shah N, Schell MJ, Zhou JM, Chung V (2019) A phase II study of nivolumab in patients with advanced refractory biliary tract cancers (BTC). J Clin Oncol 37(15_suppl):4097–4097. https://doi.org/10.1200/jco.2019.37.15_suppl.4097
Klein O, Kee D, Nagrial A, Markman B, Underhill C, Michael M, Jackett L, Lum C, Behren A, Palmer J, Tebbutt NC, Carlino MS, Cebon J (2020) Evaluation of combination nivolumab and ipilimumab immunotherapy in patients with advanced biliary tract cancers: subgroup analysis of a phase 2 nonrandomized clinical trial. JAMA Oncol 6(9):1405–1409. https://doi.org/10.1001/jamaoncol.2020.2814
Lamarca A, Palmer DH, Wasan HS, Ross PJ, Ma YT, Arora A, Falk S, Gillmore R, Wadsley J, Patel K, Anthoney A, Maraveyas A, Iveson T, Waters JS, Hobbs C, Barber S, Ryder WD, Ramage J, Davies LM, Bridgewater JA (2021) Second-line FOLFOX chemotherapy versus active symptom control for advanced biliary tract cancer (ABC-06): a phase 3, open-label, randomised, controlled trial. Lancet Oncol 0(0). https://doi.org/10.1016/S1470-2045(21)00027-9
Lamarca A, Edeline J, Goyal L (2022) How I treat biliary tract cancer. ESMO Open 7(1):100378. https://doi.org/10.1016/j.esmoop.2021.100378
Le DT, Uram JN, Wang H, Bartlett BR, Kemberling H, Eyring AD, Skora AD, Luber BS, Azad NS, Laheru D, Biedrzycki B, Donehower RC, Zaheer A, Fisher GA, Crocenzi TS, Lee JJ, Duffy SM, Goldberg RM, de la Chapelle A, Koshiji M (2015) PD-1 blockade in tumors with mismatch-repair deficiency. N Engl J Med 372(26):2509–2520. https://doi.org/10.1056/nejmoa1500596
Le DT, Durham JN, Smith KN, Wang H, Bartlett BR, Aulakh LK, Lu S, Kemberling H, Wilt C, Luber BS, Wong F, Azad NS, Rucki AA, Laheru D, Donehower R, Zaheer A, Fisher GA, Crocenzi TS, Lee JJ, Greten TF (2017) Mismatch repair deficiency predicts response of solid tumors to PD-1 blockade. Science 357(6349):409–413. https://doi.org/10.1126/science.aan6733
Lee AJ, Chun YS (2018) Intrahepatic cholangiocarcinoma: the AJCC/UICC 8th edition updates. Chin Clin Oncol 7(5):6. https://doi.org/10.21037/cco.v0i0.20321
Li W, Yu Y, Xu X, Guo X, Wang Y, Li Q, Wang Y, Cui Y, Liu H, Hao Q, Qu Y, Huang L, Liu T (2021) Toripalimab with chemotherapy as first-line treatment for advanced biliary tract tumors: update analytic results of an open-label phase II clinical study (JS001-ZS-BC001). J Clin Oncol 39(15_suppl):e16170–e16170. https://doi.org/10.1200/jco.2021.39.15_suppl.e16170

Lin J, Yang X, Zhao S, Long J, Pan J, Hu K, Zhao L, Huo L, Sang X, Wang K, Zhao H (2019) Lenvatinib plus PD-1 blockade in advanced bile tract carcinoma. Ann Oncol 30:v517. https://doi.org/10.1093/annonc/mdz253.097

Long H, Xiang T, Luo J, Li F, Lin R, Liu S, Jiang S, Hu C, Chen G, Wong E, Wan Y, Li Q-J, Zhu B (2016) The tumor microenvironment disarms CD8+T lymphocyte function via a miR-26a-EZH2 axis. OncoImmunology 5(12):e1245267. https://doi.org/10.1080/2162402x.2016.1245267

Marabelle A, Le DT, Ascierto PA, Di Giacomo AM, De Jesus-Acosta A, Delord J-P, Geva R, Gottfried M, Penel N, Hansen AR, Piha-Paul SA, Doi T, Gao B, Chung HC, Lopez-Martin J, Bang Y-J, Frommer RS, Shah M, Ghori R, Joe AK (2020a) Efficacy of pembrolizumab in patients with noncolorectal high microsatellite instability/mismatch repair–deficient cancer: results from the Phase II Keynote-158 Study. J Clin Oncol 38(1):1–10. https://doi.org/10.1200/jco.19.02105

Marabelle A, Fakih M, Lopez J, Shah M, Shapira-Frommer R, Nakagawa K, Chung HC, Kindler HL, Lopez-Martin JA, Miller WH, Italiano A, Kao S, Piha-Paul SA, Delord J-P, McWilliams RR, Fabrizio DA, Aurora-Garg D, Xu L, Jin F, Norwood K (2020b) Association of tumour mutational burden with outcomes in patients with advanced solid tumours treated with pembrolizumab: prospective biomarker analysis of the multicohort, open-label, phase 2 KEYNOTE-158 study. Lancet Oncol 21(10):1353–1365. https://doi.org/10.1016/s1470-2045(20)30445-9

Marcus L, Lemery SJ, Keegan P, Pazdur R (2019) FDA approval summary: pembrolizumab for the treatment of microsatellite instability-high solid tumors. Clin Cancer Res 25(13):3753–3758. https://doi.org/10.1158/1078-0432.ccr-18-4070

Middha S, Zhang L, Nafa K, Jayakumaran G, Wong D, Kim HR, Sadowska J, Berger MF, Delair DF, Shia J, Stadler Z, Klimstra DS, Ladanyi M, Zehir A, Hechtman JF (2017) Reliable pan-cancer microsatellite instability assessment by using targeted next-generation sequencing data. JCO Precis Oncol 1:1–17. https://doi.org/10.1200/po.17.00084

Mody K, Starr J, Saul M, Poorman K, Weinberg BA, Salem ME, VanderWalde A, Shields AF (2019) Patterns and genomic correlates of PD-L1 expression in patients with biliary tract cancers. J Gastrointest Oncol 10(6):1099–1109. https://doi.org/10.21037/jgo.2019.08.08

Oh D-Y, Lee K-H, Lee D-W, Kim TY, Bang J-H, Nam A-R, Lee Y, Zhang Q, Rebelatto M, Li W, Kim JW (2020) Phase II study assessing tolerability, efficacy, and biomarkers for durvalumab (D) ± tremelimumab (T) and gemcitabine/cisplatin (GemCis) in chemo-naïve advanced biliary tract cancer (aBTC). J Clin Oncol 38(15_suppl):4520–4520. https://doi.org/10.1200/jco.2020.38.15_suppl.4520

Oh D-Y, He AR, Qin S, Chen L-T, Okusaka T, Vogel A, Kim JW, Suksombooncharoen T, Lee MA, Kitano M, Burris HA III, Bouattour M, Tanasanvimon S, Zaucha R, Avallone A, Cundom J, Rokutanda N, Xiong J, Cohen G, Valle JW (2022) A phase 3 randomized, double-blind, placebo-controlled study of durvalumab in combination with gemcitabine plus cisplatin (GemCis) in patients (pts) with advanced biliary tract cancer (BTC): TOPAZ-1. J Clin Oncol 40(4_suppl):378–378. https://doi.org/10.1200/jco.2022.40.4_suppl.378

Oshikiri T, Miyamoto M, Shichinohe T, Suzuoki M, Hiraoka K, Nakakubo Y, Shinohara T, Itoh T, Kondo S, Katoh H (2003) Prognostic value of intratumoral CD8+ T lymphocyte in extrahepatic bile duct carcinoma as essential immune response. J Surg Oncol 84(4):224–228. https://doi.org/10.1002/jso.10321

Osipov A, Lim SJ, Popovic A, Azad NS, Laheru DA, Zheng L, Jaffee EM, Wang H, Yarchoan M (2020) Tumor mutational burden, toxicity, and response of immune checkpoint inhibitors targeting PD(L)1, CTLA-4, and combination: a meta-regression analysis. Clin Cancer Res 26(18):4842–4851. https://doi.org/10.1158/1078-0432.ccr-20-0458

Pastushenko I, Blanpain C (2019) EMT transition states during tumor progression and metastasis. Trends Cell Biol 29(3):212–226. https://doi.org/10.1016/j.tcb.2018.12.001

Patel SP, Kurzrock R (2015) PD-L1 expression as a predictive biomarker in cancer immunotherapy. Mol Cancer Ther 14(4):847–856. https://doi.org/10.1158/1535-7163.mct-14-0983

Patel H, Nilendu P, Jahagirdar D, Pal JK, Sharma NK (2017) Modulating secreted components of tumor microenvironment: A masterstroke in tumor therapeutics. Cancer Biol Ther 19(1):3–12. https://doi.org/10.1080/15384047.2017.1394538

Piha-Paul SA, Oh D, Ueno M, Malka D, Chung HC, Nagrial A, Kelley RK, Ros W, Italiano A, Nakagawa K, Rugo HS, Braud F, Varga AI, Hansen A, Wang H, Krishnan S, Norwood KG, Doi T (2020) Efficacy and safety of pembrolizumab for the treatment of advanced biliary cancer: results from the KEYNOTE -158 and KEYNOTE -028 studies. Int J Cancer 147(8):2190–2198. https://doi.org/10.1002/ijc.33013

Razumilava N, Gores GJ (2013) Notch-driven carcinogenesis: the merging of hepatocellular cancer and cholangiocarcinoma into a common molecular liver cancer subtype. J Hepatol 58(6): 1244–1245. https://doi.org/10.1016/j.jhep.2013.01.017

Rizvi S, Gores GJ (2017) Emerging molecular therapeutic targets for cholangiocarcinoma. J Hepatol 67(3):632–644. https://doi.org/10.1016/j.jhep.2017.03.026

Ross JS, Wang K, Gay L, Al-Rohil R, Rand JV, Jones DM, Lee HJ, Sheehan CE, Otto GA, Palmer G, Yelensky R, Lipson D, Morosini D, Hawryluk M, Catenacci DVT, Miller VA, Churi C, Ali S, Stephens PJ (2014) New routes to targeted therapy of intrahepatic cholangiocarcinomas revealed by next-generation sequencing. Oncologist 19(3):235–242. https://doi.org/10.1634/theoncologist.2013-0352

Sarvaria A, Madrigal JA, Saudemont A (2017) B cell regulation in cancer and anti-tumor immunity. Cell Mol Immunol 14(8):662–674. https://doi.org/10.1038/cmi.2017.35

Satiya J, Schwartz I, Tabibian JH, Kumar V, Girotra M (2020) Ablative therapies for hepatic and biliary tumors: endohepatology coming of age. Transl Gastroenterol Hepatol 5:14–14. https://doi.org/10.21037/tgh.2019.10.17

Schrimpf G (2021). Merckgroup.com. Available at: https://www.merckgroup.com/en/news/bintrafusp-topline-data-biliary-tract-cancer-16-03-2021.html. Accessed 5 June 2022

Schwartz M, Zhang Y, Rosenblatt JD (2016) B cell regulation of the anti-tumor response and role in carcinogenesis. J Immunother Cancer 4(1). https://doi.org/10.1186/s40425-016-0145-x

Sha M, Jeong S, Qiu B-J, Tong Y, Xia L, Xu N, Zhang J-J, Xia Q (2018) Isolation of cancer-associated fibroblasts and its promotion to the progression of intrahepatic cholangiocarcinoma. Cancer Med 7(9):4665–4677. https://doi.org/10.1002/cam4.1704

Shen D-Y, Zhang W, Zeng X, Liu C-Q (2013) Inhibition of Wnt/β-catenin signaling downregulates P-glycoprotein and reverses multi-drug resistance of cholangiocarcinoma. Cancer Sci 104(10): 1303–1308. https://doi.org/10.1111/cas.12223

Silva VWK, Askan G, Daniel TD, Lowery M, Klimstra DS, Abou-Alfa GK, Shia J (2016) Biliary carcinomas: pathology and the role of DNA mismatch repair deficiency. Chin Clin Oncol 5(5): 62. https://doi.org/10.21037/cco.2016.10.04

Thuwajit (2009) Alpha-smooth muscle actin-positive fibroblasts promote biliary cell proliferation and correlate with poor survival in cholangiocarcinoma. Oncol Rep 21(4). https://doi.org/10.3892/or_00000309

Tran KT, Coleman HG, McCain RS, Cardwell CR (2019) Serum biomarkers of iron status and risk of primary liver cancer: A systematic review and meta-analysis. Nutr Cancer 71(8):1365–1373. https://doi.org/10.1080/01635581.2019.1609053

TRUSELTIQ (infigratinib) capsules, for oral use (2021) Available at: Highlights of prescribing information. Available at: https://www.accessdata.fda.gov/drugsatfda_docs/label/2021/214622s000lbl.pdf

Tsai M-J, Chang W-A, Huang M-S, Kuo P-L (2014) Tumor microenvironment: a new treatment target for cancer. ISRN Biochem 2014:1–8. https://doi.org/10.1155/2014/351959.

Tyson GL, El-Serag HB (2011) Risk factors for cholangiocarcinoma. Hepatology (Baltimore, Md.) 54(1):173–184. https://doi.org/10.1002/hep.24351

Ueno M, Ikeda M, Morizane C, Kobayashi S, Ohno I, Kondo S, Okano N, Kimura K, Asada S, Namba Y, Okusaka T, Furuse J (2019) Nivolumab alone or in combination with cisplatin plus gemcitabine in Japanese patients with unresectable or recurrent biliary tract cancer: a non-randomised, multicentre, open-label, phase 1 study. Lancet Gastroenterol Hepatol 4(8): 611–621. https://doi.org/10.1016/s2468-1253(19)30086-x

Utispan K, Sonongbua J, Thuwajit P, Chau-In S, Pairojkul C, Wongkham S, Thuwajit C (2012) Periostin activates integrin α5β1 through a PI3K/AKT-dependent pathway in invasion of cholangiocarcinoma. Int J Oncol 41(3):1110–1118. https://doi.org/10.3892/ijo.2012.1530

Valle J, Wasan H, Palmer DH, Cunningham D, Anthoney A, Maraveyas A, Madhusudan S, Iveson T, Hughes S, Pereira SP, Roughton M, Bridgewater J (2010) Cisplatin plus gemcitabine versus gemcitabine for biliary tract cancer. N Engl J Med 362(14):1273–1281. https://doi.org/10.1056/nejmoa0908721.

Valle JW, Kelley RK, Nervi B, Oh D-Y, Zhu AX (2021) Biliary tract cancer. Lancet 397(10272): 428–444. https://doi.org/10.1016/S0140-6736(21)00153-7.

Vauthey JN, Blumgart LH (1994) Recent advances in the management of cholangiocarcinomas. Semin Liver Dis 14(2):109–114. https://doi.org/10.1055/s-2007-1007302

Walker C, Mojares E, del Río Hernández A (2018) Role of extracellular matrix in development and cancer progression. Int J Mol Sci 19(10):3028. https://doi.org/10.3390/ijms19103028

Wang Z, Ahmad A, Li Y, Kong D, Azmi AS, Banerjee S, Sarkar FH (2010) Emerging roles of PDGF-D signaling pathway in tumor development and progression. Biochim Biophys Acta Rev Cancer 1806(1):122–130. https://doi.org/10.1016/j.bbcan.2010.04.003

Wang M, Zhao J, Zhang L, Wei F, Lian Y, Wu Y, Gong Z, Zhang S, Zhou J, Cao K, Li X, Xiong W, Li G, Zeng Z, Guo C (2017) Role of tumor microenvironment in tumorigenesis. J Cancer 8(5): 761–773. https://doi.org/10.7150/jca.17648

Weinberg BA, Xiu J, Lindberg MR, Shields AF, Hwang JJ, Poorman K, Salem ME, Pishvaian MJ, Holcombe RF, Marshall JL, Morse MA (2019) Molecular profiling of biliary cancers reveals distinct molecular alterations and potential therapeutic targets. J Gastrointest Oncol 10(4): 652–662. https://doi.org/10.21037/jgo.2018.08.18

Yang X, Wang W, Wang C, Wang L, Yang M, Qi M, Su H, Sun X, Liu Z, Zhang J, Qin X, Han B (2014) Characterization of EGFR family gene aberrations in cholangiocarcinoma. Oncol Rep 32(2):700–708. https://doi.org/10.3892/or.2014.3261

Yao W-Y, Gong W (2021) Immunotherapy in cholangiocarcinoma: from concept to clinical trials. Surg Pract Sci 5:100028. https://doi.org/10.1016/j.sipas.2021.100028

Yoo C, Oh D-Y, Choi HJ, Kudo M, Uneo M, Kondo S, Chen L-T, Osada M, Helwig C, Dussault I, Ikeda M (2018) M7824 (MSB0011359C), a bifunctional fusion protein targeting PD-L1 and TGF-β, in Asian patients with pretreated biliary tract cancer: preliminary results from a phase I trial. Ann Oncol 29(9_suppl):viii258–viii259. https://doi.org/10.1093/annonc/mdy282.140

Yoo C, Oh D-Y, Choi HJ, Kudo M, Uneo M, Kondo S, Chen L-T, Osada M, Helwig C, Dussault I, Ikeda M (2020a) 73P Long-term follow-up of bintrafusp alfa, a bifunctional fusion protein targeting TGF-β and PD-L1, in patients with pretreated biliary tract cancer. Ann Oncol 31(4_suppl):S268–S269. https://doi.org/10.1016/j.annonc.2020.08.051

Yoo C, Oh D-Y, Choi HJ, Kudo M, Ueno M, Kondo S, Chen L-T, Osada M, Helwig C, Dussault I, Ikeda M (2020b) Phase I study of bintrafusp alfa, a bifunctional fusion protein targeting TGF-β and PD-L1, in patients with pretreated biliary tract cancer. J Immunother Cancer 8(1_suppl): e000564. https://doi.org/10.1136/jitc-2020-000564

Yoshida K, Matsuzaki K, Mori S, Tahashi Y, Yamagata H, Furukawa F, Seki T, Nishizawa M, Fujisawa J, Okazaki K (2005) Transforming growth factor-β and platelet-derived growth factor signal via c-Jun N-terminal kinase-dependent Smad2/3 phosphorylation in rat hepatic stellate cells after acute liver injury. Am J Pathol 166(4):1029–1039. https://doi.org/10.1016/s0002-9440(10)62324-3

Zhang K-S, Zhou Q, Wang Y-F, Liang L-J (2013) Inhibition of Wnt signaling induces cell apoptosis and suppresses cell proliferation in cholangiocarcinoma cells. Oncol Rep 30(3): 1430–1438. https://doi.org/10.3892/or.2013.2560

Zhang F, Li L, Yang X, Wang B, Zhao J, Lu S, Yu X (2015) Expression and activation of EGFR and STAT3 during the multistage carcinogenesis of intrahepatic cholangiocarcinoma induced by 3′-methyl-4 dimethylaminoazobenzene in rats. J Toxicol Pathol 28(2):79–87. https://doi.org/10.1293/tox.2014-0047

Zhou M, Wang C, Lu S, Xu Y, Li Z, Jiang H, Ma Y (2021) Tumor-associated macrophages in cholangiocarcinoma: complex interplay and potential therapeutic target. eBioMedicine 67. https://doi.org/10.1016/j.ebiom.2021.103375

Zhu AX, Macarulla T, Javle MM, Kelley RK, Lubner SJ, Adeva J, Cleary JM, Catenacci DVT, Borad MJ, Bridgewater JA, Harris WP, Murphy AG, Oh D-Y, Whisenant JR, Lowery MA, Goyal L, Shroff RT, El-Khoueiry AB, Chamberlain CX, Aguado-Fraile E (2021) Final overall survival efficacy results of ivosidenib for patients with advanced cholangiocarcinoma with IDH1 mutation: the phase 3 randomized clinical ClarIDHy trial. JAMA Oncol 7(11): 1669–1677. https://doi.org/10.1001/jamaoncol.2021.3836

Fibrosis and Immunotherapy in Hepatocellular Carcinoma

Sarah B. White and Dilip Rajasekhar Maddirela

Abstract

Hepatocellular carcinoma (HCC) is difficult-to-treat cancer, despite great advances in systemic therapy, with a median survival of 6–20 months, due to drug resistance, recurrence, and metastasis. Because HCC develops in patients with cirrhosis (liver fibrosis), treatment algorithms are complex, and there have yet to be discovered effective hepatic antifibrotic approaches to prevent or reverse liver fibrosis. In HCC, the background liver is cirrhotic, which contributes to oncogenesis and changes the tumor microenvironment (TME). Due to this, there are limited systemic agents that can effectively treat HCC. More recent work in advanced HCC has focused on immunotherapeutic approaches with monoclonal antibodies against CTLA-4 and PD-1 that block immune checkpoint inhibitors (ICIs). We review preclinical and clinical findings that implicate immunomodulation in HCC development and immunotherapy including checkpoint inhibitors and their positive outcomes. The intrahepatic milieu of the liver and its TME significantly affects tumor progression. Immune-based therapies have emerged as a treatment option and are relatively safe in the management of HCC. The ICIs like nivolumab and pembrolizumab have revolutionized treatment options and the combination of ICIs got accelerated FDA approval recently. A better understanding of the immune response to HCC will provide more insight into tumor characterization and identification of potential immune targets. The combination of immunotherapy with ICIs or with locoregional therapies may lead to mutually beneficial effects by increasing response rates, progression-free survival, and improving overall survival. Combining use of cellular immunotherapy in advanced HCC could increase the response rates and thereafter extend the

S. B. White (✉) · D. R. Maddirela
Department of Radiology, Division of Vascular & Interventional Radiology, Medical College of Wisconsin, Milwaukee, WI, USA
e-mail: sbwhite@mcw.edu

Interdisciplinary Cancer Research, https://doi.org/10.1007/16833_2022_122
Published online: 9 February 2023

progression-free and overall survival rate. Alternative therapies, such as checkpoint inhibitors that block CTLA-4 and PD-1, may enhance and broaden a patient's response to immunotherapy.

Keywords

Checkpoint inhibitors · Fibrosis · Hepatocellular carcinoma · Immunotherapy · Liver-directed therapies · Tumor microenvironment

1 Introduction

Chronic liver disease is a major global health problem and fibrosis is the critical factor for its progression to cirrhosis. Cirrhotic patients eventually develop hepatocellular carcinoma (HCC) in the absence of liver transplantation. Organ fibrosis contributes 45% of all mortality around cause the globe (Wynn 2004) and the development of fibrosis primarily determines the quality of life (QoL) as well as prognosis (D'Amico et al. 2018). Thus, the level of liver fibrosis affects the liver function.

Unlike other cancers, hepatocellular carcinoma (HCC) metastasizes to distant organs relatively rarely. In contrast, it routinely invades into the liver vasculature, affecting portal veins 3–10 times more often than hepatic veins. It is a characteristic feature of HCC in noncirrhotic livers (Mähringer-Kunz et al. 2021). The treatment choice between liver transplantation, resection, and liver-directed therapies (LDTs) for HCC depends on the liver function, patient's performance status, and the tumor burden and location. Systemic therapy was rarely used for HCC prior to 2008, due to lack of efficacy. In 2009, an oral multityrosine kinase inhibitor (TKI) sorafenib was shown to prolong the survival of patients in advanced HCC with preserved liver function (Child-Pugh class A) (Llovet et al. 2008). Advances in immunotherapy have now revolutionized the treatment of multiple cancers. Liver fibrosis and other serious related complications threaten the health of HCC patients and represent a serious medical burden; yet, there is still a lack of approved methods to prevent or reverse liver fibrosis. In this chapter, we summarize the overview of the immunotherapy options in the clinical settings, which are available for managing the liver TME responsible for HCC, and potential future directions. PubMed/MEDLINE was electronically searched up to 2021 to identify studies evaluating safety, efficacy, and therapeutic mechanisms of immunotherapeutic agents in HCC patients and animal models.

2 Fibrosis Is a Major Risk Factor

Structurally, hepatocytes are parenchymal cells of the liver that metabolize or detoxify all the substances which are absorbed by the portal vein from the gut and constitute 60% of total cells in the liver (Tahmasebi Birgani and Carloni 2017).

Nonparenchymal parts of the liver are composed of endothelial cells, Kupffer cells, stellate cells, and lymphocytes, which serve as an immune modulatory function of the liver against pathogens (Gao et al. 2008). Liver disease resulting in fibrosis and ultimately cirrhosis develops irrespective of the underlying etiology, which can include chronic hepatitis B virus (HBV) infections, hepatitis C virus (HCV) infections, alcoholic liver disease, nonalcoholic fatty liver disease (NAFLD), cryptogenic cirrhosis, or autoimmune hepatitis. Kanda et al. reported that viral hepatitis results in excessive liver fibrosis distribution within portal tracts and extends into the hepatic parenchyma, accompanying persistent inflammation. In addition, periportal fibrosis develops and extends across lobules, connecting mesenchymal structures. Finally, nodule formation and structure of cirrhosis are established by intrahepatic portosystemic vascular shunting and regeneration of hepatocytes (Kanda et al. 2019). During this process, the tissue remodeling and repair are associated with large depositions of collagen, fibronectin, and extracellular matrix (ECM) that results in the progressive substitution of the liver parenchyma with scar tissue (Baglieri et al. 2019). HSCs with phagocytic Kupffer cells (KCs) and liver sinusoidal endothelial cells (LSECs) also play a pivotal role in fibrosis development (Zhou et al. 2014). Therefore, the accumulation of fibrous matrix worsens normal liver function and, if left untreated, results in the development of cirrhosis. The elimination of liver fibrosis caused by NAFLD or suppression of viral replication in hepatitis B/C is the basic treatment strategy to stimulate regression or reverse fibrosis (D'Ambrosio et al. 2012). As reviewed by Change et al., control of the primary disease is the most effective strategy in the treatment of fibrosis (Chang and Li 2020).

Although liver biopsy remains the gold standard for the identification of advanced hepatic fibrosis, among the clinically available modalities in noninvasive assessment of liver disease (serum biomarkers, ultrasound-based elastography, sheerwave elastography), magnetic resonance (MR) elastography is known to have the highest diagnostic accuracy in the detection of advanced fibrosis in nonalcoholic fatty liver disease (NAFLD) (Dulai et al. 2016; Castera et al. 2019).

The incidence of HCC has tripled between 1975 and 2011, unlike other solid tumors whose incidence has remained stable. HCC differs from solid tumors due to the dense and profoundly altered microenvironment, and excessive liver fibrosis in >90% cases. Data also demonstrate that approximately one-third of patients with cirrhosis will develop HCC. Therefore, the premalignant environment with chronic hepatic cell death, inflammation, and fibrosis is characteristic feature of HCC development. Although the liver is known to regenerate after injury, persistent injury due to chronic infection and inflammation results in this elevated risk of developing HCC. Co-evolution of serological fibrotic index enhances the risk of HCC up to 15-fold and is a unique feature of liver cancer (Filliol and Schwabe 2019). Damage of hepatocytes also triggers the release of reactive oxygen species (ROS) and mediators of fibrosis inducing activation of hepatic stellate cells (HSCs), which has been reported to have a close relationship with fibrosis and HCC (Sakurai et al. 2013). In the setting of bridging fibrosis, HCC development is a progressive process in which chronic inflammation, chromosomal mutations, and eventually, malignant transformation of proliferating hepatocytes occurs (El-Serag 2011). Because the liver is the primary immune recognition site with frequent exposures of potential pathogens due to portal and arterial inflow (Guillot and Tacke 2019), the immune-mediated cell death by viral infection triggers increased production of ROS, which enhances hepatocellular oxidative stress and induces DNA mutations, contributing to fibrosis and eventual malignant transformation. The altered microenvironment fibrosis also yields altered blood flow and hypoxic hepatocytes (Mohammed et al. 2003). Areas of hypoxia in the liver parenchyma can lead to changes in molecular signaling, and upregulation of angiogenic factors, thus creating unique microenvironment that enables firm establishment of a tumor (O'Rourke et al. 2018).

Evidence suggests that HCC mainly contain fibrous nests and are stiffer than the surrounding cirrhotic liver. Fibrous nests are associated with poor prognosis and have positive correlation with HCC risk (O'Rourke et al. 2018). Furthermore, several cytokines have been identified to play significant roles in the development of cirrhosis and HCC, including platelet-derived growth factor (PDGF), transforming growth factor-β (TGF-β), tumor necrosis factor-α (TNF-α), interferon, and interleukins (ILs). PDGF-B and D have been shown to have a role in activating HSCs leading to liver fibrosis. TGF-β is known to be the most potent stimulator of fibrogenesis and is produced by a variety of cells in the liver: HSCs, KCs, LSECs, and hepatocytes (Lee and Friedman 2011). Activation of HSCs is the central driving force of hepatic fibrosis and is known to be trans-differentiated into a myofibroblast-like cell through intercellular communication between HSCs and damaged hepatocytes (Kuang et al. 2013). This activation triggers overexpression of α-smooth muscle actin (α-SMA), which are potent suppressors of hepatic immunity by affecting T-cell responses and thus play a significant role in the progression of HCC (Hautekeete and Geerts 1997). Therefore, strong expression of α-SMA is an excellent marker in the activated fibroblasts and indicator of poor survival in HCC patients (Parikh et al. 2014). Several authors have reported increased liver injury and a subsequent increase in liver fibrosis due to the absence of gut bacterial resulting in abrogation of TLR4-mediated activation of anti-apoptotic NF-κB signaling (Dapito et al. 2012; Mazagova et al. 2015). If underlying etiology is controlled or eliminated, the liver fibrosis can be restored, and risk of developing tumors can be reduced (Papatheodoridis et al. 2017).

2.1 Key Molecules and Mechanism of Liver Fibrosis Leading to HCC

Among the mechanisms of liver fibrosis, the major signaling pathways and effectors in signaling pathways include growth factor signaling (PDGF, TGF- α, EGF, VEGF), fibrogenic signaling pathway (TGF-β1), chemokine pathways (CCR5, CCR1, CXCL4, CXCL9, CXCR3), adipokine pathways (leptin, adiponectin), and neuroendocrine pathways (cannabinoid and opioid signaling, thyroid hormones, serotonin) (Aydın and Akçalı 2018). Recent reports suggest the increased expression of mesothelin in tissue fibroblasts of damaged liver causes liver, lung, and kidney fibrosis (Nishio et al. 2021).

Several alterations have been reported in HCC patients with tumor-associated microenvironment components including immune cells, fibroblast cells, endothelial cells, and extracellular matrix can augment the neoplastic cells to proliferate, growth, and invade. The main cell type in HCC that is responsible for fibrosis development is myofibroblast, which is absent, in normal liver (Mederacke et al. 2013). The myofibroblasts transform into cancer-associated fibroblasts (CAFs), through the actions of increased growth factors and continued expression of inflammatory cytokines such as PDGF, TGF-β, TNF-α, IL-6, and IL-1β (Török 2008; Tacke et al. 2009). PDGF recognizes its receptors that promote the dimerization of receptor subunits and autophosphorylation. This will lead to cellular proliferation (Wong et al. 1994) by activating Ras-mitogen protein kinase-activated pathway. Evidence suggests that blocking of these receptors is a promising potential target for the anti-fibrotic drug development. Hepatic angiogenesis is initiated by vascular endothelial growth factor (VEGF) mediated by its receptors (Schuppan et al. 2001). The growth of HCC xenografts become accelerated with the expression on TIMP-1, leading to transformation of LFs to CAFs. The secretion of CAFs induces SDF-1-mediated PI3/AKT pathway in neighboring HCC cells. This has consequently inhibited the apoptosis of HCC cells by upregulating Bcl-2/BAX ratio. Thus TIMP-1 appears as predictive factor for HCC outcome after liver resection (Song et al. 2015). Moreover, CAFs not only upregulate gene expressions of TGF-β and FAP but critically express highly potent hepatocyte growth factor (HGF) in promoting cell proliferation, migration, survival, and angiogenesis (Monvoisin et al. 1999; Efimova et al. 2004). HCC results in persistent expression of IL-6/progranulin/mTOR signaling cascade and increased phosphorylation of p70S6K, 4E-BP1, and Akt-Ser473/FoxO1 (Liu et al. 2016). PDGF-C-overexpressing hepatocytes in transgenic mice cause the activation of HSCs, which in turn secretes HGF and cytokines, resulting in the development and progression of HCC (Wright et al. 2014). Also, acquisition of an invasive phenotype is crucial for epithelial–mesenchymal transition in HCC during tumor-stroma crosstalk between TGF-β and PDGF signaling (van Zijl et al. 2009), thus regulating both tumor growth and HCC progression. In addition, MRC-5 fibroblast-conditioned medium influences multiple pathways regulating invasion in HCC. One of CAFs expressing alpha-SMA, MRC-5 fibroblast-conditioned medium promoted HCC cells in multiple ways including migration, proliferation, and apoptosis by inducing G1 phase arrest in Bel-7,402 cells in mesenchymal movement mode through activation of the α6, β3, β4, β7 integrin/FAK pathway and upregulation of MMP-2 (van Zijl et al. 2009).

Earlier studies were more focused on the mechanism of monocyte and macrophage cells in hepatic fibrogenesis and fibrosis. Macrophages play a dual role in the progression and resolution of liver fibrosis as M1 (produce inflammatory cytokine) and M2 (tissue repair) macrophages. Initial liver injury stimulates bone marrow-derived monocytes as tissue repair and turn into pro-inflammatory M1 macrophages, promoting inflammatory responses and HSC activation. M2 macrophages in turn recruit MMPs to fibrosis resolution (Barnes et al. 2015; Sun and Kisseleva 2015; Zhang et al. 2016).

Morphologically, persistent liver injury can trigger ECM accumulation and strengthen fibrosis irreversibly in HCC; thus, ECM serves as modulator of fibrogenesis and fibrinolysis (Schuppan and Kim 2013). Fibrotic livers tend to change their epithelial cell behavior by accumulating fibrillar collagens like type I and III, along with increased deposition of noncollagenous glycoproteins laminins, undulin, hyaluronic acid (HA), tenascins, elastin, and proteoglycans and fibronectin (FN) (Bataller and Brenner 2005). These events favor the extensive deposition of ECM, followed by reduced activity of ECM-degrading protein matrix metalloproteinases (MMPs), favoring the formation of the fibrotic scar (Arthur 2000). Although evidences from in vitro studies suggested the type I collagen can enhance epithelial–mesenchymal transition (EMT) (Yang et al. 2014) proliferation (Zhang et al. 2018a, b, c) and migration (Ji et al. 2010), in in vivo condition, the contribution of collagens in HCC development still remains unknown. Abnormal expression of Tenascin-C promotes defective of ECM and causes chronic liver fibrosis in HCC than liver cirrhosis (Zhao et al. 1996). Among the members of MMPs, MMP-1, MMP-8, and MMP-13 can inhibit fibrosis and promote the proliferation of hepatocyte, whereas MMP-12 and MMP-19 are fibrogenic (Robert et al. 2016; Tao et al. 2020). The unbalanced expression of TIMPs and MMPs is important contributors of liver fibrogenesis (Madtes et al. 2001). The involvement of abnormal autophagy in HSC activation can aggravate the development of fibrosis by providing energy from lipid droplets (Hernández-Gea and Friedman 2012). Autophagy and NLRP3 inflammasome promote liver injury and liver fibrosis by mediating the IL-1 pathway (Tao et al. 2020). Overexpression of IL-20 in HCC patients is associated with liver fibrosis progression and using mAb against IL-20 prevented the hepatocyte damage in vivo (Ding et al. 2018).

Most of the existing drugs are not cell-specific but still effective. The off-target effect is very high for the fibrotic liver as the uptake by fibrogenic cells is very low due to abnormal hepatic metabolism and excretion. Hence, most of the drugs tested on animals established with liver fibrosis are not effective at preclinical level (Poelstra and Schuppan 2011). The hepatic uptake of drugs occurs through hepatocytes and/or macrophages. Chemokines or cytokines have short plasma half-life and unwanted side effects result due to overexpression of their receptors, leading to liver fibrosis. Prevention of inflammation can result in the arrest of ECM turnover and consequently suppression of fibrogenesis (Poelstra and Schuppan 2011). Inhibition of inflammation may lead to inhibition of ECM turnover and thus to enhanced fibrogenesis (skewing of the immune system toward alternative macrophages, a Th2 T-cell response, or production of TGF-β).

Importantly, regression of fibrosis has also been reported to correlate with improved clinical outcomes apart from observation from in animal models and patients with chronic liver disease of diverse etiologies (Tang et al. 2019). These studies taken together indicate that understanding and detecting the dysfunction of innate and adaptive immunity in the occurrence and development of HCC (from cirrhosis, dysplastic nodules, and early HCC to advanced HCC) is necessary for uncovering the potential molecular mechanism.

Immune gene profiles are consistently downregulated during HCC progression, which leads to low tumor immunity in advanced HCC (Tang et al. 2019). A STAM™ murine model of HCC shows that alpha-TGF-β seems to enhance the fibrotic environment and increase in CD8+ T cells (Okrah et al. 2018). A recent study on gene signatures using CIBERSORT analysis by Tang et al. (2019) revealed the differential immune cell fractions in various kinds of liver tissues including healthy, cirrhotic, dysplastic nodules to HCC. Among sub-type of T-cells, CD4 and gamma delta T-cells were elevated especially in cirrhosis and dysplasia tissues and found decreased in HCC. However, Tregs and follicular helper T cells were downregulated under cirrhotic condition but elevated in HCC. The altered gene expression profile in different stages of HCC patients is represented in Table 1. T-cell activation was involved in all stages of the T-cell response when compared between HCC samples with different degrees of fibrosis. The fraction showing resting NK cells significantly related with overall survival (OS) ($P = 3.23$ E-2, HR >1) and FRS ($P = 3.93E\text{-}2$, HR > 1). The gene PVRIG (PVR-related immunoglobulin domain-containing, CD112R) is preferentially expressed in T cells and inhibited T-cell receptor-mediated signals and reported as a novel checkpoint for T cells. PVRIG is found to compete with CD226 to bind to CD112, and their disruption leads to enhanced T-cell response (Zhu et al. 2016).

3 Immunotherapy in Hepatocellular Carcinoma

Despite several advancements in systemic therapy, the prognosis of HCC is dismal with a 5-year survival rate of <30% in the United States and Asia, due to drug resistance, high recurrence rates, and metastasis (Zhang et al. 2018a, b, c; Oura et al. 2021). Effective treatment options for HCC localized to the liver include liver transplantation, surgical resection, LDTs. However, when extra-hepatic disease is present, systemic therapies remain to be the only option. Studies have shown that genetic and epigenetic alterations in cancer cells allow them to escape from the native immunity and impact the inhibitory effect of systemic therapy on tumor growth. Acquired mutations also induce resistance. The unique characteristics of hepatic immunity contribute to tumor tolerance and fibrosis in liver. Immune checkpoint inhibitors, on the other hand, are capable of handling the mutations and immune suppressive microenvironment and improve the survival of HCC patients (Nishida and Kudo 2018). In advanced HCC, immunotherapeutic approaches have increasingly focused on monoclonal antibodies against ICI pathways and these ICIs are moving into the first-line therapy (Neureiter et al. 2019). Many clinical trials (Table 1) with immunosuppressive therapies for advanced HCC have demonstrated treatment efficacy and improvements to the prognosis of HCC patients. In recent years, several clinical trials (Tables 2, 3, 4, 5 and 6) on ICIs were carried out in patients with HCC. Sorafenib is the first systemic therapy to prolong the survival of patients under unresectable HCC.

Table 1 Clinical trials and systemic treatments with regulatory approvals for hepatocellular carcinoma

Trial name	Identifier	Patient number	Primary end point	Phase	Treatment arms	Setting
SHARP (PMID: 18650514)	NCT00105443	602	OS	III	Sorafenib vs. Placebo	First line
REFLECT (PMID: 29433850)	NCT01761266	954	OS	III	Lenvatinib vs. Sorafenib	First line
IMbrave150 (PMID: 32402160)	NCT03434379	336	OS and PFS	III	Atezolizumab +bevacizumab vs. sorafenib	First line
RESORCE (PMID: 27932229)	NCT01774344	573	OS	III	Regorafenib vs. Placebo	Second line
CELESTIAL (PMID: 29972759)	NCT01908426	707	OS	III	Cabozantinib vs. Placebo	Second and third
REACH-2 (PMID: 30665869)	NCT02435433	292	OS	III	Ramucirumab vs. Placebo	Second line
CHECKMATE 040 (PMID: 28434648)	NCT01658878	262	ORR	I/II	Nivolumab	Second line
KEYNOTE 224 (PMID: 29875066)	NCT02702414	169	ORR	II	Pembrolizumab	Second line
CHECKMATE 040 (PMID: 33001135)	NCT01658878	148	ORR	I/II	Nivolumab + ipilimumab	Second line

ORR overall response rate, *OS* overall survival, *PFS* progression-free survival

Table 2 Current clinical trials for systemic treatment (first line and second line) for advanced hepatocellular carcinoma

Trial name	Identifier	Patient number	Primary end point	Phase	Treatment arms	Setting
COSMIC-312 (PMID: 32491932)	NCT03755791	740	PFS (RECIST 1.1) OS	III	Cabozantinib + atezolizumab Sorafenib	First line
LEAP-002	NCT03713593	750	PFS (RECIST 1.1) OS	III	Lenvatinib + pembrolizumab Lenvatinib	First line
HIMALAYA	NCT03298451	1,504	OS	III	Durvalumab + tremelimumab (2 regimens) Sorafenib	First line
CheckMate 9DW	NCT04039607	728	OS	III	Nivolumab + ipilimumab Sorafenib or lenvatinib	First line
N/A	NCT03764293	543	PFS (RECIST 1.1) OS	III	Camrelizumab (SHR-1210) + apatinib Sorafenib	First line
RATIONALE-301	NCT03412773	674	OS	III	Sorafenib Tislelizumab	First line
GOING	NCT04170556	78	Safety	I/II	Regorafenib (monotherapy for the first 8 weeks) + nivolumab	Second line
RENOBATE	NCT04310709	42	ORR (RECIST 1.1)	II	Regorafenib + nivolumab	First line
Bayer 19497	NCT03347292	57	Safety	Ib	Regorafenib + pembrolizumab	First line
N/A	NCT04183088	125	Part 1: Safety Part 2: PFS (RECIST 1.1) ORR (RECIST 1.1)	II	Part 1: Regorafenib + tislelizumab Part 2: Regorafenib + tislelizumab vs Regorafenib	First line
N/A	NCT04442581	29	ORR (RECIST 1.1)	II	Cabozantinib + pembrolizumab	First line
N/A	NCT03941873	111	Phase 1: Safety Phase 2: ORR	I/II	Phase I: Sitravatinib Sitravatinib + tislelizumab Phase II: Sitravatinib Sitravatinib + tislelizumab	First- and later line

ORR overall response rate, *OS* overall survival, *PFS* progression-free survival

Table 3 Current clinical trials for systemic treatment in combination therapy with TACE for hepatocellular carcinoma

Trial name	Identifier	Patient number	Primary end point	Phase	Treatment arms	Setting
LEAP-012	NCT04246177	950	PFS (RECIST 1.1) OS	III	Lenvatinib + pembrolizumab + TACE TACE	First line
EMERALD-1	NCT03778957	724	PFS (RECIST 1.1)	III	Durvalumab + TACE Durvalumab + bevacizumab + TACE TACE	First line
CheckMate 74 W	NCT04340193	26	Time to TACE progression OS	III	Nivolumab + ipilimumab + TACE Nivolumab + TACE TACE	First line
ABC-HCC	NCT04803994	434	Time to failure of treatment strategy	III	Atezolizumab + bevacizumab TACE	First line
RENOTACE	NCT04777851	496	PFS (RECIST 1.1)	III	Regorafenib + nivolumab TACE	First line

ORR overall response rate, *OS* overall survival, *PFS* progression-free survival

Table 4 Current clinical trials for systemic treatments (adjuvant setting) after surgery or ablation for hepatocellular carcinoma

Trial name	Identifier	Patient number	Primary end point	Phase	Treatment arms
CheckMate 9DX	NCT03383458	545	RFS	III	Nivolumab Placebo
KEYNOTE-937	NCT03867084	950	RFS OS	III	Pembrolizumab Placebo
IMbrave050	NCT04102098	668	RFS	III	Atezolizumab + bevacizumab Active surveillance
EMERALD-2	NCT03847428	888	RFS	III	Atezolizumab + bevacizumab TACE

OS overall survival, *RFS* recurrence-free survival

Table 5 Current clinical trials for systemic treatments (neoadjuvant setting) after surgery or ablation for hepatocellular carcinoma

Trial name	Identifier	Patient number	Primary end point	Phase	Treatment arms
NIVOLEP	NCT03630640	43	Local RFS	II	Nivolumab + electroporation
CaboNivo	NCT03299946	15	Safety	Ib	Cabozantinib + nivolumab + surgery
N/A	NCT03337841	50	One-year RFS rate		Pembrolizumab + surgery/ ablation
PLENTY202001	NCT04425226	192	Pembrolizumab + Lenvatinib + liver transplantation No intervention + liver transplantation		Pembrolizumab + Lenvatinib + liver transplantation No intervention + liver transplantation

OS overall survival, *RFS* recurrence-free survival

3.1 Immune Checkpoint Inhibitors and Novel Immunotherapy Targets

Immune checkpoint inhibitors maintain immunotherapeutic balance and prevent uncontrolled immunity against overactivation of T cells. ICIs counteract the escape mechanism and allow the immune system to reactivate and generate tumor-specific T cells by blocking checkpoint-mediated signals (Ozer et al. 2021). The systematic development of treatment for advanced HCC over the last decade is presented in Fig. 1. The pattern in HCC is similar to other solid tumors like presentation of recognized cancer cell antigen to antigen-presenting cells (APC) and moving to

Table 6 Current clinical trials on gene and cell-based systemic treatments (palliative)

Identifier	Patient number	Primary end point	Phase	Treatment arms
NCT02905188	14	Safety	I	CAR-GPC3 T cells
NCT03980288	6	Safety	I	CAR-GPC3 T cells
NCT04011033	144	OS; PFS; DCR	I	iNKT cells + TACE
NCT03319459	44	Safety	I	FATE-NK100
NCT03841110	37	Safety	I	FT500 (allogeneic NK cells)
NCT03071094	14	Safety; ORR	I/II	Pexastimogene devacirepvec + nivolumab

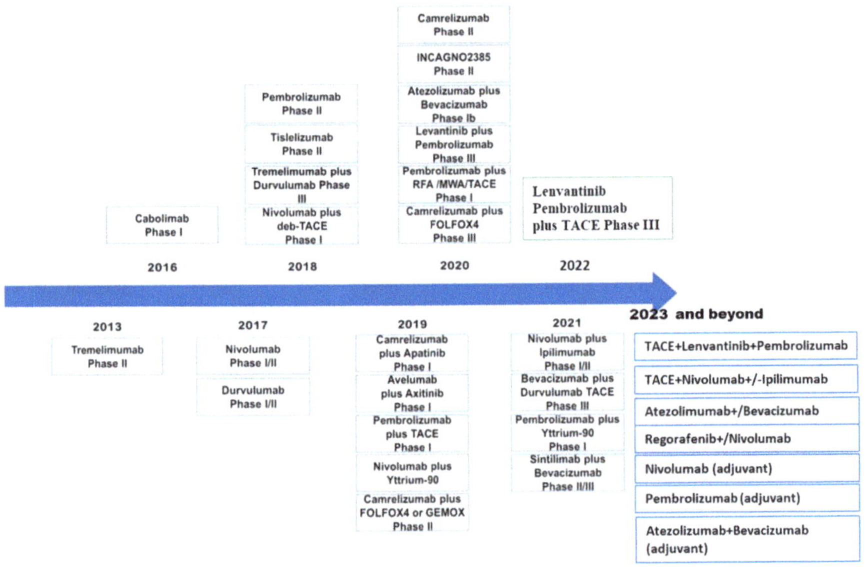

Fig. 1 Development of clinical trials and different treatment strategies for hepatocellular carcinoma
The figure represents immune checkpoint inhibitors and their clinical trials with different treatment strategies for HCC from 2013 to 2023 and beyond (partially adopted from Leone et al. PMID: 34065489)

lymph node to initiate T-cell activation. The cytotoxicity is exerted when $CD8^+$ T cells are mobilized to the tumor site. In addition, the uniqueness of the dual blood supply, as mentioned earlier, makes the liver more immune-tolerant. Notably, this is due to the upregulation of cytokines (Lawal et al. 2021), increased expression of co-inhibitory molecules like programmed cell death-1 (PD-1), and cytotoxic T-lymphocyte associated antigen-4 (CTLA-4), and increased differentiation of Treg cells (Ghavimi et al. 2020; Zhang et al. 2020) Tumor growth can be suppressed by using inhibitors of these proteins which can enhance tumor immunity. Despite recent failures with response rates of 20 to 25%, checkpoint-based immunotherapy is considered to be one of the most promising strategies for HCC treatment. To enhance the overall response rates, checkpoint inhibitors are now being tested in combination with other systemic treatments. The recent review on checkpoint inhibitors of HCC and their clinical trials suggested the main ICIs under investigation include programmed cell death protein (PDI-1) programmed cell death protein-ligand 1 (PD-L1), B and T lymphocyte attenuators, cytotoxic T-lymphocyte associated antigen 4 (CTLA-4), V-domain immunoglobulin suppressor of T-cell activation, T-cell immunoglobulin and mucin domain (TIM-3), lymphocyte activation gene-3 (LAG-3), and TNF receptor superfamily member 4 (Ozer et al. 2021). The use of ICIs, used as single agents, or combination therapy regulates the balance of immune homeostasis and indicates as landmark in clinical research. The various trials (completed and ongoing) with ICIs, used as single agents or combination therapy to alter the imbalances in the TME are represented in Tables 2, 3, 4, 5 and 6.

Nivolumab, a PD-1 blocker and fully humanized Ig4 antibody, enhances the activity of effector T cells in the HCC microenvironment. This humanized antibody is the first FDA-approved second-line therapy. A multicenter, open-label, phase I/II clinical trial CheckMate 040 (NCT01658878) enrolled 262 patients and demonstrated favorable results and a good safety profile. The dose escalation showed durable objective response rates of 15–20%, a median survival of 15.6 months, and tumor reduction. In addition, among 22 responders, the drug response lasted for at least 6 months in 91% and 12 months in 55% (El-Khoueiry et al. 2017). Ipilimumab is another promising anti-CTLA4 antibody in HCC therapy, mostly administrated as an adjuvant therapy with nivolumab. A recent approval by FDA for the combination of nivolumab and ipilimumab for the treatment of patients as a second-line therapy, revealed an overall response rate of 35% with a complete response of 12% and partial response of 22% (Crocenzi et al. 2017). Another anti-PD1 pembrolizumab (KEYNOTE-224) was demonstrated in a nonrandomized study to have an objective response rate (ORR) of 17%, one patient (1%) had a complete response, 17 patients (16%) had a partial response, 46 patients (44%) had stable disease, 33% showed disease progression. 73% of patients experienced adverse events related to treatments, with only 15% reported severe events (Zhu et al. 2018). The subsequent randomized, double-blind, phase III study, KEYNOTE-240, did not reach the desired statistical outcome when evaluating pembrolizumab vs placebo in the second-line setting (HR for OS (0.78; 95% CI, 0.611 to 0.998; P 5.02) and PFS (HR 0.72; 95% CI, 0.57 to 0.90)). A similar phase III trial in Asian is actively recruiting and will assess the safety and efficacy of pembrolizumab vs best supportive care in the 2nd line.

Another active ongoing phase III trial evaluates anti-PD-L1 antibodies, such as durvalumab (NCT03298451), atezolizumab (NCT03434379), as well as BGB-A317 (NCT03412773). In phase I/II study, durvalumab, in advanced HCC patients having prior treatment with sorafenib, demonstrated 10.3% ORR (NCT01693562) (Yang et al. 2018). Avelumab showed moderate efficacy after a median follow-up of 13.7 months (no CR, 10.0% PR, and 63.3% SD) and was well tolerated without any grade 4 adverse events and only 23% of patients experienced grade 3 adverse events (Lee et al. 2021). Other PD-1/PD-L1 trials, including tislelizumab (NCT03412773, NCT03419897), are currently in progress as well. The Rationale-301 trial is a randomized, open-label, multicenter study that will evaluate the efficacy and safety of tislelizumab vs. sorafenib, in patients with advanced unresectable HCC (Qin et al. 2019). The best response rate to date was observed with the combination of atezolizumab plus bevacizumab with OS rate of 67.2% vs 54.6% at 12 months and PFS was 6.8 vs 4.3 months compared to sorafenib alone in unresectable HCC in the IMbrave150 trial (HR for disease progression or death, 0.59; 95% CI, 0.47 to 0.76; $P < 0.001$) (Finn et al. 2020).

A phase II, noncontrolled, open-label, multicenter study of the anti-CTLA-4 antibody tremelimumab (NCT01008358) revealed a safety profile, and promising effect as antitumor and antiviral effects, associated with an enhanced specific anti-HCV immune response in patients with HCC and HCV infection. It showed a 17.6% partial response rate, and a 76% disease control rate in patients with advanced HCC whose disease progressed under sorafenib treatment (Sangro et al. 2013).

The combination of TKIs and ICIs is also being investigated in ongoing clinical trials to obtain a better understanding of survival rates, although they have independent proven efficacy. Kudo et al. reported the antitumor activity and the effect of the combination of avelumab plus axitinib as first-line treatment in patients with advanced HCC (NCT03289533). The results have shown ORR of 31.8% [95% CI: 13.9–54.9%] as per mRECIST, with manageable toxicity profile (Kudo et al. 2021). Similarly, investigations studying combinations of pembrolizumab with either lenvatinib or regorafenib. The preclinical studies with the combination of pembrolizumab and lenvatinib decreased TGF-β and IL-10 by suppressing tumor-associated macrophages, Tregs, and related TME constituents (Kato et al. 2016). Second-generation immunotherapies such as TIM-3, LAG-3, and killer cell immunoglobulin-like receptor (KIR) as described earlier, combined with anti-PD-1 antibody therapies to activate T-cell infiltration against TME. These include NCT03005782, NCT01968109, and NCT03099109 studies and are at early phase of clinical trials in treating advanced HCC.

An ongoing open-label, multicenter, randomized phase III HIMALAYA study (NCT03298451) evaluating the efficacy of tremelimumab (anti-CTLA-4 mAb) plus durvalumab (anti-PD-L1 mAb) in the treatment of patients with no prior systemic therapy for unresectable HCC. The study showed promising results regarding the survival and antitumor response (Abou-Alfa et al. 2018b).

Sorafenib is widely considered as more potent antiangiogenic/vascular drug through inhibition of vascular endothelial growth factor (VEGF) receptors (VEGFRs) and platelet-derived growth factor receptors (PDGFRs) and is the only available standard of care for advanced HCC for a decade. The transient delay in HCC growth is the major advantage of sorafenib in HCC patients. The well-established clinical trial for patients with advanced-stage HCC is the SHARP study (Llovet et al. 2008), which demonstrated the superiority of sorafenib vs placebo with an OS of 10.7 months vs 7.9 months. The stromal-derived key pathways, hypoxia, may however play a role in the resistance against sorafenib. The antifibrotic effects were observed in cirrhotic livers with sorafenib and vatalanib

(Liu et al. 2009; Bai et al. 2012). The activity of PDGF-B, as profibrotic factor, can be effectively blocked by sorafenib. Conversion of HSCs to myofibroblasts, during hepatic fibrogenesis or activation of fibroblasts, is actively mediated by profibrotic factors like SDF-1α, via MAPK pathway activation after sorafenib treatment (O'Shanick 1986). SDF-1α can counteract the antifibrotic effects of sorafenib through PDGFR-inhibition, thus leading to tumor-associated fibrosis (O'Shanick 1986). The combination therapy inhibited HCC growth, compared to sorafenib alone.

The REFLECT study demonstrated the efficacy of lenvatinib, a small molecular type V TKI, with more potent activity against VEGF receptors and FGFR family. The study indicated the superiority of lenvatinib vs sorafenib with OS 13.6 months vs 12.3 months. The combination of atezolizumab and bevacizumab in NCT03434379 trial resulted in improved OS and progression-free survival (PFS), effectively replacing sorafenib, with near-normal liver function (Finn et al. 2020). Alternative treatments including nivolumab or sorafenib or lenvatinib may be considered. As a second-line treatment for patients who received sorafenib with HCC, regorafenib and cabozantinib were shown to prolong the survival compared with placebo and this combination was approved recently by FDA (Bruix et al. 2017; Abou-Alfa et al. 2018a). The analysis of ninety-nine HCC tissue samples by Meng-yun et al. showed low levels of FIB-4 values (a noninvasive liver fibrosis biomarker) and high CD8+ T-cell infiltration correlated with prolonged survival (Ke et al. 2021). The Golgi membrane protein 1 (GOLMI) is a serum biomarker for liver diseases and HCC and considered as the most upregulated protein. The GOLMI1 promotes immunosuppression of HCC. Targeting EGFR and the PD-1/PD-L1 axis together improved the treatment efficacy in HCC (Ke et al. 2021). The application of different therapeutic options for HCC is limited owing to patient's overall health status, as >80% of cases develop cirrhosis. In principle, patients with early-stage HCC tumors recommended for resection or local ablation and/or at intermediate stages TACE or advanced disease systemic therapies. As the QOL is the major end point for cancer research, the pivotal clinical trials investigating multikinase inhibitors such as sorafenib, lenvatinib, or regorafenib, indicated the incidence of treatment-related adverse events (grade ≥3) exceeded more than 50% (Bruix et al. 2017; Villanueva 2019).

The retrospective studies of Chapiro et al., and Jin et al., respectively, did not find the treatment efficacy when Sorafenib combined with TACE in randomized clinical trial and in advanced liver cancer patients in European, American, or Asian populations (Chapiro et al. 2014; Jin et al. 2018). Another randomized, multicenter, prospective trial improved the PFS in the group treated with sorafenib and TACE than TACE alone (25.2 months vs. 13.5 months, respectively; $P = 0.006$) (Kudo et al. 2020). A systematic review and meta-analysis of Li et al., and Wei et al., respectively, on the combined use of TACE and sorafenib treatment in unresectable HCC condition, indicated the prolong TTP and DCR in patients (Li et al. 2018; Wei et al. 2019). Combining LRTs and ICIs can enhance the antitumor immune response and combination of different drugs and TACE, including nivolumab-TACE

(NCT03572582), nivolumab-DEB-TACE (NCT03143270), and pembrolizumab-TACE (NCT03397654). Nakamoto (Nakamoto et al. 2011) demonstrated an anti-HCC tumor effect in a combination therapy using immunotherapy strategies based on dendritic cells with TAE (Nakamoto et al. 2011) and exerted prolonged recurrence-free survival.

Several trials are ongoing to assess the effectiveness of Cobolimab (TSR-022), (NCT02817633), anti-Tim-3-like MBG453, and Symo23 (a recombinant human Ab). Cobolimab is an anti-Tim-3 mAb and the investigation is currently ongoing with combination of anti-PD-1, anti- LAG-3, or anti PD-L1 mAbs in a multicenter, open-label, first in human study, in patients with advanced HCC (Murtaza et al. 2016). INCAGN02385 is an anti-LAG-3 mAb, evaluated in cynomolgus monkeys shown enhanced T-cell responsiveness to TCR stimulation or in combination with PD-1/PD-L1 axis blockade. The preliminary investigation at phase I, open-label, dose escalation trial (NCT03538028) for safety, tolerability, and preliminary efficacy has yet to be published.

Earlier studies suggested that locoregional therapy may not often result less than ideal therapeutic outcomes due to hypoxia generated after treatment in the progression of HCC (Makary et al. 2021). To overcome the therapeutic bottlenecks like different treatments, and efficacies at different stages of HCC, combination of therapies may lead to effective tumor responses and longer survival.

3.2 Adoptive Cell Transfer Therapy

Adoptive cell transfer (ACT) therapy refers to the extraction of immune cells from the tumor or from the patient to use them in vitro culture and millions of copies of these copies re-inject them better able to target the patient's cancer cells. Although it does not work for every patient, it is widely the most effective method for treating patients with solid tumors. ACT application include tumor-infiltrating lymphocytes (TILs), dendritic cells (DCs), natural killer cells, cytokine-induced killer cells (CIK) cells, T-cell receptor T Cells (TCR-T Cell), and chimeric antigen receptor T cell (CAR-T Cell). A retrospective study (Zhang et al. 2018a, b, c) evaluated the prognosis and factors influencing DC-CIK cell therapy following the use of TACE for HCC. The study group (TACE+DC-CIK cell therapy) showed significantly longer OS than control group (TACE only) in HBV-infected HCC patients. Although solid tumors are less sensitive to CAR-T cell therapy, probably due to the reduced aggregation of therapeutic T cells at tumor sites (Liu et al. 2020), in the application of liver cancer treatment demonstrated promising results (Batra et al. 2020). The efficacy of antiangiogenic drug apatinib in a phase II trial (NCT03046979) in patients with advanced HCC as a first-line therapy resulted in the median OS and PFS of 13.8 months and 8.7 months, respectively. The treatment-related adverse events include proteinuria (39/1%), hypertension (34.8%), and hand-foot skin reactions (34.8%). Immunotherapy based on dendritic cells combined with TAE had shown promising anti-HCC tumor effect (Nakamoto et al. 2011).

3.3 Liver-Directed Therapies in Combination Therapy

Liver-directed therapies, TACE and ablation, can result in liver tumor necrosis, inducing T-cell responses (Ayaru et al. 2007; Mizukoshi et al. 2013; Yang et al. 2019) as well as PD-1/PD-L1 expression in the TME (Noman et al. 2014; Shi et al. 2016). The LDTs, surgical approaches, and sub-lethal incomplete treatments might trigger oncogenesis by metastasis, microenvironmental changes by angiogenesis or several other pathways (Tanis et al. 2014; Rozenblum et al. 2015; Ahmed et al. 2016). Radiofrequency ablation induces both local and systemic immune responses in HCC treatment. Following RFA, an increase in inflammatory cytokines such as IL-1β, IL-6, TNF-α (Ali et al. 2005; Ahmad et al. 2010; Erinjeri et al. 2013; Huang et al. 2019), tumor-specific antibodies CD4+ T cells, CD8+ T cells (Dromi et al. 2009), central memory lymphocytes (CD45RA-/CCR7+) (Mizukoshi et al. 2013; Rochigneux et al. 2019), and in infiltrating CD45RO+ memory T cells (Huang et al. 2019) were observed. The increase in survival of patients is correlated with a higher count of tumor-specific CD8+ T cells (Wissniowski et al. 2003; Dromi et al. 2009). This is due to the thermal and nonthermal effect on the liver and other organs and correlated to more aggressive nature for tumor progression. This is also attributed to local and systemic inflammation vial IL-6, IL-8, HIF-1α, and other factors as described earlier. In recent years, LDTs have played a significant role in shaping the tumor immunity by altering the composition of TME (Biondetti et al. 2021) in HCC. A meta-analysis by Guoming conducted the prognostic impact of tumor-infiltrating CD45RO+ memory T lymphocytes (CD45RO+ T cells) in solid tumors and found that higher the density of intratumoral CD45RO+ T cells inversely correlated with TNM stage. The enhanced infiltration of CD45RO+ T cells is considered as marker of improved clinical outcome and overall survival in all types of solid tumors (Hu and Wang 2017). A decrease in TGF-β, IL-10, and Tregs is associated following RFA treatment (Fietta et al. 2009; Widenmeyer et al. 2011; Huang et al. 2019).

Cryotherapy, which involves cellular destruction of hepatic tumor by freezing at temperatures below −40 °C via Joule-Thompson effect, makes the tumor antigens available to the host immune system. This therapy mediates immunostimulatory response and higher post-ablative levels of serum IL-1, NF-kβ, and TNF-α (Mehta et al. 2016). This therapy also offers both local and potential immunostimulatory effects, based on whether there is more necrosis or apoptosis (Sabel et al. 2010). Microwave ablation (MWA), compared to RFA, delivers higher intratumor temperatures, but exhibits relatively weak immune response (Slovak et al. 2017). In HCC lesions that have undergone ablation, there is increased infiltration of lymphocytes especially CD3+ T cells and CD56+ NK cells but not B cells (Dong et al. 2003). A combination of MWA and immunotherapy indicated an increase in CD8+ T cells after a month follow-up, and reduction in HBV load (Zhou et al. 2011).

TACE and drug-eluting bead TACE (DEB-TACE) induce local cell death and trigger tumor immune response (Greten et al. 2019). After TACE, CD4+/CD8+ ratio, and NK cells increased in HCC patients but on the other hand, CD8+ cells were reduced (Huang et al. 2015). Min Ju Kim et al. reported post-TACE causes changes in various cytokines levels. The early changes in IL-6 correlate acute-phase response and late-phase elevation of Th2 cytokine levels reflect immune suppression (Kim et al. 2013). Yttrium-90 (90Y) microsphere radioembolization delivers a high of dose of radiation to the tumor and known to control the disease effectively (Taylor et al. 2019). After surgical removal of the downstaged tumor by TARE, various signs of immune activation have been described. Accumulating studies reported significant elevation of TNF-α in both CD8+ and CD4+ T cells; APC cells, IL-6 and IL-8, induction of endothelial injury markers (vW factor and PAI-I) (Fernandez-Ros et al. 2015; Seidensticker et al. 2017; Domouchtsidou et al. 2018). However, compared to TACE, stimulation of HIF-1α and VEGF seems negligible with TARE (Zhang et al. 2015). Another important and emerging noninvasive hyperthermic ablative therapy is HIFU, which causes an increase in temperature to 60–85 °C. In HCC, after treatment with HIFU, this is an increase in CD4+ lymphocytes and decrease in immune suppressive cytokines, VEGF, TGF-β1, and TGF-β2 (Wu et al. 2004; Zhou et al. 2008; Ma et al. 2019). Baofeng Ma et al. recently reported a significant elevation of IL-2 and IFN-γ after 3-month follow-up of HIFU treatment (Ma et al. 2019).

Laser ablation is another noninvasive locally ablative technique, currently under use to treat tumor lesions, especially focusing on the target region with laser optical fibers, leading to above 60 °C, causing coagulative necrosis (Di Costanzo et al. 2014). The potential immunological effect of laser ablation is associated with a rise in the number of macrophages and CD8 T cells; expression of HSP70, HSP90, and HSP27, IL-6 and TNF receptor (Ivarsson et al. 2005). Stereotactic body radiation therapy (SBRT) delivers high doses of radiation to the liver tumor, with less risk of damaging normal adjacent tissues. HCC patients upon exposure to SBRT demonstrated to changes in peripheral NK and CD3+, CD56+, NKT cells, and associated with increased OS (Li et al. 2021).

Irreversible electroporation is a percutaneous nonthermal ablation technique that delivers electrical pulses, altering the electrochemical potential of the cell membrane, and causing necrosis and apoptosis (Scheffer et al. 2014). Animal studies reveal the promotion of infiltration of inflammatory cells adjacent to the ablation volume and secrete cytokines, and revert the abnormal Th2 status promoted by HCC. In HCC patients, the macrophage migration inhibitory factor, responsible for maintaining the inflammatory response, was enhanced significantly after IRE, compared to RFA (Sugimoto et al. 2019).

Several recent studies reported the advantage in combination therapy with LDT and immunotherapy to magnify the immunological response (Biondetti et al. 2021) against tumor antigens and intensify even more by addition of MKIs/anti-VEGF agents, to remodel the TME in favor antitumor effect. The combination therapy with ICIs and liver-directed therapies, such as TACE, the well-preserved liver status should be at BCLC B stage. Studies suggested that combining systemic therapies and LRTs with ICIs may represent a future useful strategy.

Combination of ablative procedures with immunotherapy not only induces the release of tumor antigens, but increases the secretion and release of cytokines and even more enhanced if it is followed by an adjuvant immunotherapy (Biondetti et al. 2021). Though the strategy is not for curative purpose, it is aimed to increase the effectiveness of immunotherapy by transformation from "cold" tumor into a "hot" one. Several studies are underway, evaluating the combination of nivolumab vs SIRT, pembrolizumab vs SIRT, nivolumab vs TACE, pembrolizumab vs TACE, TACE vs. TACE with lenvatinib and pembrolizumab (LEAP-012), TACE with durvalumab +/− bevacizumab (EMERALD-1) with trial numbers NCT03380130, NCT03033446, NCT02837029, NCT03099564, NCT03143270, NCT03572582, and NCT03397654, NCT04246117, NCT03778957, respectively. The combination of RFA and cellular therapy in HCC mononuclear cells significantly reduced HCC recurrence compared to RFA alone. The mononuclear cells on harvesting, induced into NK cells, γδT cells, and CIK cells and re-infusion of those induced cells into the patients improved PFS (Cui et al. 2014). This combination therapy may be helpful in preventing the recurrence rate for the patients with HCC after RFA.

4 Conclusion

In conclusion, we attempted to explore and describe the recently published and ongoing clinical studies on the various immune cells responsible for chronic liver disease. With chronic inflammation, and intrahepatic immunosuppressive microenvironment, liver fibrosis is firmly established and contributes the development of HCC. Immunotherapy is the emerging mainstream treatment for advanced HCC, with challenges include low response rate and acquired resistance. Systemic therapy/ immune checkpoint inhibitors and/or LRTs, in combination therapy, can alter and counteract the original tumor immune status and enhance the efficacy of immunotherapy and prolong the OS. The importance of this approach is demonstrated by various ongoing studies, based on this combination, and would be beneficial to improve the quality of life for HCC patients.

References

Abou-Alfa GK, Meyer T, Cheng AL, El-Khoueiry AB, Rimassa L, Ryoo BY, Cicin I, Merle P, Chen Y, Park JW, Blanc JF, Bolondi L, Klümpen HJ, Chan SL, Zagonel V, Pressiani T, Ryu MH, Venook AP, Hessel C, Borgman-Hagey AE, Schwab G, Kelley RK (2018a) Cabozantinib in patients with advanced and progressing hepatocellular carcinoma. N Engl J Med 379:54–63

Abou-Alfa GK, Chan SL, Furuse J, Galle PR, Kelley RK, Qin SK, Armstrong J, Darilay A, Vlahovic G, Negro A et al (2018b) A randomized, multicenter phase 3 study of durvalumab (D) and tremelimumab (T) as first-line treatment in patients with unresectable hepatocellular carcinoma (HCC): HIMALAYA study. J Clin Oncol 36:TPS4144

Ahmad F, Gravante G, Bhardwaj N, Strickland A, Basit R, West K, Sorge R, Dennison AR, Lloyd DM (2010) Changes in interleukin-1β and 6 after hepatic microwave tissue ablation compared with radiofrequency, cryotherapy and surgical resections. Am J Surg 200:500–506

Ahmed M, Kumar G, Moussa M, Wang Y, Rozenblum N, Galun E, Goldberg SN (2016) Hepatic radiofrequency ablation-induced stimulation of distant tumor growth is suppressed by c-Met inhibition. Radiology 279:103–117

Ali MY, Grimm CF, Ritter M, Mohr L, Allgaier HP, Weth R, Bocher WO, Endrulat K, Blum HE, Geissler M (2005) Activation of dendritic cells by local ablation of hepatocellular carcinoma. J Hepatol 43:817–822
Arthur MJ (2000) Fibrogenesis II. Metalloproteinases and their inhibitors in liver fibrosis. Am J Physiol Gastrointest Liver Physiol 279:G245–G249
Ayaru L, Pereira SP, Alisa A, Pathan AA, Williams R, Davidson B, Burroughs AK, Meyer T, Behboudi S (2007) Unmasking of alpha-fetoprotein-specific CD4(+) T cell responses in hepatocellular carcinoma patients undergoing embolization. J Immunol 178:1914–1922
Aydın MM, Akçalı KC (2018) Liver fibrosis. Turk J Gastroenterol 29:14–21
Baglieri J, Brenner DA, Kisseleva T (2019) The role of fibrosis and liver-associated fibroblasts in the pathogenesis of hepatocellular carcinoma. Int J Mol Sci 20:1723
Bai Q, An J, Wu X, You H, Ma H, Liu T, Gao N, Jia J (2012) HBV promotes the proliferation of hepatic stellate cells via the PDGF-B/PDGFR-β signaling pathway in vitro. Int J Mol Med 30: 1443–1450
Barnes MA, McMullen MR, Roychowdhury S, Madhun NZ, Niese K, Olman MA, Stavitsky AB, Bucala R, Nagy LE (2015) Macrophage migration inhibitory factor is required for recruitment of scar-associated macrophages during liver fibrosis. J Leukoc Biol 97:161–169
Bataller R, Brenner DA (2005) Liver fibrosis. J Clin Invest 115:209–218
Batra SA, Rathi P, Guo L, Courtney AN, Fleurence J, Balzeau J, Shaik RS, Nguyen TP, Wu MF, Bulsara S, Mamonkin M, Metelitsa LS, Heczey A (2020) Glypican-3-specific CAR T cells Coexpressing IL15 and IL21 have superior expansion and antitumor activity against hepatocellular carcinoma. Cancer Immunol Res 8:309–320
Biondetti P, Saggiante L, Ierardi AM, Iavarone M, Sangiovanni A, Pesapane F, Fumarola EM, Lampertico P, Carrafiello G (2021) Interventional radiology image-guided locoregional therapies (LRTs) and immunotherapy for the treatment of HCC. Cancers (Basel) 13:5797
Bruix J, Qin S, Merle P, Granito A, Huang YH, Bodoky G, Pracht M, Yokosuka O, Rosmorduc O, Breder V, Gerolami R, Masi G, Ross PJ, Song T, Bronowicki JP, Ollivier-Hourmand I, Kudo M, Cheng AL, Llovet JM, Finn RS, LeBerre MA, Baumhauer A, Meinhardt G, Han G (2017) Regorafenib for patients with hepatocellular carcinoma who progressed on sorafenib treatment (RESORCE): a randomised, double-blind, placebo-controlled, phase 3 trial. Lancet 389:56–66
Castera L, Friedrich-Rust M, Loomba R (2019) Noninvasive assessment of liver disease in patients with nonalcoholic fatty liver disease. Gastroenterology 156:1264–1281.e1264
Chang Y, Li H (2020) Hepatic antifibrotic pharmacotherapy: are we approaching success? J Clin Transl Hepatol 8:222–229
Chapiro J, Duran R, Geschwind JF (2014) Combination of intra-arterial therapies and sorafenib: is there a clinical benefit? Radiol Med 119:476–482
Crocenzi TS, El-Khoueiry AB, Yau TC, Melero I, Sangro B, Kudo M, Hsu C, Trojan J, Kim T-Y, Choo S-P, Meyer T, Kang Y-K, Yeo W, Chopra A, Baakili A, Cruz CMD, Lang L, Neely J, Welling T (2017) Nivolumab (nivo) in sorafenib (sor)-naive and sorafenib-experienced pts with advanced hepatocellular carcinoma (HCC): CheckMate 040 study. J Clin Oncol 35(15_suppl):4013
Cui J, Wang N, Zhao H, Jin H, Wang G, Niu C, Terunuma H, He H, Li W (2014) Combination of radiofrequency ablation and sequential cellular immunotherapy improves progression-free survival for patients with hepatocellular carcinoma. Int J Cancer 134:342–351
D'Ambrosio R, Aghemo A, Rumi MG, Ronchi G, Donato MF, Paradis V, Colombo M, Bedossa P (2012) A morphometric and immunohistochemical study to assess the benefit of a sustained virological response in hepatitis C virus patients with cirrhosis. Hepatology 56:532–543
D'Amico G, Morabito A, D'Amico M, Pasta L, Malizia G, Rebora P, Valsecchi MG (2018) New concepts on the clinical course and stratification of compensated and decompensated cirrhosis. Hepatol Int 12:34–43
Dapito DH, Mencin A, Gwak GY, Pradere JP, Jang MK, Mederacke I, Caviglia JM, Khiabanian H, Adeyemi A, Bataller R, Lefkowitch JH, Bower M, Friedman R, Sartor RB, Rabadan R, Schwabe RF (2012) Promotion of hepatocellular carcinoma by the intestinal microbiota and TLR4. Cancer Cell 21:504–516

Di Costanzo GG, Francica G, Pacella CM (2014) Laser ablation for small hepatocellular carcinoma: state of the art and future perspectives. World J Hepatol 6:704–715

Ding WZ, Han GY, Jin HH, Zhan CF, Ji Y, Huang XL (2018) Anti-IL-20 monoclonal antibody suppresses hepatocellular carcinoma progression. Oncol Lett 16:6156–6162

Domouchtsidou A, Barsegian V, Mueller SP, Best J, Ertle J, Bedreli S, Horn PA, Bockisch A, Lindemann M (2018) Impaired lymphocyte function in patients with hepatic malignancies after selective internal radiotherapy. Cancer Immunol Immunother 67:843–853

Dong BW, Zhang J, Liang P, Yu XL, Su L, Yu DJ, Ji XL, Yu G (2003) Sequential pathological and immunologic analysis of percutaneous microwave coagulation therapy of hepatocellular carcinoma. Int J Hyperth 19:119–133

Dromi SA, Walsh MP, Herby S, Traughber B, Xie J, Sharma KV, Sekhar KP, Luk A, Liewehr DJ, Dreher MR, Fry TJ, Wood BJ (2009) Radiofrequency ablation induces antigen-presenting cell infiltration and amplification of weak tumor-induced immunity. Radiology 251:58–66

Dulai PS, Sirlin CB, Loomba R (2016) MRI and MRE for non-invasive quantitative assessment of hepatic steatosis and fibrosis in NAFLD and NASH: clinical trials to clinical practice. J Hepatol 65:1006–1016

Efimova EA, Glanemann M, Liu L, Schumacher G, Settmacher U, Jonas S, Langrehr JM, Neuhaus P, Nüssler AK (2004) Effects of human hepatocyte growth factor on the proliferation of human hepatocytes and hepatocellular carcinoma cell lines. Eur Surg Res 36:300–307

El-Khoueiry AB, Sangro B, Yau T, Crocenzi TS, Kudo M, Hsu C, Kim TY, Choo SP, Trojan J, Welling THR, Meyer T, Kang YK, Yeo W, Chopra A, Anderson J, Dela Cruz C, Lang L, Neely J, Tang H, Dastani HB, Melero I (2017) Nivolumab in patients with advanced hepatocellular carcinoma (CheckMate 040): an open-label, non-comparative, phase 1/2 dose escalation and expansion trial. Lancet 389:2492–2502

El-Serag HB (2011) Hepatocellular carcinoma. N Engl J Med 365:1118–1127

Erinjeri JP, Thomas CT, Samoilia A, Fleisher M, Gonen M, Sofocleous CT, Thornton RH, Siegelbaum RH, Covey AM, Brody LA, Alago W Jr, Maybody M, Brown KT, Getrajdman GI, Solomon SB (2013) Image-guided thermal ablation of tumors increases the plasma level of interleukin-6 and interleukin-10. J Vasc Interv Radiol 24:1105–1112

Fernandez-Ros N, Iñarrairaegui M, Paramo JA, Berasain C, Avila MA, Chopitea A, Varo N, Sarobe P, Bilbao JI, Dominguez I, D'Avola D, Herrero JI, Quiroga J, Sangro B (2015) Radioembolization of hepatocellular carcinoma activates liver regeneration, induces inflammation and endothelial stress and activates coagulation. Liver Int 35:1590–1596

Fietta AM, Morosini M, Passadore I, Cascina A, Draghi P, Dore R, Rossi S, Pozzi E, Meloni F (2009) Systemic inflammatory response and downmodulation of peripheral CD25+Foxp3+ T-regulatory cells in patients undergoing radiofrequency thermal ablation for lung cancer. Hum Immunol 70:477–486

Filliol A, Schwabe RF (2019) Contributions of fibroblasts, extracellular matrix, stiffness, and mechanosensing to hepatocarcinogenesis. Semin Liver Dis 39:315–333

Finn RS, Qin S, Ikeda M, Galle PR, Ducreux M, Kim TY, Kudo M, Breder V, Merle P, Kaseb AO, Li D, Verret W, Xu DZ, Hernandez S, Liu J, Huang C, Mulla S, Wang Y, Lim HY, Zhu AX, Cheng AL (2020) Atezolizumab plus bevacizumab in unresectable hepatocellular carcinoma. N Engl J Med 382:1894–1905

Gao B, Jeong WI, Tian Z (2008) Liver: an organ with predominant innate immunity. Hepatology 47:729–736

Ghavimi S, Apfel T, Azimi H, Persaud A, Pyrsopoulos NT (2020) Management and treatment of hepatocellular carcinoma with immunotherapy: a review of current and future options. J Clin Transl Hepatol 8:168–176

Greten TF, Mauda-Havakuk M, Heinrich B, Korangy F, Wood BJ (2019) Combined locoregional-immunotherapy for liver cancer. J Hepatol 70:999–1007

Guillot A, Tacke F (2019) Liver macrophages: old dogmas and new insights. Hepatol Commun 3: 730–743

Hautekeete ML, Geerts A (1997) The hepatic stellate (Ito) cell: its role in human liver disease. Virchows Arch 430:195–207

Hernández-Gea V, Friedman SL (2012) Autophagy fuels tissue fibrogenesis. Autophagy 8:849–850

Hu G, Wang S (2017) Tumor-infiltrating CD45RO(+) memory T lymphocytes predict favorable clinical outcome in solid tumors. Sci Rep 7:10376

Huang M, Wang X, Bin H (2015) Effect of transcatheter arterial chemoembolization combined with argon-helium cryosurgery system on the changes of NK cells and T cell subsets in peripheral blood of hepatocellular carcinoma patients. Cell Biochem Biophys 73:787–792

Huang KW, Jayant K, Lee PH, Yang PC, Hsiao CY, Habib N, Sodergren MH (2019) Positive immuno-modulation following radiofrequency assisted liver resection in hepatocellular carcinoma. J Clin Med 8:385

Ivarsson K, Myllymäki L, Jansner K, Stenram U, Tranberg KG (2005) Resistance to tumour challenge after tumour laser thermotherapy is associated with a cellular immune response. Br J Cancer 93:435–440

Ji J, Zhao L, Budhu A, Forgues M, Jia HL, Qin LX, Ye QH, Yu J, Shi X, Tang ZY, Wang XW (2010) Let-7g targets collagen type I alpha2 and inhibits cell migration in hepatocellular carcinoma. J Hepatol 52:690–697

Jin PP, Shao SY, Wu WT, Zhao XY, Huang BF, Fu QH, Que RS, Hu QD (2018) Combination of transarterial chemoembolization and sorafenib improves outcomes of unresectable hepatocellular carcinoma: an updated systematic review and meta-analysis. Jpn J Clin Oncol 48:1058–1069

Kanda T, Goto T, Hirotsu Y, Moriyama M, Omata M (2019) Molecular mechanisms driving progression of liver cirrhosis towards hepatocellular carcinoma in chronic hepatitis B and C infections: a review. Int J Mol Sci 20:1358

Kato Y, Bao X, Macgrath S, Tabata K, Hori Y, Tachino S et al (2016) Lenvatinib Mesylate (LEN) enhanced antitumor activity of a PD-1 blockade agent by potentiating Th1 immune response. Ann Oncol 27(suppl_6):1–14. https://doi.org/10.1093/annonc/mdw362.02

Ke MY, Xu T, Fang Y, Ye YP, Li ZJ, Ren FG, Lu SY, Zhang XF, Wu RQ, Lv Y, Dong J (2021) Liver fibrosis promotes immune escape in hepatocellular carcinoma via GOLM1-mediated PD-L1 upregulation. Cancer Lett 513:14–25

Kim MJ, Jang JW, Oh BS, Kwon JH, Chung KW, Jung HS, Jekarl DW, Lee S (2013) Change in inflammatory cytokine profiles after transarterial chemotherapy in patients with hepatocellular carcinoma. Cytokine 64:516–522

Kuang P, Zhao W, Su W, Zhang Z, Zhang L, Liu J, Ren G, Yin Z, Wang X (2013) 18-β-glycyrrhetinic acid inhibits hepatocellular carcinoma development by reversing hepatic stellate cell-mediated immunosuppression in mice. Int J Cancer 132:1831–1841

Kudo M, Ueshima K, Ikeda M, Torimura T, Tanabe N, Aikata H, Izumi N, Yamasaki T, Nojiri S, Hino K, Tsumura H, Kuzuya T, Isoda N, Yasui K, Aino H, Ido A, Kawabe N, Nakao K, Wada Y, Yokosuka O, Yoshimura K, Okusaka T, Furuse J, Kokudo N, Okita K, Johnson PJ, Arai Y (2020) Randomised, multicentre prospective trial of transarterial chemoembolisation (TACE) plus sorafenib as compared with TACE alone in patients with hepatocellular carcinoma: TACTICS trial. Gut 69:1492–1501

Kudo M, Motomura K, Wada Y, Inaba Y, Sakamoto Y, Kurosaki M, Umeyama Y, Kamei Y, Yoshimitsu J, Fujii Y, Aizawa M, Robbins PB, Furuse J (2021) Avelumab in combination with axitinib as first-line treatment in patients with advanced hepatocellular carcinoma: results from the phase 1b VEGF liver 100 trial. Liver Cancer 10:249–259

Lawal G, Xiao Y, Rahnemai-Azar AA, Tsilimigras DI, Kuang M, Bakopoulos A, Pawlik TM (2021) The immunology of hepatocellular carcinoma. Vaccines (Basel) 9:1184

Lee UE, Friedman SL (2011) Mechanisms of hepatic fibrogenesis. Best Pract Res Clin Gastroenterol 25:195–206

Lee DW, Cho EJ, Lee JH, Yu SJ, Kim YJ, Yoon JH, Kim TY, Han SW, Oh DY, Im SA, Kim TY, Lee Y, Kim H, Lee KH (2021) Phase II study of avelumab in patients with advanced hepatocellular carcinoma previously treated with sorafenib. Clin Cancer Res 27:713–718

Li L, Zhao W, Wang M, Hu J, Wang E, Zhao Y, Liu L (2018) Transarterial chemoembolization plus sorafenib for the management of unresectable hepatocellular carcinoma: a systematic review and meta-analysis. BMC Gastroenterol 18:138

Li TT, Sun J, Wang Q, Li WG, He WP, Yang RC, Duan XZ (2021) The effects of stereotactic body radiotherapy on peripheral natural killer and CD3(+)CD56(+) NKT-like cells in patients with hepatocellular carcinoma. Hepatobiliary Pancreat Dis Int 20:240–250

Liu Y, Lui EL, Friedman SL, Li L, Ye T, Chen Y, Poon RT, Wo J, Kok TW, Fan ST (2009) PTK787/ZK22258 attenuates stellate cell activation and hepatic fibrosis in vivo by inhibiting VEGF signaling. Lab Investig 89:209–221

Liu F, Zhang W, Yang F, Feng T, Zhou M, Yu Y, Yu X, Zhao W, Yi F, Tang W, Lu Y (2016) Interleukin-6-stimulated progranulin expression contributes to the malignancy of hepatocellular carcinoma cells by activating mTOR signaling. Sci Rep 6:21260

Liu G, Rui W, Zheng H, Huang D, Yu F, Zhang Y, Dong J, Zhao X, Lin X (2020) CXCR2-modified CAR-T cells have enhanced trafficking ability that improves treatment of hepatocellular carcinoma. Eur J Immunol 50:712–724

Llovet JM, Ricci S, Mazzaferro V, Hilgard P, Gane E, Blanc JF, de Oliveira AC, Santoro A, Raoul JL, Forner A, Schwartz M, Porta C, Zeuzem S, Bolondi L, Greten TF, Galle PR, Seitz JF, Borbath I, Häussinger D, Giannaris T, Shan M, Moscovici M, Voliotis D, Bruix J (2008) Sorafenib in advanced hepatocellular carcinoma. N Engl J Med 359:378–390

Ma B, Liu X, Yu Z (2019) The effect of high intensity focused ultrasound on the treatment of liver cancer and patients' immunity. Cancer Biomark 24:85–90

Madtes DK, Elston AL, Kaback LA, Clark JG (2001) Selective induction of tissue inhibitor of metalloproteinase-1 in bleomycin-induced pulmonary fibrosis. Am J Respir Cell Mol Biol 24: 599–607

Mähringer-Kunz A, Meyer FI, Hahn F, Müller L, Düber C, Pinto Dos Santos D, Galle PR, Weinmann A, Kloeckner R, Schotten S (2021) Hepatic vein tumor thrombosis in patients with hepatocellular carcinoma: prevalence and clinical significance. United European Gastroenterol J 9:590–597

Makary MS, Ramsell S, Miller E, Beal EW, Dowell JD (2021) Hepatocellular carcinoma locoregional therapies: outcomes and future horizons. World J Gastroenterol 27:7462–7479

Mazagova M, Wang L, Anfora AT, Wissmueller M, Lesley SA, Miyamoto Y, Eckmann L, Dhungana S, Pathmasiri W, Sumner S, Westwater C, Brenner DA, Schnabl B (2015) Commensal microbiota is hepatoprotective and prevents liver fibrosis in mice. FASEB J 29:1043–1055

Mederacke I, Hsu CC, Troeger JS, Huebener P, Mu X, Dapito DH, Pradere JP, Schwabe RF (2013) Fate tracing reveals hepatic stellate cells as dominant contributors to liver fibrosis independent of its aetiology. Nat Commun 4:2823

Mehta A, Oklu R, Sheth RA (2016) Thermal ablative therapies and immune checkpoint modulation: can Locoregional approaches effect a systemic response? Gastroenterol Res Pract 2016: 9251375

Mizukoshi E, Yamashita T, Arai K, Sunagozaka H, Ueda T, Arihara F, Kagaya T, Yamashita T, Fushimi K, Kaneko S (2013) Enhancement of tumor-associated antigen-specific T cell responses by radiofrequency ablation of hepatocellular carcinoma. Hepatology 57:1448–1457

Mohammed NA, Abd El-Aleem S, Appleton I, Maklouf MM, Said M, McMahon RF (2003) Expression of nitric oxide synthase isoforms in human liver cirrhosis. J Pathol 200:647–655

Monvoisin A, Neaud V, De Lédinghen V, Dubuisson L, Balabaud C, Bioulac-Sage P, Desmoulière A, Rosenbaum J (1999) Direct evidence that hepatocyte growth factor-induced invasion of hepatocellular carcinoma cells is mediated by urokinase. J Hepatol 30:511–518

Murtaza A, Laken H, Correia JDS, McNeeley P, Altobell L, Zhang J, Vancutsem P, Wilcoxen K, Jenkins D (2016) Discovery of TSR-022, a novel, potent anti-human TIM-3 therapeutic antibody. Eur J Cancer 69:S102

Nakamoto Y, Mizukoshi E, Kitahara M, Arihara F, Sakai Y, Kakinoki K, Fujita Y, Marukawa Y, Arai K, Yamashita T, Mukaida N, Matsushima K, Matsui O, Kaneko S (2011) Prolonged recurrence-free survival following OK432-stimulated dendritic cell transfer into hepatocellular carcinoma during transarterial embolization. Clin Exp Immunol 163:165–177

Neureiter D, Stintzing S, Kiesslich T, Ocker M (2019) Hepatocellular carcinoma: therapeutic advances in signaling, epigenetic and immune targets. World J Gastroenterol 25:3136–3150
Nishida N, Kudo M (2018) Immune checkpoint blockade for the treatment of human hepatocellular carcinoma. Hepatol Res 48:622–634
Nishio T, Koyama Y, Liu X, Rosenthal SB, Yamamoto G, Fuji H, Baglieri J, Li N, Brenner LN, Iwaisako K, Taura K, Hagood JS, LaRusso NF, Bera TK, Pastan I, Brenner DA, Kisseleva T (2021) Immunotherapy-based targeting of MSLN(+) activated portal fibroblasts is a strategy for treatment of cholestatic liver fibrosis. Proc Natl Acad Sci U S A 118:e2101270118
Noman MZ, Desantis G, Janji B, Hasmim M, Karray S, Dessen P, Bronte V, Chouaib S (2014) PD-L1 is a novel direct target of HIF-1α, and its blockade under hypoxia enhanced MDSC-mediated T cell activation. J Exp Med 211:781–790
O'Rourke JM, Sagar VM, Shah T, Shetty S (2018) Carcinogenesis on the background of liver fibrosis: implications for the management of hepatocellular cancer. World J Gastroenterol 24: 4436–4447
Okrah K, Tarighat S, Liu B, Koeppen H, Wagle MC, Cheng G, Sun C, Dey A, Chang MT, Sumiyoshi T, Mounir Z, Cummings C, Hampton G, Amler L, Fridlyand J, Hegde PS, Turley SJ, Lackner MR, Huang SM (2018) Transcriptomic analysis of hepatocellular carcinoma reveals molecular features of disease progression and tumor immune biology. NPJ Precis Oncol 2:25
O'Shanick GJ (1986) Neuropsychiatric complications in head injury. Adv Psychosom Med 16: 173–193
Oura K, Morishita A, Tani J, Masaki T (2021) Tumor immune microenvironment and immunosuppressive therapy in hepatocellular carcinoma: a review. Int J Mol Sci 22:5801
Ozer M, George A, Goksu SY, George TJ, Sahin I (2021) The role of immune checkpoint blockade in the hepatocellular carcinoma: a review of clinical trials. Front Oncol 11:801379
Papatheodoridis GV, Idilman R, Dalekos GN, Buti M, Chi H, van Boemmel F, Calleja JL, Sypsa V, Goulis J, Manolakopoulos S, Loglio A, Siakavellas S, Keskın O, Gatselis N, Hansen BE, Lehretz M, de la Revilla J, Savvidou S, Kourikou A, Vlachogiannakos I, Galanis K, Yurdaydin C, Berg T, Colombo M, Esteban R, Janssen HLA, Lampertico P (2017) The risk of hepatocellular carcinoma decreases after the first 5 years of entecavir or tenofovir in Caucasians with chronic hepatitis B. Hepatology 66:1444–1453
Parikh JG, Kulkarni A, Johns C (2014) α-smooth muscle actin-positive fibroblasts correlate with poor survival in hepatocellular carcinoma. Oncol Lett 7:573–575
Poelstra K, Schuppan D (2011) Targeted therapy of liver fibrosis/cirrhosis and its complications. J Hepatol 55:726–728
Qin S, Finn RS, Kudo M, Meyer T, Vogel A, Ducreux M, Macarulla TM, Tomasello G, Boisserie F, Hou J, Li X, Song J, Zhu AX (2019) RATIONALE 301 study: tislelizumab versus sorafenib as first-line treatment for unresectable hepatocellular carcinoma. Future Oncol 15:1811–1822
Robert S, Gicquel T, Victoni T, Valença S, Barreto E, Bailly-Maître B, Boichot E, Lagente V (2016) Involvement of matrix metalloproteinases (MMPs) and inflammasome pathway in molecular mechanisms of fibrosis. Biosci Rep 36:e00360
Rochigneux P, Nault JC, Mallet F, Chretien AS, Barget N, Garcia AJ, Del Pozo L, Bourcier V, Blaise L, Grando-Lemaire V, N'Kontchou G, Nahon P, Seror O, Ziol M, Ganne-Carrié N, Olive D (2019) Dynamic of systemic immunity and its impact on tumor recurrence after radiofrequency ablation of hepatocellular carcinoma. Onco Targets Ther 8:1615818
Rozenblum N, Zeira E, Scaiewicz V, Bulvik B, Gourevitch S, Yotvat H, Galun E, Goldberg SN (2015) Oncogenesis: an "off-target" effect of radiofrequency ablation. Radiology 276:426–432
Sabel MS, Su G, Griffith KA, Chang AE (2010) Rate of freeze alters the immunologic response after cryoablation of breast cancer. Ann Surg Oncol 17:1187–1193
Sakurai T, Kudo M, Umemura A, He G, Elsharkawy AM, Seki E, Karin M (2013) p38α inhibits liver fibrogenesis and consequent hepatocarcinogenesis by curtailing accumulation of reactive oxygen species. Cancer Res 73:215–224
Sangro B, Gomez-Martin C, de la Mata M, Iñarrairaegui M, Garralda E, Barrera P, Riezu-Boj JI, Larrea E, Alfaro C, Sarobe P, Lasarte JJ, Pérez-Gracia JL, Melero I, Prieto J (2013) A clinical

trial of CTLA-4 blockade with tremelimumab in patients with hepatocellular carcinoma and chronic hepatitis C. J Hepatol 59:81–88
Scheffer HJ, Nielsen K, de Jong MC, van Tilborg AA, Vieveen JM, Bouwman AR, Meijer S, van Kuijk C, van den Tol PM, Meijerink MR (2014) Irreversible electroporation for nonthermal tumor ablation in the clinical setting: a systematic review of safety and efficacy. J Vasc Interv Radiol 25:997–1011; quiz 1011
Schuppan D, Kim YO (2013) Evolving therapies for liver fibrosis. J Clin Invest 123:1887–1901
Schuppan D, Ruehl M, Somasundaram R, Hahn EG (2001) Matrix as a modulator of hepatic fibrogenesis. Semin Liver Dis 21:351–372
Seidensticker M, Powerski M, Seidensticker R, Damm R, Mohnike K, Garlipp B, Klopffleisch M, Amthauer H, Ricke J, Pech M (2017) Cytokines and (90)Y-Radioembolization: relation to liver function and overall survival. Cardiovasc Intervent Radiol 40:1185–1195
Shi L, Chen L, Wu C, Zhu Y, Xu B, Zheng X, Sun M, Wen W, Dai X, Yang M, Lv Q, Lu B, Jiang J (2016) PD-1 blockade boosts radiofrequency ablation-elicited adaptive immune responses against tumor. Clin Cancer Res 22:1173–1184
Slovak R, Ludwig JM, Gettinger SN, Herbst RS, Kim HS (2017) Immuno-thermal ablations – boosting the anticancer immune response. J Immunother Cancer 5:78
Song T, Dou C, Jia Y, Tu K, Zheng X (2015) TIMP-1 activated carcinoma-associated fibroblasts inhibit tumor apoptosis by activating SDF1/CXCR4 signaling in hepatocellular carcinoma. Oncotarget 6:12061–12079
Sugimoto K, Kakimi K, Takeuchi H, Fujieda N, Saito K, Sato E, Sakamaki K, Moriyasu F, Itoi T (2019) Irreversible electroporation versus radiofrequency ablation: comparison of systemic immune responses in patients with hepatocellular carcinoma. J Vasc Interv Radiol 30:845–853.e846
Sun M, Kisseleva T (2015) Reversibility of liver fibrosis. Clin Res Hepatol Gastroenterol 39(Suppl 1):S60–S63
Tacke F, Luedde T, Trautwein C (2009) Inflammatory pathways in liver homeostasis and liver injury. Clin Rev Allergy Immunol 36:4–12
Tahmasebi Birgani M, Carloni V (2017) Tumor microenvironment, a paradigm in hepatocellular carcinoma progression and therapy. Int J Mol Sci 18:405
Tang X, Shu Z, Zhang W, Cheng L, Yu J, Zhang M, Zheng S (2019) Clinical significance of the immune cell landscape in hepatocellular carcinoma patients with different degrees of fibrosis. Ann Transl Med 7:528
Tanis E, Nordlinger B, Mauer M, Sorbye H, van Coevorden F, Gruenberger T, Schlag PM, Punt CJ, Ledermann J, Ruers TJ (2014) Local recurrence rates after radiofrequency ablation or resection of colorectal liver metastases. Analysis of the European Organisation for Research and Treatment of Cancer #40004 and #40983. Eur J Cancer 50:912–919
Tao Y, Wang N, Qiu T, Sun X (2020) The role of autophagy and NLRP3 inflammasome in liver fibrosis. Biomed Res Int 2020:7269150
Taylor AC, Maddirela D, White SB (2019) Role of radioembolization for biliary tract and primary liver cancer. Surg Oncol Clin N Am 28:731–743
Török NJ (2008) Recent advances in the pathogenesis and diagnosis of liver fibrosis. J Gastroenterol 43:315–321
van Zijl F, Mair M, Csiszar A, Schneller D, Zulehner G, Huber H, Eferl R, Beug H, Dolznig H, Mikulits W (2009) Hepatic tumor-stroma crosstalk guides epithelial to mesenchymal transition at the tumor edge. Oncogene 28:4022–4033
Villanueva A (2019) Hepatocellular carcinoma. N Engl J Med 380:1450–1462
Wei Y, Liu J, Yan M, Zhao S, Long Y, Zhang W (2019) Effectiveness and safety of combination therapy of transarterial chemoembolization and apatinib for unresectable hepatocellular carcinoma in the Chinese population: a meta-analysis. Chemotherapy 64:94–104
Widenmeyer M, Shebzukhov Y, Haen SP, Schmidt D, Clasen S, Boss A, Kuprash DV, Nedospasov SA, Stenzl A, Aebert H, Wernet D, Stevanović S, Pereira PL, Rammensee HG, Gouttefangeas C

(2011) Analysis of tumor antigen-specific T cells and antibodies in cancer patients treated with radiofrequency ablation. Int J Cancer 128:2653–2662

Wissniowski TT, Hänsler J, Neureiter D, Frieser M, Schaber S, Esslinger B, Voll R, Strobel D, Hahn EG, Schuppan D (2003) Activation of tumor-specific T lymphocytes by radio-frequency ablation of the VX2 hepatoma in rabbits. Cancer Res 63:6496–6500

Wong L, Yamasaki G, Johnson RJ, Friedman SL (1994) Induction of beta-platelet-derived growth factor receptor in rat hepatic lipocytes during cellular activation in vivo and in culture. J Clin Invest 94:1563–1569

Wright JH, Johnson MM, Shimizu-Albergine M, Bauer RL, Hayes BJ, Surapisitchat J, Hudkins KL, Riehle KJ, Johnson SC, Yeh MM, Bammler TK, Beyer RP, Gilbertson DG, Alpers CE, Fausto N, Campbell JS (2014) Paracrine activation of hepatic stellate cells in platelet-derived growth factor C transgenic mice: evidence for stromal induction of hepatocellular carcinoma. Int J Cancer 134:778–788

Wu F, Wang ZB, Lu P, Xu ZL, Chen WZ, Zhu H, Jin CB (2004) Activated anti-tumor immunity in cancer patients after high intensity focused ultrasound ablation. Ultrasound Med Biol 30:1217–1222

Wynn TA (2004) Fibrotic disease and the T(H)1/T(H)2 paradigm. Nat Rev Immunol 4:583–594

Yang MC, Wang CJ, Liao PC, Yen CJ, Shan YS (2014) Hepatic stellate cells secretes type I collagen to trigger epithelial mesenchymal transition of hepatoma cells. Am J Cancer Res 4: 751–763

Yang H, Shen K, Zhu C, Li Q, Zhao Y, Ma X (2018) Safety and efficacy of durvalumab (MEDI4736) in various solid tumors. Drug Des Devel Ther 12:2085–2096

Yang X, Guo Y, Guo Z, Si T, Xing W, Yu W, Wang Y (2019) Cryoablation inhibition of distant untreated tumors (abscopal effect) is immune mediated. Oncotarget 10:4180–4191

Zhang Y, Li Y, Ji H, Zhao X, Lu H (2015) Transarterial Y90 radioembolization versus chemoembolization for patients with hepatocellular carcinoma: a meta-analysis. Biosci Trends 9:289–298

Zhang CY, Yuan WG, He P, Lei JH, Wang CX (2016) Liver fibrosis and hepatic stellate cells: etiology, pathological hallmarks and therapeutic targets. World J Gastroenterol 22:10512–10522

Zhang G, Li R, Deng Y, Zhao L (2018a) Conditional survival of patients with hepatocellular carcinoma: results from the surveillance, epidemiology, and end results registry. Expert Rev Gastroenterol Hepatol 12:515–523

Zhang J, Li H, Gao D, Zhang B, Zheng M, Lun M, Wei M, Duan R, Guo M, Hua J, Liu Q, Bai J, Liu H, Zheng J, Yao H (2018b) A prognosis and impact factor analysis of DC-CIK cell therapy for patients with hepatocellular carcinoma undergoing postoperative TACE. Cancer Biol Ther 19:475–483

Zhang R, Ma M, Lin XH, Liu HH, Chen J, Chen J, Gao DM, Cui JF, Ren ZG, Chen RX (2018c) Extracellular matrix collagen I promotes the tumor progression of residual hepatocellular carcinoma after heat treatment. BMC Cancer 18:901

Zhang L, Ding J, Li HY, Wang ZH, Wu J (2020) Immunotherapy for advanced hepatocellular carcinoma, where are we? Biochim Biophys Acta Rev Cancer 1874:188441

Zhao M, Laissue JA, Zimmermann A (1996) Tenascin and type IV collagen expression in liver cell dysplasia and in hepatocellular carcinoma. Histol Histopathol 11:323–333

Zhou Q, Zhu XQ, Zhang J, Xu ZL, Lu P, Wu F (2008) Changes in circulating immunosuppressive cytokine levels of cancer patients after high intensity focused ultrasound treatment. Ultrasound Med Biol 34:81–87

Zhou P, Liang P, Dong B, Yu X, Han Z, Xu Y (2011) Phase I clinical study of combination therapy with microwave ablation and cellular immunotherapy in hepatocellular carcinoma. Cancer Biol Ther 11:450–456

Zhou WC, Zhang QB, Qiao L (2014) Pathogenesis of liver cirrhosis. World J Gastroenterol 20: 7312–7324

Zhu Y, Paniccia A, Schulick AC, Chen W, Koenig MR, Byers JT, Yao S, Bevers S, Edil BH (2016) Identification of CD112R as a novel checkpoint for human T cells. J Exp Med 213:167–176

Zhu AX, Finn RS, Edeline J, Cattan S, Ogasawara S, Palmer D, Verslype C, Zagonel V, Fartoux L, Vogel A, Sarker D, Verset G, Chan SL, Knox J, Daniele B, Webber AL, Ebbinghaus SW, Ma J, Siegel AB, Cheng AL, Kudo M (2018) Pembrolizumab in patients with advanced hepatocellular carcinoma previously treated with sorafenib (KEYNOTE-224): a non-randomised, open-label phase 2 trial. Lancet Oncol 19:940–952

Liver Cancer and the Curative Potential of Nanomedicine

Adrian Kuzmanović, Cheng Lin, and Matthias Bartneck

Abstract

Hepatocellular carcinoma (HCC) is one of the most common cancers in the world and ranks third in the world cancer mortality rate. Due to the latent growth of HCC, most patients are diagnosed at the late stage of the disease where treatment options are scarce. HCC is initiated through mutations caused by carcinogen exposure, smoking, alcohol abuse, and viral infections. Genetic and environmental factors such as nutrition contribute to the risk of developing HCC. Inflammation plays a key role for driving HCC progression, what is owed to the fact that exaggerated inflammation can trigger cell death of the sensitive hepatocytes, the most important parenchymal cells of the liver. Despite the high potency for regeneration, hepatocytes are frequently replaced by scar tissue. This can lead to hepatic fibrosis as well as cirrhosis, and ultimately to cancer. One key problem of HCC therapy is that the diagnosis occurs frequently at a late stage, where treatment options are limited. Immunotherapy based on targeting checkpoint inhibitors, which unleashes the break of antitumor immunity, works only for a minority of patients. This might partly be owed to the tolerogenic function of the liver. Nanotechnology opens novel perspectives for liver tumor treatment, what is owed to the hepatic clearance of nanoparticles by the liver. Molecular heterogeneity and accompanying liver dysfunction make the transformation of nanomedicines for therapy of liver cancer complicated. In this article, we

A. Kuzmanović · M. Bartneck (✉)
Department of Internal Medicine III, University Hospital RWTH Aachen, Aachen, Germany
e-mail: mbartneck@ukaachen.de

C. Lin
Department of Internal Medicine III, University Hospital RWTH Aachen, Aachen, Germany

Department of Rheumatology and Shanghai Institute of Rheumatology, Renji Hospital, School of Medicine, Shanghai Jiao Tong University, Shanghai, People's Republic of China

Interdisciplinary Cancer Research, https://doi.org/10.1007/16833_2023_129
Published online: 14 February 2023

summarize the current treatment options as well as challenges and future prospects of nanomedicine for treatment of liver cancer.

Keywords

Antigen · Checkpoint inhibitors · Cirrhosis · Clinical trials · Co-delivery · Cytokines · Gene therapy · HCC · Hepatic clearance · Hepatic fibrosis · Hepatocellular carcinoma · Hepatocytes · Immune cells · Immunotherapy · Inflammation · Ionizable lipids · Lipid nanoparticles · Liposomes · Liver cancer · LNP · Macrophages · mRNA vaccines · Nanomedicine · Nanotechnology · RNA silencing · siRNA · Therapeutics · Tumor · Vaccine

1 Hepatocellular Carcinoma

1.1 Prevalence and Initiation of HCC

Clinicians distinguish different types of primary liver cancer, several of them very common, such as hepatocellular carcinoma (HCC), intra- and extrahepatic cholangiocarcinoma, fibrolamellar carcinoma, hepatoblastomas, angiosarcoma of the liver, and epitheloid hemangioendothelioma. HCC is the most prevalent type of primary liver cancer, accounting for 85–90% of all primary liver cancers. It is one of the deadliest types of cancer and represents a massive health burden to the world (Bray et al. 2018). Within the male population, it is the fifth most common type of cancer, while ranking seventh in women. About 600,000 deaths globally are attributed to HCC, with a 5-year survival rate being only 6.9% and the median survival time is only 11 months – one of the probable causes is that only a minority of patients are diagnosed at early stage. Geographical distribution of HCC prevalence also varies, with most cases occurring in Asia and Africa, particularly within middle- and low-income countries. The highest incidence of HCC can be observed in Asian countries – 99 per 100,000 people in Mongolia, and 29 per 100,000 persons in Japan. In China, the incidence is 35 per 100,000 people, totals to 395,000 cases annually, i.e., half of the world's total number of cases (Ozakyol 2017).

In addition, liver cancer can be caused by metastases that originate from primary tumors of other organs. The most common type of secondary liver cancer is derived from disseminating breast, colorectal, and lung cancer cells (Ananthakrishnan et al. 2006).

Generally, mutations in the genome of HC lead to HCC. These can be caused by several processes. In particular, persistent viral infections attribute to HCC, with 50–55% of cases attributable to hepatitis B virus (HBV), and 25–30% to hepatitis C virus (HCV) (Chan and Sung 2006). Excessive consumption of alcohol, especially in combination with HBV or HCV, is also considered an important risk factor for HCC development (Chakraborty and Sarkar 2022). Alcohol metabolization in the liver produces metabolites highly toxic to hepatocytes, such as acetaldehyde, which leads to the formation of adducts, inactivating the glutathione peroxidase system and resulting in mitochondrial damage. Further effects lead to DNA synthesis and repair defects, generation of reactive oxygen species, and the induction of inflammatory signaling in the liver (Setshedi et al. 2010; Tunissiolli et al. 2017). Also, diabetes significantly attributes to HCC progression, with a 2.5-fold higher risk for HCC development, independent of viral hepatitis and alcohol abuse (Zhou et al. 2014).

Similar to viruses and alcohol abuse, exposure and ingestion of fungal aflatoxins, particularly of aflatoxin B1, were shown to result in mutations of proto-oncogene genes and tumor suppressor genes such as p53 (Tunissiolli et al. 2017). In 2011, a study revealed food, such as plant oil, corn, rice and wheat flour from three different areas in China were contaminated with aflatoxin (Sun et al. 2011). Aflatoxin is problematic because it has high heat stability, and therefore, cooking and most industrial processes cannot inactivate or destroy it (Mahato et al. 2019).

1.2 Treatment of HCC

The choice of an appropriate treatment for HCC, as presented in Fig. 1a, requires the assessment of multiple factors like patient's tumor stage, and clinical manifestations. For early-stage HCC patients, hepatectomy, liver transplantation, ablation, interventional therapy, radiotherapy, and chemotherapy provide multiple treatment options (Lau et al. 2001). The presence of different immune cell populations in liver tissue also presents a factor that needs to be considered when designing an appropriate strategy (Fig. 1b), as reviewed before (Heymann and Tacke 2016). Unfortunately, for advanced-stage HCC patients, systemic therapies and transarterial radioembolization (TACE) – an adjuvant treatment for HCC patients after surgery are the only available methods (Chen et al. 2020; Kudo 2018; Raoul et al. 2019). Surgical resection is still the most effective way but only 15% patients have the opportunity to perform surgery, the rest have to switch to other treatment methods like chemotherapy (Leung et al. 2003; Zhou et al. 2009; Kudo 2018).

The very low early-stage discovery rate leads to the fact that the main therapeutic strategies primarily rely on surgical interventions – liver resection and transplantation. Other methods for nonresectable tumors include radiotherapy for small tumors, and chemotherapy. The latter one is limited by off-target toxicity and chemoresistance (Younis et al. 2019). Throughout the last two decades, immune- and gene-based therapies have emerged, with a gain in popularity in recent years. Encouragingly, nanotechnology, immune checkpoint blockade immunotherapy, adoptive cell therapy (ACT), and cancer vaccines bring new hope for patients (Khalil et al. 2016; Ribas and Wolchok 2018; El-Mayta et al. 2021; Li et al. 2021).

In recent years, the drug treatment of liver cancer has made more progress. Targeted drugs and immunotherapy have the leading role in the treatment of HCC and intrahepatic cholangiocarcinoma (ICC). Many clinical studies have obtained positive results and entered the clinical frontline application. In the process of targeted therapy for liver cancer, sorafenib was the first drug approved for clinical application to advanced-stage patients in 2007. In the following 10 years, it was widely recognized as a first-line therapy of HCC (Wilhelm et al. 2004; Chang et al.

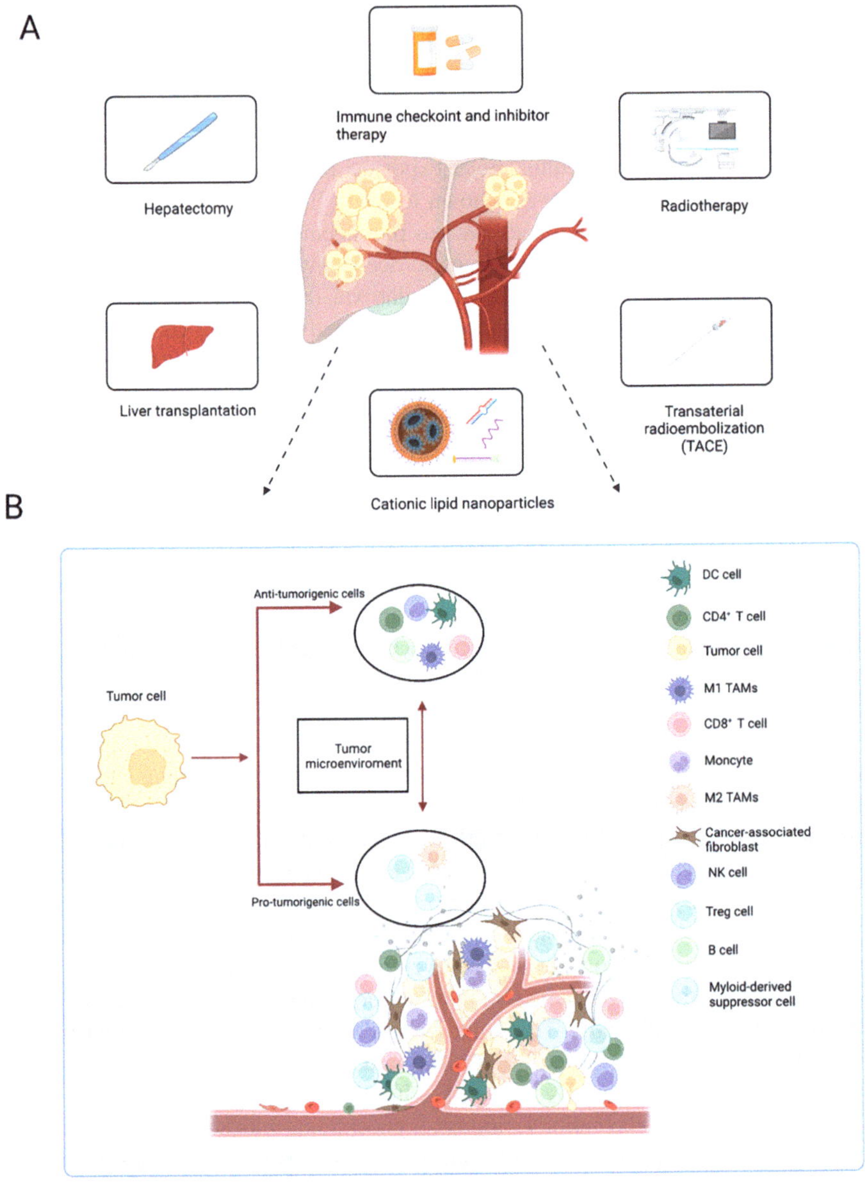

Fig. 1 Treatment options and cellular microenvironment of hepatocellular carcinoma. (**a**) Current treatment options for hepatocellular carcinoma include surgical excision, chemotherapy, radiotherapy, liver transplantation, transarterial radioembolization (TACE), small molecules, and nanomedicines. During all the methods, hepatectomy is still the most efficient treatment. (**b**) Infiltration of different immune cells is one of the reasons for the complex tumor microenvironment, Macrophages play an important role in tumor microenvironment and can be polarized by diverse stimuli into many different specialized subpopulations (M1 and M2). Outbalances between M1 and M2 can lead to different situations. (Created with BioRender.com)

2007; Llovet et al. 2008; Marisi et al. 2018). After decades, more targeted drugs have emerged, lenvatini, regorafenib, carbozantinib also received the FDA approval but only as second-line treatment post-sorafenib (Lee et al. 2019; Finn et al. 2020a, b).

To further improve the efficacy in advanced-stage HCC, combination treatments show significant progress. For example, compared to sorafenib single-use treatment, advanced HCC patients with preserved liver function are suggested to receive atezolizumab and bevacizumab over sorafenib, which showed superiority for the combination with an immune checkpoint inhibitor (Finn et al. 2020a, b). However, the limits of chemotherapy caused a lot of side effects. Due to the lack of specific tumor targeting function of chemotherapy drugs, the biodistribution in the body is uncontrolled. In addition to killing tumor cells, they can also cause serious damage to normal tissues. Besides, the insufficient amount of drugs and frequent emergence of multidrug resistance (MDR) also led to the poor effect or failure of treatment (Song et al. 2010; Hu and Zhang 2012).

1.2.1 Immunotherapy for Treatment of Liver Cancer

Immunotherapy aims at mobilizing the patient's own immune system to eliminate tumor cells (Liu et al. 2022). Several clinical trials have in recent years used ICIs on patients with advanced tumors. Programmed cell death 1 protein (PD1) has been described as a breaking system for immune responses – inhibition of the PD1 binding to its ligand causes an enhanced immune activation and antitumor response (Kudo 2017). Phase I/II clinical trial, CheckMate 040 has used Nivolumab, an anti-programmed cell death protein (anti-PD1) antibody, with a disease control rate of 64% in patients with advanced tumors, while another phase II trial using Pembrolizumab, also an anti-PD1 antibody, showed an objective response rate of 17% in patients showing tumor progression on Sorafenib (Zhu et al. 2016; El-Khoueiry et al. 2017). Nivolumab has already been approved for the treatment of melanoma from 2014 in Japan (Kudo 2017). Unfortunately, the phase III trials with Pembrolizumab showed the drug didn't have a favorable impact on patient survival (Finn et al. 2020a, b). Another method that harnesses the patient's immune system is through ex vivo stimulation of T cells against tumor-associated antigens. Naive T cells are harnessed, primed, and expanded against tumor antigens. In the case of HCC, tumor-associated antigens most researched are AFP, glypican-3, NY-ESO-1, and MAGE-A1 (Mizukoshi and Kaneko 2019).

The liver itself is described as an immune-tolerant tissue, due to its physiological properties (Heymann and Tacke 2016). Several cell types work directly in coherence with the adaptive immune system. Liver sinusoidal endothelial cells act as antigen-presenting cells, preventing acute responses to bacterial antigens from the portal circulation. Kupffer cells are specialized liver-specific macrophages responsible for bacteria removal, cytokine production, and $CD4^+$ regulatory T cell proliferation, through Forkhead box P3 activation (Kole et al. 2020). Tumor cells can upregulate negative immune costimulation and alter aspects of the tumor microenvironment, enhancing immune evasion and tumor progression (Llovet et al. 2022). Immune checkpoint inhibitors (ICI), such as Atezolizumab and Bevacizumab, are currently in use for treatment of solid malignancies. ICI molecules disable inhibitory signals of

T-cell activation, and thus enable tumor-reactive T cells to facilitate an effective antitumor response and induce cytokine release, slowing down tumor progression (Wei et al. 2018; Robert 2020).

The limited success of immunotherapy for treatment of liver cancer might be related to the complex function of the liver in regard to its role in tolerance induction. Improved understanding of the tolerogenic immunity might also help to pave novel roads for liver immunotherapy.

2 Nanomedicines for Treatment of Liver Cancer

Nanomedicines are nanotechnologically generated drugs that are applicable for medical treatment and that cover a broad range of sizes of a few up to several hundred nanometers. Depending on the nature of the material, they can be assigned into two basic categories: inorganic or organic. In many cases, the organic particulates are clinically usable and the inorganic formulations are mostly used in research (Bartneck 2017). Gold nanoparticles (AuNP) probably are the most intensively studied type of inorganic nanoparticle. AuNP can easily be altered in size, shape, and functionalization such as nanorods (Nikoobakht et al. 2002), nanocages (Chen et al. 2005), or nanostars (Krichevski and Markovich 2007). The metallic nature allows for optical and magnetic properties being traceable in whole body-based and cellular imaging (Soo Choi et al. 2007; Bartneck et al. 2012). However, a drawback of inorganic nanoparticles is given by their temporal accumulation in the body, owed to their missing degradability. Earlier own studies have demonstrated that gold nanorods reside in the liver to a similar extent after 7 days compared to the level after 1 day (Bartneck et al. 2012).

Organic nanoparticles cover a broad spectrum, such as those based on polymers, i.e., N-(2-Hydroxypropyl) methacrylamide (HPMA) (Bartneck et al. 2017), or lipid-based nanoparticles (Bartneck et al. 2015). Many organic nanoparticles are biodegradable; however, this does not apply for all organic nanoparticles, for instance, fullerenes or carbon nanotubes (Kümmerer et al. 2011) are not biodegradable.

In this chapter, we put the focus on lipid nanoparticles that are capable of delivering genetic medicines, particularly nucleic acids.

2.1 Viral and Nonviral Systems for Delivery of Nucleic Acids

The biggest hurdle is the delivery of mRNA to target cells. Larger size of mRNA (300–5,000 kDa), as compared to smaller oligonucleotides, such as siRNA (~14 kDa), negative charge and degradability, present as a challenge for successful passing of mRNA through the cell membrane. For this reason, a delivery system is required (Kauffman et al. 2015). Several virus-based gene therapy approaches are being explored in respective clinical studies. Recombinant adeno-associated virus (AAV) vector can be used as a delivery system for exogenous genes, which can then be persistently expressed in an episomal state, while also being able to infect nondividing and dividing cells, with low toxicity (Smith 2008; Wang et al. 2016a, b). Oncolytic viruses are a group of potential viruses with the ability to infect and lyse cancer cells, without any toxicity to normal cells. Their function in simultaneous addition is being observed in several clinical trials (Hemminki et al. 2020). Several gene therapy-based drugs currently on the market, such as Luxturna, Zolgensma, Oncorine, and Imlygic, use viral vectors. But, because of the expensive nature of these drugs and potential immunogenicity and oncogenic issues, other vectors are explored (Sainz-Ramos et al. 2021).

Viral delivery is costly and bears the risk of delivering the cargo to unintended sites. Thus, a lot of research has been done on nonviral delivery systems. Nanoparticles for nucleic acid delivery enable to target molecular key points that previously were undruggable. This applies for instance for molecular targets for which there is no usable small molecule available. The combination of gene therapy and nanodelivery systems has broadened the therapeutic and biomedical applications of these molecules, such as biological analysis, gene silencing, protein substitution, and vaccines. A very large number of systems for nucleic acid delivery have been explored in clinical studies. Lipid nanoparticle delivery systems are the most widely used nonviral nucleic acid carriers. Upon uptake by cells, they fuse with the negatively charged endocytoid membrane, allowing effective nucleic acid delivery. RNA delivery technology using lipid nanoparticles (LNP) plays an important role in the practical application of mRNA vaccines and RNA-based therapies. Encapsulating RNA into LNP can prevent RNA from being degraded by nucleases, particularly by RNases in blood.

There are many different types of RNA that have been used as therapeutic. To mention only a few, therapeutics have included messenger RNA (mRNA), antisense RNA (asRNA), RNA aptamers that enable to target a specific cell types, and various types of RNA interference (RNAi). RNAi can be achieved by siRNA, which is very specific, but also by the less specific micro-RNAs. CRISPR guide RNAs (gRNAs) make gene editing possible, but at the moment are to our mind not really usable in vivo, due to a lack of precision owed to mismatch repair. While most types of therapeutic RNA can inhibit molecular targets, only mRNA can introduce a protein encoded by the specific mRNA. This has been explored in the course of vaccine development.

Generally, RNA drugs can control the expression of specific genes by inhibition (mediated by RNAi) or by induction of a protein (by mRNA) (Chakraborty and Sarkar 2022). siRNAs are double-stranded RNA that specifically target complementary mRNA for degradation, causing gene silencing (Dana et al. 2017). siRNA-mediated gene silencing has the advantage over other molecular therapeutics and monoclonal antibody drugs in the much simpler Watson-Crick base pairing with mRNA, as opposed to antibody drugs that need to recognize the specific and complicated spatial conformation of targeted proteins. Naked, unmodified siRNA is very unstable, has poor pharmacokinetic behavior, and can possibly have off-target effects. Thus, a carrier is needed for successful targeted delivery to specific cell types (Hu et al. 2020). mRNA-based therapies rely on several advantages they provide against plasmid DNA: mRNA functions in the cytoplasm, it will not insert itself into the host genome because it is not carried by viral vectors, mRNA is also nontoxic and immunologically inert (Yang et al. 2020a, b). Since the 1990s, immense progress has been made in the preparation of mRNA therapeutics. There are several ways mRNA expression can be enhanced through modifications, such as inhibiting RNA de-capping and improving enzymatic degradation resistance. Presence of 5′ and 3′ untranslated regions is important for RNA-binding protein and microRNA recruitment (Kowalski et al. 2019).

2.2 Lipid-Based Delivery Systems

Lipid nanoparticles have many advantages – biocompatibility, low toxicity, enhanced physical stability, ease of scalability, and low production cost. The biggest advantage being the ease of introducing changes in the composition and ratio of lipids and other components, which can all influence encapsulation efficiency and specific targeting and release (Fig. 2a) (Mahmoud et al. 2022). Lipids are amphiphilic molecules, composed of three domains: a hydrophobic tail region, coupled by a linker with a polar head group (Hou et al. 2021). Because of their structure, lipids can form vesicles, such as lipid nanoparticles, liposomes, and lipid emulsions, and thus have been researched as potential vessels for all kinds of molecular cargo (Yang et al. 2020a, b).

2.2.1 Cationic Lipid Nanoparticles

Cationic lipids in their structure have a head group with a permanent positive charge. 1,2-di-O-octadecenyl-3-trimethylammonium-propane, 1,2-dioleoyl-sn-glyce- ro-3-phosphoethanolamine (DOPE), and 1,2-dioleoyl-3-trimethylammonium-propane (DOTAP) – a biodegradable analogue of DOTMA, were all studied for mRNA delivery (Hou et al. 2021). DOTAP-based nanoparticle delivery of mRNA was already observed in several types of cancer, including colorectal, ovarian, and hepatocellular carcinoma (Kornek et al. 2008; Lara et al. 2012; Lei et al. 2020). A 2017 study by Liu et al. constructed lipid nanoparticles consisting of 2-dioleyloxy-N,N-dimethyl-3-aminopropane (DODMA), egg phosphatidylcholine, cholesterol and cholesterol-polyethylene glycol for delivery of miR-122, a microRNA (miRNA) – a small RNA molecule, consisting of around 20 nucleotides that regulates gene expression at posttranscriptional level, with tumor-suppressive properties, highly downregulated in HCC. Successful administration of these LNPs resulted in a ~ 50% growth suppression of HCC xenografts within 30 days, correlated with the suppression of target genes and angiogenesis (Hsu et al. 2013). Further along, Makita et al. in Makita et al. 2017 used lipid nanoparticles composed of 3-((5-(dimethylamino) pentanoyl)oxy)2,2-bis(((3-pentyloctanoyl)oxy)methyl) propyl 3-pentyloctanoate, DPPC, cholesterol and GS-020 – a PEG diacylglycerol, used for application of drugs sensitive to ionic charges, for delivery of siRNA targeting the kinetochore-associated protein 2 (KNTC2). Repeated administration of the LNPs containing the siRNA-inhibited tumor tissue growth, possibly by disturbing the mitosis process in tumor cells by causing an accumulation of

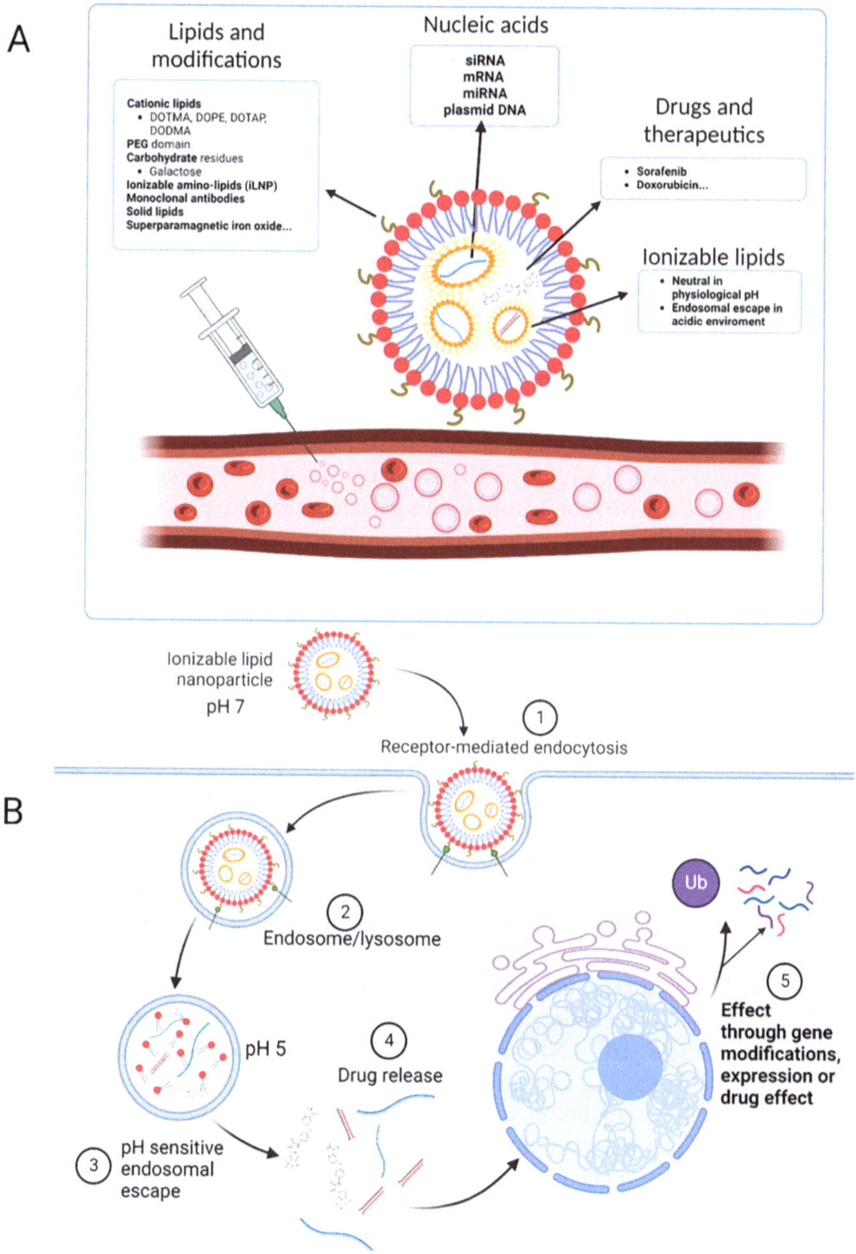

Fig. 2 Composition and mechanism of release of ionizable lipid nanoparticles for nucleic acids. (**a**) Lipid-based drugs are an important transport vehicle, and have demonstrated long-term success for instance in chemotherapeutics. The ease of which different compounds can be introduced into the structure of the lipid nanoparticle proves as an advantage over many other carrier particles. (**b**) The cargo, i.e. a nucleic acid, has to get out of the carrier before it is degraded by enzymes of the lysosome which form in late endosomes. Endosomal escape can for instance be realized via incorporation of pH-sensitive lipids, leading to enhanced release of nucleic acids in lysosomal regions where the pH is around 5. (Created with BioRender.com)

phosphorylated HH3, correlated with chromosome condensation, leading to a G2/M cell cycle arrest (Makita et al. 2017).

Furthermore, advances have been made with utilizing the specific targeting opportunities of modified LNPs with gene modifications and drug-based therapeutics or immunotherapies. The end result is a combinational therapy method that maximizes therapeutic efficacy, while overpowering the cells' innate resistance mechanisms with reduced side effects. Vascular endothelial growth factor (VEGF) is necessary for tumor angiogenesis, also for hepatocellular carcinoma (Zhang et al. 2012). Intraperitoneal treatment with VEGF-A siRNA had no effect, until coupled with an immunostimulatory motif in combination with the DOTAP formulation, which significantly reduced hepatic VEGF-A expression and increased intrahepatic interferon type 1 response, thus enhancing the antitumor response (Zhang et al. 2012). Another study that utilized the VEGF pathway was done by Huang et al. in Huang et al. 2018, where VEGF siRNAs were delivered by lipid/calcium/phosphate nanoparticles (LCP-LNP) conjugated with a galactose derivative. Galactose residues are recognized by the asialoglycoprotein receptor, which is found predominantly on hepatocytes and HCC cells. The study demonstrated how phenyl β-d-galactoside-decorated LCP-LNPS show significantly increased siRNA delivery into HCC cells compared to normal hepatocytes (Huang et al. 2018).

2.2.2 Ionizable Lipid Nanoparticles for siRNA Delivery

Ionizable lipid nanoparticles (iLNP) contain ionizable amino lipids that can be combined with various other modifications to the lipid structure and different cargo (Fig. 2a). They are neutral in physiological conditions, but following endocytosis, the acidic environment of the endosome changes the nanoparticle charge, causing a significantly enhanced endosomal escape of the nucleic acid and accumulation in the cytoplasm (Fig. 2b) (Jayaraman et al. 2012). Nanoparticles with pKa values between 6.0 and 6.5, can trap nucleic acid at acidic pH value, while maintaining a neutral zeta potential at physiological pH. This leads to an increased safety by reducing nonspecific protein binding in circulation (Yang et al. 2020a, b). LNP characteristics, including lipid composition, nitrogen/phosphate molar ratio (N/P), LNP particle size, LNP particle size distribution, Zeta potential and RNA encapsulation efficiency, affect the biological distribution and the therapeutic effect, indicating that these conditions should be optimized to maximize the therapeutic effect. Nanoparticles (NPs) used in current research carry drugs to HCC tumor sites with their physical and chemical properties, and then inhibit tumor growth (Baetke et al. 2015).

The genetic material, such as siRNA can also be condensed by addition of polyethyleneimine (PEI) or protamine, forming a nucleic acid core and neutralizing the siRNAs charge, reducing the amount of cationic lipids needed for coating (Younis et al. 2019). Specific targeting can also be improved by the addition of specific targeting ligands. SP94 is a synthetic 12-amino-acid peptide, with a high success rate for recognition of an unknown cell membrane molecule on hepatocellular carcinoma surgical specimens (Lo et al. 2008). It is also known that iLNPs bind Apoliprotein E (ApoE) to their surface, which interacts with low-density lipoprotein receptor (LDLR), highly expressed on the outer membrane of hepatocytes, thus enhancing their uptake into hepatocytes (Woitok et al. 2020a, b). Patisiran, an FDA-approved siRNA-based drug that stops the production of misfolded proteins in hepatocytes, exploits the uptake of lipid nanoparticles via the LDLR receptor uptake and is used as a treatment of polyneuropathy in people with hereditary transthyretin-mediated amyloidosis (Adams et al. 2019).

Woitok et al. (2020a, b) generated LNP containing Jnk2 siRNA, with aminolipid KL-52, 2-distearoyl-3-phosphatidylcholine (DSPC), cholesterol, and α-[3′-(1,2-dimyristoyl-3-propanoxy)-carboxamide-propyl]-ω-meth-oxy-polyoxyethylene (PEG-c-DOMG). c-Jun N-terminal kinase-2 (JNK) activation is prompted by the downregulation of the regulatory subunit IKKγ (NEMO) and NF-κB inactivation in murine hepatocytes and has several functions in HCC (Cubero et al. n.d.; Zoubek et al. 2019; Yang et al. 2020a, b). The formulated Jnk2-siRNA-LNP successfully accumulated in the liver and efficiently silenced Jnk2 in hepatocytes. Chronic treatment with the siRNA LNPs reduced premalignant nodules, suppressing tumor initiation, and it also prevented HCC progression at 44 weeks of age. On the other hand, the study demonstrated the importance of careful stage dependent intervention – downregulation of Jnk2 in 12-week-old mice caused a proinflammatory response, resulting in hepatic damage, impaired liver fibrosis, immune cell infiltration, and apoptosis (Woitok et al. 2020a, b).

2.2.3 Ionizable Lipid Nanoparticles for mRNA Delivery

In 2020, the mRNA vaccine targeting COVID-19 was authorized for emergency use by the US Food and Drug Administration (FDA). The mRNA vaccines BNT16b2 and mRNA-1,273 produced by Pfizer BioNTech and Moderna can prevent severe cases of COVID-19. One of the remarkable cases are the mRNA-based COVID-19 vaccines that showed the priority of LNPs as a nanocarrier delivery system (Mitchell et al. 2021). Other mRNA-based vaccines targeting malaria, human immunodeficiency virus, influenza, and cancer are also under development (FDA 2021; Oliver et al. 2021; Ruiz-Fresneda et al. 2021).

With the success of the covid vaccine, tumor vaccine is one of the most promising therapeutic strategies in tumor immunotherapy. Messenger RNA (mRNA) vaccine is a new type of vaccine. It can activate the immune system to achieve the goal of immunotherapy by delivering specific antigen mRNA sequences to the body and expressing corresponding antigen proteins. Compared with traditional vaccines, mRNA vaccines have the advantages of short production cycle, high effectiveness, and strong immunogenicity. In recent years, the application of mRNA vaccines in tumor immunotherapy has attracted extensive attention, but the instability and low delivery efficiency of mRNA limit its application. Nanoparticles can effectively solve the problem of mRNA vaccine delivery, greatly promote the research process and clinical application of mRNA tumor vaccine, and has become a hot spot in mRNA vaccine research. Messenger RNA (mRNA) vaccine refers to an immune preparation that synthesizes mRNA sequences encoding specific antigens in vitro, delivers them to the body through a specific delivery system and expresses corresponding antigen proteins, thereby stimulating the body to produce specific immune response to achieve the purpose of immunotherapy (Sahin et al. 2014).

The ApoE-LDLR axis was also explored for applications with mRNA. For instance, Yang et al. (2020a, b) produced lipid nanoparticles composed of three lipids – iBL0713, cholesterol and C16-PEG, using a codon-optimized and polyadenylated mRNA for erythropoietin, with anti-reverse-cap analog added. This iLNP formulation, combined with a modified mRNA, not only greatly improved the translational efficiency in hepatocellular carcinoma cells, but also the specific delivery to hepatocytes, possibly through ApoE binding and LDLR interaction (Yang et al. 2020a, b).

In 1990, Wolff et al. injected mRNA into the skeletal muscle of mice through intramuscular injection, and found that the corresponding protein was expressed in mice, revealing for the first time the possibility of the application of mRNA technology in the field of vaccine research (Wolff et al. 1992). In 1995, Conry et al. constructed mRNA splicers encoding luciferase and human carcinoembryonic antigen (Conry et al. 1995). Mice received anti-tumor immune response after injection of liposome-based mRNA. They first explored and proposed the use of mRNA vaccine as a means of tumor treatment. In the field of tumor therapy, mRNA vaccines have also emerged from a variety of types of vaccines, and the research development is relatively rapid. Compared with DNA, mRNA has the following advantages: it is safer, can be expressed without entering the nucleus, and has no risk of infection or cancer (Patel et al. 2019); it is more efficient. It can improve the efficiency and stability of mRNA translation through sequence optimization and various carrier modifications; the preparation is faster. For the target protein with known arbitrary sequence, its mRNA can be synthesized in a short time, saving drug development time (Pardi et al. 2018).

The mRNA tumor vaccine is gradually favored by researchers, which also broadens the path of personalized vaccine development. mRNA can be divided into nonreplicating mRNA (NRM) and self-replicating mRNA (SAM). NRM is synthesized by in vitro transcription using linear plasmid DNA or polymerase chain reaction template, with simple structure, short RNA sequence, and no additional coding protein. SAM directly inserts the sequence encoding antigen into a single-stranded RNA virus, where the target gene replaces the gene encoding structural protein, while the gene encoding mRNA replication is complete. SAM can produce a higher level of antigen expression and sustainable expression, but its molecular weight is far greater than NRM, leading to its production process being more complex than NRM. In addition, the immunogenicity of additional proteins encoded in SAM is difficult to control and may induce unexpected immune responses. Its safety remains to be verified (Sahin et al. 2014; Pardi et al. 2018).

The mRNA tumor vaccine generally uses the mRNA encoding tumor-derived antigen to translate the specific tumor antigen protein as the "target" through the human protein synthesis system, to induce the body to generate the immune response against the "target," and then recognize and eliminate tumor cells. At present, tumor-derived antigens encoded by vaccines mainly include tumor-associated antigen (TAA) and neoantigen (Neo Ag) (Stenzl et al. 2017). In addition, some mRNA vaccines can also encode immune stimulators such as cytokines and chemokines for tumor treatment. Compared with other types of tumor vaccines, the production process of mRNA tumor vaccines is simple, the preparation speed is faster, and the mass production is easy; It has high effectiveness, dual immune mechanism, and strong immunogenicity; the research and development cycle is short, and new candidate vaccines can be rapidly developed for precise and personalized treatment of tumors. In addition, mRNA can not only encode antigens, but also provide adjuvant activity in some cases. By activating Toll-like receptors (TLR), such as TLR3, TLR7 and TLR8, it can provide costimulatory signals to promote the maturation of dendritic cells (DCs) and enhance the immune response intensity (Amin et al. 2015). Tumor vaccines have been demonstrated to effectively present tumor antigens and activate $CD8^+$ and $CD4^+$ T cells, leading to an antigen-specific response (Aurisicchio et al. 2018). RNA vaccines bypass MHC-based restrictions, without a need for an adjuvant and without a risk of integrating into the recipient's genome (McNamara et al. 2015).

A large advantage of mRNA-based vaccines is the capability to encode various antigens – full proteins and peptides, integrating all tumor antigens together (Fan et al. 2021). They also induce a stronger humoral and cellular response as compared to protein-based vaccines (Chen et al. 2022). In 2021, Zhang et al. constructed a DC-targeted RNA LNP cancer vaccine, using isolated RNA from murine hepatoma cells. Tumor RNA LNPs were ingested by dendritic cells (DC), which led to DC maturation and translation of tumor antigen. DCs then activate tumor antigen specific $CD8^+$ and $CD4^+$ T cells. These cells effectively triggered a specific antitumor immune response, preventing and inhibiting HCC growth.

2.2.4 Combination of Nucleic Acid Gene Therapy with Small Molecules

Several studies have included a co-delivery of chemotherapeutics and siRNA in using iLNP. In 2016, Zhang et al. have devised a polymer of N-succinyl chitosan-poly-L-lysine-palmitic acid (NSC-PLL-PA) for co-delivering doxorubicin and siRNA-P-glycoprotein (P-gp) for treatment of HepG2 human liver cancer cells (Dang et al. 2019). P-gp, also called the multidrug resistance protein (MDR1) is an ATP binding cassette transporter and has been described frequently as being able to prevent the effective employment of a large number of compounds, including cancer therapeutics (Seelig 2020).

The downregulation of P-gp using RNA interference successfully increased the effectiveness of doxorubicin, ultimately significantly inhibiting tumor growth (Dang et al. 2019). Another drug very often used for synergistic treatment with gene-based interaction is Sorafenib. Sorafenib was the first molecule-targeted drug approved by the FDA for the clinical treatment of HCC (Kim and Park 2011). It is a multi-kinase inhibitor, with several mechanisms of action. It can block the Raf/MEK/ERK-mediated signaling pathway, but also block tumor angiogenesis by inhibiting the vascular endothelial growth factor receptor (VEGFR) and platelet-derived growth factor receptor (PDGFR) (Gong et al. 2017). On the other hand, treatment with Sorafenib presents several problems, such as its poor solubility in water, potential side effects, like skin rash, diarrhea, and elevated blood pressure. Congenital resistance and/or resistance acquired posttreatment have also been reported (Kong et al. 2021). Targeting through nanoparticles, such as LNPs could overcome tissue off-targeting and drug resistance, so several different LNP formulations and combinations with gene modifications have been studied. Younis et al. (2019) combined Sorafenib treatment with midkine gene siRNA (MK-siRNA). Midkine is a heparin-binding growth factor highly expressed during mouse embryogenesis, but also in various diseases, such as cancer, including HCC (Zhu et al. 2013; Filippou et al. 2020). It is involved in several physiological processes, such as growth, survival, migration, reproduction, and repair, and thus has a very important role in pathogenesis (Muramatsu 2010). The synthesized LNP was also modified by the addition of the SP94 peptide for targeted delivery to HCC cells. The SP94-LNPs showed high specificity for HCC, and the addition of MK-siRNA successfully increased the success of Sorafibin treatment, ultimately producing a treatment with an efficient uptake by the HCC cells, selective cytotoxicity only to HCC cells, efficient gene silencing, and high biosafety (Younis et al. 2019).

In 2019, Wang et al. used LNPs modified with the addition of anti-GPC3 monoclonal antibodies for GPC3 targeting, which is highly expressed in HCC cells. Also, instead of regular gene silencing using siRNAs, in this study anti-miR-27a was used as a potential therapeutic, with the addition of Sorafenib (Wang et al. 2019). MiR-27a has been found to inhibit aspects of adipogenesis, promote myeloblast differentiation and tumor development. It also regulates chemotherapy resistance in several cancers, including HCC. Previous studies have shown that miR-27a could, through downregulation of transcription factors, such as FOXO-1 and PPAR-γ, inhibit expression of genes involved in tumor suppression p21, p27, Bax and caspase-3, and increase liver cell proliferation (Chen et al. 2013). The therapy with an LNP containing the combination of Sorafenib and anti-miRNA27a resulted in a lower cell viability and an increase in the apoptosis cell proportion, while it showed a suppression of the tumor burden in a liver cancer xenograft model. Importantly, it induced no toxicity (Wang et al. 2019).

Lipid/calcium/phosphate (LCP) nanoparticles were also used in a treatment with Sorafenib. In the study, LCPs were used for targeted delivery of TNF-related apoptosis-inducing ligand TRAIL plasmid DNA, which shows a high cytotoxic effect in cancer cells. The plasmid DNA was encapsulated in a calcium phosphate core, with protamine added to facilitate nuclear delivery. The combination of TRAIL plasmid DNA and Sorafenib treatment successfully attenuated progression of HCC (Liu et al. 2018). Solid colloidal LNPs (SLN) have also been used for the delivery of Sorafenib. SLNs are composed of solid lipids such as lecithin and triacylglycerol, and display specific targeting, good stability, and can encapsulate lipophilic and hydrophilic drugs (Kong et al. 2021). Grillone et al. (2015) used cetyl palmitate lipid with the addition of superparamagnetic iron oxide nanoparticles within the lipid matrix by hot homogenization. Sorafenib-loaded magnetic lipid nanoparticles were then delivered to the disease site with the help of a remote magnetic field. The produced nanoparticles had a Sorafenib-loading efficiency of about 90% and were shown to have good stability in aqueous environment. They were also very convenient for imaging and could be tracked through MRI. The Sorafenib-loaded magnetic LNPs were able to use both the Sorafenib's cytotoxic effect to inhibit HepG2 cancer cell proliferation, and also focus on a desired area through magnetically driven accumulation (Grillone et al. 2015).

2.3 Immunomodulatory Gene Therapy with LNP

Cytokines are potential therapeutic targets, such as interleukin-12 (IL-12). It has been shown to express antitumor activity through stimulating proliferation and the cytotoxic function of NK, NKT, and $CD8^+$ cells, promoting cell-mediated immunity through macrophages, lymphocytes, and endothelial cells, differentiation of naïve $CD4^+$ T, and stimulating the production of interferon-γ (IFNγ). IFNγ can subsequently destroy cancer cells directly or through the recruitment of macrophages and NK cells (Presky et al. 1996; Salem et al. 2004; Berraondo et al. 2018). Previous studies using intratumoral gene therapy with IL-12 plasmid DNA have already shown that IL-12 gene therapy could be an effective therapeutic strategy for HCC (Harada et al. 2004). After treatment with lipid nanoparticles containing IL-12 mRNA, both a reduction in tumor growth, tumor burden, and an increase in overall survival in HCC-bearing mice was observed. An increase in infiltration of immune cells ($CD3^+$ and $CD4^+$ T cells) into the tumor was also shown (Lai et al. 2018). In 2016, Wang et al. have shown that IL12 leads to a significant tumor growth inhibition by polarizing macrophages to their M1 phenotype through the downregulation of Stat-3 (Wang et al. 2016a, b). M1 macrophages, also called tumoricidal macrophages, are activated by LPS or IFNγ, and produce IL-1, IL-12, IL-23, TNF-α, and potentially can terminate microorganisms and cancer cells (Porta et al. 2011; Wang et al. 2016a, b). Kupffer cells are resident macrophages in the liver and present the largest population of resident macrophages in the body. They are involved in cell communication and homeostasis maintenance, but primarily in immune processes – particle engulfment, antigen presentation, and T cell stimulation (Dixon et al. 2013).

Kupffer cells also play an important part in several diseases such as acute liver injury, fibrosis, cirrhosis, fatty liver diseases, and liver tumors (Ritz et al. 2018). Because of this reason, macrophages have been targeted for several therapeutic agents. Kupffer cells can receive nanoparticles through mannose and Fc receptors (Grainger 2015). Specific uptake is determined by size and charge – larger nanoparticles being more deposited in liver (>200 nm). Positively charged nanoparticles seem to have a greater uptake, compared to those with a neutral charge (Colino et al. 2020). Nanoparticles, such as liposomes have the advantage of being normally phagocytized by macrophages in circulation, thus passive targeting can be used for drug administration for some infectious diseases, such as Leishmania (Torrado et al. 2008). Surface modifications, such as the addition of mannose to a PEGylated liposome can also aid in specific targeting. In 2016, Hagimori et al. designed a mannosylated lipid with an increased mannose receptor binding through serine-glycine repeats, which successfully associated with mouse peritoneal macrophages (Hagimori et al. 2018).

A promising approach to macrophage interaction is through repolarization of M2 pro-tumorigenic macrophages to anti-tumorigenic M1 macrophages. Recently, in 2022, Galbraith et al. have created lipid nanoparticles carrying a Cas12a/RICTOR-crRNA complex in order to block macrophages from differentiating into the M2 phenotype. The results showed an increase in the number of M1 macrophages in ration to M2 macrophages, as well as an increased number of immune cells-macrophages, $CD8^+$ and $CD8^+/PD1^+$ T cells. This approach changes the tumor microenvironment from supporting tumor growth into a tumor suppressive environment, which aided by accompanying immunotherapy could lead to synergistic effects (Galbraith et al. 2022). Inorganic particles could also be used in a synergistic fashion with vaccines. Calcium phosphate polyetilenimine/SiO2 nanoparticles carrying a Toll-like-3 ligand were used to induct an immune response, with pronounced immunostimulatory effects in vitro with Kupffer cells and liver sinusoidal endothelial cells (Sokolova et al. 2017). On the other hand, toll-like receptor 4 (TLR4) activation blocks mRNA translation, without reducing LNP uptake, with TLR4 inhibition improving delivery (Lokugamage et al. 2020). This indicates a need for a greater understanding of specific mRNA delivery for inflammatory diseases.

3 Approved Nanomedicines and Those at Clinical Trials

Until December, 2021, the FDA has approved only a few anticancer nanomedicines of which none is specific for HCC: Doxil, Onivyde, Abraxane, Eligard, and Vyxeos (Bakrania et al. 2021). Nonetheless, several trials have tried utilizing the nanomedicine approach in HCC. In 2018, Tak et al., published the results of a phase III "HEAT" trial (NCT00617981) combining the effects of thermosensitive liposomal doxorubicin with radiofrequency ablation (RFA). HCC lesions in size of up to 3 cm can be cured through surgery or RFA, but intermediate-size multinodular lesions larger than 3 cm, palliative care treatments are suggested, such as doxorubicin monotherapy (Ye and Chen 2011). Thermodox is a lyso-thermosensitive liposomal doxorubicin (LTLD) contained in a heat-sensitive liposome. The liposomes accumulate around the tumor tissue and remain encapsulated at normal body temperature. Using RFA to apply a local source of heating through microwaves, upon heating to ≥40 °C, doxorubicin gets released and diffuses into the tissue. It has been found that heated LTLD results in 25-fold greater doxorubicin tumor concentrations than nonliposomal doxorubicin (Kong et al. 2000). The results showed that inadequate heating duration led to doxorubicin concentrations in local tumor tissue that showed no therapeutic effect. The study called for further research to expand the heating duration time, but the second Phase III clinical "OPTIMA" trial was abandoned (NCT02112656) (Tak et al. 2018; Regenold et al. 2022).

Omega Therapeutics is currently deploying a Phase I/II open-label study to determine the antitumor activity of OTX-2002 (NCT05497453). OTX-2002 is described as a lipid nanoparticle containing a biscistronic mRNA that codes for two independent epigenomic controllers, used to regulate the expression of c-Myc (MYC) oncogene through epigenetic modifications. MYC is a transcription factor controlling almost 15% of expressed genes. Dysregulation of the said oncogene was observed in several types of cancer (Ahmadi et al. 2021). Several promising findings have been stated by the company through preclinical data – a reduction of MYC mRNA levels in liver in nonhuman primates, decreased MYC mRNA, protein levels, and cell viability in several HCC cell lines, with no toxicity to normal cells, and a reduced tumor growth in HCC xenograft models. Further clinical trials will focus on OTX-2002 as a single therapeutic agent and in combination with standard care of patients with HCC (NCT05497453).

The EphA2 receptor is a member of receptor tyrosine kinases. It is primarily expressed in epithelial cells, and seems to have a role in cell proliferation, migration, and angiogenesis (Lamorte and Park 2001). It is also overexpressed in several types of cancer, including prostate, lung, breast, ovarian, colorectal, and skin cancer. Higher EphA2 expression is also associated with poor prognosis, higher chance of metastasis occurrence, and reduced patient survival (Xiao et al. 2020). Enriched EphA2 expression and signaling has been also observed in HCC. CRISPR/CAS9-mediated EphA2 inhibition resulted in suppression of two oncogenic pathways – AKT and JAK1/STAT3, and a significantly delayed tumor development in a murine HCC model (Wang et al. 2021). Although not directly aimed at HCC, a new 1,2-dioleoyl-sn-glycero-3-phosphatidylcholine (DOPC) neutral liposome incorporated EphA2 siRNA (EPHARNA) is currently in the recruitments stage for a phase I trial study (NCT01591356) of the side effects and best dose for solid tumor treatment. The DOPC neutral liposome has in previous studies been shown to deliver inhibitory nucleic acids with tenfold better efficiency, as when compared to cationic liposomes (Landen et al. 2005). The current study is underway and will hopefully have a completion date sometime in 2024.

4 Conclusion

The success of the COVID-19 mRNA vaccine has inspired the confidence of scientists to apply the technology for tumor treatment. However, despite promising results from preclinical and clinical studies, the use of nanomaterials still faces some challenges. One of the challenges in the delivery of nucleic acid carriers is organ selectivity and cell selectivity, due to many factors affecting the biological distribution of nanoparticles. A lot of research shows most of the nanoparticles accumulate in the liver, but how to lead them to tumor cells is a critical part for further development. Another challenge is the standardization of nanoparticle production. In case of lipid nanoparticles, the specific mixing conditions vary with different types of processing instruments. In order to ensure the accuracy of clinical results, the homogeneity of nucleic acid nanocarriers is crucial, and purification and characterization methods should be standardized in the community. Tumorigenesis is a complex process associated with multiple changes in molecular biology. In case of the liver, the functions in tolerance make the situation even more complicated. However, we are confident that, although there are still many things to overcome, precision nanomedicine is the most promising tool for cancer treatment.

References

Adams D, Koike H, Slama M, Coelho T (2019) Hereditary transthyretin amyloidosis: a model of medical progress for a fatal disease. Nat Rev Neurol 15:387–404

Ahmadi SE, Rahimi S, Zarandi B, Chegeni R, Safa M (2021) MYC: a multipurpose oncogene with prognostic and therapeutic implications in blood malignancies. J Hematol Oncol 14:1–49

Amin A, Dudek AZ, Logan TF, Lance RS, Holzbeierlein JM, Knox JJ, Master VA, Pal SK, Miller WH, Karsh LI (2015) Survival with AGS-003, an autologous dendritic cell–based immunotherapy, in combination with sunitinib in unfavorable risk patients with advanced renal cell carcinoma (RCC): phase 2 study results. J Immunother Cancer 3:1–13

Ananthakrishnan A, Gogineni V, Saeian K (2006) Epidemiology of primary and secondary liver cancers. In: Seminars in interventional radiology. Copyright© 2006 by Thieme Medical Publishers, New York

Aurisicchio L, Pallocca M, Ciliberto G, Palombo F (2018) The perfect personalized cancer therapy: cancer vaccines against neoantigens. J Exp Clin Cancer Res 37:1–10

Baetke SC, Lammers T, Kiessling F (2015) Applications of nanoparticles for diagnosis and therapy of cancer. Br J Radiol 88:20150207

Bakrania A, Zheng G, Bhat M (2021) Nanomedicine in hepatocellular carcinoma: a new frontier in targeted cancer treatment. Pharmaceutics 14:41

Bartneck M (2017) Immunomodulatory nanomedicine. Macromol Biosci 17:1700021

Bartneck M, Ritz T, Keul HA, Wambach M, Bornemann J, Gbureck U, Ehling J, Lammers T, Heymann F, Gassler N (2012) Peptide-functionalized gold nanorods increase liver injury in hepatitis. ACS Nano 6:8767–8777

Bartneck M, Scheyda KM, Warzecha KT, Rizzo LY, Hittatiya K, Luedde T, Storm G, Trautwein C, Lammers T, Tacke F (2015) Fluorescent cell-traceable dexamethasone-loaded liposomes for the treatment of inflammatory liver diseases. Biomaterials 37:367–382

Bartneck M, Schlosser CT, Barz M, Zentel R, Trautwein C, Lammers T, Tacke F (2017) Immunomodulatory therapy of inflammatory liver disease using selectin-binding glycopolymers. ACS Nano 11:9689–9700

Berraondo P, Etxeberria I, Ponz-Sarvise M, Melero I (2018) Revisiting interleukin-12 as a cancer immunotherapy agentIL12: back to the future. Clin Cancer Res 24:2716–2718

Bray F, Ferlay J, Soerjomataram I, Siegel RL, Torre LA, Jemal A (2018) Global cancer statistics 2018: GLOBOCAN estimates of incidence and mortality worldwide for 36 cancers in 185 countries. CA Cancer J Clin 68:394–424

Chakraborty E, Sarkar D (2022) Emerging therapies for hepatocellular carcinoma (HCC). Cancers 14:2798

Chan HL-Y, Sung JJ-Y (2006) Hepatocellular carcinoma and hepatitis B virus. In: Seminars in liver disease. Copyright© 2006 by Thieme Medical Publishers, Inc., New York

Chang YS, Adnane J, Trail PA, Levy J, Henderson A, Xue D, Bortolon E, Ichetovkin M, Chen C, McNabola A (2007) Sorafenib (BAY 43-9006) inhibits tumor growth and vascularization and induces tumor apoptosis and hypoxia in RCC xenograft models. Cancer Chemother Pharmacol 59:561–574

Chen J, Saeki F, Wiley BJ, Cang H, Cobb MJ, Li Z-Y, Au L, Zhang H, Kimmey MB, Li X (2005) Gold nanocages: bioconjugation and their potential use as optical imaging contrast agents. Nano Lett 5:473–477

Chen Z, Ma T, Huang C, Zhang L, Lv X, Xu T, Hu T, Li J (2013) MiR-27a modulates the MDR1/P-glycoprotein expression by inhibiting FZD7/β-catenin pathway in hepatocellular carcinoma cells. Cell Signal 25:2693–2701

Chen J-h, Lu L, Wen T-F, Huang Z-Y, Zhang T, Zeng Y-Y, Li X-C, Xiang B-D, Lu C-d, Xu X (2020) Adjuvant lenvatinib in combination with TACE for hepatocellular carcinoma patients with high risk of postoperative relapse (LANCE): interim results from a muticenter prospective cohort study. Proc Am Soc Clin Oncol 38(15_suppl):4580

Chen J, Ye Z, Huang C, Qiu M, Song D, Li Y, Xu Q (2022) Lipid nanoparticle-mediated lymph node–targeting delivery of mRNA cancer vaccine elicits robust CD8+ T cell response. Proc Natl Acad Sci 119:e2207841119

Colino CI, Lanao JM, Gutierrez-Millan C (2020) Targeting of hepatic macrophages by therapeutic nanoparticles. Front Immunol 11:218

Conry RM, LoBuglio AF, Wright M, Sumerel L, Pike MJ, Johanning F, Benjamin R, Lu D, Curiel DT (1995) Characterization of a messenger RNA polynucleotide vaccine vector. Cancer Res 55: 1397–1400

Cubero F, Zhao G, Nevzorova Y, Masaoudi MA, Verdier J, Peng J, Schaefer F, Hatting M, Hermanns N, Boekschoten M (n.d.) Hematopoietic cells-derived Jnk1 drives hepatic injury and hepatocellular carcinoma in IKKγ/NEMO-deleted livers. Zeitschrift für Gastroenterologie

Dana H, Chalbatani GM, Mahmoodzadeh H, Karimloo R, Rezaiean O, Moradzadeh A, Mehmandoost N, Moazzen F, Mazraeh A, Marmari V (2017) Molecular mechanisms and biological functions of siRNA. Int J Biomed Sci 13:48

Dang W, Xie Y, Cao P, Xin S, Wang J, Li S, Li Y, Lu J (2019) N6-methyladenosine and viral infection. Front Microbiol 10:417

Dixon L, Barnes M, Tang H, Pritchard M, Laura E, Clinic C (2013) Kupffer cells in the liver. Compr Physiol 3:785–797

El-Khoueiry A, Sangro B, Yau T, Crocenzi T, Kudo M, Hsu C, Kim T, Choo S, Trojan J, Welling T, dela Cruz C, Lang L, Neely J, Tang H, Dastani HB, Melero I (2017) Nivolumab in patients with advanced hepatocellular carcinoma (CheckMate 040): an open-label, non-comparative, phase 1/2 dose escalation and expansion trial. Lancet 389:2492–2502

El-Mayta R, Zhang Z, Hamilton AG, Mitchell MJ (2021) Delivery technologies to engineer natural killer cells for cancer immunotherapy. Cancer Gene Ther 28:947–959

Fan C, Qu H, Wang X, Sobhani N, Wang L, Liu S, Xiong W, Zeng Z, Li Y (2021) Cancer/testis antigens: from serology to mRNA cancer vaccine. In: Seminars in cancer biology. Elsevier

FDA F (2021) Approves first COVID-19 vaccine. US Food and Drug Administration

Filippou PS, Karagiannis GS, Constantinidou A (2020) Midkine (MDK) growth factor: a key player in cancer progression and a promising therapeutic target. Oncogene 39:2040–2054

Finn R, Ryoo B, Merle P, Kudo M, Bouattour M, Lim H, Breder V, Edeline J, Chao Y, Ogasawara S (2020a) KEYNOTE-240 investigators. Pembrolizumab as second-line therapy in patients with advanced hepatocellular carcinoma in KEYNOTE-240: a randomized, double-blind, phase III trial. J Clin Oncol 38:193–202

Finn RS, Qin S, Ikeda M, Galle PR, Ducreux M, Kim T-Y, Kudo M, Breder V, Merle P, Kaseb AO (2020b) Atezolizumab plus bevacizumab in unresectable hepatocellular carcinoma. N Engl J Med 382:1894–1905

Galbraith T, Cook H, Ziemys A, Leonard F (2022) Abstract PR04: CRISPR lipid nanoparticle modulates the tumor immune microenvironment of liver metastases by genetically eliminating M2 polarization pathway in macrophages. Clin Cancer Res 28:PR04-PR04

Gong L, Giacomini MM, Giacomini C, Maitland ML, Altman RB, Klein TE (2017) PharmGKB summary: sorafenib pathways. Pharmacogenet Genomics 27:240

Grainger GHH-CD (2015) The phagocyte problem. Nano Today 10:487–510

Grillone A, Riva ER, Mondini A, Forte C, Calucci L, Innocenti C, de Julian FC, Cappello V, Gemmi M, Moscato S (2015) Active targeting of sorafenib: preparation, characterization, and in vitro testing of drug-loaded magnetic solid lipid nanoparticles. Adv Healthc Mater 4: 1681–1690

Hagimori M, Chinda Y, Suga T, Yamanami K, Kato N, Inamine T, Fuchigami Y, Kawakami S (2018) Synthesis of high functionality and quality mannose-grafted lipids to produce macrophage-targeted liposomes. Eur J Pharm Sci 123:153–161

Harada N, Shimada M, Okano S, Suehiro T, Soejima Y, Tomita Y, Maehara Y (2004) IL-12 gene therapy is an effective therapeutic strategy for hepatocellular carcinoma in immunosuppressed mice. J Immunol 173:6635–6644

Hemminki O, Dos Santos JM, Hemminki A (2020) Oncolytic viruses for cancer immunotherapy. J Hematol Oncol 13:1–15
Heymann F, Tacke F (2016) Immunology in the liver--from homeostasis to disease. Nat Rev Gastroenterol Hepatol 13:88–110
Hou X, Zaks T, Langer R, Dong Y (2021) Lipid nanoparticles for mRNA delivery. Nat Rev Mater 6:1078–1094
Hsu S-h, Yu B, Wang X, Lu Y, Schmidt CR, Lee RJ, Lee LJ, Jacob ST, Ghoshal K (2013) Cationic lipid nanoparticles for therapeutic delivery of siRNA and miRNA to murine liver tumor. Nanomed Nanotechnol Biol Med 9:1169–1180
Hu C-MJ, Zhang L (2012) Nanoparticle-based combination therapy toward overcoming drug resistance in cancer. Biochem Pharmacol 83:1104–1111
Hu B, Zhong L, Weng Y, Peng L, Huang Y, Zhao Y, Liang X-J (2020) Therapeutic siRNA: state of the art. Signal Transduct Target Ther 5:1–25
Huang K-W, Lai Y-T, Chern G-J, Huang S-F, Tsai C-L, Sung Y-C, Chiang C-C, Hwang P-B, Ho T-L, Huang R-L (2018) Galactose derivative-modified nanoparticles for efficient siRNA delivery to hepatocellular carcinoma. Biomacromolecules 19:2330–2339
Jayaraman M, Ansell SM, Mui BL, Tam YK, Chen J, Du X, Butler D, Eltepu L, Matsuda S, Narayanannair JK (2012) Maximizing the potency of siRNA lipid nanoparticles for hepatic gene silencing in vivo. Angew Chem 124:8657–8661
Kauffman KJ, Dorkin JR, Yang JH, Heartlein MW, DeRosa F, Mir FF, Fenton OS, Anderson DG (2015) Optimization of lipid nanoparticle formulations for mRNA delivery in vivo with fractional factorial and definitive screening designs. Nano Lett 15:7300–7306
Khalil DN, Smith EL, Brentjens RJ, Wolchok JD (2016) The future of cancer treatment: immunomodulation, CARs and combination immunotherapy. Nat Rev Clin Oncol 13:273–290
Kim HY, Park J-W (2011) Molecularly targeted therapies for hepatocellular carcinoma: sorafenib as a stepping stone. Dig Dis 29:303–309
Kole C, Charalampakis N, Tsakatikas S, Vailas M, Moris D, Gkotsis E, Kykalos S, Karamouzis MV, Schizas D (2020) Immunotherapy for hepatocellular carcinoma: a 2021 update. Cancers 12:2859
Kong G, Anyarambhatla G, Petros WP, Braun RD, Colvin OM, Needham D, Dewhirst MW (2000) Efficacy of liposomes and hyperthermia in a human tumor xenograft model: importance of triggered drug release. Cancer Res 60:6950–6957
Kong F-H, Ye Q-F, Miao X-Y, Liu X, Huang S-Q, Xiong L, Wen Y, Zhang Z-J (2021) Current status of sorafenib nanoparticle delivery systems in the treatment of hepatocellular carcinoma. Theranostics 11:5464
Kornek M, Lukacs-Kornek V, Limmer A, Raskopf E, Becker U, Klöckner M, Sauerbruch T, Schmitz V (2008) 1, 2-dioleoyl-3-trimethylammonium-propane (DOTAP)-formulated, immune-stimulatory vascular endothelial growth factor a small interfering RNA (siRNA) increases antitumoral efficacy in murine orthotopic hepatocellular carcinoma with liver fibrosis. Mol Med 14:365–373
Kowalski PS, Rudra A, Miao L, Anderson DG (2019) Delivering the messenger: advances in technologies for therapeutic mRNA delivery. Mol Ther 27:710–728
Krichevski O, Markovich G (2007) Growth of colloidal gold nanostars and nanowires induced by palladium doping. Langmuir 23:1496–1499
Kudo M (2017) Immune checkpoint blockade in hepatocellular carcinoma: 2017 update. Liver Cancer 6:1–12
Kudo M (2018) Systemic therapy for hepatocellular carcinoma: latest advances. Cancers 10:412
Kümmerer K, Menz J, Schubert T, Thielemans W (2011) Biodegradability of organic nanoparticles in the aqueous environment. Chemosphere 82:1387–1392
Lai I, Swaminathan S, Baylot V, Mosley A, Dhanasekaran R, Gabay M, Felsher DW (2018) Lipid nanoparticles that deliver IL-12 messenger RNA suppress tumorigenesis in MYC oncogene-driven hepatocellular carcinoma. J Immunother Cancer 6:1–11

Lamorte L, Park M (2001) The receptor tyrosine kinases: role in cancer progression. Surg Oncol Clin N Am 10:271–288

Landen C Jr, Chavez-Reyes A, Bucana C, Schmandt R, Deavers MT, Lopez-Berestein G, Sood AK (2005) Therapeutic EphA2 gene targeting in vivo using neutral liposomal small interfering RNA delivery. Cancer Res 65:6910–6918

Lara H, Wang Y, Beltran AS, Juárez-Moreno K, Yuan X, Kato S, Leisewitz AV, Fredes MC, Licea AF, Connolly DC (2012) Targeting serous epithelial ovarian cancer with designer zinc finger transcription factors. J Biol Chem 287:29873–29886

Lau W-y, Leung TW, Liew C-t, Ho SK, Simon C, Tang AM (2001) Preoperative systemic chemoimmunotherapy and sequential resection for unresectable hepatocellular carcinoma. Ann Surg 233:236

Lee M, Ryoo B-Y, Hsu C-H, Numata K, Stein S, Verret W, Hack S, Spahn J, Liu B, Abdullah H (2019) Randomised efficacy and safety results for atezolizumab (Atezo)+ bevacizumab (Bev) in patients (pts) with previously untreated, unresectable hepatocellular carcinoma (HCC). Ann Oncol 30:v875

Lei S, Zhang X, Men K, Gao Y, Yang X, Wu S, Duan X, Wei Y, Tong R (2020) Efficient colorectal cancer gene therapy with IL-15 mRNA nanoformulation. Mol Pharm 17:3378–3391

Leung TW, Yu S, Johnson P, Geschwind J, Vogl T, Engelmann K, Gores G, Giovannini M, O'Grady J, Heneghan M (2003) Phase II study of the efficacy and safety of cisplatin-epinephrine injectable gel administered to patients with unresectable hepatocellular carcinoma. J Clin Oncol 21:652–658

Li Q, Zhao Z, Qin X, Zhang M, Du Q, Li Z, Luan Y (2021) A checkpoint-regulatable immune niche created by injectable hydrogel for tumor therapy. Adv Funct Mater 31:2104630

Liu CH, Chern GJ, Hsu FF, Huang KW, Sung YC, Huang HC, Qiu JT, Wang SK, Lin CC, Wu CH (2018) A multifunctional nanocarrier for efficient TRAIL-based gene therapy against hepatocellular carcinoma with desmoplasia in mice. Hepatology 67:899–913

Liu JK, Irvine AF, Jones RL, Samson A (2022) Immunotherapies for hepatocellular carcinoma. Cancer Med 11:571–591

Llovet JM, Ricci S, Mazzaferro V, Hilgard P, Gane E, Blanc J-F, De Oliveira AC, Santoro A, Raoul J-L, Forner A (2008) Sorafenib in advanced hepatocellular carcinoma. N Engl J Med 359:378–390

Llovet JM, Castet F, Heikenwalder M, Maini MK, Mazzaferro V, Pinato DJ, Pikarsky E, Zhu AX, Finn RS (2022) Immunotherapies for hepatocellular carcinoma. Nat Rev Clin Oncol 19: 151–172

Lo A, Lin C-T, Wu H-C (2008) Hepatocellular carcinoma cell-specific peptide ligand for targeted drug delivery. Mol Cancer Ther 7:579–589

Lokugamage MP, Gan Z, Zurla C, Levin J, Islam FZ, Kalathoor S, Sato M, Sago CD, Santangelo PJ, Dahlman JE (2020) Mild innate immune activation overrides efficient nanoparticle-mediated RNA delivery. Adv Mater 32:1904905

Mahato DK, Lee KE, Kamle M, Devi S, Dewangan KN, Kumar P, Kang SG (2019) Aflatoxins in food and feed: an overview on prevalence, detection and control strategies. Front Microbiol 10: 2266

Mahmoud K, Swidan S, El-Nabarawi M, Teaima M (2022) Lipid based nanoparticles as a novel treatment modality for hepatocellular carcinoma: a comprehensive review on targeting and recent advances. J Nanobiotechnol 20:1–42

Makita Y, Murata S, Katou Y, Kikuchi K, Uejima H, Teratani M, Hoashi Y, Kenjo E, Matsumoto S, Nogami M (2017) Anti-tumor activity of KNTC2 siRNA in orthotopic tumor model mice of hepatocellular carcinoma. Biochem Biophys Res Commun 493:800–806

Marisi G, Cucchetti A, Ulivi P, Canale M, Cabibbo G, Solaini L, Foschi FG, De Matteis S, Ercolani G, Valgiusti M (2018) Ten years of sorafenib in hepatocellular carcinoma: are there any predictive and/or prognostic markers? World J Gastroenterol 24:4152

McNamara MA, Nair SK, Holl EK (2015) RNA-based vaccines in cancer immunotherapy. J Immunol Res 2015:794528

Mitchell MJ, Billingsley MM, Haley RM, Wechsler ME, Peppas NA, Langer R (2021) Engineering precision nanoparticles for drug delivery. Nat Rev Drug Discov 20:101–124
Mizukoshi E, Kaneko S (2019) Immune cell therapy for hepatocellular carcinoma. J Hematol Oncol 12:1–11
Muramatsu T (2010) Midkine, a heparin-binding cytokine with multiple roles in development, repair and diseases. Proc Jpn Acad Ser B 86:410–425
Nikoobakht B, Burda C, Braun M, Hun M, El-Sayed MA (2002) The quenching of CdSe quantum dots photoluminescence by gold nanoparticles in solution. Photochem Photobiol 75:591–597
Oliver SE, Gargano JW, Marin M, Wallace M, Curran KG, Chamberland M, McClung N, Campos-Outcalt D, Morgan RL, Mbaeyi S (2021) The advisory committee on immunization practices' interim recommendation for use of moderna COVID-19 vaccine—United States, December 2020. Morb Mortal Wkly Rep 69:1653
Ozakyol A (2017) Global epidemiology of hepatocellular carcinoma (HCC epidemiology). J Gastrointest Cancer 48:238–240
Pardi N, Hogan MJ, Porter FW, Weissman D (2018) mRNA vaccines—a new era in vaccinology. Nat Rev Drug Discov 17:261–279
Patel S, Athirasala A, Menezes PP, Ashwanikumar N, Zou T, Sahay G, Bertassoni LE (2019) Messenger RNA delivery for tissue engineering and regenerative medicine applications. Tissue Eng A 25:91–112
Porta C, Riboldi E, Totaro MG, Strauss L, Sica A, Mantovani A (2011) Macrophages in cancer and infectious diseases: the 'good'and the 'bad'. Immunotherapy 3:1185–1202
Presky D, Yang H, Minetti L, Chua AO, Nabavi N, Wu CY, Gately MK, Gubler U (1996) A functional interleukin 12 receptor complex is composed of two beta-type cytokine receptor subunits. Proc Natl Acad Sci U S A 93:14002–14007
Raoul J-L, Forner A, Bolondi L, Cheung TT, Kloeckner R, de Baere T (2019) Updated use of TACE for hepatocellular carcinoma treatment: how and when to use it based on clinical evidence. Cancer Treat Rev 72:28–36
Regenold M, Bannigan P, Evans JC, Waspe A, Temple MJ, Allen C (2022) Turning down the heat: the case for mild hyperthermia and thermosensitive liposomes. Nanomedicine 40:102484
Ribas A, Wolchok JD (2018) Cancer immunotherapy using checkpoint blockade. Science 359: 1350–1355
Ritz T, Krenkel O, Tacke F (2018) Dynamic plasticity of macrophage functions in diseased liver. Cell Immunol 330:175–182
Robert C (2020) A decade of immune-checkpoint inhibitors in cancer therapy. Nat Commun 11:1–3
Ruiz-Fresneda MA, Jiménez-Contreras E, Ruiz-Fresneda C, Ruiz-Pérez R (2021) Bibliometric analysis of international scientific production on pharmacologic treatments for SARS-CoV-2/ COVID-19 during 2020. Front Public Health 9:778203
Sahin U, Karikó K, Türeci Ö (2014) mRNA-based therapeutics—developing a new class of drugs. Nat Rev Drug Discov 13:759–780
Sainz-Ramos M, Gallego I, Villate-Beitia I, Zarate J, Maldonado I, Puras G, Pedraz JL (2021) How far are non-viral vectors to come of age and reach clinical translation in gene therapy? Int J Mol Sci 22:7545
Salem ML, Kadima AN, Zhou Y, Nguyen CL, Rubinstein MP, Demcheva M, Vournakis JN, Cole DJ, Gillanders WE (2004) Paracrine release of IL-12 stimulates IFN-γ production and dramatically enhances the antigen-specific T cell response after vaccination with a novel peptide-based cancer vaccine. J Immunol 172:5159–5167
Seelig A (2020) P-Glycoprotein: one mechanism, many tasks and the consequences for pharmacotherapy of cancers. Front Oncol 10:576559
Setshedi M, Wands JR, de la Monte SM (2010) Acetaldehyde adducts in alcoholic liver disease. Oxidative Med Cell Longev 3:178–185
Smith R (2008) Adeno-associated virus integration: virus versus vector. Gene Ther 15:817–822

Sokolova V, Shi Z, Huang S, Du Y, Kopp M, Frede A, Knuschke T, Buer J, Yang D, Wu J (2017) Delivery of the TLR ligand poly (I: C) to liver cells in vitro and in vivo by calcium phosphate nanoparticles leads to a pronounced immunostimulation. Acta Biomater 64:401–410

Song XR, Zheng Y, He G, Yang L, Luo YF, He ZY, Li SZ, Li JM, Yu S, Luo X (2010) Development of PLGA nanoparticles simultaneously loaded with vincristine and verapamil for treatment of hepatocellular carcinoma. J Pharm Sci 99:4874–4879

Soo Choi H, Liu W, Misra P, Tanaka E, Zimmer JP, Itty Ipe B, Bawendi MG, Frangioni JV (2007) Renal clearance of quantum dots. Nat Biotechnol 25:1165–1170

Stenzl A, Feyerabend S, Syndikus I, Sarosiek T, Kübler H, Heidenreich A, Cathomas R, Grüllich C, Loriot Y, Gracia SP (2017) Results of the randomized, placebo-controlled phase I/IIB trial of CV9104, an mRNA based cancer immunotherapy, in patients with metastatic castration-resistant prostate cancer (mCRPC). Ann Oncol 28:v408–v409

Sun G, Wang S, Hu X, Su J, Zhang Y, Xie Y, Zhang H, Tang L, Wang J-S (2011) Co-contamination of aflatoxin B1 and fumonisin B1 in food and human dietary exposure in three areas of China. Food Addit Contam 28:461–470

Tak WY, Lin S-M, Wang Y, Zheng J, Vecchione A, Park SY, Chen MH, Wong S, Xu R, Peng C-Y (2018) Phase III HEAT study adding lyso-thermosensitive liposomal doxorubicin to radiofrequency ablation in patients with unresectable hepatocellular carcinoma lesions RFA and lyso-thermosensitive liposomal doxorubicin in HCC. Clin Cancer Res 24:73–83

Torrado J, Espada R, Ballesteros M, Torrado-Santiago S (2008) Amphotericin B formulations and drug targeting. J Pharm Sci 97:2405–2425

Tunissiolli NM, Castanhole-Nunes MMU, Biselli-Chicote PM, Pavarino ÉC, da Silva RF, Goloni-Bertollo EM (2017) Hepatocellular carcinoma: a comprehensive review of biomarkers, clinical aspects, and therapy. Asian Pac J Cancer Prev 18:863

Wang Q, Cheng F, Ma T-t, Xiong H-Y, Li Z-W, Xie C-L, Liu C-Y, Tu Z-G (2016a) Interleukin-12 inhibits the hepatocellular carcinoma growth by inducing macrophage polarization to the M1-like phenotype through downregulation of Stat-3. Mol Cell Biochem 415:157–168

Wang Y-G, Huang P-P, Zhang R, Ma B-Y, Zhou X-M, Sun Y-F (2016b) Targeting adeno-associated virus and adenoviral gene therapy for hepatocellular carcinoma. World J Gastroenterol 22:326

Wang Z, Zhao K, Zhang Y, Duan X, Zhao Y (2019) Anti-GPC3 antibody tagged cationic switchable lipid-based nanoparticles for the co-delivery of anti-miRNA27a and sorafenib in liver cancers. Pharm Res 36:1–13

Wang H, Hou W, Perera A, Bettler C, Beach JR, Ding X, Li J, Denning MF, Dhanarajan A, Cotler SJ (2021) Targeting EphA2 suppresses hepatocellular carcinoma initiation and progression by dual inhibition of JAK1/STAT3 and AKT signaling. Cell Rep 34:108765

Wei SC, Duffy CR, Allison JP (2018) Fundamental mechanisms of immune checkpoint blockade therapy. Cancer Discov 8:1069–1086

Wilhelm SM, Carter C, Tang L, Wilkie D, McNabola A, Rong H, Chen C, Zhang X, Vincent P, McHugh M (2004) BAY 43-9006 exhibits broad spectrum oral antitumor activity and targets the RAF/MEK/ERK pathway and receptor tyrosine kinases involved in tumor progression and angiogenesis. Cancer Res 64:7099–7109

Woitok MM, Zoubek ME, Doleschel D, Bartneck M, Mohamed MR, Kießling F, Lederle W, Trautwein C, Cubero FJ (2020a) Lipid-encapsulated siRNA for hepatocyte-directed treatment of advanced liver disease. Cell Death Dis 11:1–14

Woitok MM, Zoubek ME, Doleschel D, Bartneck M, Mohamed MR, Kiessling F, Lederle W, Trautwein C, Cubero FJ (2020b) Lipid-encapsulated siRNA for hepatocyte-directed treatment of advanced liver disease. Cell Death Dis 11:343

Wolff J, Dowty ME, Jiao S, Repetto G, Berg RK, Ludtke JJ, Williams P, Slautterback DB (1992) Expression of naked plasmids by cultured myotubes and entry of plasmids into T tubules and caveolae of mammalian skeletal muscle. J Cell Sci 103:1249–1259

Xiao T, Xiao Y, Wang W, Tang YY, Xiao Z, Su M (2020) Targeting EphA2 in cancer. J Hematol Oncol 13:1–17

Yang T, Li C, Wang X, Zhao D, Zhang M, Cao H, Liang Z, Xiao H, Liang X-J, Weng Y (2020a) Efficient hepatic delivery and protein expression enabled by optimized mRNA and ionizable lipid nanoparticle. Bioact Mater 5:1053–1061

Yang T, Li C, Wang X, Zhao D, Zhang M, Cao H, Liang Z, Xiao H, Liang XJ, Weng Y (2020b) Efficient hepatic delivery and protein expression enabled by optimized mRNA and ionizable lipid nanoparticle. Bioact Mater 5:9

Ye S, Chen R (2011) Comments on management of hepatocellular carcinoma: an update. Zhonghua gan zang bing za zhi= Zhonghua ganzangbing zazhi=. Chin J Hepatol 19:251–253

Younis MA, Khalil IA, Abd Elwakil MM, Harashima H (2019) A multifunctional lipid-based nanodevice for the highly specific codelivery of sorafenib and midkine siRNA to hepatic cancer cells. Mol Pharm 16:4031–4044

Zhang L, Wang J-N, Tang J-M, Kong X, Yang J-Y, Zheng F, Guo L-Y, Huang Y-Z, Zhang L, Tian L (2012) VEGF is essential for the growth and migration of human hepatocellular carcinoma cells. Mol Biol Rep 39:5085–5093

Zhou Q, Sun X, Zeng L, Liu J, Zhang Z (2009) A randomized multicenter phase II clinical trial of mitoxantrone-loaded nanoparticles in the treatment of 108 patients with unresected hepatocellular carcinoma. Nanomedicine 5:419–423

Zhou YM, Zhang XF, Wu LP, Sui CJ, Yang JM (2014) Risk factors for combined hepatocellular-cholangiocarcinoma: a hospital-based case-control study. World J Gastroenterol 20: 12615–12620

Zhu W-W, Guo J-J, Guo L, Jia H-L, Zhu M, Zhang J-B, Loffredo CA, Forgues M, Huang H, Xing X-J (2013) Evaluation of midkine as a diagnostic serum biomarker in hepatocellular carcinoma. Clin Cancer Res 19:3944–3954

Zhu A, Knox J, Kudo M, Chan S, Finn R, Siegel A, Ma J, Watson P, Cheng A-L (2016) Pembrolizumab in patients with previously treated advanced hepatocellular carcinoma: phase 2 KEYNOTE-224 study. Ann Oncol 27:vi242

Zoubek ME, Woitok MM, Sydor S, Nelson LJ, Bechmann LP, Lucena MI, Andrade RJ, Bast A, Koek GH, Trautwein C (2019) Protective role of c-Jun N-terminal kinase-2 (JNK2) in ibuprofen-induced acute liver injury. J Pathol 247:110–122

Locoregional Therapies for Hepatocellular Carcinoma

Alexander E. Hare and Mina S. Makary

Abstract

Hepatocellular cancer (HCC) is the fifth most common cancer and is the most common primary cancer of the liver. HCC has a very poor prognosis, because while current surgical treatment methods are curative for a small portion of patients, they are not appropriate for the majority of patients due to local and nonlocal metastases and patient comorbidities. To expand treatment options for nonsurgical patients, catheter-directed and percutaneous locoregional therapies have been developed. Locoregional therapies provide treatment options for patients with early- and late-stage disease, and treatment goals include cure, downstaging, palliation, and bridging patients to transplantation. These approaches can also be combined with each other and non-locoregional therapies, like systemic chemotherapy, in treatments. In this chapter, we cover current transarterial embolization (bland), chemoembolization, radioembolization, and ablation-based treatment methods in hepatocellular cancer, along with the clinical indications for each and a brief summary of their outcomes, along with a discussion of emerging treatment approaches.

Keywords

Ablation · Hepatocellular carcinoma · Locoregional therapy · Outcomes · TACE · TARE

A. E. Hare · M. S. Makary (✉)
Division of Interventional Radiology, Department of Radiology, The Ohio State University Wexner Medical Center, Columbus, OH, USA
e-mail: mina.makary@osumc.edu

Interdisciplinary Cancer Research, https://doi.org/10.1007/16833_2022_42
Published online: 6 October 2022

1 Introduction

Hepatocellular carcinoma (HCC) is the most common primary liver malignancy, making up 90% of liver tumors, and is the fifth most common cancer overall (Wong et al. 2017; Kim and El-Serag 2019; Llovet et al. 2021). Mean survival rates vary between 6 and 20 months with a 5-year relative survival rate estimated at 19.6% (Byam et al. 2013; Lewis et al. 2017). Estimates attribute approximately 700,000 deaths to liver cancers annually (Ferlay et al. 2015). No more than 10% of patient have surgical indications at the time of diagnosis, when for most patients, the disease has progressed past the early stages (Roayaie et al. 2015; Villanueva 2019). The combination of unfavorable prognoses with limited surgical indications supports the need for enhanced treatment modalities and methods. Liver-directed locoregional therapies are critical components of treatment and continue to see development.

There are a variety of targeted locoregional therapies approaches that have recently been developed (Makary et al. 2020). There are image-guided techniques that utilize locoregional delivery of radiotherapeutic, chemotherapeutic, or ablative therapy methods. These techniques include transarterial chemoembolization (TACE), transarterial embolization (TAE), transarterial radioembolization (TARE), and ablation as minimally invasive approaches, dependent on patient and tumor characteristics. These approaches may have curative intent, but have other goals in addition including tumor cytoreduction and downstaging, hypertrophy of hepatic tissue, and palliation (Inchingolo et al. 2019). Tumor cytoreduction and hepatic hypertrophy may be performed to bridge to resection surgeries or transplantation. A key advantage of locoregional treatments is decreased morbidity than surgical-based therapies (Kis et al. 2017). In this chapter, we will discuss the classification of HCC, management with a focus on targeted treatment approaches (TACE, TAE, TARE, ablation), and the various risks, benefits, and clinical indications/goals of each approach (Table 1).

Portal vein thrombosis (PVT) is a specific complication of HCC with ramifications for embolotherapy due to the vulnerability of the liver's blood supply (Zane and Makary 2021). With decreased blood flow through the portal vein, the liver comes to rely greatly on the hepatic artery for perfusion, which is the target of targeted embolotherapy. HCC's propensity for vascular invasion and promotion of a hypercoagulable state, like many cancers, also leads to high rates of PVT, with approximately half of patients developing the complication (Cerrito et al. 2019). In non-diseased livers, PVT is compensated via both the hepatic arteries and venous collaterals adjacent to the portal vein, known as cavernous transformation. These collaterals lead to increased pressure in the portal system to maintain perfusion, and further side effects that delineate BCLC stage C disease in these patients. Transplant is contraindicated, with only locoregional and system therapy available.

2 Classification Systems in HCC

Staging of HCC is dependent on observed levels of liver dysfunction. Several staging systems and clinical tools exist for use in prognostication (Levy et al. 2002; Grieco et al. 2005; Yau et al. 2014; Kim et al. 2020). The most widely used

Table 1 Comparison of locoregional therapies used in hepatocellular carcinoma

Modality	Method	Clinical indication	Risks	Benefits
TAE	Microparticles	Disease control, downstaging, bridging to transplantation	Liver failure/abscess/biloma, PES	Improved OS vs. supportive. Decreased chemotherapeutic toxicity. Cheaper than TACE
TACE	Emulsified chemotherapeutic agents or drug-eluting beads	Disease control, downstaging, bridging to transplantation	Liver failure/abscess/biloma, PES	Improved OS vs. supportive. Synergy of embolic and chemotherapeutic effects
TARE	Yt^{90} radioisotopes on microspheres	Disease control, downstaging, bridging to transplantation Radiation segmentectomy with curative intent	Radiation-induced liver disease/pneumonitis, PES, liver failure/abscess/biloma	Improved QoL/TTP vs. TACE Segmentectomy outcomes equal to resection at 5 years
Ablation	Laser, radiofrequency, microwaves, cryoablation	Early stage, nonsurgical HCC (<2–3 cm). Adjuvant to TACE for tumors up to 5 cm	Post-ablation syndrome, bleeding, local organ injury	Similar outcomes to surgery for small tumors (<3 cm)

PES post-embolization syndrome, *OS* overall survival, *QoL* quality of life, *TTP* time to progression

of these is the Barcelona Clinic Liver Cancer (BCLC) classification scheme (Llovet et al. 1999; Kulik and El-Serag 2019; Mauro and Forner 2022). The BCLC has been used internationally for recommending therapy using liver dysfunction, tumor burden, and performance status to stratify patients into five stages, from stage 0 (early disease) to stage D (highly advanced disease) as shown in Fig. 1 (Clavien et al. 2012; Mauro and Forner 2022). The Milan criteria are a set of criteria relevant to patients with HCC to determine appropriateness of liver transplantation. They include (1) one lesion smaller than 5 cm or three lesions that are smaller than 3 cm, (2) no extrahepatic manifestations, and (3) no evidence of gross vascular invasion (Mazzaferro et al. 1996; Yao 2008). While not directly engaged in the staging of HCC, the Milan criteria are utilized in medical decision-making for identification of patients who will have good outcomes following liver transplantation.

In cases of very early disease (BCLC stages 0 and A) with preserved liver function, transplantation is the preferred approach, dependent on the patient's clinical status. If the patient is not a transplant candidate, resection or ablation can yield long-term benefits with overall survival greater than 5 years. In cases of unresectable early HCC, due to spread of disease or the location, locoregional therapies are utilized for potential downstaging or bridging to transplant.

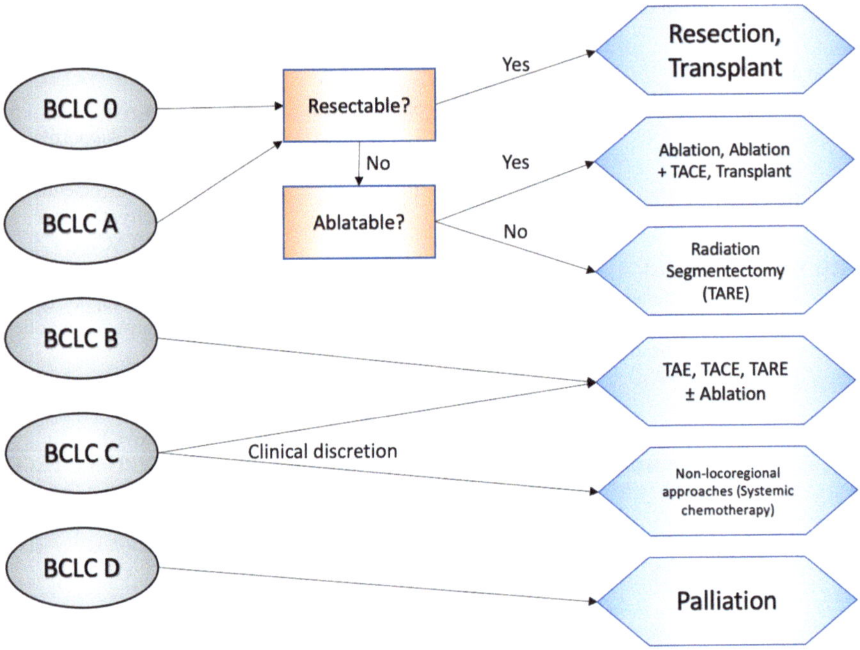

Fig. 1 BCLC stages are used to determine treatment approaches. BCLC: Barcelona Clinic Liver Cancer. TACE: transarterial chemoembolization. TARE: transarterial radioembolization. TAE: transarterial embolization

Locoregional therapies are also used for stage B and more progressed disease states. Dependent on the location and spread of disease within the liver, catheter-directed locoregional approaches can be used for stage B patients. Well-defined nodules, preserved portal flow, and good access to the tumor are all traits that favor use of locoregional therapies in these patients versus diffuse, bilobar disease that makes targeting therapy ineffective or impossible. Beyond stage C, the extent of disease involvement makes locoregional therapy no longer preferred in favor of systemic therapy. Current systemic therapies include sorafenib and lenvatinib (Kudo et al. 2018). Locoregional approaches can also be used to support systemic therapies for local disease control.

Beyond survival measures, HCC necessitates measurement of tumor response outcomes because the complexity of the disease generally indicates treatment regimens that utilize multiple therapies. To measure this tumor response, the modified Response Evaluation Criteria in Solid Tumors (mRECIST) was developed in 2010 to satisfy the controversy caused by the difficulty in adequately assessing treatment approaches (Lencioni and Llovet 2010; Kim et al. 2015). Prior to the mRECIST was the RECIST model, which evaluated tumor size reduction; however, it was incapable of assessing necrosis and devascularization within a tumor, which are the primary mechanisms of action of locoregional therapies, making it unsuitable for application to them (Forner et al. 2009). Therefore, the American Association for the Study of Liver Diseases developed mRECIST, which includes arterial enhancement in its measures and has been validated (Tsochatzis et al. 2014; Vincenzi et al. 2015). In contrast, quantitative EASL (qEASL) and volumetric RECIST have been introduced along with the propagation of three-dimensional imaging techniques, but mRECIST is still the most broadly used measure of tumor response (Tacher et al. 2016).

3 Transarterial Embolization (TAE)

The key principle underlying TAE is that HCC recruits arterial blood flow from the hepatic arteries to supply the growing tumor, whereas the healthy liver parenchyma receives blood supply primarily from the portal vein, and only secondarily from the hepatic artery (Breedis and Young 1954; Gbolahan et al. 2017). For this reason, TAE is targeted at preventing the hepatic artery from providing blood supply to the tumor. The tumor then undergoes cell death via cellular membrane disruption brought out by the ischemic state formed by embolism of the hepatic artery blood supply (Brown et al. 1998). In contrast with TACE or TARE, TAE is "bland" embolization due to the lack of additional chemotherapeutic or radioactive agents provided alongside the embolic function. The embolic agents used include PVA particles, gelatin sponge, or spherical embolic agents (Brown et al. 1998; Tsochatzis et al. 2014).

As with locoregional approaches generally, TAE is most used for BCLC class B and C patients, though it can also be used in class A patients to extend transplantation candidate status with respect to the Milan criteria (Lammer et al. 2010). The goal in class B and C patients is in downstaging the disease to prepare patients for transplant. TAE is not recommended to patients with poor liver function, defined as a Child-Pugh score of B8 or higher, nor for certain comorbidities, such as elevated LFTs, decompensated cirrhosis, low creatinine clearance, and uncontrolled varices (Raoul et al. 2011). Thus, pre-treatment testing will include varied testing and evaluation, including clinical indices in addition to the Child-Pugh, such as the ALBI (albumin-bilirubin), MELD (Model for End-stage Liver Disease), and ECOG (Eastern Cooperative Oncology Group to determine patient status (Levy et al. 2002; Johnson et al. 2015; Guerrini et al. 2018). Generally, TAE is for patients who are not surgical candidates whose disease is also predominantly located in the liver. Imaging studies are also utilized prior to procedure to evaluate the vascular anatomy and patency of the portal vein and access site (Sieghart et al. 2015).

A key challenge of TAE is identifying the specific branches of the hepatic artery that have been recruited for the tumor supply, and thus that are to be embolized in the procedure, along with precise placement of the catheter for accurate delivery of embolic agents to maximize treatment effect and minimize any collateral damage from ischemia in adjacent liver tissues. Side effects and complications of TAE generally arise from these regional ischemic effects, and because of that fact, TAE is excellent in preserving healthy liver tissue not affected by the embolized vessel versus the systemic effect of chemotherapy. The advantage of not using chemotherapeutic agents with potential toxic effects is the strongest factor promoting usage of

TAE vs. TACE. The risk of post-embolization syndrome, while high in TAE (68.7%), is lower than that of embolization using chemotherapeutic agents (TACE, 74.7%) (Agrawal et al. 2021). In addition, hospitalizations due to PES were shorter for TAE patients (1.47 days) than TACE patients (1.12 days).

The underlying purpose of locoregional therapies is to increase overall survival in patients. Compared to supportive care, initial studies on unresectable HCC showed a survival benefit to TAE (hazard ratio 0.47, $p = 0.025$) (Llovet et al. 2002). However, in the same study, 1-year survival was 75% and 2-year survival was 50% for TAE, compared to 82% and 63%, respectively, for TACE. Both were significant improvements on supportive care showing 63% and 27%, respectively ($p = 0.009$, TACE vs. control), though there was no significant difference between the TAE and TACE ($p = 0.009$, TACE vs. control). A more recent study that compared TAE to doxorubicin TACE demonstrated an insignificant survival difference between treatments (hazard ratio 1.11; 95% CI 0.71–1.76; $p = 0.64$) (Brown et al. 2016). That study also demonstrated a similar rate of adverse events, tumor response (via RECIST), and progression-free survival.

4 Transarterial Chemoembolization (TACE)

In contrast to TAE, TACE utilizes chemotherapeutic agents alongside embolic agents. The underlying principles remain the same, including the importance of the differing blood supply between tumor (hepatic artery) and healthy tissue (portal venous system). This dichotomy is more critical in TACE than TAE due to the importance of targeting chemotherapeutic agents to tumor tissue (Breedis and Young 1954). Chemotherapeutic agents targeted through catheters in the hepatic artery will minimize the toxic effects on healthy liver tissue.

Longer-term outcomes of TACE are analyzed with respect to the patient's baseline staging, performance status, extrahepatic disease, and tumor burden. The Assessment for Retreatment with TACE (ART) score was developed to guide decision-making for assessing possible treatment benefit of TACE treatments beyond the initial treatment and is independent of Child-Pugh score (Sieghart et al. 2013; Hucke et al. 2014). The ART score is a significant predictor of overall survival, and using a score of 2.5 as a cutoff prior to second TACE treatment identified patients who would not receive any benefit from further TACE treatment sessions. An additional measure applied to TACE patients is the albumin-bilirubin (ALBI) grade (Johnson et al. 2015). It has been used to predict survival of HCC patients who have been treated with either TACE or TARE (Young et al. 2022). It has been shown to have greater capacity to discriminate than Child-Pugh in patients treated with DEB-TACE with their BCLC stage.

There are two primary variants of TACE in use today. Conventional TACE (cTACE) introduces chemotherapeutic agents through the hepatic artery. These agents are emulsified with Lipiodol oil to increase effectiveness of the treatment through decreased washout and increased concentrations. In contrast, drug-eluting beads (also called drug-eluting microspheres) have been used (DEB-TACE) to provide greater targeting of agents to the tumor versus healthy tissues (Poon et al.

2007; Malagari et al. 2008; Woo and Heo 2015). Advantages of DEB-TACE over cTACE also include increased ischemic effect due to improved embolic activity, less systemic delivery of chemotherapeutic agents, and consistent dosing of chemotherapeutic agents. In recent studies, DEB-TACE has been shown to have superior tumor response to cTACE (via mRECIST) and lower rates of side effects like PES than cTACE (Razi et al. 2022). However, other studies suggest cTACE yields greater symptomatic improvement, though those studies agree that DEB-TACE is associated with lesser LFT elevation and PES rates, which suggests improvements in toxicity with DEB-TACE (Makary et al. 2016; de Baere et al. 2020). Overall survival improvements in DEB-TACE have not been observed vs. cTACE through multiple studies (Facciorusso et al. 2015).

Lastly, balloon occlusion methods (B-TACE) are another newer variant of TACE, which recent data suggesting improved oncological response over DEB-TACE (Hatanaka et al. 2018; Lucatelli et al. 2021). In addition, B-TACE treatment approaches have a lower retreatment rate than cTACE or DEB-TACE, though side effect rates like post-embolization syndrome were higher (Golfieri et al. 2021). At this time, there is unclear survival benefit to B-TACE over other TACE approaches.

With an increase in TACE-based treatment approaches, there is a need for ever-increasing data and analysis of that data to understand potential treatments to develop algorithms for improved clinical care. Unfortunately, despite a proliferation of studies, there is a dearth of availability to those studies (Grégory et al. 2020). Only 39% of identified randomized controlled trials assessing TACE for HCC have made their results publicly available. In addition to concerns of inappropriate use of research funding, this raises concerns about bias in scientific publications and harms our collective ability to develop clinically appropriate treatments.

5 Transarterial Radioembolization (TARE)

Transarterial radioembolization (TARE) relies on the same process of targeted embolization via the hepatic artery, leading to local ischemia and tumor death, as other embolization methods previously discussed. Also called selective internal radiotherapy (SIRT), the process utilizes a radioisotope of yttrium, called Y^{90}, bound to microspheres instead of the TACE approach of chemotherapeutic agents bound to microspheres (Salem et al. 2016). Beta decay of Y^{90} irradiates the tumor over a 2-week period (Makary et al. 2018). The treatment procedure is similar in approach to TAE or TACE, in which a catheter is directed into the hepatic artery to direct the therapeutic agent (Y^{90}) in a specific lobar or segmental distribution. In contrast to external beam radiation (EBRT), Y^{90} therapy allows much higher radiation dosing and correlated increased efficacy in antitumor activity (Dawson et al. 2002). Due to the longer time period necessary for irradiation, there is an extended latency compared to TAE or TACE before complete visualization of the effect of treatment (Lewandowski et al. 2009). This time period generally requires multiple months before posttreatment diagnostic imaging is done to assess efficacy.

In comparison with TACE, the delivery of off-target Y^{90} radioisotopes poses even greater risk of off-target radiation, which is mitigated through a planning arteriography procedure performed 1–2 weeks prior to Y^{90} treatment. During the planning arteriography, the tumor's vascular anatomy is investigated, and extrahepatic vessels that may lead to off-target Y^{90} deposition are identified. Any vessels that may lead to this deposition are embolized (Gaba 2015). In addition, the degree of lung shunting is determined. Due to the nature of the vascular return from the hepatic veins into the right side of the heart and into the lungs along with the capillary beds of the lung providing a filter to accumulate Y^{90}, it is necessary to identify that the catheter site chosen will not cause radioactive particles to accumulate in the lung, leading to lung fibrosis. One last key advantage of this preprocedural planning is careful modulating treatment dosing to personalize for each patient to a degree not present in non-radioembolization techniques (Kallini et al. 2016).

Y^{90} therapy as a treatment approach is traditionally used for palliation in patients with unresectable HCC and is performed on the lobar scale within the liver (Salem et al. 2010; Makary et al. 2018). TACE has been the standard of care for patients with BCLC stage B disease, but Y^{90} has become more common, with possible superiority in time to progression measures. Unfortunately, Y^{90} therapy is not appropriate for patients with high total bilirubin (greater than 2 mg/dL), whereas TACE remains a potential treatment option for patients up to 4 mg/dL (Makary et al. 2018). Y^{90} therapy is also the preferred locoregional treatment approach in patients who have developed portal vein thrombosis.

Radiation segmentectomy is a specific application of TARE with the goal of applying an ablative dose of radiation to HCC-bearing parenchyma (Vouche et al. 2014). By using the lobar volume, a specific dose, generally 120–150 Gy, is introduced via the segmental blood supply to replicate the effect of ablative therapy within only the targeted segment or segments in a catheter-directed technique. Usually no more than two hepatic segments are targeted, which also has the advantage of limited nontarget irradiation. Unlike other catheter-directed therapies, radiation segmentectomy (RS) is performed with curative intent, as a stand-in for the surgical resection that it mimics. In a retrospective study of 70 patients treated with RS, response rates, tumor, and survival were comparable to other curative-intent treatments, such as resection, transplantation, and radiofrequency ablation (Lewandowski et al. 2018).

Outcomes in Y^{90} radioembolization vary based on patient characteristics, but studies have shown survival benefits approximately equal to TACE. The SIRTACE study, whose goal was to compare TARE to TACE in BCLC stages A, B, and C, showed similar quality of life between treatments, with TARE only requiring a single procedure. Both treatments also showed similar rates of tumor response via RECIST and disease control (Kolligs et al. 2015). Another study comparing quality of life between patients undergoing TARE or TACE (which had lower tumor and disease burdens) showed superior social and functional well-being, along with higher embolotherapy-specific scores in the TARE group, despite the lesser disease in the TACE patient group. Long-term outcome studies in patients with Child-Pugh A and B diseases have shown that patients with Child-Pugh disease received greater benefits from treatment, regardless of portal vein thrombosis, whereas patients with Child-Pugh B disease had a worse overall outcome, exacerbated by portal vein thrombosis (Salem et al. 2010). Side effects associated with Y^{90} therapy include fatigue, pain, nausea/vomiting, and radioembolization-induced liver disease (REILD), which is a form of sinusoidal obstruction syndrome. REILD appears 4–8 weeks following Y^{90} treatment as jaundice and ascites along with GGT and alkaline phosphatase increases (Salem et al. 2013; Bruix et al. 2016).

6 Ablation

Ablation is a locoregional technique traditionally used in small (<3 cm) tumors in patients for which surgical resection would be recommended, but are poor surgical candidates, though it also has utility in higher-stage disease for somewhat larger tumors (<5 cm) as an adjuvant to other methods (Heimbach et al. 2018). The initial techniques utilized percutaneous ethanol injection, a method of chemical ablation, but this method was surpassed and was shown inferior to newer methods in the 1990s (Lee and Lee 2018). The most common techniques utilized in locoregional treatment of HCC are microwave ablation (MWA), radiofrequency ablation (RFA), and cryoablation (Inchingolo et al. 2019). MWA uses local heat to lead to cell death via an oscillating microwave field (915/2450 MHz), which has a reduced heat sink effect compared to RFA, a similar approach that substitutes microwave field heating with alternating electric current to create heating, leading to cellular necrosis. RFA's principal advantage over MWA is improved precision of the energy to the target site, avoiding off-target necrosis (Yang et al. 2015). This trade-off between precise targeting and less heat sink indicates MWA as a preferred method in proximity to large vessels, large tumors (3–5 cm), or multiple nodule disease (Xu et al. 2019). Compared to the heat-based methods, cryoablation (CA) uses argon and helium gases to freeze and thaw the target tissues alternately, based on the Joule-Thomson effect (Song 2016). This destroys the tissue through direct tissue destruction and vascular-related injury (Gage and Baust 1998; Hoffmann and Bischof 2002). However, the side effects of cryoablation are severe, leading it to be a less commonly chosen technique than other ablative methods for HCC (Wu et al. 2015). Like thermal ablation, the heat sink effect is at play, wherein adjacent vascular structures dampen the ablative effect by conducting heat into the ablated region.

One common side effect of HCC is post-ablation syndrome (PAS), a flu-like illness with an onset of approximately 3 days post-procedure, which lasts 5 days. It is associated with tumor volumes of >50 cm^3, ablated tissue volumes of >150 cm^3, large volumes ablated outside the tumor (>125 cm^3), or elevated aspartate aminotransferase levels (>350 IU/L) post-ablation (Dodd et al. 2005). The most common symptoms reported are fever, malaise, chills, delayed pain, and nausea. Approximately 25–35% of patients receiving ablation for HCC experience PAS, and it is most common after RFA ablation procedures.

Studies on outcomes of ablation in HCC are generally performed with surgical resection as a comparison group, not other locoregional therapies, because ablation is most commonly used in patients with BCLC stage 0 and A diseases who are not good surgical candidates instead of more advanced disease, like catheter-directed therapies are. Like resection, ablation is often performed with curative intent. Several studies have been performed to evaluate RFA versus resection. In meta-analyses comparing these treatments, hepatic resection may have superior overall survival and 2-year survival than RFA; however, RFA is superior to resection in decreasing complications and decreasing length of hospital stays (Weis et al. 2013; Xu et al. 2014, 2018). However, in those studies, small tumors (<2 cm) showed similar overall survival, suggesting a difference in survival between these groups stems from larger tumors. Other studies that have targeted RFA vs. hepatic resection have specifically interrogated the intermediate size tumors (2–3 cm) over a longer period of time posttreatment and found no significant difference in survival or recurrence (Pompili et al. 2013).

MWA has similar application to RFA, and multiple meta-analyses have directly compared these treatment approaches. For smaller tumors, meta-analyses show no significant difference in tumor response and survival between both treatments groups (Chinnaratha et al. 2016; Facciorusso et al. 2016). However, both studies agreed that MWA showed significant improved response rates in larger tumors than RFA. There was not sufficient data to stratify results by tumor stage, which may only correlate with tumor size, and not necessarily explain the difference. There was a significant difference in adverse events noted, with MWA responsible for greater numbers of adverse events, which is a possible mechanism for explaining the similar overall response between treatments. MWA is a superior ablation approach for larger tumors with a higher rate of complications.

Cryoablation studies are generally performed with comparison to both other well-established methods of ablation, RFA and MWA. One study showed significant decreased tumor progression at 1, 2, and 3 years in the cryoablation group than the RFA treatment group in analysis limited to small (<3 cm) (Wang et al. 2015). Overall survival and tumor-free survival showed no significant difference in either group, and a similar rate of major complications was observed. One study, specifically targeted at safety of cryoablation and RFA in cirrhotic patients, found no significant difference in overall safety between these techniques, with similar complication rates of 39.4% in cryoablation and 25.7% in RFA ($p = 0.26$) (Dunne et al. 2014). A separate study had differing results, concluding that RFA was safer (3.3% complication rate) than cryoablation (40.7% complication rate, 1 postoperative death) (Pearson et al. 1999). In addition to the study population in Dunne et al. including specifically cirrhotic patients, whereas Pearson et al. excluded Child's class C cirrhotic patients, this discrepancy is also partially explained by the rates of intrahepatic abscesses and pleural effusions in the cryoablation group in Pearson et al., which were not measured in Dunne et al.

Ablation therapies have also been used as a component of combination therapies for HCC, used for larger tumors (>3 cm) that are BCLC stage A or B that are unresectable. These combined treatments include RFA with TACE and have shown improved response and disease-free survival than TACE alone in tumors 2–3 cm in size, along with improved tumor control than RFA alone treatments (Kim et al.

2012). A study targeting RFA with and without TACE investigated HCC up to 7 cm has that TACE-RFA was superior to RFA in improving overall survival (Peng et al. 2013). A more recent retrospective study has investigated the time period between RFA and TACE treatments and its impact on survival measures and complete response rate (Liu et al. 2019). In it, the TACE plus RFA within 30-day group had significantly improved overall survival, progression-free survival, and complete response than the TACE plus RFA outside 30-day group, emphasizing that closely spaced treatments are important for improving patient outcomes.

7 Future Directions

Locoregional therapies are under constant development and novel therapies and in the immediate future will likely rely on novel modifications to improve current methods. One novel modification in the realm of TACE is improvement in catheter technology to address one of the limiting factors in TACE treatment: careful targeting and dosing of chemotherapeutic agents. Microvalve infusion (MVI) catheters allow treatment to prevent retrograde flow of chemotherapeutic agents into healthy liver tissue and increase uptake into target sites (Titano et al. 2019). Recent studies have shown improved tumor response and increased deposition of DEB-TACE microspheres into tumor tissue compared to standard end-hole catheters. Current clinical trials are undergoing to improve this technology.

Ablation is defined by its varied approaches to delivering energy to the target site. In addition to the well-established modalities of RFA, MWA, and cryoablation, newer ablation modalities have begun to be applied to HCC. Irreversible electroporation (IRE) is an ablation technique relying on high-current electrical pulses to promote the formation of pores in the cellular membrane, which leads to cell death (Zimmerman et al. 2017). Electroporation has also been combined with immunotherapies, which has promising data showing efficacy (Trotovšek et al. 2021). More data is currently needed to identify the role of IRE in the future treatment of HCC. High-intensity focused ultrasound (HIFU) is a novel noninvasive ablation technique that utilizes sonication to deliver energy to microbubbles and has been shown to have similar outcomes to TACE (Huang et al. 2019). Like IRE, there is not enough data to fully draw conclusions as to the utility of HIFU in HCC treatment.

Histotripsy is an ablation-based treatment, with similarities to HIFU, which has been used in multiple applications, including recently HCC. Histotripsy uses focused ultrasound to destroy tissues through cavitation (Xu et al. 2021). Like HIFU, histotripsy has the advantage of being nonthermal and uses no ionizing radiation, which are the root causes of complications of other ablation approaches. The key difference between HIFU and histotripsy is the reliance on mechanical cavitation in histotripsy to induce damage, compared to HIFU delivering ultrasound waves to create heat. Initial studies of histotripsy in HCC have been applied to a xenograft murine model, which showed potential for histotripsy for HCC treatment (Worlikar et al. 2018). Current clinical trials are in process for using histotripsy for liver tumors

(The HistoSonics System for Treatment of Primary and Metastatic Liver Tumors Using Histotripsy (#HOPE4LIVER US) n.d.).

Combination of locoregional therapies with immunotherapies has been a steadily growing field in the treatment of HCC (Zongyi and Xiaowu 2020; Sangro et al. 2021). The targeted nature of catheter-directed locoregional therapies allows for the delivery of immunologic agents, such as Toll-like receptor agonists, cGAS-STING agonists, RIGI or MDA5 agonists, and inflammasome agonists, with toxicity too great for systemic administration, but intratumor delivery potentiates response and decreases the toxicity to acceptable levels. In addition to delivering immunotherapies via catheter-directed methods, more conventional locoregional therapies, most commonly TACE, TARE, and RFA, can be performed alongside systemic immunotherapies in a treatment regimen (Duffy et al. 2017; Rodríguez-Ruiz et al. 2018).

8 Conclusions

HCC is the third most common cause of cancer-related death across the world (Kim and El-Serag 2019). Prognosis is poor for patients with advanced disease, and while surgical resection and transplantation are appropriate treatments for many patients with early disease, numerous clinical complications make that an impossibility for many others. Locoregional therapies have applications for cure in early-stage patients, along with effective palliation or downstaging to curative therapies for late-stage patients. TAE, TACE, and TARE have all proven to be catheter-directed therapies that can effectively improve survival in patients with disparate clinical comorbidities and tumor burdens, while ablation has been shown to be useful as a noninvasive alternative for surgical intervention in patients who are not surgical candidates along with possessing synergistic function in combination therapies. New refinements on each of these techniques are being developed constantly, and researchers have been studying each before making these opportunities available for clinical practice.

Compliance with Ethical Standards

Disclosure of Interests All authors declare they have no conflict of interest.

Funding None.

References

Agrawal R et al (2021) Identifying predictors and evaluating the role of steroids in the prevention of post-embolization syndrome after transarterial chemoembolization and bland embolization. Ann Gastroenterol 34(2):241–246. https://doi.org/10.20524/aog.2020.0566

Breedis C, Young G (1954) The blood supply of neoplasms in the liver. Am J Pathol 30(5):969–977

Brown KT et al (1998) Particle embolization for hepatocellular carcinoma. J Vasc Interv Radiol 9(5):822–828. https://doi.org/10.1016/s1051-0443(98)70398-7

Brown KT et al (2016) Randomized trial of hepatic artery embolization for hepatocellular carcinoma using doxorubicin-eluting microspheres compared with embolization with microspheres alone. J Clin Oncol 34(17):2046–2053. https://doi.org/10.1200/JCO.2015.64.0821
Bruix J, Reig M, Sherman M (2016) Evidence-based diagnosis, staging, and treatment of patients with hepatocellular carcinoma. Gastroenterology 150(4):835–853. https://doi.org/10.1053/j.gastro.2015.12.041
Byam J, Renz J, Millis JM (2013) Liver transplantation for hepatocellular carcinoma. Hepatobil Surg Nutrit 2(1):22–30. https://doi.org/10.3978/j.issn.2304-3881.2012.11.03
Cerrito L et al (2019) Treatment of hepatocellular carcinoma in patients with portal vein tumor thrombosis: beyond the known frontiers. World J Gastroenterol 25(31):4360–4382. https://doi.org/10.3748/wjg.v25.i31.4360
Chinnaratha MA et al (2016) Percutaneous thermal ablation for primary hepatocellular carcinoma: a systematic review and meta-analysis. J Gastroenterol Hepatol 31(2):294–301. https://doi.org/10.1111/jgh.13028
Clavien P-A et al (2012) Recommendations for liver transplantation for hepatocellular carcinoma: an international consensus conference report. Lancet Oncol 13(1):e11–e22. https://doi.org/10.1016/S1470-2045(11)70175-9
Dawson LA et al (2002) Analysis of radiation-induced liver disease using the Lyman NTCP model. Int J Radiat Oncol Biol Phys 53(4):810–821. https://doi.org/10.1016/s0360-3016(02)02846-8
de Baere T et al (2020) Real life prospective evaluation of new drug-eluting platform for chemoembolization of patients with hepatocellular carcinoma: PARIS registry. Cancers 12(11). https://doi.org/10.3390/cancers12113405
Dodd GD et al (2005) Percutaneous radiofrequency ablation of hepatic tumors: postablation syndrome. Am J Roentgenol 185(1):51–57. https://doi.org/10.2214/ajr.185.1.01850051
Duffy AG et al (2017) Tremelimumab in combination with ablation in patients with advanced hepatocellular carcinoma. J Hepatol 66(3):545–551. https://doi.org/10.1016/j.jhep.2016.10.029
Dunne RM et al (2014) Percutaneous treatment of hepatocellular carcinoma in patients with cirrhosis: a comparison of the safety of cryoablation and radiofrequency ablation. Eur J Radiol 83(4):632–638. https://doi.org/10.1016/j.ejrad.2014.01.007
Facciorusso A et al (2015) Transarterial chemoembolization: evidences from the literature and applications in hepatocellular carcinoma patients. World J Hepatol 7(16):2009–2019. https://doi.org/10.4254/wjh.v7.i16.2009
Facciorusso A, di Maso M, Muscatiello N (2016) Microwave ablation versus radiofrequency ablation for the treatment of hepatocellular carcinoma: a systematic review and meta-analysis. Int J Hyperthermia 32(3):339–344. https://doi.org/10.3109/02656736.2015.1127434
Ferlay J et al (2015) Cancer incidence and mortality worldwide: sources, methods and major patterns in GLOBOCAN 2012. Int J Cancer 136(5):E359–E386. https://doi.org/10.1002/ijc.29210
Forner A et al (2009) Evaluation of tumor response after locoregional therapies in hepatocellular carcinoma: are response evaluation criteria in solid tumors reliable? Cancer 115(3):616–623. https://doi.org/10.1002/cncr.24050
Gaba RC (2015) Planning arteriography for Yttrium-90 microsphere Radioembolization. Semin Interv Radiol 32(4):428–438. https://doi.org/10.1055/s-0035-1564708
Gage AA, Baust J (1998) Mechanisms of tissue injury in cryosurgery. Cryobiology 37(3):171–186. https://doi.org/10.1006/cryo.1998.2115
Gbolahan OB et al (2017) Locoregional and systemic therapy for hepatocellular carcinoma. J Gastrointest Oncol 8(2):215–228. https://doi.org/10.21037/jgo.2017.03.13
Golfieri R et al (2021) Retrospective European multicentric evaluation of selective Transarterial chemoembolisation with and without balloon-occlusion in patients with hepatocellular carcinoma: a propensity score matched analysis. Cardiovasc Intervent Radiol 44(7):1048–1059. https://doi.org/10.1007/s00270-021-02805-5

Grégory J et al (2020) Results of trials assessing transarterial chemoembolization for treating hepatocellular carcinoma are critically underreported. Eur Radiol 30(10):5633–5640. https://doi.org/10.1007/s00330-020-06900-y

Grieco A et al (2005) Prognostic factors for survival in patients with early-intermediate hepatocellular carcinoma undergoing non-surgical therapy: comparison of Okuda, CLIP, and BCLC staging systems in a single Italian Centre. Gut 54(3):411–418. https://doi.org/10.1136/gut.2004.048124

Guerrini GP et al (2018) Value of HCC-MELD score in patients with hepatocellular carcinoma undergoing liver transplantation. Prog Transplant (Aliso Viejo, Calif.) 28(1):63–69. https://doi.org/10.1177/1526924817746686

Hatanaka T, Arai H, Kakizaki S (2018) Balloon-occluded transcatheter arterial chemoembolization for hepatocellular carcinoma. World J Hepatol 10(7):485–495. https://doi.org/10.4254/wjh.v10.i7.485

Heimbach JK et al (2018) AASLD guidelines for the treatment of hepatocellular carcinoma. Hepatology 67(1):358–380. https://doi.org/10.1002/hep.29086

Hoffmann NE, Bischof JC (2002) The cryobiology of cryosurgical injury. Urology 60(2 Suppl 1): 40–49. https://doi.org/10.1016/s0090-4295(02)01683-7

Huang L et al (2019) Efficacy and safety of high-intensity focused ultrasound ablation for hepatocellular carcinoma by changing the acoustic environment: microbubble contrast agent (SonoVue) and transcatheter arterial chemoembolization. Int J Hyperth 36(1):243–251. https://doi.org/10.1080/02656736.2018.1558290

Hucke F et al (2014) The ART-strategy: sequential assessment of the ART score predicts outcome of patients with hepatocellular carcinoma re-treated with TACE. J Hepatol 60(1):118–126. https://doi.org/10.1016/j.jhep.2013.08.022

Inchingolo R et al (2019) Locoregional treatments for hepatocellular carcinoma: current evidence and future directions. World J Gastroenterol 25(32):4614–4628. https://doi.org/10.3748/wjg.v25.i32.4614

Johnson PJ et al (2015) Assessment of liver function in patients with hepatocellular carcinoma: a new evidence-based approach-the ALBI grade. J Clin Oncol 33(6):550–558. https://doi.org/10.1200/JCO.2014.57.9151

Kallini JR et al (2016) Transarterial Radioembolization with Yttrium-90 for the treatment of hepatocellular carcinoma. Adv Ther 33(5):699–714. https://doi.org/10.1007/s12325-016-0324-7

Kim H-S, El-Serag HB (2019) The Epidemiology of Hepatocellular Carcinoma in the USA. Curr Gastroenterol Rep 21(4):17. https://doi.org/10.1007/s11894-019-0681-x

Kim JW et al (2012) Hepatocellular carcinomas 2-3 cm in diameter: transarterial chemoembolization plus radiofrequency ablation vs. radiofrequency ablation alone. Eur J Radiol 81(3):e189–e193. https://doi.org/10.1016/j.ejrad.2011.01.122

Kim MN et al (2015) Evolution from WHO to EASL and mRECIST for hepatocellular carcinoma: considerations for tumor response assessment. Expert Rev Gastroenterol Hepatol 9(3):335–348. https://doi.org/10.1586/17474124.2015.959929

Kim KM et al (2020) Child-Pugh, MELD, MELD-Na, and ALBI scores: which liver function models best predicts prognosis for HCC patient with ascites? Scand J Gastroenterol 55(8): 951–957. https://doi.org/10.1080/00365521.2020.1788139

Kis B et al (2017) Liver-directed therapies for hepatocellular carcinoma and intrahepatic cholangiocarcinoma. Cancer Control 24(3):1073274817729244. https://doi.org/10.1177/1073274817729244

Kolligs FT et al (2015) Pilot randomized trial of selective internal radiation therapy vs. chemoembolization in unresectable hepatocellular carcinoma. Liver Int 35(6):1715–1721. https://doi.org/10.1111/liv.12750

Kudo M et al (2018) Lenvatinib versus sorafenib in first-line treatment of patients with unresectable hepatocellular carcinoma: a randomised phase 3 non-inferiority trial. Lancet (London, England) 391(10126):1163–1173. https://doi.org/10.1016/S0140-6736(18)30207-1

Kulik L, El-Serag HB (2019) Epidemiology and management of hepatocellular carcinoma. Gastroenterology 156(2):477–491.e1. https://doi.org/10.1053/j.gastro.2018.08.065
Lammer J et al (2010) Prospective randomized study of doxorubicin-eluting-bead embolization in the treatment of hepatocellular carcinoma: results of the PRECISION V study. Cardiovasc Intervent Radiol 33(1):41–52. https://doi.org/10.1007/s00270-009-9711-7
Lee DH, Lee JM (2018) Recent advances in the image-guided tumor ablation of liver malignancies: radiofrequency ablation with multiple electrodes, real-time multimodality fusion imaging, and new energy sources. Korean J Radiol 19(4):545–559. https://doi.org/10.3348/kjr.2018.19.4.545
Lencioni R, Llovet JM (2010) Modified RECIST (mRECIST) assessment for hepatocellular carcinoma. Semin Liver Dis 30(1):52–60. https://doi.org/10.1055/s-0030-1247132
Levy I, Sherman M, Liver Cancer Study Group of the University of Toronto (2002) Staging of hepatocellular carcinoma: assessment of the CLIP, Okuda, and child-Pugh staging systems in a cohort of 257 patients in Toronto. Gut 50(6):881–885. https://doi.org/10.1136/gut.50.6.881
Lewandowski RJ et al (2009) A comparative analysis of transarterial downstaging for hepatocellular carcinoma: chemoembolization versus radioembolization. Am J Transplant 9(8):1920–1928. https://doi.org/10.1111/j.1600-6143.2009.02695.x
Lewandowski RJ et al (2018) Radiation Segmentectomy: potential curative therapy for early hepatocellular carcinoma. Radiology 287(3):1050–1058. https://doi.org/10.1148/radiol.2018171768
Lewis DR et al (2017) Early estimates of SEER cancer incidence, 2014. Cancer 123(13):2524–2534. https://doi.org/10.1002/cncr.30630
Liu D et al (2019) Transarterial chemoembolization followed by radiofrequency ablation for hepatocellular carcinoma: impact of the time interval between the two treatments on outcome. J Vasc Interv Radiol 30(12):1879–1886. https://doi.org/10.1016/j.jvir.2019.07.029
Llovet JM, Brú C, Bruix J (1999) Prognosis of hepatocellular carcinoma: the BCLC staging classification. Semin Liver Dis 19(3):329–338. https://doi.org/10.1055/s-2007-1007122
Llovet JM et al (2002) Arterial embolisation or chemoembolisation versus symptomatic treatment in patients with unresectable hepatocellular carcinoma: a randomised controlled trial. Lancet (London, England) 359(9319):1734–1739. https://doi.org/10.1016/S0140-6736(02)08649-X
Llovet JM et al (2021) Hepatocellular carcinoma. Nat Rev Dis Primers 7(1):6. https://doi.org/10.1038/s41572-020-00240-3
Lucatelli P et al (2021) Balloon occluded TACE (B-TACE) vs DEM-TACE for HCC: a single center retrospective case control study. BMC Gastroenterol 21(1):51. https://doi.org/10.1186/s12876-021-01631-w
Makary MS et al (2016) Conventional versus drug-eluting bead Transarterial chemoembolization for neuroendocrine tumor liver metastases. J Vasc Interv Radiol 27(9):1298–1304. https://doi.org/10.1016/j.jvir.2016.05.014
Makary MS et al (2018) Yttrium-90 microsphere selective internal radiation therapy for liver metastases following systemic chemotherapy and surgical resection for metastatic adrenocortical carcinoma. World J Clin Oncol 9(1):20–25. https://doi.org/10.5306/wjco.v9.i1.20
Makary MS et al (2020) Locoregional therapy approaches for hepatocellular carcinoma: recent advances and management strategies. Cancers 12(7). https://doi.org/10.3390/cancers12071914
Malagari K et al (2008) Transarterial chemoembolization of unresectable hepatocellular carcinoma with drug eluting beads: results of an open-label study of 62 patients. Cardiovasc Intervent Radiol 31(2):269–280. https://doi.org/10.1007/s00270-007-9226-z
Mauro E, Forner A (2022) Barcelona clinic liver cancer 2022 update: linking prognosis prediction and evidence-based treatment recommendation with multidisciplinary clinical decision-making. Liver Int 42(3):488–491. https://doi.org/10.1111/liv.15180
Mazzaferro V et al (1996) Liver transplantation for the treatment of small hepatocellular carcinomas in patients with cirrhosis. N Engl J Med 334(11):693–699. https://doi.org/10.1056/NEJM199603143341104
Pearson AS et al (1999) Intraoperative radiofrequency ablation or cryoablation for hepatic malignancies. Am J Surg 178(6):592–598. https://doi.org/10.1016/S0002-9610(99)00234-2

Peng Z-W et al (2013) Radiofrequency ablation with or without transcatheter arterial chemoembolization in the treatment of hepatocellular carcinoma: a prospective randomized trial. J Clin Oncol 31(4):426–432. https://doi.org/10.1200/JCO.2012.42.9936
Pompili M et al (2013) Long-term effectiveness of resection and radiofrequency ablation for single hepatocellular carcinoma ≤3 cm. Results of a multicenter Italian survey. J Hepatol 59(1):89–97. https://doi.org/10.1016/j.jhep.2013.03.009
Poon RTP et al (2007) A phase I/II trial of chemoembolization for hepatocellular carcinoma using a novel intra-arterial drug-eluting bead. Clin Gastroenterol Hepatol 5(9):1100–1108. https://doi.org/10.1016/j.cgh.2007.04.021
Raoul J-L et al (2011) Evolving strategies for the management of intermediate-stage hepatocellular carcinoma: available evidence and expert opinion on the use of transarterial chemoembolization. Cancer Treat Rev 37(3):212–220. https://doi.org/10.1016/j.ctrv.2010.07.006
Razi M et al (2022) Comparison of tumor response following conventional versus drug-eluting bead transarterial chemoembolization in early- and very early-stage hepatocellular carcinoma. J Intervent Med 5(1):10–14. https://doi.org/10.1016/j.jimed.2021.12.004
Roayaie S et al (2015) The role of hepatic resection in the treatment of hepatocellular cancer. Hepatology (Baltimore, Md.) 62(2):440–451. https://doi.org/10.1002/hep.27745
Rodríguez-Ruiz ME et al (2018) Combined immunotherapy encompassing intratumoral poly-ICLC, dendritic-cell vaccination and radiotherapy in advanced cancer patients. Ann Oncol 29(5):1312–1319. https://doi.org/10.1093/annonc/mdy089
Salem R et al (2010) Radioembolization for hepatocellular carcinoma using Yttrium-90 microspheres: a comprehensive report of long-term outcomes. Gastroenterology 138(1): 52–64. https://doi.org/10.1053/j.gastro.2009.09.006
Salem R, Mazzaferro V, Sangro B (2013) Yttrium 90 radioembolization for the treatment of hepatocellular carcinoma: biological lessons, current challenges, and clinical perspectives. Hepatology (Baltimore, Md.) 58(6):2188–2197. https://doi.org/10.1002/hep.26382
Salem R et al (2016) Y90 radioembolization significantly prolongs time to progression compared with chemoembolization in patients with hepatocellular carcinoma. Gastroenterology 151(6): 1155–1163.e2. https://doi.org/10.1053/j.gastro.2016.08.029
Sangro B et al (2021) Advances in immunotherapy for hepatocellular carcinoma. Nat Rev Gastroenterol Hepatol 18(8):525–543. https://doi.org/10.1038/s41575-021-00438-0
Sieghart W et al (2013) The ART of decision making: retreatment with transarterial chemoembolization in patients with hepatocellular carcinoma. Hepatology 57(6):2261–2273. https://doi.org/10.1002/hep.26256
Sieghart W, Hucke F, Peck-Radosavljevic M (2015) Transarterial chemoembolization: modalities, indication, and patient selection. J Hepatol 62(5):1187–1195. https://doi.org/10.1016/j.jhep.2015.02.010
Song KD (2016) Percutaneous cryoablation for hepatocellular carcinoma. Clin Mol Hepatol 22(4): 509–515. https://doi.org/10.3350/cmh.2016.0079
Tacher V et al (2016) Comparison of existing response criteria in patients with hepatocellular carcinoma treated with Transarterial chemoembolization using a 3D quantitative approach. Radiology 278(1):275–284. https://doi.org/10.1148/radiol.2015142951
The HistoSonics System for Treatment of Primary and Metastatic Liver Tumors Using Histotripsy (#HOPE4LIVER US) (n.d.). ClinicalTrials.gov Identifier: NCT04572633. Available at: https://clinicaltrials.gov/ct2/show/NCT04572633. Accessed 31 May 2022
Titano JJ et al (2019) End-hole versus microvalve infusion catheters in patients undergoing drug-eluting microspheres-TACE for solitary hepatocellular carcinoma tumors: a retrospective analysis. Cardiovasc Intervent Radiol 42(4):560–568. https://doi.org/10.1007/s00270-018-2150-6
Trotovšek B et al (2021) New era of electrochemotherapy in treatment of liver tumors in conjunction with immunotherapies. World J Gastroenterol 27(48):8216–8226. https://doi.org/10.3748/wjg.v27.i48.8216

Tsochatzis EA et al (2014) Transarterial chemoembolization and bland embolization for hepatocellular carcinoma. World J Gastroenterol 20(12):3069–3077. https://doi.org/10.3748/wjg.v20.i12.3069

Villanueva A (2019) Hepatocellular carcinoma. N Engl J Med 380(15):1450–1462. https://doi.org/10.1056/NEJMra1713263

Vincenzi B et al (2015) Prognostic relevance of objective response according to EASL criteria and mRECIST criteria in hepatocellular carcinoma patients treated with loco-regional therapies: a literature-based meta-analysis. PloS One 10(7):e0133488. https://doi.org/10.1371/journal.pone.0133488

Vouche M et al (2014) Unresectable solitary hepatocellular carcinoma not amenable to radiofrequency ablation: multicenter radiology-pathology correlation and survival of radiation segmentectomy. Hepatology 60(1):192–201. https://doi.org/10.1002/hep.27057

Wang C et al (2015) Multicenter randomized controlled trial of percutaneous cryoablation versus radiofrequency ablation in hepatocellular carcinoma. Hepatology 61(5):1579–1590. https://doi.org/10.1002/hep.27548

Weis S et al (2013) Radiofrequency (thermal) ablation versus no intervention or other interventions for hepatocellular carcinoma. Cochrane Database Syst Rev 12:CD003046. https://doi.org/10.1002/14651858.CD003046.pub3

Wong MCS et al (2017) International incidence and mortality trends of liver cancer: a global profile. Sci Rep 7:45846. https://doi.org/10.1038/srep45846

Woo HY, Heo J (2015) Transarterial chemoembolization using drug eluting beads for the treatment of hepatocellular carcinoma: now and future. Clin Mol Hepatol 21(4):344–348. https://doi.org/10.3350/cmh.2015.21.4.344

Worlikar T et al (2018) Histotripsy for non-invasive ablation of Hepatocellular Carcinoma (HCC) tumor in a subcutaneous xenograft murine model. Annual International Conference of the IEEE Engineering in Medicine and Biology Society. IEEE Engineering in Medicine and Biology Society. Annual International Conference, 2018, pp 6064–6067. https://doi.org/10.1109/EMBC.2018.8513650

Wu S et al (2015) Cryoablation versus radiofrequency ablation for hepatic malignancies: a systematic review and literature-based analysis. Medicine 94(49):e2252. https://doi.org/10.1097/MD.0000000000002252

Xu Q et al (2014) Comparison of hepatic resection and radiofrequency ablation for small hepatocellular carcinoma: a meta-analysis of 16,103 patients. Sci Rep 4:7252. https://doi.org/10.1038/srep07252

Xu X-L et al (2018) Radiofrequency ablation versus hepatic resection for small hepatocellular carcinoma: systematic review of randomized controlled trials with meta-analysis and trial sequential analysis. Radiology 287(2):461–472. https://doi.org/10.1148/radiol.2017162756

Xu Z et al (2019) The combination strategy of Transarterial chemoembolization and radiofrequency ablation or microwave ablation against hepatocellular carcinoma. Anal Cell Pathol (Amst) 2019:8619096. https://doi.org/10.1155/2019/8619096

Xu Z et al (2021) Histotripsy: the first noninvasive, non-ionizing, non-thermal ablation technique based on ultrasound. Int J Hyperthermia 38(1):561–575. https://doi.org/10.1080/02656736.2021.1905189

Yang W et al (2015) Radiofrequency ablation of hepatocellular carcinoma in difficult locations: strategies and long-term outcomes. World J Gastroenterol 21(5):1554–1566. https://doi.org/10.3748/wjg.v21.i5.1554

Yao FY (2008) Liver transplantation for hepatocellular carcinoma: beyond the Milan criteria. Am J Transplant 8(10):1982–1989. https://doi.org/10.1111/j.1600-6143.2008.02351.x

Yau T et al (2014) Development of Hong Kong liver cancer staging system with treatment stratification for patients with hepatocellular carcinoma. Gastroenterology 146(7):1691–700.e3. https://doi.org/10.1053/j.gastro.2014.02.032

Young LB et al (2022) Survival analysis using Albumin-Bilirubin (ALBI) grade for patients treated with drug-eluting embolic transarterial chemoembolization for hepatocellular carcinoma. J Vasc Intervent Radiol 33(5):510–517.e1. https://doi.org/10.1016/j.jvir.2022.02.005
Zane KE, Makary MS (2021) Locoregional therapies for hepatocellular carcinoma with portal vein tumor thrombosis. Cancers (Basel). 13(21):5430. https://doi.org/10.3390/cancers13215430
Zimmerman A, Grand D, Charpentier KP (2017) Irreversible electroporation of hepatocellular carcinoma: patient selection and perspectives. J Hepatocell Carcinoma 4:49–58. https://doi.org/10.2147/JHC.S129063
Zongyi Y, Xiaowu L (2020) Immunotherapy for hepatocellular carcinoma. Cancer Lett 470:8–17. https://doi.org/10.1016/j.canlet.2019.12.002

Drug Resistance in Hepatocellular Carcinoma

Xinxin Chen, Jin Li, Yuhong Huang, and Chao Liang

Abstract

At present, the burden of cancer morbidity and mortality is increasing all over the world. Among all kinds of cancer, primary liver cancer is discovered to be the seventh most common cancer and has an extremely high death rate, affecting countless patients. Globally, the main histology of primary liver cancer is hepatocellular carcinoma (HCC), characterized by pathological changes in parenchymal cells of the liver, accounting for over 70% of liver cancer. Currently, HCC treatment mainly includes surgical intervention, local ablative therapy, and local therapy. Sorafenib, a multityrosine kinase inhibitor, is approved for use in patients with unresectable HCC. Tyrosine kinase inhibitors (TKIs) (lenvatinib, regorafenib, and cabozantinib) and a VEGF receptor inhibitor (ramucirumab) have also been used for HCC treatment. Although drugs for HCC treatment have been under further development, the acquired drug resistance increasingly affects the management of HCC patients. Researchers have focused on drug resistance for many years and identified complex regulatory networks involved in this issue.

Xinxin Chen and Jin Li contributed equally to this work.

X. Chen · J. Li · Y. Huang
Department of Biology, School of Life Sciences, Southern University of Science and Technology, Shenzhen, China

C. Liang (✉)
Department of Biology, School of Life Sciences, Southern University of Science and Technology, Shenzhen, China

Institute of Integrated Bioinfomedicine and Translational Science (IBTS), School of Chinese Medicine, Hong Kong Baptist University, Hong Kong SAR, China

State Key Laboratory of Proteomics, National Center for Protein Sciences (Beijing), Beijing Institute of Lifeomics, Beijing, China
e-mail: liangc@sustech.edu.cn

Interdisciplinary Cancer Research, https://doi.org/10.1007/16833_2022_110
Published online: 3 February 2023

In this chapter, the significant research progress regarding mechanisms underlying HCC drug resistance is summarized, which may benefit the treatment of HCC patients and give future direction for new drug development for HCC.

Keywords

Autophagy · Drug resistance · Hepatocellular carcinoma · Noncoding RNA · Sorafenib

1 Introduction

The incidence of liver cancer is higher in eastern and southeastern Asia, but it is also on the rise in India and some countries in Europe, America, and Oceania (McGlynn et al. 2015; Petrick et al. 2020). HCC is the most common primary malignant tumor of the liver. Risk factors such as chronic viral hepatitis and alcoholic/nonalcoholic graded steatohepatitis can cause cirrhosis and hepatocellular lesions, which may eventually lead to the occurrence of HCC (Ogunwobi et al. 2019). Early HCC can be treated with radiofrequency ablation (RFA), hepatic resection, or liver transplantation. Up to now, chemotherapy is still applied as a mainstream treatment for advanced HCC. However, due to tumor heterogeneity, liver cirrhosis, and other possible reasons, few drugs have shown significant efficacy. Sorafenib is recognized globally as an advanced standard therapy for HCC. Although sorafenib has a certain therapeutic effect, it can only temporarily prolong the survival of patients, who are prone to develop sorafenib resistance and predict poor prognosis. Some other drugs, such as lenvatinib and regorafenib, have also shown some positive effects. Generally speaking, so far, the effective treatment of advanced HCC is still a major problem perplexing doctors and researchers.

Sorafenib, a multi-target tyrosine kinase inhibitor, shows multiple targeting for vascular endothelial growth factor receptor (VEGFR), platelet-derived growth factor receptor (PDGFR), rearrange during transfection (RET), c-Kit, and so on. With validated antiangiogenic and antiproliferative functions, it has become the first mature drug for systemic treatment of advanced HCC patients. Furthermore, sorafenib increases median overall survival of HCC patients from 8 months to 11 months without an uncontrolled toxicity profile (Llovet et al. 2018; Ko et al. 2020). Lenvatinib has also been shown to be a multi-receptor tyrosine kinase inhibitor against VEGFR-1/VEGFR-2/VEGFR-3, fibroblast growth factor receptor (FGFR) 1/2/3/4, PDGFRα, RET, and c-Kit, which has also shown some activity in HCC (Kudo et al. 2018; Matsuki et al. 2018). Studies have shown that lenvatinib is not inferior to sorafenib in efficacy and also has controllable tolerability, which has been approved in the America, European Union, China, and Japan for unresectable HCC treatment (Al-Salama et al. 2019). Similar to sorafenib, cabozantinib and regorafenib are tyrosine kinase inhibitors that are approved for the treatment of partial advanced HCC patients (Kelley et al. 2020). Regorafenib inhibits a variety of cancer-promoting pathways, such as the activation of several kinases involved in angiogenesis, tumorigenesis, metastasis, and tumor immunity. Studies have shown that regorafenib can extend survival in patients who have developed tumor progression after taking sorafenib (Personeni et al. 2018; Facciorusso et al. 2019). Although these drugs have been used to treat HCC patients and extend their survival to some extent, the development of multidrug resistance (MDR) to these drugs is the inherent weakness of clinical oncology and often leads to poor prognosis.

The outflow of anticancer drugs from tumor cells is caused by multiple reasons, including upregulation of drug efflux transporters, altered drug redistribution in cells, apoptotic signaling pathways abolished, DNA damage repair ability increasing, drug metabolism changing, and cancer stem cells (CSCs) generating and activating are involved. For instance, the mechanism of sorafenib acquired resistance may involve the interaction of PI3K/Akt and JAK-Stat pathways, enablement of hypoxia-induced pathways, and participation of epithelial-mesenchymal transformation (EMT) (Zhu et al. 2017; Chen et al. 2019a, b; Wei et al. 2019a, b). In this chapter, the mechanisms affecting drug resistance of HCC are covered.

2 Classical Pathways and Proteins Involved in HCC Drug Resistance

Proteins are the main undertaker of life activities. Various proteins and other signal molecules in cells form a huge signal transduction network to regulate various life processes in cells and maintain the homeostasis of the internal environment. Some proteins have been proven to be involved in regulating the response of HCC cells to various treatments and play a certain role in the development of HCC drug resistance. So do some classical signaling pathways, such as PI3K/Akt, Notch, etc. In this section, some proteins and classical signaling pathways associated with HCC resistance are mentioned (Table 1).

Song et al. found that collagen 1A1 (COL1A1) content could affect the spheroid formation ability and chemotherapy resistance of HCC. They found that connective tissue growth factor (CTGF) upregulation in dense HCC spheroids independent of TGF-β leads to COL1A1 accumulation and drug resistance. Losartan, an inhibitor of type 1 collagen synthesis, affects spheroid formation by disrupting cell contact and has the potential to increase the efficacy of anticancer therapy in hepatocellular carcinoma cell lines (Song et al. 2017). Fucosyltransferase (FUT) family is involved in the synthesis of cell surface antigens. Cheng et al. analyzed the expression of FUT family proteins in drug-resistant cell lines and found that FUT4, FUT6, and FUT8 expression is involved in HCC drug resistance by regulating the activity of PI3K/AKT signaling pathway and the expression of tumor multidrug-resistant related protein 1 (MRP1) (Cheng et al. 2013). Tian et al. found that the Lin28 expression, which is related to tumor relapse after chemotherapy, was closely related to paclitaxel resistance. Abnormal expression of Lin28 can inhibit the expression of let-7

Table 1 Pathways and related proteins involved in HCC drug resistance

Pathways	Related proteins	Annotation	References
	CTGF	Upregulation leading to COL1A1 accumulation and drug resistance	Song et al. (2017)
PI3K/AKT	FUT4/6/8	Regulating MRP1 expression	Cheng et al. (2013)
	USP22	Regulating SIRT1 expression	Ling et al. (2017)
	HIF-1α	Regulating ADAM17 expression	Wang et al. (2013)
	KPNA3	Upregulating TWIST expression	Nwosu et al. (2020)
Lin28/let-7/Bcl-xL	Lin28	Upregulating Bcl/xL	Tian et al. (2014)
Notch	NOTCH1	Downregulated by VPA to inhibit HCC MDR	Sun et al. (2015)
FZD2-AXL-NUAK1/2		Promoting EMT and drug resistance	Golkowski et al. (2020)
IL-6/STAT3		Activated by M2 promoting HCC proliferation	Yin et al. (2018)
	HIF-1	Upregulation leading to HCC proliferation and viability	Luo et al. (2014)
	MT-1G	Key regulator of sorafenib resistance	Sun et al. (2016)
MAPK/ERK		Activation leading to HCC drug resistance	Nwosu et al. (2020)
	KPNA3	Upregulating TWIST expression	Hu et al. (2019)
Hedgehog		Activation leading to sorafenib resistance	Wang et al. (2020a, b)
PDCD5/p53	Lgr5	Upregulation leading to HCC drug resistance	Ma et al. (2019)
SSP	PHGDH	Key driver of sorafenib resistance	Wei et al. (2019a, b)
NF-κB/EMT	TNF-α	Promote sorafenib resistance	Tan et al. (2019)

family microRNA and upregulate antiapoptotic protein Bcl-xL, suggesting that Lin28/let-7/Bcl-xL pathway is the fundamental of Hep3B cell drug resistance (Tian et al. 2014). Many previous studies have shown that Notch signal is abnormally expressed in HCC and plays a carcinogenic role. Sun and colleagues found that the anticonvulsant drug valproic acid (VPA) worked as a histone deacetylase (HDAC) inhibitor to reduce HDAC4 expression and thereby inhibit HCC cell proliferation. Moreover, VPA can downregulate Notch pathway by inhibiting Notch1 and its target gene HES1. Therefore, they propose that VPA may have anti-HCC therapeutic activity (Sun et al. 2015). Ling et al. found that the deubiquitinating enzyme ubiquitin-specific protease 22 (USP22), a marker of cancer stem cells, may be involved in HCC multidrug resistance. Their work revealed that USP22 affects the AKT pathway and MRP1 expression by regulating SIRT1 expression and may be a potential target of HCC MDR (Ling et al. 2017). Wang et al. found that upregulated A disintegrin and metalloproteinase-17 (ADAM17) expression is involved in HIF-1α induced HCC cisplatin resistance. HIF-1α regulates the expression of ADAM17 and promotes hypoxia-induced cisplatin resistance through EGFR/PI3K/AKT signaling pathway (Wang et al. 2013). Martin Golkowski and colleagues analyzed HCC cell lines and HCC samples by pharmacoproteomics and revealed that FZD2-AXL-NUAK1/2 signaling module is involved in promoting EMT and drug resistance (Golkowski et al. 2020). Yin et al. found that IL-6/STAT3 signaling pathway is activated in M2-type macrophages, but the reverse is true in M1-type macrophages. However, M2 macrophages promote HCC cell proliferation and migration, which is beneficial to drug resistance to HCC cells. Their study revealed that IL-6/STAT3 pathway is involved in the polarization of M1/M2 macrophage and may be associated with drug resistance in HCC chemotherapy (Yin et al. 2018). Hypoxia promotes tumor progression and is associated with tumor radiation and chemotherapy resistance. HIF-1 level in HCC microenvironment shows significantly increase, results in increased viability of HCC cells and promoted HCC cell proliferation. HIF-1 can change angiogenesis, glucose metabolism, cell proliferation, cell apoptosis, invasion, and metastasis features in the tumor site by regulating the expression of VEGF, GLUT1, NOS, p53, MMP2, and other genes (Luo et al. 2014).

3 Noncoding RNAs Associated with HCC Drug Resistance

In the central dogma, genetic information flows among three molecules, DNA, RNA, and proteins, thus completing the transcription and translation of genetic information. In humans, although protein-coding sequences make up only about 1.5% of the genome, for a long time, RNA is considered as the bridge for the transfer of genetic information from DNA to protein or just intermediate products of transcription that does not have any function (Fu 2014). Certainly, what we already known is some RNAs also play a role as functional RNAs, such as ribosomal RNA (rRNA), transfer RNA (tRNA), and so on. For the past few years, with the deepening of research on RNA, more and more functional RNAs have been discovered. Noncoding RNA (ncRNA) refers to RNAs incapable to encode proteins. A variety of RNAs with known functions including rRNA, tRNA, snRNA, small nucleolar RNA (snoRNA), microRNA, and some RNAs with unknown functions are included. These ncRNAs often perform their respective biological functions at the RNA level (Fu 2014; Saw et al. 2021). Many ncRNAs are considered to be key regulatory factors in various signaling networks. For example, some ncRNAs are involved in important life process such as chromosome maintenance, segregation, and cell cycle regulation (e.g., U1 RNA), which leads to specific cellular responses and fates

(Mattick and Makunin 2006). Not only that, ncRNAs have been identified to be a fat part in disease and drug resistance in cancer (Matsui and Corey 2017; Ding et al. 2018a, b). In this section, some ncRNAs associated with drug resistance in hepatocellular carcinoma will be covered (Table 2).

3.1 MicroRNAs Associated with HCC Drug Resistance

MicroRNAs (miRNAs), a class of small RNA molecules encoded by endogenous genes with a length of about 20–24 nucleotides, are characterized by the regulation of posttranscriptional gene expression. They have a variety of important regulatory functions in cells involved in the occurrence, development, and drug resistance of tumors as potential oncogenes or tumor suppressor genes (Iorio and Croce 2012; Lu and Rothenberg 2018).

The miR-30 family is a group of miRNAs that may have tumor suppressor or tumor promoting effects. Zhang et al. found that sorafenib can reduce the miR-30a-5p expression, and miR-30a-5p directly targets the pro-inflammatory cytokine Cardiotrophin-like cytokine factor 1 (CLCF1). The reduced miR-30a-5p expression promoted the upregulation of CLCF1, and then CLCF1 affected the glycolysis of HCC cells through activating PI3K/Akt pathway. They proposed that the decreased miR-30a-5p expression was conducive to the metastasis and invasion of cancer cells and the generation of drug resistance to sorafenib (Zhang et al. 2020a, b, c, d). The work of Gramantieri et al. revealed that the maturation of miR-30e-3p required wild-type TP53. miR-30e-3p targets MDM2, PTEN, and CDKN1B/P27 to affect tumor cell proliferation and invasion and change the stem cell characteristics of HCC cells by targeting epithelial cell adhesion molecule. As the circulating level of miR-30e-3p was elevated in sorafenib-resistant populations, the authors hypothesized that it could be a predictor for HCC sorafenib resistance (Gramantieri et al. 2020). Ji et al. reported that the miR-486-3p expression was decreased in sorafenib-resistant HCC tissues and cells, indicating that it might be involved in the mechanism of sorafenib resistance in HCC. Moreover, miR-486-3p could downregulate FGFR4 and EGFR expression by targeting their 3'UTR. Therefore, the authors suggest that miR-486-3p may serve as a novel combination therapy to treat sorafenib resistance (Ji et al. 2020). Xiao Fu et al. found that the expression of miR-32-5p was significantly increased in multidrug-resistant HCC cells. The increased miR-32-5p directly acted on PTEN and downregulated its expression, eventually activating the PI3K/Akt pathway, and inducing MDR by promoting angiogenesis and EMT. Moreover, drug-resistant cells also delivered miR-32-5p to drug-sensitive cells through exosomes and diffusing drug resistance (Fu et al. 2018). miR-204 is a member of the miR-183 family. Yuhui et al. reported that miR-204 can affect the response of HCC cells to drugs by acting directly on NUAK1. NUAK1 expression was elevated in HCC, and miR-204 enhanced drug sensitivity by downregulating NUAK1 expression (Yu et al. 2019). Liu et al. found that miR-222 promoted HCC cell proliferation, migration, and invasion. It can also reduce apoptosis and enhanced sorafenib resistance by activating PI3K/Akt pathway (Liu et al. 2014). The study of

Table 2 Noncoding RNAs involved in HCC drug resistance

Factors	Tendency	Target(s)	Pathway	References
MicroRNA				
miR-30a-5p	Downregulated	CLCF1	PI3K/Akt	Zhang et al. (2020a, b, c, d)
miR-30e-3p	Upregulated	MDM2, PTEN, and CDKN1B/P27		Gramantieri et al. (2020)
miR-486-3p	Downregulated	FGFR4, EGFR		Ji et al. (2020)
miR-32-5p	Upregulated	PTEN	PI3K/Akt	Fu et al. (2018)
miR-204		NUAK1		Yu et al. (2019)
miR-222			PI3K/Akt	Liu et al. (2014)
miR-223	Downregulated	FOXO3a		Zhou et al. (2019a, b, c)
miR-591			PI3K/Akt	Wang et al. (2019)
miR-7-5p		TYRO3	TYRO3/phosphoinositide 3-kinase/protein kinase B signal transduction pathway	Kabir et al. (2018)
miR-31		NDRG3		Du et al. (2017)
miR-541	Downregulated	ATG2A, RAB1B		Xu et al. (2020a, b, c, d)
miR-125b	Downregulated		Mir-125b-loss / HIF1α axis	
LncRNA				
LINC01234		miR-31-5p		Chen et al. (2020)
LINC00160		miR-132	LINC00160/miR-132/PIK3R3 axis	Zhang et al. (2020a, b, c, d)
SNHG3		miR128	miR128/CD151/Akt pathway	Zhang et al. (2019)
NR2F1-AS1		miR-363		Huang et al. (2018)
NRAL		miR-340-5p	Nrf2-dependent antioxidant pathway	Wu et al. (2019)
NEAT1		miR-204		Li et al. (2020a, b, c, d)
FOXS2-AS1		miR-150-5p	Nrf2 signaling pathway	Sui et al. (2019)
PDIA3P1		miR-125 a/b/miR-124	NF-κB signaling pathway	Xie et al. (2020)

(continued)

Table 2 (continued)

Factors	Tendency	Target(s)	Pathway	References
TTN-AS1		miR-16-5p	PTEN/Akt pathway	Zhou et al. (2021)
MALAT1		miR-3129-5p		Cao et al. (2021)
		miR-140-5p		Fan et al. (2020)
GAS5	Downregulated	miR-21	miR-21-PTEN pathway	Wang et al. (2020a, b)
SNHG1	Upregulated	SLC3A2	Akt pathway	Li et al. (2019)
lncASAR	Upregulated	PTEN mRNA	PI3K/Akt	Li et al. (2017)
H19	Upregulated		MAPK/ERK pathway	Ding et al. (2018a, b)
KRAL		miR-141	NFR pathway	Wu et al. (2018)
KCNQ1OT1		miR-7-5p	miR-7-5p/ABCC1 axis	Hu et al. (2018)
		miR-338-3		Zhong et al. (2020)
		miR-506	PD-L1	Zhang et al. (2020a, b, c, d)
HULC		miR-6825-5p, miR-6845-5p, miR-6886-3p	"HULC/USP22/ SIRT1/protective autophagy" pathway	Xiong et al. (2017)
circRNA				
circMET	Upregulated		miR-30-5p/Snail/ DPP4 axis	Huang et al. (2020)
cirCFBXO11	Upregulated	miR605	miR605/FOXO3/ ABCB1 axis	Li et al. (2020a, b, c, d)
circRNA-SORE	Upregulated	YBX1		Xu et al. (2020a, b, c, d)
	Upregulated	miR-103a-2-5p and miR-620-3p	Wnt/β-catenin pathway	Xu et al. (2020a, b, c, d)
circUHRF1	Upregulated			Zhang et al. (2020a, b, c, d)
Circ_0031242		miR-924	Circ_0031242/ miR-924/ POU3F2 axis	Fan et al. (2021)
circ_0003418	Downregulated		Wnt/β -catenin pathway	Chen et al. (2019a, b)
circFN1	Upregulated	miR-1205	miR-1205/E2F1 signaling pathway	Yang et al. (2020)
Circ_0003998	Upregulated	miR-218-5p		Li et al. (2020a, b, c, d)

Zhou et al. showed that miR-223 was expressed at a low level in HCC cells. Overexpression of miR-223 inhibited autophagy and reversed chemoresistance of HCC cells through direct action on the transcription factor FOXO3a (Zhou et al. 2019a, b, c). Wang et al. found that miR-591 could inhibit FBO2 expression by blocking the PI3K/Akt/mammalian target of rapamycin axis. Their study showed that miR-591 inhibited the development of HCC by reducing CSC drug resistance (Wang et al. 2019). Kabir et al. found that miR-7-5p (miR-7), as an effective tumor suppressor of human HCC, could overcome the resistance of sorafenib in HCC caused by the abnormal expression of TYRO3/PI3K/protein kinase B (PKB) signal transduction pathway by directly acting on TYRO3 (Kabir et al. 2018). Du et al. showed that miR-31 is a key regulator of HCC and its target gene NDRG3 is essential for HCC cell survival. Their findings suggested that MicroRNA31-NDRG3 regulatory axis is critical for HCC cell survival and drug resistance (Du et al. 2017). Studies have shown that downregulated miR-541 in human HCC tissues is associated with poor prognosis. Xu and colleagues found that autophagy associated gene 2A (ATG2A) and Ras-related protein RabB-1B (RAB1B) are direct targets of miR-541, and miR-541-ATG2A/RAB1B axis plays an important part in patients' response to sorafenib. Combination of miR-541 and sorafenib could further inhibit cancer cell proliferation (Xu et al. 2020a, b, c, d). Wei et al. found that the loss of Mir-125b significantly activated the hypoxia-inducible factor 1α (HIF1α)/pAKT loop in relapsed liver cancer. miR-125b attenuates HIF1α translation and blocks the HF1 α/platelet-derived growth factor β (PDGFβ)/pAKT/HIF1α loop by targeting PDGFβ receptors. They suggest that miR-125b-loss /HIF1α axis induces CD24 expression and erythropoietin (EPO) is involved in the generation of transarterial chemoembolization resistance (Wei et al. 2021).

3.2 LncRNA Associated with HCC Drug Resistance

Long noncoding RNAs (lncRNA) are noncoding RNAs containing more than 200 nucleotides. lncRNA participates in epigenetic regulation, cell cycle regulation, and cancer drug resistance (Chen et al. 2017). It is now believed that lncRNA is associated with the regulation of cancer cell proliferation, metastasis, and drug resistance through a variety of mechanisms, and its abnormal expression may cause the occurrence of drug resistance (Pan et al. 2015).

3.2.1 LncRNA Associated with HCC Drug Resistance Through ceRNA Mechanism

Salmena et al. described a "competitive endothelial RNA" (ceRNA) hypothesis (Salmena et al. 2011). MiRNAs induce gene silencing by binding with mRNAs, while ceRNAs regulate gene expression by binding with microRNAs through microRNA response elements (MREs). By binding with miRNAs competitively, ceRNAs change the availability of miRNAs binding to their targets and thus play a regulatory role in protein synthesis. lncRNAs, circular RNAs (circRNAs), and other RNAs compete with miRNAs by acting as natural miRNA sponges that regulate the expression of protein-coding genes (Salmena et al. 2011; Ergun and Oztuzcu 2015). Studies have shown that ceRNAs have unique regulatory mechanisms and functional roles in HCC (Long et al. 2019; Zhao et al. 2019; Li et al. 2020a; b, c, d).

Chen et al. found that lncRNA LINC01234 can inhibit the downregulation of melanoma-associated antigen A3 (MAGEA3) by miR-31-5p and participate in HCC cisplatin resistance through this pathway (Chen et al. 2020). LncRNA LINC00160, a highly expressed lncRNA in HCC cells, can downregulate miR-132 by acting as ceRNA, leading to upregulation of PIK3R3, which ultimately inhibits autophagy and viability of HCC cells, reduces sorafenib resistance, and promotes apoptosis. Animal experiments have also shown that silencing LINC00160 can suppress tumorigenicity of HCC cells in vivo through LINC00160/miR-132/PIK3R3 axis (Zhang et al. 2020a, b, c, d). Small nucleolar RNA host gene 3 (SNHG3) is an overexpressed lncRNA involved in sorafenib resistance in HCC cells. The study of Zhang et al. showed that SNHG3 shares a miR128 target sequence with CD151 and can competitively bind with miR128. Overexpression of SNHG3 in HCC cells promotes sorafenib resistance by regulating miR128/CD151/Akt pathway (Zhang et al. 2019). Oxaliplatin (OXA) is an inhibitor of DNA synthesis that causes DNA cross-linking damage by preventing DNA replication and transcription which leads to cell death. OXA resistance is also a major bottleneck in HCC chemotherapy. Similarly, Huang et al. reported that lncRNA NR2F1-AS1 could function as a ceRNA to bind miR-363 through its 3'UTR, thereby upregulating the expression of a drug-resistant protein multidrug resistance-associated protein 1 (ABCC1) in HCC cells, leading to the production of OXA resistance (Huang et al. 2018). Similarly, Nrf2 regulation-associated lncRNA (NRAL) interacts with miR-340-5p and regulates Nrf2 expression through ceRNA mechanism. NRAL activated Nrf2-dependent antioxidant pathways and promoted HCC cisplatin resistance (Wu et al. 2019). Autophagy can regulate metabolic processes in harsh microenvironments to protect cancer cells from death and is found to be associated with tumor drug resistance. Li et al demonstrated that LncRNA Nuclear Enriched Transcript1 (NEAT1) functioned as a miR-204 sponge and upregulated ATG3 expression by inhibiting the function of miR-204, so as to promote autophagy and inhibit the efficacy of sorafenib (Li et al. 2020a, b, c, d). Another NEAT1-related study indicated that the expression of P-glycoprotein (P-gp) and multidrug resistance protein 1 (MRP1) were promoted in HCC tissues and cells through directly targeting NEAT1 by high expression of Bcl-2-associated transcription factor 1 (BCLAF1) to promote 5-FU resistance (Mou et al. 2020). Sui et al. proposed that lncRNA FOXS2-AS1 regulated the expression of transmembrane protein 9 (TEMEM9) through ceRNA mechanism as a miR-150-5p sponge. This pathway furthermore regulated HCC sorafenib resistance by acting on the Nrf2 pathway (Sui et al. 2019). Xie et al. found that lncRNA protein disulfate isomerase family A member 3 pseudogene 1 (PDIA3P1) inhibited tumor necrosis factor receptor-associated factor 6 (TRAF6) by binding with miR-125a/b/miR124 and further enhanced NF-κB signaling pathway, ultimately conferring chemoresistance on HCC cells (Xie et al. 2020). Zhou et al. proposed that lncRNA TTN-AS1 sponges miR-16-5p and ultimately regulates sorafenib resistance through TTN-AS1/miR-16-5p/Cyclin E1 axis and PTEN/Akt pathway (Zhou et al. 2021). Dysregulation of lncRNA MALAT1 is involved in doxorubicin resistance in HCC. Cao et al. proposed that lncRNA MALAT1 regulated the miR-3129-5p/NOVA1 axis and mediates doxorubicin resistance in HCC by acting as a sponge for miR-3129-5p (Cao et al. 2021). Lei Fan et al. found that lncRNA MALAT1 regulates Aurora A expression through sponging miR-140-5p, thereby promoting sorafenib resistance (Fan et al. 2020). A lncRNA called growth crim-specific 5 (GAS5) which is downregulated in HCC cells and tumor tissues was found to modulate PTEN expression by sponging miR-21. Since GAS5 plays a tumor suppressive role in HCC through regulating miR-21-PTEN pathway, the decreased expression of GAS5 could enhance HCC cell proliferation and increase doxorubicin resistance (Wang et al. 2020a, b).

3.2.2 LncRNAs in HCC Drug Resistance Through Other Mechanisms

Li et al. showed that sorafenib induced the translocation of miR-21 to nucleus in sorafenib-resistant HCC cells, which led to upregulation of a lncRNA called small nucleolar RNA host gene 1 (SNHG1), and SNHG1 upregulated the expression of SLC3A2, thereby activating Akt pathway and promoting the generation of sorafenib resistance (Li et al. 2019). PTEN, a tumor suppressor gene, is closely related to tumorigenesis. PTEN protein plays an important role in cell growth, apoptosis, adhesion, and migration. Studies have shown that the upregulated lncARSR in HCC is involved in tumor invasion and predicts poor prognosis. The study of Li et al. found that lncARSR could directly interact with PTEN mRNA and promote its degradation, thereby reducing PTEN expression and activating PI3K/Akt pathway and finally leading to the development of doxorubicin resistance (Li et al. 2017). Cancer stem cells (CSCs) are cells with high tumorigenic potential and strong self-renewal ability. CSCs have greater environmental adaptability, such as high levels of reactive oxygen species (ROS) scavengers, and are therefore more likely to develop drug resistance (Najafi et al. 2019). H19 is a lncRNA related to embryonic development and growth control. Studies have shown that the H19 expression is upregulated in HCC tissues, suggesting that it may be related to epigenetic mechanism of tumor genesis and development. Kai et al. showed that H19 appeared to participate in oxidative stress (OS) regulation in HCC CSCs by activating MAPK/ERK pathway and promoting the proliferation of CSCs and tumor progression. Correspondingly, inhibition of H19 may induce OS by blocking the MAPK/ERK signaling pathway, thereby reducing chemotherapy resistance of CSCs (Ding et al. 2018a, b). Besides, Xu et al. also raised that H19 promotes the EMT of HCC cells as a precursor of miR-675, thus contributing to HCC sorafenib resistance (Xu et al. 2020a, b, c, d). Keap1 (Kelch-like ECH-associated protein 1)-NFE2 L2 or Nrf2 (nuclear factor erythroid 2-related factor 2) functions as "cellular defense machinery" to cope with oxidative/electrophilic stimuli or chemical insults. LncRNA KRAL was found to be a sponge for miR-141, which can induce Keap expression and thus inhibit nuclear factor erythroid 2-related factor 2 (NFR) pathway and then attenuate 5-FU resistance (Wu et al. 2018). Hu et al. found that lncRNA KCNQ1OT1 regulated ABCC1 expression via sponging miR-7-5p and regulated the drug resistance of HCC to OXA through miR-7-5p/ABCC1 axis (Hu et al. 2018). Zhong et al.

found that lncRNA KCNQ1OT1 was involved in autophagy regulation and cisplatin resistance by downregulating miR-338-3 (Zhong et al. 2020). Zhang et al. showed that lncRNA KCNQ1OT1 functioned as a ceRNA of miR-506. And the KCNQ1OT1 axis may be involved in the immune escape of sorafenib-resistant HCC cells by regulating PD-L1 expression (Zhang et al. 2020a, b, c, d). Guo et al. proved that the expression of small nucleolar RNA host gene 16 (SNHG16) was upregulated in HCC and may act on tumor cell proliferation, migration, invasion, and sorafenib resistance (Guo et al. 2019). The work of Xiong et al. suggested that protective autophagy is important for HCC resistance. They found three miRNAs — miR-6825-5p, miR6845-5p and miR-6886-3p that could reduce USP22 expression by binding to the 3'-untranslated region of USP22 mRNA and they were downregulated by lncRNA HULC. LncRNA HULC attenuated the responsibility of HCC cells to chemotherapy drugs through "HULC/USP22/SIRT1/protective autophagy" pathway (Xiong et al. 2017).

3.3 circRNA Associated with HCC Drug Resistance

Circular RNA (circRNA) is noncoding RNA molecules mainly generated by noncanonical splicing and has a closed ring structure. circRNAs are generally localized in the cytoplasm. In the past, they were mostly taken for "junk" as the product of noncanonical splicing. They are not affected by RNA exonuclease leading to higher stability consequently. It has found that circRNAs can bind to microRNAs and act as microRNA sponges. More and more circRNAs have been found to be associated with human diseases (such as cancer) (Kristensen et al. 2019; Zang et al. 2020).

Studies have shown that the amplification of chromosome 7q21-7q31 is related to tumor recurrence and MDR. Huang et al. found that the expression of a circular RNA MET in this region was significantly increased in HCC cells. Their research shows that increased circMET expression induces EMT through the miR-30-5p/Snail/DPP4 axis and leads to compromised tumor immune microenvironment (Huang et al. 2020). The significantly upregulated circRNA cirCFBXO11 in HCC tissues was demonstrated to regulate downstream FOXO3 expression through the interaction with miR605. cirCFBXO11 participated in HCC OXA resistance via miR605/FOXO3/ABCB1 axis (Li et al. 2020a, b, c, d). Xu et al. reported that upregulated circRNA-SORE prevented the interaction between YBX1 and E3 ubiquitin ligase PRP19 by binding the main oncogenic protein YBX1, thus abrogating YBX1 degradation. In addition, circRNA-SORE can also spread sorafenib resistance in HCC cells through exosomal transport (Xu et al. 2020a, b, c, d). They also found that circRNA-SORE stability was increased in sorafenib-resistant HCC cells as a result of increased levels of specific adenosine N6-methyladenosine (m6A) in circRNA-SORE. circRNA-SORE sponges miR-103a-2-5p and miR-620-3p through the ceRNA mechanism. As a result, Wnt/β-catenin pathway was activated and sorafenib resistance was induced (Xu et al. 2020a, b, c, d). The study of Zhang et al. found that highly expressed circUHRF1 HCC suggests poor prognosis and NK cell dysfunction. They proposed that exosome circUHRF1 secreted by HCC cells inhibits interferon (IFN)-γ and tumor necrosis factor (TNF)-α secretion and upregulates TIM-3 expression by degrading miR-449c-5p to inhibit NK cell function. They found that high levels of plasma exosome circUHRF1 were associated with a decreased proportion of NK cells and reduced NK cell infiltration and may lead to PD1 immunotherapy resistance (Zhang et al. 2020a, b, c, d). Fan et al. revealed that Circ_0031242 sponges miR-924 and influences cisplatin (DDP) resistance of HCC through the Circ_0031242/ miR-924/POU3F2 (POU class 3 homeobox 2) axis (Fan et al. 2021). Chen et al. proposed that circ_0003418 was downregulated in tissues and cell lines of HCC and associated with disease progression. Their study showed that circ_0003418 inhibits HCC cisplatin resistance by inhibiting Wnt/β-catenin pathway (Chen et al. 2019a, b). Yang et al. observed that circFN1 was upregulated in HCC cell lines and HCC tissues from patients and was markedly associated with the aggressiveness of HCC. They proposed that circFN1 could serve as a sponge for miR-1205 to promote HCC sorafenib resistance by promoting E2F1 expression through miR-1205/E2F1 signaling pathway (Yang et al. 2020). Li et al. reported an elevated expression of Circ_0003998 HCC tissues, doxorubicin (DOX) resistant tissues, and cells. Further studies found that Circ_0003998 sponges miR-218-5p through ceRNA mechanism to affect EIF5A2 expression and promote HCC DOX resistance (Li et al. 2020a, b, c, d).

4 Epigenetics

Different from the changes in gene function caused by intrinsic reshape in DNA sequences (such as mutations), epigenetics refers to the alternation such as DNA methylation or histone modification without changes in DNA sequence itself to gain heritable changes in phenotypes. Studies suggest that epigenetic changes may also participate in HCC drug resistance. For example, some studies have suggested that methylation changes in EMT are involved in the common mechanism of drug resistance to cancer therapy. Galle et al. found that DNA methylation changes in EMT drive sorafenib resistance (Galle et al. 2020).

Lin et al. found that m6A methyltransferase METTL3 promoted the 3'UTR region of FOXO3 mRNA methylation and enhanced its stability in a YTHDF1-dependent manner. Downregulation of METTL3 expression in HCC enhances sorafenib resistance by activating pathways related to angiogenesis and autophagy (Lin et al. 2020). ADAMTSL5 is secreted by liver cancer cells and plays a role in maintaining the function of key carcinogenesis signaling pathways. Arechederra et al. found that ADAMTSL5 abrogation led to a reduction of several oncogenic inputs relevant to HCC, including reduced expression and/or phosphorylation levels of receptor tyrosine kinases c-Met, EGFR, PDGFRβ, IGF1Rβ, or FGFR4. As an epigenetic regulated oncogene in HCC, ADAMTSL5 may act as a major regulator of tumorigenic and drug resistance of HCC and affect the efficacy of drugs used in cancer treatment (Arechederra et al. 2021). TNF-α is an important inflammatory cytokine, which promotes tumor invasion and metastasis. A G-actin monomer binding protein thymosin β4 (Tβ4) was found to be overexpressed in drug-resistant

HCC cells. Abnormal expression of Tβ4 caused by epigenetic changes may induce resistance to antiangiogenic therapy through stemness acquisition. Ohata et al. proposed that epigenetic drugs recruited at the Tβ4 promoter site antagonize chromatin regulators and make drug-resistant HCC cells sensitive to angiogenic therapy (Ohata et al. 2017).

5 Autophagy

Metabolic homeostasis is the basis of all biological activities. Biosynthesis and degradation are two important processes to maintain metabolic homeostasis. Autophagy is the process that cytoplasmic substances enter the lysosome through the autophagosome and are degraded in the lysosome. Autophagy is the main degradation system in cells. It can not only remove substances but also participate in cellular material renewal and homeostasis as a dynamic circulation system, playing a certain role in metabolic adaptation, intracellular quality control, and other life processes. Current studies suggest that autophagy is also closely related to the development of many diseases and may be involved in the generation of cancer drug resistance.

Jun et al. reported a highly expressed Rage (the receptor for advanced glycation end products) in HCC cells. It regulates autophagy by acting on AMPK/mTOR signaling pathway to promote HCC cell proliferation and induce sorafenib resistance (Li et al. 2018). Xu et al. found that cisplatin could induce the downregulation of miR-199a-5p and further increase cisplatin resistance by targeting autophagy-related gene 7 (ATG7) to activate autophagy (Xu et al. 2012). YAP is a key effector of the Hippo pathway and associates with cancer progression, metastasis, and invasion. Zhou et al. found that YAP overexpression in some HCC cells induced multidrug resistance and inhibited autophagy-related cell death by acting on the RAC1-ROS-mTOR pathway (Zhou et al. 2019a, b, c). Fu et al. found that ADRB2 signaling downstream of adrenaline negatively regulates autophagy by destroying Beclin1/VPS34/Atg14 complex in an Akt-dependent manner, leading to HIF1α stabilization and glucose metabolism reprogramming and further inducing sorafenib resistance (Wu et al. 2016). Lui et al. reported that overexpression of the cell surface molecule CD24 resulted in increased expression of PP2A protein and induced inactivation of the mTOR/AKT pathway, thereby regulating sorafenib resistance by increasing autophagy levels of HCC cells (Lu et al. 2018). Studies have shown that endoplasmic reticulum stress-related autophagy plays a crucial role in apoptosis resistance. Zhou et al. found that low concentrations of melatonin can inhibit autophagy through PERK-ATF4-Beclin1 pathway and increase sorafenib responsibility (Zhou et al. 2019a, b, c). The study of Fu et al. showed that tumor associated macrophages (TAM) in the tumor microenvironment inhibited OXA induced cell death in HCC by inducing autophagy and led to OXA resistance (Fu et al. 2019). ManTong et al. found that overexpressed Annexin A3 (ANXA3) in HCC cells and patient-derived xenografts promoted HCC cell survival by activating autophagy and inhibiting PKCδ/P38 dependent apoptosis (Tong et al. 2018). Transactivation response element RNA-binding protein 2 (TRABR2) is a double-strand RNA-binding protein that stabilizes Dicer and induces miRNA maturation. Lai et al. found that TARBP2 protein in HCC cells becomes unstable via the autophagy lysosomal proteolysis pathway, thus stabilizing Nanog expression (a CSC marker), which contributes to the promotion of sorafenib resistance (Lai et al. 2019). Zhang et al. reported that chloroquine improved the therapeutic effect of cisplatin chemotherapy in HepG2 cells by blocking autophagy and deteriorating mitochondrial dysfunction. They proposed that chloroquine might be a potential adjuvant to overcome chemotherapy resistance in HCC (Zhang et al. 2017). Su et al. found that galectin-1, a β-galactoside binding lectin, is upregulated in HCC patients. Upregulated galectin-1 promoted tumor development by facilitating cancer cell adhesion, migration, and proliferation. Further studies showed that increased mitochondrial autophagy induced by galectin-1 could reduce the loss of mitochondrial potential and apoptosis induced by cisplatin therapy; their study demonstrated that using thiodigalactoside to inhibit galectin-1 could significantly enhance the therapeutic effect of cisplatin (Su et al. 2016). PERK stimulation activates C-Jun-terminal kinase (JNK), leading to gathering unfused autophagosomes that enhance autophagic death through cleavage of poly-ADP-ribotransferase (PARP-1). Shao et al. found that Crizo plus Dox stimulate autophagosome formation by activating PERK-p-eIF2 α-MDR1 and kill HCC cells by activating PERK-JNK-autophagic cell death pathways (Shao et al. 2021). Jin et al. found that miR-26a/b can promote apoptosis and sensitive HCC cells to chemotherapy by downregulating autophagy promoter ULK1 (Jin et al. 2017). Barot et al. researched the processes related to glycogen metabolism in HCC cells and found that inhibiting glycogen phosphorylase (GP) induces cancer cell death through intrinsic apoptotic pathways, such as mitochondrial dysfunction and autophagy adaptation. They also found that inhibiting GP made HCC more sensitive to sorafenib and regorafenib (Barot et al. 2019). Yu et al. found that lycorine has the ability to induce apoptosis and autophagy in vitro and in vivo. Lycorine inducing apoptosis and autophagy are associated with decreased expression of tongue cancer resistance-associated protein 1 (TCRP1). Their study demonstrated that lycorine promoted apoptosis and autophagy in HCC cells through the TCRP1/Akt/mTOR pathway (Yu et al. 2017). Ding et al. found that the expression of calcium-/calmodulin-dependent serine protein kinase (CASK) in HCC cells was positively correlated with sorafenib resistance and poor prognosis of the disease. Their work showed that CASK depletion enhanced the chemical sensitivity of HCC by inducing apoptosis and enhancing autophagy death mediated by the JNK/C-Jun pathway. As a potential therapeutic target for HCC, inhibition of CASK can enhance sorafenib efficiency to a certain extent by promoting apoptosis and autophagy (Ding et al. 2021). The stress-inducible nuclear protein 1 (NUPR1) is reported to be a transcription factor that plays a critical role in chromatin remodeling, cell cycle control, DNA damage response, aging, autophagy, apoptosis, and lipid metabolism. Augello et al. demonstrated that NUPR1 regulates autophagy in HCC cells at baseline and after chemotherapy and that NUPR1 KD inhibits autophagy flow and induces apoptosis sensitivity to sorafenib. Their work demonstrated that the combination of the novel P73 small molecule activator NSC59984 and sorafenib is a new promising treatment for HCC (Augello et al. 2021). Han et al. found that sorafenib-resistant HCC cells accumulate hepatocyte growth factor (HGF) excessively and overexpress c-Met kinase and its phosphorylated form, leading to activation of Akt and ERK pathways. The anticancer activity of the Akt inhibitor MK2206 primarily depends on induction of apoptosis and autophagy death, whereas capmatinib treatment resulted in cell cycle stagnation at the G1 phase. Their study showed that the dual inhibition of the Akt inhibitor Mk2206 and the c-Met inhibitor capmatinib could help in sorafenib-resistant HCC treatment (Han et al. 2017). Li et al showed that asiaticoside induces apoptosis and G1 cell cycle arrest in QGY-7703 and BEL-7402 cells by inhibiting the PI3K/Akt and MAPK/ERK signaling pathways. In addition, asiaticoside also considerably antagonized P-GP mediated HCC cell MDR (Ma et al. 2020).

6 Other Mechanisms in HCC Drug Resistance

In addition to mechanisms mentioned above, some other mechanisms have also been reported to be involved in the regulation of HCC drug resistance. For example, mitophagy, ferroptosis, oxidative stress response, metabolic pathway changes and so on. ATPase family AAA domain containing 3A (ATAD3A) is the core regulatory gene of mitophagy. Wu et al. found that hypoxia induction was involved in the regulation of mitophagy through the miR-210-5p-ATAD3A axis and ultimately participated in HCC sorafenib resistance (Wu et al. 2020). Ferroptosis is a novel regulatory form of cell death characterized by iron-dependent accumulation of lipid hydroperoxides. This form of cell death has certain prospects in cancer therapy (Nie et al. 2018). Sun et al. prompted that metallothionein (MT)-1G was involved in sorafenib resistance by suppressing ferroptosis through affecting glutathione consumption and lipid peroxidation of HCC cells. Their results revealed that MT-1G is a key regulator of sorafenib resistance in human HCC cells and a promising therapeutic target (Sun et al. 2016). In the study of Nwosu et al., overactivation of the ERK pathway in HCC cells was found to be associated with HCC drug resistance. They found that severe metabolic damage leads to the induction of the ERK pathway and some HCC cells developed a pro-survival phenotype after the ERK pathway was blocked, providing new insights into understanding, predicting, and overcoming drug resistance in HCC (Nwosu et al. 2020). In HCC, both CD44 and Hedgehog signaling pathways play an important role in the characterization of tumor-initiating cells. Wang et al. found that in CD44 positive HCC, Hedgehog signaling inhibitor (GANT61) can effectively inhibit the activity of patient-derived organoid cells and reverse sorafenib drug resistance. They proposed that sorafenib and GANT61 had a strong synergistic inhibitory effect on cell activity and malignant properties in vitro and in vivo (Wang et al. 2020a, b). Karyopherin subunit alpha 3 (KPNA3), a member of the α3 subfamily of KPNA proteins, has been implicated in HCC sorafenib resistance. Hu et al. found that overexpression of KPNA3 significantly enhanced PI3K/Akt and MAPK/ERK signaling pathway activities in HCC drug-resistant cells, and phosphorylation of ERK induced the expression of TWIST, leading to EMT and sorafenib resistance ultimately (Hu et al. 2019). Leucine-rich repeat containing G protein-coupled receptor 5 (Lgr5) is a member of the G protein-coupled receptor family. Study has shown that it promoted drug resistance by regulating PDCD5/p53 signaling axis, inducing EMT and inhibiting cell apoptosis in HCC (Ma et al. 2019). Through genome-wide CRISPR/Cas9 library screening, Wei et al. reported that glycerol phosphate dehydrogenase (PHGDH), the first enzyme involved in the serine synthesis pathway (SSP), was a key driving force of sorafenib drug resistance. Sorafenib treatment activated SSP by inducing PHGDH expression in HCC cells. Activated PHGDH and SSP produce antioxidants and α-ketoglutarate in response to sorafenib-induced oxidative stress and promote HCC cell survival. Using RNAi and CRISPR/Cas9 knockdown models, they found that PHGDH inactivation paralyzed SSP and reduced αKG, serine, and NADPH production while increasing ROS levels and inducing apoptosis of HCC cells (Wei et al. 2019a, b). Researches showed that TNF-α in liver cancer tissue was significantly upregulated than that in normal liver tissue. Wenliang Tan et al. suggested that high TNF-α expression in HCC indicated poor prognosis upon sorafenib treatment. Their work reveals that TNF-α promotes sorafenib resistance in HCC cells by promoting EMT and ulinastatin can enhance sorafenib's effect on HCC cells even in the case of high TNF-α expression by acting on NF-κB/EMT signaling pathway (Tan et al. 2019).

7 Conclusion

For the treatment of unresectable HCC, drugs like sorafenib are still the primary choice at present. However, the emergence of multidrug resistance has brought serious adverse impact on the efficacy of drugs and suggests a poor prognosis. Therefore, it is of great practical significance to study the mechanism of HCC drug resistance. In this chapter, we summarized the mechanisms involved in HCC drug resistance, including noncoding RNAs, epigenetic factors, and autophagy. The initiation of MDR involves complex mechanisms and many signaling pathways. For example, many noncoding RNAs have been reported to be associated with HCC drug resistance. Autophagy, changes in epigenetic modifications, and signaling pathways such as PI3K/Akt and MAPK/ERK also seem to contribute. With the deepening of research, more and more drug resistance mechanisms have been reported, providing one solution after another to solve the problems of HCC drug resistance. However, due to the heterogeneity of cancer and the complexity of drug resistance mechanisms, no real clinical solutions have been proposed so far. No matter finding new drugs or increasing adjuvant therapy through mechanism exploration, the solution of HCC drug resistance still needs further research.

Acknowledgment This review is supported by the National Natural Science Foundation of China (Grant No. 82172386 and 81922081), the Department of Education of Guangdong Province (Grant No. 2021KTSCX104), the Guangdong Basic and Applied Basic Research Foundation (Grant No. 2022A1515012164), and Shenzhen Science and Technology Program (Grant No. JCYJ20210324104201005).

Author Contributions C.L. supervises and revises the manuscript. X.C. and J.L. write and edit the manuscript and make the tables. Y.H. provides the professional expertise.

Competing Interests The authors declare no competing interests.

References

Al-Salama ZT, Syed YY, Scott LJ (2019) Lenvatinib: a review in hepatocellular carcinoma. Drugs 79:665–674

Arechederra M, Bazai SK, Abdouni A, Sequera C, Mead TJ, Richelme S, Daian F, Audebert S, Dono R, Lozano A, Gregoire D, Hibner U, Allende DS, Apte SS, Maina F (2021) ADAMTSL5 is an epigenetically activated gene underlying tumorigenesis and drug resistance in hepatocellular carcinoma. J Hepatol 74:893–906

Augello G, Emma MR, Azzolina A, Puleio R, Condorelli L, Cusimano A, Giannitrapani L, McCubrey JA, Iovanna JL, Cervello M (2021) The NUPR1/p73 axis contributes to sorafenib resistance in hepatocellular carcinoma. Cancer Lett 519:250–262

Barot S, Abo-Ali EM, Zhou DL, Palaguachi C, Dukhande VV (2019) Inhibition of glycogen catabolism induces intrinsic apoptosis and augments multikinase inhibitors in hepatocellular carcinoma cells. Exp Cell Res 381:288–300

Cao Y, Zhang F, Wang H, Bi C, Cui J, Liu F, Pan H (2021) LncRNA MALAT1 mediates doxorubicin resistance of hepatocellular carcinoma by regulating miR-3129-5p/Noval axis. Mol Cell Biochem 476:279–292

Chen Q-N, Wei C-C, Wang Z-X, Sun M (2017) Long non-coding RNAs in anti-cancer drug resistance. Oncotarget 8:1925–1936

Chen S, Cao Q, Wen W, Wang H (2019a) Targeted therapy for hepatocellular carcinoma: challenges and opportunities. Cancer Lett 460:1–9

Chen H, Liu S, Li M, Huang P, Li X (2019b) circ_0003418 inhibits tumorigenesis and cisplatin chemoresistance through Wnt/β-catenin pathway in hepatocellular carcinoma. Onco Targets Ther 12:9539–9549

Chen Y, Zhao H, Li H, Feng X, Tang H, Qiu C, Zhang J, Fu B (2020) LINC01234/MicroRNA-31-5p/MAGEA3 axis mediates the proliferation and chemoresistance of hepatocellular carcinoma cells. Mol Ther Nucleic Acids 19:168–178

Cheng L, Luo S, Jin C, Ma H, Zhou H, Jia L (2013) FUT family mediates the multidrug resistance of human hepatocellular carcinoma via the PI3K/Akt signaling pathway. Cell Death Dis 4: e923–e923

Ding K, Liao Y, Gong D, Zhao X, Ji W (2018a) Effect of long non-coding RNA H19 on oxidative stress and chemotherapy resistance of CD133+ cancer stem cells via the MAPK/ERK signaling pathway in hepatocellular carcinoma. Biochem Biophys Res Commun 502:194–201

Ding B, Lou W, Xu L, Fan W (2018b) Non-coding RNA in drug resistance of hepatocellular carcinoma. Biosci Rep 38:BSR20180915

Ding B, Bao C, Jin L, Xu L, Fan W, Lou W (2021) CASK silence overcomes sorafenib resistance of hepatocellular carcinoma through activating apoptosis and autophagic cell death. Front Oncol 11:681683–681683

Du Z, Niu S, Xu X, Xu Q (2017) MicroRNA31-NDRG3 regulation axes are essential for hepatocellular carcinoma survival and drug resistance. Cancer Biomark 19:221–230

Ergun S, Oztuzcu S (2015) Oncocers: ceRNA-mediated cross-talk by sponging miRNAs in oncogenic pathways. Tumour Biol 36:3129–3136

Facciorusso A, Abd El Aziz MA, Sacco R (2019) Efficacy of regorafenib in hepatocellular carcinoma patients: a systematic review and meta-analysis. Cancer 12:36

Fan L, Huang X, Chen J, Zhang K, Gu YH, Sun J, Cui SY (2020) Long noncoding RNA MALAT1 contributes to sorafenib resistance by targeting miR-140-5p/Aurora-A signaling in hepatocellular carcinoma. Mol Cancer Ther 19:1197–1209

Fan W, Chen L, Wu X, Zhang T (2021) Circ_0031242 silencing mitigates the progression and drug resistance in DDP-resistant hepatoma cells by the miR-924/POU3F2 axis. Cancer Manag Res 13:743–755

Fu X-D (2014) Non-coding RNA: a new frontier in regulatory biology. Natl Sci Rev 1:190–204

Fu X, Liu M, Qu S, Ma J, Zhang Y, Shi T, Wen H, Yang Y, Wang S, Wang J, Nan K, Yao Y, Tian T (2018) Exosomal microRNA-32-5p induces multidrug resistance in hepatocellular carcinoma via the PI3K/Akt pathway. J Exp Clin Cancer Res 37:52–52

Fu X-T, Song K, Zhou J, Shi Y-H, Liu W-R, Shi G-M, Gao Q, Wang X-Y, Ding Z-B, Fan J (2019) Tumor-associated macrophages modulate resistance to oxaliplatin via inducing autophagy in hepatocellular carcinoma. Cancer Cell Int 19:71–71

Galle E, Thienpont B, Cappuyns S, Venken T, Busschaert P, Van Haele M, Van Cutsem E, Roskams T, van Pelt J, Verslype C, Dekervel J, Lambrechts D (2020) DNA methylation-driven EMT is a common mechanism of resistance to various therapeutic agents in cancer. Clin Epigenetics 12:27–27

Golkowski M, Lau HT, Chan M, Kenerson H, Vidadala VN, Shoemaker A, Maly DJ, Yeung RS, Gujral TS, Ong SE (2020) Pharmacoproteomics identifies kinase pathways that drive the epithelial-mesenchymal transition and drug resistance in hepatocellular carcinoma. Cell Syst 11:196–207.e197

Gramantieri L, Pollutri D, Gagliardi M, Giovannini C, Quarta S, Ferracin M, Casadei-Gardini A, Callegari E, De Carolis S, Marinelli S, Benevento F, Vasuri F, Ravaioli M, Cescon M, Piscaglia F, Negrini M, Bolondi L, Fornari F (2020) MiR-30e-3p influences tumor phenotype through MDM2/TP53 axis and predicts sorafenib resistance in hepatocellular carcinoma. Cancer Res 80:1720–1734

Guo Z, Zhang J, Fan L, Liu J, Yu H, Li X, Sun G (2019) Long noncoding RNA (lncRNA) small nucleolar RNA host Gene 16 (SNHG16) predicts poor prognosis and sorafenib resistance in hepatocellular carcinoma. Med Sci Monit 25:2079–2086

Han P, Li H, Jiang X, Zhai B, Tan G, Zhao D, Qiao H, Liu B, Jiang H, Sun X (2017) Dual inhibition of Akt and c-Met as a second-line therapy following acquired resistance to sorafenib in hepatocellular carcinoma cells. Mol Oncol 11:320–334

Hu H, Yang L, Li L, Zeng C (2018) Long non-coding RNA KCNQ1OT1 modulates oxaliplatin resistance in hepatocellular carcinoma through miR-7-5p/ ABCC1 axis. Biochem Biophys Res Commun 503:2400–2406

Hu B, Cheng J-W, Hu J-W, Li H, Ma X-L, Tang W-G, Sun Y-F, Guo W, Huang A, Zhou K-Q, Gao P-T, Cao Y, Qiu S-J, Zhou J, Fan J, Yang X-R (2019) KPNA3 confers sorafenib resistance to advanced hepatocellular carcinoma via TWIST regulated epithelial-mesenchymal transition. J Cancer 10:3914–3925

Huang H, Chen J, Ding C-M, Jin X, Jia Z-M, Peng J (2018) LncRNA NR2F1-AS1 regulates hepatocellular carcinoma oxaliplatin resistance by targeting ABCC1 via miR-363. J Cell Mol Med 22:3238–3245

Huang X-Y, Zhang P-F, Wei C-Y, Peng R, Lu J-C, Gao C, Cai J-B, Yang X, Fan J, Ke A-W, Zhou J, Shi G-M (2020) Circular RNA circMET drives immunosuppression and anti-PD1 therapy resistance in hepatocellular carcinoma via the miR-30-5p/snail/DPP4 axis. Mol Cancer 19:92–92

Iorio MV, Croce CM (2012) MicroRNA dysregulation in cancer: diagnostics, monitoring and therapeutics. A comprehensive review. EMBO Mol Med 4:143–159

Ji L, Lin Z, Wan Z, Xia S, Jiang S, Cen D, Cai L, Xu J, Cai X (2020) miR-486-3p mediates hepatocellular carcinoma sorafenib resistance by targeting FGFR4 and EGFR. Cell Death Dis 11:250–250

Jin F, Wang Y, Li M, Zhu Y, Liang H, Wang C, Wang F, Zhang C-Y, Zen K, Li L (2017) MiR-26 enhances chemosensitivity and promotes apoptosis of hepatocellular carcinoma cells through inhibiting autophagy. Cell Death Dis 8:e2540–e2540

Kabir TD, Ganda C, Brown RM, Beveridge DJ, Richardson KL, Chaturvedi V, Candy P, Epis M, Wintle L, Kalinowski F, Kopp C, Stuart LM, Yeoh GC, George J, Leedman PJ (2018) A microRNA-7/growth arrest specific 6/TYRO3 axis regulates the growth and invasiveness of sorafenib-resistant cells in human hepatocellular carcinoma. Hepatology 67:216–231

Kelley RK, Mollon P, Blanc J-F, Daniele B, Yau T, Cheng A-L, Valcheva V, Marteau F, Guerra I, Abou-Alfa GK (2020) Comparative efficacy of cabozantinib and regorafenib for advanced hepatocellular carcinoma. Adv Ther 37:2678–2695

Ko K-L, Mak L-Y, Cheung K-S, Yuen M-F (2020) Hepatocellular carcinoma: recent advances and emerging medical therapies. F1000Research 9:1620

Kristensen LS, Andersen MS, Stagsted LVW, Ebbesen KK, Hansen TB, Kjems J (2019) The biogenesis, biology and characterization of circular RNAs. Nat Rev Genet 20:675–691

Kudo M, Finn RS, Qin S, Han K-H, Ikeda K, Piscaglia F, Baron A, Park J-W, Han G, Jassem J, Blanc JF, Vogel A, Komov D, Evans TRJ, Lopez C, Dutcus C, Guo M, Saito K, Kraljevic S, Tamai T, Ren M, Cheng A-L (2018) Lenvatinib versus sorafenib in first-line treatment of patients with unresectable hepatocellular carcinoma: a randomised phase 3 non-inferiority trial. Lancet 391:1163–1173

Lai H-H, Li C-W, Hong C-C, Sun H-Y, Chiu C-F, Ou D-L, Chen P-S (2019) TARBP2-mediated destabilization of Nanog overcomes sorafenib resistance in hepatocellular carcinoma. Mol Oncol 13:928–945

Li Y, Ye Y, Feng B, Qi Y (2017) Long noncoding RNA lncARSR promotes doxorubicin resistance in hepatocellular carcinoma via modulating PTEN-PI3K/Akt pathway. J Cell Biochem 118: 4498–4507

Li J, Wu P-W, Zhou Y, Dai B, Zhang P-F, Zhang Y-H, Liu Y, Shi X-L (2018) Rage induces hepatocellular carcinoma proliferation and sorafenib resistance by modulating autophagy. Cell Death Dis 9:225–225

Li W, Dong X, He C, Tan G, Li Z, Zhai B, Feng J, Jiang X, Liu C, Jiang H, Sun X (2019) LncRNA SNHG1 contributes to sorafenib resistance by activating the Akt pathway and is positively regulated by miR-21 in hepatocellular carcinoma cells. J Exp Clin Cancer Res 38:183–183

Li X, Ding J, Wang X, Cheng Z, Zhu Q (2020a) NUDT21 regulates circRNA cyclization and ceRNA crosstalk in hepatocellular carcinoma. Oncogene 39:891–904

Li X, He J, Ren X, Zhao H, Zhao H (2020b) Circ_0003998 enhances doxorubicin resistance in hepatocellular carcinoma by regulating miR-218-5p/EIF5A2 pathway. Diagn Pathol 15:141

Li J, Qin X, Wu R, Wan L, Zhang L, Liu R (2020c) Circular RNA circFBXO11 modulates hepatocellular carcinoma progress and oxaliplatin resistance through miR-605/FOXO3/ABCB1 axis. J Cell Mol Med 24:5152–5161

Li X, Zhou Y, Yang L, Ma Y, Peng X, Yang S, Li H, Liu J (2020d) LncRNA NEAT1 promotes autophagy via regulating miR-204/ATG3 and enhanced cell resistance to sorafenib in hepatocellular carcinoma. J Cell Physiol 235:3402–3413

Lin Z, Niu Y, Wan A, Chen D, Liang H, Chen X, Sun L, Zhan S, Chen L, Cheng C, Zhang X, Bu X, He W, Wan G (2020) RNA m(6) A methylation regulates sorafenib resistance in liver cancer through FOXO3-mediated autophagy. EMBO J 39:e103181–e103181

Ling S, Li J, Shan Q, Dai H, Lu D, Wen X, Song P, Xie H, Zhou L, Liu J, Xu X, Zheng S (2017) USP22 mediates the multidrug resistance of hepatocellular carcinoma via the SIRT1/AKT/MRP1 signaling pathway. Mol Oncol 11:682–695

Liu K, Liu S, Zhang W, Ji B, Wang Y, Liu Y (2014) miR-222 regulates sorafenib resistance and enhance tumorigenicity in hepatocellular carcinoma. Int J Oncol 45:1537–1546

Llovet JM, Montal R, Sia D, Finn RS (2018) Molecular therapies and precision medicine for hepatocellular carcinoma. Nat Rev Clin Oncol 15:599–616

Long J, Bai Y, Yang X, Lin J, Yang X, Wang D, He L, Zheng Y, Zhao H (2019) Construction and comprehensive analysis of a ceRNA network to reveal potential prognostic biomarkers for hepatocellular carcinoma. Cancer Cell Int 19:90–90

Lu TX, Rothenberg ME (2018) MicroRNA. J Allergy Clin Immunol 141:1202–1207

Lu S, Yao Y, Xu G, Zhou C, Zhang Y, Sun J, Jiang R, Shao Q, Chen Y (2018) CD24 regulates sorafenib resistance via activating autophagy in hepatocellular carcinoma. Cell Death Dis 9: 646–646

Luo D, Wang Z, Wu J, Jiang C, Wu J (2014) The role of hypoxia inducible factor-1 in hepatocellular carcinoma. Biomed Res Int 2014:409272

Ma Z, Guo D, Wang Q, Liu P, Xiao Y, Wu P, Wang Y, Chen B, Liu Z, Liu Q (2019) Lgr5-mediated p53 repression through PDCD5 leads to doxorubicin resistance in hepatocellular carcinoma. Theranostics 9:2967–2983

Ma Y, Wen J, Wang J, Wang C, Zhang Y, Zhao L, Li J, Feng X (2020) Asiaticoside antagonizes proliferation and chemotherapeutic drug resistance in hepatocellular carcinoma (HCC) cells. Med Sci Monit 26:e924435–e924435

Matsui M, Corey DR (2017) Non-coding RNAs as drug targets. Nat Rev Drug Discov 16:167–179

Matsuki M, Hoshi T, Yamamoto Y, Ikemori-Kawada M, Minoshima Y, Funahashi Y, Matsui J (2018) Lenvatinib inhibits angiogenesis and tumor fibroblast growth factor signaling pathways in human hepatocellular carcinoma models. Cancer Med 7:2641–2653

Mattick JS, Makunin IV (2006) Non-coding RNA. Hum Mol Genet 15:R17–R29

McGlynn KA, Petrick JL, London WT (2015) Global epidemiology of hepatocellular carcinoma: an emphasis on demographic and regional variability. Clin Liver Dis 19:223–238

Mou S-J, Yang P-F, Liu Y-P, Xu N, Jiang W-W, Yue W-J (2020) BCLAF1 promotes cell proliferation, invasion and drug-resistance though targeting lncRNA NEAT1 in hepatocellular carcinoma. Life Sci 242:117177

Najafi M, Mortezaee K, Majidpoor J (2019) Cancer stem cell (CSC) resistance drivers. Life Sci 234: 116781

Nie J, Lin B, Zhou M, Wu L, Zheng T (2018) Role of ferroptosis in hepatocellular carcinoma. J Cancer Res Clin Oncol 144:2329–2337

Nwosu ZC, Piorońska W, Battello N, Zimmer AD, Dewidar B, Han M, Pereira S, Blagojevic B, Castven D, Charlestin V, Holenya P, Lochead J, De La Torre C, Gretz N, Sajjakulnukit P, Zhang L, Ward MH, Marquardt JU, di Magliano MP, Lyssiotis CA, Sleeman J, Wölfl S, Ebert MP, Meyer C, Hofmann U, Dooley S (2020) Severe metabolic alterations in liver cancer lead to ERK pathway activation and drug resistance. EBioMedicine 54:102699–102699

Ogunwobi OO, Harricharran T, Huaman J, Galuza A, Odumuwagun O, Tan Y, Ma GX, Nguyen MT (2019) Mechanisms of hepatocellular carcinoma progression. World J Gastroenterol 25: 2279–2293

Ohata Y, Shimada S, Akiyama Y, Mogushi K, Nakao K, Matsumura S, Aihara A, Mitsunori Y, Ban D, Ochiai T, Kudo A, Arii S, Tanabe M, Tanaka S (2017) Acquired resistance with epigenetic alterations under long-term antiangiogenic therapy for hepatocellular carcinoma. Mol Cancer Ther 16:1155–1165

Pan JJ, Xie XJ, Li X, Chen W (2015) Long non-coding RNAs and drug resistance. Asian Pac J Cancer Prev 16:8067–8073

Personeni N, Pressiani T, Santoro A, Rimassa L (2018) Regorafenib in hepatocellular carcinoma: latest evidence and clinical implications. Drugs Context 7:212533–212533

Petrick JL, Florio AA, Znaor A, Ruggieri D, Laversanne M, Alvarez CS, Ferlay J, Valery PC, Bray F, McGlynn KA (2020) International trends in hepatocellular carcinoma incidence, 1978–2012. Int J Cancer 147:317–330

Salmena L, Poliseno L, Tay Y, Kats L, Pandolfi PP (2011) A ceRNA hypothesis: the Rosetta Stone of a hidden RNA language? Cell 146:353–358

Saw PE, Xu X, Chen J, Song EW (2021) Non-coding RNAs: the new central dogma of cancer biology. Sci China Life Sci 64:22–50

Shao M, Shi R, Gao Z-X, Gao S-S, Li J-F, Li H, Cui S-Z, Hu W-M, Chen T-Y, Wu G-R, Zhang J, Xu J, Sy M-S, Li C (2021) Crizotinib and doxorubicin cooperatively reduces drug resistance by mitigating MDR1 to increase hepatocellular carcinoma cells death. Front Oncol 11: 650052–650052

Song Y, Kim JS, Choi EK, Kim J, Kim KM, Seo HR (2017) TGF-β-independent CTGF induction regulates cell adhesion mediated drug resistance by increasing collagen I in HCC. Oncotarget 8: 21650–21662

Su Y-C, Davuluri GVN, Chen C-H, Shiau D-C, Chen C-C, Chen C-L, Lin Y-S, Chang C-P (2016) Galectin-1-induced autophagy facilitates cisplatin resistance of hepatocellular carcinoma. PLoS One 11:e0148408–e0148408

Sui C, Dong Z, Yang C, Zhang M, Dai B, Geng L, Lu J, Yang J, Xu M (2019) LncRNA FOXD2-AS1 as a competitive endogenous RNA against miR-150-5p reverses resistance to sorafenib in hepatocellular carcinoma. J Cell Mol Med 23:6024–6033

Sun G, Mackey LV, Coy DH, Yu C-Y, Sun L (2015) The histone deacetylase inhibitor vaproic acid induces cell growth arrest in hepatocellular carcinoma cells via suppressing notch signaling. J Cancer 6:996–1004

Sun X, Niu X, Chen R, He W, Chen D, Kang R, Tang D (2016) Metallothionein-1G facilitates sorafenib resistance through inhibition of ferroptosis. Hepatology 64:488–500

Tan W, Luo X, Li W, Zhong J, Cao J, Zhu S, Chen X, Zhou R, Shang C, Chen Y (2019) TNF-α is a potential therapeutic target to overcome sorafenib resistance in hepatocellular carcinoma. EBioMedicine 40:446–456

Tian N, Han Z, Li Z, Zhou M, Fan C (2014) Lin28/let-7/Bcl-xL pathway: the underlying mechanism of drug resistance in Hep3B cells. Oncol Rep 32:1050–1056

Tong M, Che N, Zhou L, Luk ST, Kau PW, Chai S, Ngan ES, Lo C-M, Man K, Ding J, Lee TK, Ma S (2018) Efficacy of annexin A3 blockade in sensitizing hepatocellular carcinoma to sorafenib and regorafenib. J Hepatol 69:826–839

Wang XJ, Feng CW, Li M (2013) ADAM17 mediates hypoxia-induced drug resistance in hepatocellular carcinoma cells through activation of EGFR/PI3K/Akt pathway. Mol Cell Biochem 380:57–66

Wang H, Tang Y, Yang D, Zheng L (2019) MicroRNA-591 functions as a tumor suppressor in hepatocellular carcinoma by lowering drug resistance through inhibition of far-upstream element-binding protein 2-mediated phosphoinositide 3-Kinase/Akt/Mammalian target of rapamycin axis. Pharmacology 104:173–186

Wang C, Ke S, Li M, Lin C, Liu X, Pan Q (2020a) Downregulation of LncRNA GAS5 promotes liver cancer proliferation and drug resistance by decreasing PTEN expression. Mol Gen Genomics 295:251–260

Wang S, Wang Y, Xun X, Zhang C, Xiang X, Cheng Q, Hu S, Li Z, Zhu J (2020b) Hedgehog signaling promotes sorafenib resistance in hepatocellular carcinoma patient-derived organoids. J Exp Clin Cancer Res 39:22–22

Wei L, Lee D, Law C-T, Zhang MS, Shen J, Chin DW-C, Zhang A, Tsang FH-C, Wong CL-S, Ng IO-L, Wong CC-L, Wong C-M (2019a) Genome-wide CRISPR/Cas9 library screening identified PHGDH as a critical driver for Sorafenib resistance in HCC. Nat Commun 10: 4681–4681

Wei L, Wang X, Lv L, Liu J, Xing H, Song Y, Xie M, Lei T, Zhang N, Yang M (2019b) The emerging role of microRNAs and long noncoding RNAs in drug resistance of hepatocellular carcinoma. Mol Cancer 18:147–147

Wei X, Zhao L, Ren R, Ji F, Xue S, Zhang J, Liu Z, Ma Z, Wang XW, Wong L, Liu N, Shi J, Guo X, Roessler S, Zheng X, Ji J (2021) MiR-125b loss activated HIF1α/pAKT loop, leading to transarterial chemoembolization resistance in hepatocellular carcinoma. Hepatology 73: 1381–1398

Wu F-Q, Fang T, Yu L-X, Lv G-S, Lv H-W, Liang D, Li T, Wang C-Z, Tan Y-X, Ding J, Chen Y, Tang L, Guo L-N, Tang S-H, Yang W, Wang H-Y (2016) ADRB2 signaling promotes HCC progression and sorafenib resistance by inhibiting autophagic degradation of HIF1α. J Hepatol 65:314–324

Wu L, Pan C, Wei X, Shi Y, Zheng J, Lin X, Shi L (2018) lncRNA KRAL reverses 5-fluorouracil resistance in hepatocellular carcinoma cells by acting as a ceRNA against miR-141. Cell Commun Signal 16:47–47

Wu L-L, Cai W-P, Lei X, Shi K-Q, Lin X-Y, Shi L (2019) NRAL mediates cisplatin resistance in hepatocellular carcinoma via miR-340-5p/Nrf2 axis. J Cell Commun Signal 13:99–112

Wu H, Wang T, Liu Y, Li X, Xu S, Wu C, Zou H, Cao M, Jin G, Lang J, Wang B, Liu B, Luo X, Xu C (2020) Mitophagy promotes sorafenib resistance through hypoxia-inducible ATAD3A dependent Axis. J Exp Clin Cancer Res 39:274–274

Xie C, Zhang L-Z, Chen Z-L, Zhong W-J, Fang J-H, Zhu Y, Xiao M-H, Guo Z-W, Zhao N, He X, Zhuang S-M (2020) A hMTR4-PDIA3P1-miR-125/124-TRAF6 regulatory axis and its function in NF kappa B signaling and chemoresistance. Hepatology 71:1660–1677

Xiong H, Ni Z, He J, Jiang S, Li X, He J, Gong W, Zheng L, Chen S, Li B, Zhang N, Lyu X, Huang G, Chen B, Zhang Y, He F (2017) LncRNA HULC triggers autophagy via stabilizing Sirt1 and attenuates the chemosensitivity of HCC cells. Oncogene 36:3528–3540

Xu N, Zhang J, Shen C, Luo Y, Xia L, Xue F, Xia Q (2012) Cisplatin-induced downregulation of miR-199a-5p increases drug resistance by activating autophagy in HCC cell. Biochem Biophys Res Commun 423:826–831

Xu J, Ji L, Liang Y, Wan Z, Zheng W, Song X, Gorshkov K, Sun Q, Lin H, Zheng X, Chen J, Jin R-A, Liang X, Cai X (2020a) CircRNA-SORE mediates sorafenib resistance in hepatocellular carcinoma by stabilizing YBX1. Signal Transduct Target Ther 5:298–298

Xu WP, Liu JP, Feng JF, Zhu CP, Yang Y, Zhou WP, Ding J, Huang CK, Cui YL, Ding CH, Zhang X, Lu B, Xie WF (2020b) miR-541 potentiates the response of human hepatocellular carcinoma to sorafenib treatment by inhibiting autophagy. Gut 69:1309–1321

Xu Y, Liu Y, Li Z, Li H, Li X, Yan L, Mao J, Shen J, Chen W, Xue F (2020c) Long non-coding RNA H19 is involved in sorafenib resistance in hepatocellular carcinoma by upregulating miR-675. Oncol Rep 44:165–173

Xu J, Wan Z, Tang M, Lin Z, Jiang S, Ji L, Gorshkov K, Mao Q, Xia S, Cen D, Zheng J, Liang X, Cai X (2020d) N(6)-methyladenosine-modified CircRNA-SORE sustains sorafenib resistance in hepatocellular carcinoma by regulating β-catenin signaling. Mol Cancer 19:163–163

Yang C, Dong Z, Hong H, Dai B, Song F, Geng L, Lu J, Yang J, Sui C, Xu M (2020) circFN1 mediates sorafenib resistance of hepatocellular carcinoma cells by sponging miR-1205 and regulating E2F1 expression. Mol Ther Nucleic Acids 22:421–433

Yin Z, Ma T, Lin Y, Lu X, Zhang C, Chen S, Jian Z (2018) IL-6/STAT3 pathway intermediates M1/M2 macrophage polarization during the development of hepatocellular carcinoma. J Cell Biochem 119:9419–9432

Yu H, Qiu Y, Pang X, Li J, Wu S, Yin S, Han L, Zhang Y, Jin C, Gao X, Hu W, Wang T (2017) Lycorine promotes autophagy and apoptosis via TCRP1/Akt/mTOR axis inactivation in human hepatocellular carcinoma. Mol Cancer Ther 16:2711

Yu Y, Wang Y, Xiao X, Cheng W, Hu L, Yao W, Qian Z, Wu W (2019) MiR-204 inhibits hepatocellular cancer drug resistance and metastasis through targeting NUAK1. Biochem Cell Biol 97:563–570

Zang J, Lu D, Xu A (2020) The interaction of circRNAs and RNA binding proteins: an important part of circRNA maintenance and function. J Neurosci Res 98:87–97

Zhang N, Xie H, Lu W, Li F, Li J, Guo Z (2017) Chloroquine sensitizes hepatocellular carcinoma cells to chemotherapy via blocking autophagy and promoting mitochondrial dysfunction. Int J Clin Exp Pathol 10:10056–10065

Zhang PF, Wang F, Wu J, Wu Y, Huang W, Liu D, Huang XY, Zhang XM, Ke AW (2019) LncRNA SNHG3 induces EMT and sorafenib resistance by modulating the miR-128/CD151 pathway in hepatocellular carcinoma. J Cell Physiol 234:2788–2794

Zhang P-F, Gao C, Huang X-Y, Lu J-C, Guo X-J, Shi G-M, Cai J-B, Ke A-W (2020a) Cancer cell-derived exosomal circUHRF1 induces natural killer cell exhaustion and may cause resistance to anti-PD1 therapy in hepatocellular carcinoma. Mol Cancer 19:110–110

Zhang W, Liu Y, Fu Y, Han W, Xu H, Wen L, Deng Y, Liu K (2020b) Long non-coding RNA LINC00160 functions as a decoy of microRNA-132 to mediate autophagy and drug resistance in hepatocellular carcinoma via inhibition of PIK3R3. Cancer Lett 478:22–33

Zhang Z, Tan X, Luo J, Yao H, Si Z, Tong J-S (2020c) The miR-30a-5p/CLCF1 axis regulates sorafenib resistance and aerobic glycolysis in hepatocellular carcinoma. Cell Death Dis 11: 902–902

Zhang J, Zhao X, Ma X, Yuan Z, Hu M (2020d) KCNQ1OT1 contributes to sorafenib resistance and programmed death-ligand-1-mediated immune escape via sponging miR-506 in hepatocellular carcinoma cells. Int J Mol Med 46:1794–1804

Zhao L, Hu K, Cao J, Wang P, Li J, Zeng K, He X, Tu PF, Tong T, Han L (2019) lncRNA miat functions as a ceRNA to upregulate sirt1 by sponging miR-22-3p in HCC cellular senescence. Aging (Albany NY) 11:7098–7122

Zhong W, Dai Q, Huang Q (2020) Effect of lncRNA KCNQ1OT1 on autophagy and drug resistance of hepatocellular carcinoma cells by targeting miR-338-3p. Cell Mol Biol 66: 191–196

Zhou Y, Chen E, Tang Y, Mao J, Shen J, Zheng X, Xie S, Zhang S, Wu Y, Liu H, Zhi X, Ma T, Ni H, Chen J, Chai K, Chen W (2019a) miR-223 overexpression inhibits doxorubicin-induced autophagy by targeting FOXO3a and reverses chemoresistance in hepatocellular carcinoma cells. Cell Death Dis 10:843–843

Zhou B, Lu Q, Liu J, Fan L, Wang Y, Wei W, Wang H, Sun G (2019b) Melatonin increases the sensitivity of hepatocellular carcinoma to sorafenib through the PERK-ATF4-Beclin1 pathway. Int J Biol Sci 15:1905–1920

Zhou Y, Wang Y, Zhou W, Chen T, Wu Q, Chutturghoon VK, Lin B, Geng L, Yang Z, Zhou L, Zheng S (2019c) YAP promotes multi-drug resistance and inhibits autophagy-related cell death in hepatocellular carcinoma via the RAC1-ROS-mTOR pathway. Cancer Cell Int 19:179–179

Zhou Y, Huang Y, Dai T, Hua Z, Xu J, Lin Y, Han L, Yue X, Ho L, Lu J, Ai X (2021) LncRNA TTN-AS1 intensifies sorafenib resistance in hepatocellular carcinoma by sponging miR-16-5p and upregulation of cyclin E1. Biomed Pharmacother 133:111030

Zhu Y-J, Zheng B, Wang H-Y, Chen L (2017) New knowledge of the mechanisms of sorafenib resistance in liver cancer. Acta Pharmacol Sin 38:614–622

Liver Stereotactic Body Radiotherapy (SBRT)

Olwen Leaman Alcibar, Fernando López Campos,
José Antonio Blanco, Patricia Tavera Pomata, and Carmen Rubio

Abstract

The liver organ fulfills many important functions that go from vitamin reservoir to protein production or immune cell homeostasis in the blood. This chapter concentrates on its immune function, and specifically on its propensity to immune tolerance when antigens get in contact with resident immune cells in the liver. Although these antigens generally come from intestinal blood through the portal vein and are mostly not pathogenic, in a cancer scenario these antigens can come from primary tumors that metastasize to the liver or primary tumor of the liver per se. Whether the tumor microenvironment of the liver triggers an effective killing immune response depends on the balance between inhibitory and activating signals present in the tissue. Today, there is robust clinical evidence indicating a worse prognosis of patients with liver metastases compared to patients with metastases in other organs. In addition, these patients have had worse tumor outcomes when new immunotherapies have been tested in clinical trials, together suggesting there is something especially immunosuppressive about the liver tissue. Initial preclinical and translational studies have brought some light into this tunnel and have explained, at least partly, where the problem resides. Early work on the combination of immunotherapy and liver Stereotactic Body

O. Leaman Alcibar (✉) · P. Tavera Pomata
Radiation Oncology Department, Gregorio Marañón General University Hospital, Madrid, Spain
e-mail: olweneriete.leaman@salud.madrid.org

F. López Campos
Radiation Oncology Department, Ramón y Cajal University Hospital, Madrid, Spain

J. A. Blanco
Radiation Oncology Department, Marqués de Valdecilla University Hospital, Santander, Cantabria, Spain

C. Rubio
Radiation Oncology Department, HM University Hospital Sanchinarro, Madrid, Spain

Interdisciplinary Cancer Research, https://doi.org/10.1007/16833_2022_94
Published online: 30 December 2022

Radiotherapy (SBRT), both in primary liver cancer and liver metastases, have come forward as a promising solution to the immune tolerance of the liver.

Keywords

Immunotherapy · Liver · Stereotactic body radiotherapy · Tumor microenvironment

1 Introduction

1.1 General Functions and Blood Flow of the Liver

The human liver is essential for the elimination of toxins, the maintenance of metabolic homeostasis, the production of certain proteins and the regulation of immunity. It is composed of functional units called lobules, and it is capable of renewing its tissue due to the plasticity of resident hepatocytes (Li et al. 2020).

The liver organ serves as a filter for the approximately 1.5 liters of blood per minute that crosses its circuit. About 80% of this blood flow comes from the portal vein that drains the gastrointestinal tract and the rest comes from the hepatic artery. This blood travels loaded with antigens originated from food that has been digested and commensal antigens, which are largely harmless and therefore, should not activate a systemic immune response nor a local liver immune response. This immune tolerance in the liver is accomplished by a variety of immunomodulating antigen-presenting cells (APCs) that reside in its sinusoids (Doherty 2015).

1.2 Liver Histology and Cell Composition

The hepatic sinusoids are specialized vascular channels through which portal vein blood travels toward the liver's central vein. Sinusoids have endothelial cells surrounding them as a first barrier and a small Disse space between the endothelial cells and the hepatocytes that serve as the second barrier.

Liver sinusoidal endothelial cells are aligned leaving gaps between them that make blood travel to the space of Disse, and therefore enable direct contact of antigen-enriched blood with hepatocytes. In these small but specialized vasculature, cells that belong both to the innate and adaptive immunity play an important role in the everyday decision of immune tolerance versus immune attack.

Multiple types of innate liver leucocytes accumulate in the liver and function as pathogen sensors that respond rapidly to microbial pathogens, cells infected by pathogens and tumor cells. They selectively activate the adaptive immune response, such as those that involve T helper type 1 (Th1), Th2, Th17, regulatory T (Treg) cell, and antibody responses. This innate leukocyte population is integrated by NK cells, CD56+ T cells, natural killer T (NKT) cells, gamma/delta T cells, and mucosal associated invariant T (MAIT) cells. Using contact-dependent interactions and release of cytokines they enhance antigen presentation and activate CD4+ and CD8+ T cells, B cells, and macrophages (Doherty 2015).

In addition, immune receptors for pathogen molecules or microorganism components such as toll-like receptors (TLR), nucleotide-binding oligomerization domain-like receptors (NRL), and retinoic acid inducible gene 1-like receptors (RLR) are spread out in parenchymal and non-parenchymal liver cells, making most cells in the sinusoid capable of initiating an immune activating cascade.

Hepatocytes, on the other hand, produce C-reactive protein, complement, opsonizing proteins, cytokines like IL-6, TNFα, and TGFβ, which will be key factors in acute phase inflammation and posterior liver tissue repair (Protzer et al. 2012).

Hepatic APCs, on the other hand, come in different forms and flavors, but overall have the capacity to favor a tolerogenic response against exposed antigens. The mechanisms used by them for this purpose are T cell killing or inactivation and influencing maturation of naïve T cells into inhibitory regulatory T cells.

1.3 Mechanisms That May Explain Immunotherapy Failure in the Liver

Antigen presentation by APCs in the liver preferentially leads to immune tolerance. This is convenient for our body in the context of healthy liver and the absence of pathogens or tumor cells in the blood flow or hepatic tissue. Nevertheless, our immune system needs to overcome the liver immune tolerance to activate an immune response when pathogens and tumor cells are present. The success of such immune attack will depend on the balance between activating and inhibiting factors and cells in the liver. Analyzing the cell population that enables a regulatory immunity niche in the liver can bring light to this unresolved problem.

Dendritic cells (DC) in the liver have been categorized into myeloid DC (mDC) and plasmacytoid DC. mDC are positive for CD11b, CD11c, and CD1c and produce IL-10 in abundance as well as inducing its production on T cells, generating a hypoproliferation and hyporesponsiveness of CD4+ antigen specific T cells (Bamboat et al. 2009). In addition, preclinical studies have shown that liver stroma is capable of synthesizing macrophage colony-stimulating factor (M-CSF) locally, which in turn enhances the inhibitory nature of these DCs (Xia et al. 2008). Plasmacytoid DC are positive for CD-123 (IL-3 receptor) and BDCA-2 (C type lectin) on their cell surface and are known to upregulate PD-L1 in response to TLR and NLR activation, which in turn promotes regulatory T cell differentiation via IL-27 release (Moseman et al. 2004). Taken together, compared with DCs in other organs, liver DCs are less likely to activate and produce IL-12 to help T cell activation, and are more prone to produce IL-10, IL-27, and IDO during activation of their pathogen receptors (Eckert et al. 2015). Interestingly, mouse studies brought to light the knowledge that the lipid content of DCs can influence their immune capacity; being the low lipid DCs tolerogenic and vice versa (Ibrahim et al. 2012).

Kupffer cells (KC) are also APCs and are liver-resident macrophages. They have pathogen receptors such as TLR and NRL and can internalize antigens and damaged apoptotic cells by phagocytosis. On their surface they have many MHC molecules and through them and other costimulatory molecules can activate naïve T cells. While this is true for immunogenic M1 macrophages or M1 KC, it is not for alternatively activated M2 macrophages or M2 KC, that produce high levels of IL-10 and TGFβ which are both inhibitory molecules. Cell studies state that upon TLR activation KC preferentially act as M2 macrophages (Zhang et al. 2011). In addition, when antigen presentation happens such as in tumors, PD-L1 appears in KC of the TME, interacting with PD-1+CD8+ T cells that lose their proliferative and effector functions (Wu et al. 2009).

Hepatic stellate cells (HSC) are found in the space of Disse and are known for vitamin A storage. They help in the regulation of blood flow across the sinusoids. They too are capable of presenting antigens to T cells, upregulating PD-L1 and this can lead to allograft acceptance (Charles et al. 2013). HSC can also specifically present glycolipids to CD1-restricted T cells and promote the appearance of myeloid-derived suppressor cells (MDSC) that inhibit T cells in many ways (Höchst et al. 2013).

Finally, liver sinusoidal endothelial cells (LSEC), although belonging to the liver parenchyma, have the potential of antigen presentation. Antigens are presented in their surface with MHC molecules to CD8+ T cells that become tolerogenic (von Oppen et al. 2009) and to CD4+ T cells that become regulatory T cells CD4+ FOXP3+ (Kruse et al. 2009). Moreover, LSEC can secrete IL-10, PGE, TGFβ, and PD-L1 (Uhrig et al. 2005; Diehl et al. 2008) that all in all promote liver immune tolerogenic niche.

2 Data of Immunotherapy Failure in Patients with Liver Metastasis

For years it has been unknown whether metastatic liver progression of the different tumor types could limit the efficacy of immunomodulatory drugs (Mehlen and Puisieux 2006) and therefore determine both the prognosis and the response to established immunological therapies (Reck et al. 2019). Likewise, several recently published studies have postulated that liver metastases decrease the systemic efficacy of immunotherapy both in preclinical models and in patients (Yu et al. 2021; Lee et al. 2020), the latter obtaining a limited benefit, regardless of other established response biomarkers such as tumor mutational load or PD-L1 expression.

Immunomodulatory drugs take advantage of tumor infiltrating CD8+ effector T cells to induce long-lasting responses to immune therapy in a wide variety of tumor pathologies (Cohen et al. 2019; Antonia et al. 2017; Larkin et al. 2019; Bajorin et al. 2021; Kelly et al. 2021). Unfortunately, the causes of the absence of immune response remain poorly defined (Chen and Mellman 2017), and it is not clear whether the location of the metastases influences the efficacy of these treatments.

Currently, we have data from studies such as the one by Tumeh et al. (2017), where a reduced response and shorter progression-free survival were observed in a cohort of patients diagnosed with melanoma treated with the anti-PD-1 pembrolizumab [objective response rate (ORR), 30.6%; median PFS, 5.1 months] in patients with liver metastases compared with patients without such lesions (ORR, 56.3%; median PFS, 20.1 months) $p \leq 0.0001$, these data being confirmed in a validation cohort of the KEYNOTE-002 trial ($p = 0.0006$).

Similarly, a reduced response rate and shorter PFS were observed in non-small cell lung cancer patients with liver metastases treated with pembrolizumab [median PFS, 1.8 months; 95% confidence interval (CI), 1.4–2.0], compared with those without liver metastases ($n = 119$, median PFS, 4.0 months; 95% CI, 2.1–5.1), $p = 0.0094$. Liver biopsies of a subgroup of these patients ($n = 62$) demonstrated a reduction in the density of CD8 + T cells in the margin of the tumor analyzed (Tumeh et al. 2017).

These findings are in line with those published later by other groups, which reported overall survival data for patients treated with another anti-PD1, nivolumab, within the phase I trial CA209-003, in patients with metastatic melanoma, kidney clear cell carcinoma, and non-small cell lung carcinoma, describing in the multivariate analysis that the presence of liver metastasis is independently associated with a reduced probability of survival at 5 years in these patients (odds ratio [OR], 0.31; CI 95%, 0.12–0.83, $p = 0.02$) (Topalian et al. 2019).

Similarly, the presence of liver metastasis in patients treated with nivolumab, diagnosed with advanced gastric cancer, has been significantly associated with the development of hyperprogressive disease, presenting a shorter overall survival and PFS (Sasaki et al. 2020).

Different retrospective studies published to date, which include patients treated with immunotherapy with different primary tumors and liver metastases, corroborate that the presence of these lesions can be considered a poor prognostic factor, suggesting the need to develop up-to-date prognostic models that would need to be validated in prospective cohorts (Bilen et al. 2019; Schmid et al. 2018).

The mechanism of immune scape induced by liver metastases is unclear. One of the hypotheses proposed, developed from murine models, advocates that the presence of tumor antigens within the liver leads to systemic suppression of antitumor immunity, together with regulatory T cell and intratumoral MDSC (CD11b+ monocytes) induction, causing a dysfunctional immune state that is difficult to reverse with anti-PD-1 monotherapy unless Treg cells are depleted (anti-CTLA-4) or destabilized (inhibitor of EZH2) with combination immunotherapies (Lee et al. 2020).

However, combination studies of these treatments (anti CTLA-4 and anti PD-1) in patients diagnosed with metastatic melanoma, continue to show us that those individuals with liver metastases have worse clinical results in terms of ORR (OR, 0.33; $p = 0.02$), PFS (HR, 4.03; $p < 0.01$), and overall survival (HR, 3.17; $p = 0.01$) than those who did not present liver metastatic lesions, leaving room for much improvement (Gide et al. 2020).

The results of a cohort of 718 patients with metastatic melanoma, non-small cell lung cancer, urothelial carcinoma, and renal cell tumors who received immune-checkpoint inhibitors were recently reported, analyzing the clinical benefit of patients with liver metastases and these being significantly lower compared to patients without them. These happened regardless of the histology of the disease, observing a strong inverse correlation of hepatic metastatic tropism with the efficacy of immunotherapy, after adjusting for tumor mutational load and PD-L1 expression (Yu et al. 2021). In that same publication, murine models were treated in the same conditions and those with liver metastases did not develop an adequate immune response, diverting activated CD8+ T cells from the systemic circulation and reducing their function. Specifically at the liver level, activated antigen-specific Fas+ CD8+ T cells underwent apoptosis after their interaction with FasL+ CD11b + F4/80+ monocyte-derived macrophages, recruited by these hepatic lesions (Yu et al. 2021). This preclinical work established the mechanisms on which the clinical results could be explained. This work, despite being a retrospective and monocentric analysis and considering that the preclinical model of liver metastasis used may not fully represent human pathophysiology, suggests that liver metastasis may be an important determinant of the efficacy of immunotherapy and supports other clinical studies that advocate that different anatomical locations of metastases show different response patterns, and therefore unique mechanisms of resistance that need to be accounted for in future personalized treatments (Lee et al. 2020).

3 Role of Liver SBRT

Historically, the use of external beam radiation therapy has had a limited role in the treatment of liver primary tumors and metastases because of the risk of liver toxicity induced by high doses of radiation delivered to normal liver tissue. Radiation-induced liver disease (RILD), classically manifesting as a triad of anicteric hepatomegaly, ascites, and elevation of alkaline phosphatase, has been a feared complication (Marks et al. 2010).

However, recent technological advances have contributed to the development of SBRT (Stereotactic Body Radiation Therapy) as a precisely focused radiation technique that allows the treatment of liver lesions with an ablative intent while significantly limiting the dose to the healthy liver and surrounding tissues. The emergence of sophisticated software, the use of high dose gradient achieved through highly conformed dosimetry, and quality control in the delivery with image guidance systems such as integrating intrafraction control of liver motion have all contributed to an accurate tumor-directed treatment, with a reduction on treatment uncertainty, increasing the potential to deliver higher doses of radiation and thereby improving local control minimizing toxicity.

The safety and effectiveness of SBRT has been evaluated with encouraging results in retrospective and prospective clinical studies of liver metastases, showing minimal toxicity, high rates of local control (LC), and promising overall survival (OS) as we will explain in the following paragraphs.

First results of SBRT published in 1995 showed the feasibility of the technique (Blomgren et al. 1995). Phase I studies reported the dose escalation benefit in a single fraction (14-26 Gy), reaching a 66% LC at 18 months (Herfarth et al. 2004), and a multicenter phase I/II study demonstrated the safety and efficacy of high doses of SBRT, administering 60 Gy in 3 fractions, with a 2-year LC of 92% and only 2% of grade 3 toxicity (Rusthoven et al. 2009). Together, SBRT is a very well-tolerated treatment with a low toxicity profile (G1-2 of 0–28%), grade 3 late complications of less than 5%, and exceptional severe toxicity. Equally, long-term results with SBRT have shown that the technique is safe and gives extended survival rates (Goodman et al. 2016).

As far as LC is concerned, it seems mainly influenced by lesion size (<5 cm) and radiation dose (BED > 100 Gy). Dose escalation appears to be particularly important. When dose-escalated regimens are analyzed, 1- and 2-year LC rates range from 90–100% and 81–100%, respectively (Herfarth et al. 2004; Mendez Romero et al. 2006; Rusthoven et al. 2009), and phase II studies with even higher doses (75 Gy in 3 fractions) have reported 2-year LC of 91% and high OS (2-year OS of 65%) (Scorsetti et al. 2015). This data supports that there is a dose–response relationship for radiation and local control outcomes, and that SBRT adds a novel mechanism of radiation-induced damage. Higher doses per fraction, in addition to direct cytotoxicity, also induce endothelial apoptosis and microvascular damage that in conjunction lead to death of the tissue supplied by that vasculature, which in turn produces a substantial effect on the tumor cell kill (Fuks and Kolesnick 2005).

A major limitation of the published studies is the lack of randomized studies comparing SBRT with other local techniques. Nevertheless, the studies show that SBRT offers an effective and secure alternative for the treatment of liver metastases. Regarding patient characteristics, most of these studies have treated patients with a limited number of liver metastases (1–3 lesions) and have recruited mainly inoperable patients or those with unresectable metastases and consequently, with the fewest local therapeutic options. SBRT therefore, must be studied in a prospective setting in combination with systemic therapies and immunotherapy, just as other local options are, to demonstrate higher OS in the same patient and cancer settings.

Recently, SBRT has been increasingly used for the management of hepatocellular carcinoma (HCC) in patients who previously relapsed or are not candidates for surgery, ablation, or embolization. SBRT provides noninvasive, curative local therapy for HCC, and multiple retrospective and prospective studies have also demonstrated that SBRT is an effective bridge for liver transplant (Sapisochin et al. 2017; Wong et al. 2021) as well as being effective for local therapy in patients with inoperable early, intermediate, or advanced-stage HCC disease. All in all, SBRT in HCC is associated with excellent local control, and despite the increased risk of hepatic decompensation and poor functional status of patients with advanced-stage HCC, SBRT is generally well-tolerated, and a recent meta-analysis of retrospective and prospective studies has reported a grade 3 or higher toxicity < 10% (Thomas and Feng 2021).

HCC type of cancer has been shown to be associated with liver inflammation and an immunosuppressive microenvironment (Prieto et al. 2015). In the same manner, high expression of programmed death-ligand 1 (PD-L1) in tumors correlates with a poorer prognosis in patients with resected HCC (Gao et al. 2009). Upregulation of programmed cell death protein-1 (PD-1) and PD-L1 on T cells is often associated with more advanced disease stages and higher risks of recurrence suggesting that immunotherapy may benefit HCC patients (Shi et al. 2011). Two recently published phase II trials testing antiPD-1 inhibitors for advanced HCC, Nivolumab in the CheckMate 040 study (El-Khoueiry et al. 2017), and Pembrolizumab in the KEYNOTE-224 study (Zhu et al. 2018) have shown substantial tumor responses (15–20%) and promising duration of response (median: 16 months), with both favorable survival-rates and manageable toxicity profiles.

4 Data on the Benefit of Adding Liver SBRT to ICI Treatment

Although we have previously talked about how immunotherapy has not achieved the expected results in liver metastases, it is also true that recent studies have shown that, thanks to techniques such as liver SBRT, we can significantly influence the "immunosensitivity" of these tumors.

This immunostimulation is achieved by several routes, the first and most studied being tumor DNA damage (lethal and/or sublethal damage) (Sato et al. 2021). According to this first pathway, damage produced to the tumor derived from the administration of radiotherapy treatment achieves direct and indirect cell death that leads to the exposure of antigens and, therefore, a greater effect on the immune system with or without systemic immunotherapy (Formenti and Demaria 2009). This is the mechanism that is also involved in the effect known as "abscopal," through which distant lesions to the tumor treated with radiation also respond to such treatment (Ashrafizadeh et al. 2020; Demaria and Formenti 2020). Today we have preclinical data that explains that this activation of the immune system occurs through the activation of cGAS/STING pathway and an overexpression of INF-I, INF-β, PD-L1, and increased exposure by type I MHC tumor cells (Chen et al. 2016).

The second pathway through which radiotherapy influences immune stimulation would be the one that is activated by influencing molecular changes in the tumor microenvironment (TME). In this pathway, the underlying immunological mechanism is the so-called immunogenic cell death (ICD). Due to the acute molecular damage produced by radiation, activation and migration of dendritic cells is promoted, as well as antigenic exposure to CD8+ T lymphocytes. This induces the expression of HMGB1 protein, calreticulin, and ATP. HMGB 1, when released extracellularly, acts as an agonist on dendritic cell receptors, activating them and inducing an antigen-specific response by T lymphocytes. Calreticulin is a chaperone that is expressed at the membrane of the endoplasmic reticulum and its expression at the cell membrane induces phagocytosis of irradiated tumor cells by dendritic cells. ATP acts by activating dendritic cells via binding to their P2X7 receptor, which induces the production of INF-gamma (Sia et al. 2020).

Interestingly, there is an in silico study published by *Basler et al* that is able to calculate circulating lymphocytes (CL) according to different liver SBRT treatment parameters used. For example, radiation modality (3D, VMAT, 10MV-FFF) and *beam-on time* were related to CL. On the other hand, tumor location in the liver was shown to impact CL, being the central metastases the ones that had worse CL outcome. Finally, as previously expected, larger PTV volumes impacted negatively on CL (Basler et al. 2018).

In the era of immunotherapy with checkpoint inhibitors, the combination of these treatments with radiotherapy is rapidly expanding both in palliative settings and within clinical trials. Although robust clinical evidence is not yet available to administer these combinations in the daily practice, there is sufficient preclinical data indicating a synergy between both treatments as previously explained, and initial clinical experiences that indicate likewise with minimal effects on toxicity that will be explained in the next paragraphs.

The first case of synergy between liver SBRT and anti-CTLA4 was published in 2013 (Golden et al. 2013). The patient in question was diagnosed with metastatic lung cancer adenocarcinoma. He had previously received pemetrexed, gemcitabine, and thoracic radiation therapy and suffered a systemic progression in multiple organs. At this point, he started ipilimumab (anti-CTLA4) × 3 cycles and concomitantly received five fractions of liver SBRT that were administered on alternate days, reaching a maximum dose of 30Gy. Subsequently, a multiple-level regression was observed by means of PET-CT, as well as a reduction in the CEA tumor marker. When blood samples were analyzed, an increase in total leukocytes and eosinophils was detected post-treatment, both considered predictive biomarkers of a favorable response to anti-CTLA4 drugs.

Following with blood biomarkers of response to liver SBRT, the Mayo Clinic found in one study, a significant reduction in the populations of CD3+ T cells and immature NK cells (CD56br and CD16- phenotype) in the blood of patients treated with liver SBRT compared to controls, regardless of the local treatment being applied for primary liver cancer or metastatic cancer (Gustafson et al. 2017). Similarly, patients with liver metastases were evaluated in a study in Michigan and showed reduced T cell numbers and lower T cell diversity and function than controls. Parallel translational and preclinical work demonstrated that this was due to a cell to cell contact of F4/80+ macrophages and CD8+ T cells in the TME that eliminated T cells, further supporting the previous hypothesis, and highlighting one of the mechanisms of immunotherapy failure (Yu et al. 2021). This study followed by testing on preclinical models liver radiotherapy and demonstrated how this treatment can eliminate immunosuppressive hepatic macrophages and increase hepatic T cell survival and blood circulation.

In 2019, Liu et al (2019) published the first positive clinical responses, both local and abscopal, of the synergy between an anti-PD-L1 agent and liver SBRT in 3 patients with intrahepatic cholangiocarcinoma. This result was achieved despite low expression (<1%) of PD-L1 in the tumors, thus potentially expanding the indications of immunotherapy in this cancer scenario.

Also in 2019, a retrospective series of 5 patients with unresectable hepatocellular carcinoma treated with SBRT and anti-PD-1 therapy demonstrated spectacular results with local control rates and 1-year overall survival of 100% (Chiang et al. 2019).

Regarding optimal doses of radiation and how they can influence tumor response at the immune level, a consensus has not yet been reached beyond admitting that the standard fractionation at 2Gy per fraction does not seem to be the most appropriate. Thus, most authors are more inclined towards fractionation with more ablative doses (Pérez-Romasanta et al. 2021). Nevertheless, there is a study in which these ablative doses were evaluated in mice, and it was established that doses below 10Gy repeated in 2–3 fractions are adequate to produce an ICD and consequently an abscopal response, since the canonical cGAS-STING pathway is inhibited with higher doses and does not achieve the activation of IFN genes necessary to amplify the immune response (Vanpouille-Box et al. 2017).

Surprisingly, and to overcome the toxicity produced by radiation's ablative doses in organs in the proximity of the tumor, there are authors who describe how low doses at the tumor margin can also achieve good tumor responses when combined with ablative doses to the center of the tumor in combination with immunotherapy (Luke et al. 2018; Onderdonk 2019).

In conclusion, there is initial preclinical and clinical evidence of the synergy between liver SBRT and immunotherapy, both in primary liver cancer and liver metastases. Most importantly, this combination strategy seems to be able to overcome the natural preferential immune tolerance of the liver, and as a result, there is an increasing use of this combination in hospitals. Some authors argue that biological knowledge in this field is still in its infancy in comparison to our knowledge in radiation technology and safety (Kamran et al. 2021). In the same line, we currently have state-of-the-art linear accelerators at our disposal that allow us to perform highly precise treatments with ablative doses of radiation (Rokni et al. 2021), we can administer targeted treatments, and, on the other hand, we are not yet able to obtain the data that will help us decide, among other things, the sequence and fractionation of these treatments, and the most appropriate combination in each type of metastatic tumor or organ (Demaria et al. 2021). All these and some other questions remain unanswered, and clinical trials are needed to obtain the necessary data on the topic. Meanwhile, efforts should be made to extrapolate and adapt the results of existing clinical trials to daily clinical practice in order to offer the best and most innovative therapeutic alternatives to patients.

5 Conclusions

Advances in the recent years, both in the field of immunotherapy and in the implementation of complex local cancer treatments, have allowed us to optimize the clinical results of our patients, carrying out increasingly personalized treatments. In this context, it is vital to deepen into the pathophysiological mechanisms that underlie each cancer and that condition different response patterns depending on the underlying histology and anatomical location. Specifically, the presence of tumor lesions at the liver level configures a subgroup of patients with a particularly poor prognosis, whose treatment today remains a challenge. The identification of new biomarkers and the significant preclinical advances are helping to understand more and more this scenario of the disease, laying the foundations for a multimodal treatment in which the use of oncological treatments in combination allows to enhance the results. Therefore, combination trials are urgently needed to define the optimal strategy in each clinical scenario that will incorporate effective local liver SBRT and systemic immunotherapy to improve patient outcome.

Acknowledgments None.

Compliance with Ethical Standards The authors declare that there is no conflict of interest.

References

Antonia SJ, Villegas A, Daniel D, Vicente D, Murakami S, Hui R et al (2017) Durvalumab after chemoradiotherapy in stage III non-small-cell lung cancer. N Engl J Med 377(20):1919–1929

Ashrafizadeh M, Farhood B, Musa AE, Taeb S, Rezaeyan A, Najafi M (2020) Abscopal effect in radioinmunotherapy. Int Immunopharmacol 85:106663

Bajorin DF, Witjes JA, Gschwend JE, Schenker M, Valderrama BP, Tomita Y et al (2021) Adjuvant nivolumab versus placebo in muscle-invasive urothelial carcinoma. N Engl J Med 384 (22):2102–2114

Bamboat ZM, Stableford JA, Plitas G, Burt BM, Nguyen HM, Welles AP, Gonen M, Young JW, DeMatteo RP (2009) Human liver dendritic cells promote T cell hyporesponsiveness. J Immunol 182:1901–1911

Basler L, Andratschke N, Ehrbar S, Guckenberger M, Tanadini-Lang S (2018) Modelling the immunosuppressive effect of liver SBRT by simulating the dose to circulating lymphocytes: an in-silico planning study. Radiat Oncol 13(1):10

Bilen MA, Shabto JM, Martini DJ, Liu Y, Lewis C, Collins H et al (2019) Sites of metastasis and association with clinical outcome in advanced stage cancer patients treated with immunotherapy. BMC Cancer 19(1):857

Blomgren H, Lax I, Naslund I, Svanstrom R (1995) Stereotactic high dose fraction radiation therapy of extracranial tumors using an accelerator. Clinical experience of the first thirty-one patients. Acta Oncol 34(6):861–870

Charles R, Chou H-S, Wang L, Fung JJ, Lina L, Qian S (2013) Human hepatic stellate cells inhibit T cell response through B7-H1 pathway. Transplantation 96(1):17–24

Chen DS, Mellman I (2017) Elements of cancer immunity and the cancer-immune set point. Nature 541(7637):321–330

Chen Q, Sun L, Chen ZJ (2016) Regulation and function of the cGAS-STING pathway of cytosolic DNA sensing. Nat Immunol 17:1142–1149

Chiang C-L, Chan ACY, Chiu KWH, Kong F-MS (2019) Front Oncol 12(9):115

Cohen EEW, Bell RB, Bifulco CB, Burtness B, Gillison ML, Harrington KJ et al (2019) The Society for Immunotherapy of Cancer consensus statement on immunotherapy for the treatment of squamous cell carcinoma of the head and neck (HNSCC). J Immunother Cancer 7(1):184

Demaria S, Formenti SC (2020) The abscopal effect 67 years later: from a side story to center stage. Br J Radiol 93:20200042

Demaria S, Guha C, Schoenfeld J, Morris Z, Monjazeb A, Sikora A, Crittenden M, Shiao S, Khleif S, Gupta S, Formenti SC, Vikram B, Coleman CN, Ahmed MM (2021) Radiation dose and fraction in immunotherapy: one-size régimen does not fit all settings, so how does one choose? J Immunother Cancer 9(4):e002038

Diehl L, Schurich A, Grochtmann R, Hegenbarth S, Chen L, Knolle PA (2008) Tolerogenic maturation of liver sinusoidal endothelial cells promotes B7-homolog 1-dependent CD8+ T cell tolerance. Hepatology 47:296–305

Doherty DG (2015) Immunity, tolerance and autoimmunity in the liver: a comprehensive review. J Autoimmun. https://doi.org/10.1016/j.jaut.2015.08.020
Eckert C, Klein N, Kornek M, Lukacs-Kornek V (2015) The complex myeloid network of the liver with diverse functional capacity at steady state and in inflammation. Front Immunol 20(6):179
El-Khoueiry AB, Sangro B, Yau T, Crocenzi TS, Kudo M, Hsu C, Kim TY, Choo SP, Trojan J, Rd WTH, Meyer T, Kang YK, Yeo W, Chopra A, Anderson J, Dela Cruz C, Lang L, Neely J, Tang H, Dastani HB, Melero I (2017) Nivolumab in patients with advanced hepatocellular carcinoma (CheckMate 040): an open-label, non-comparative, phase 1/2 dose escalation and expansion trial. Lancet 389(10088):2492–2502
Formenti SC, Demaria S (2009) Systemic effects of local radiotherapy. Lancet Oncol 10:718–726
Fuks Z, Kolesnick R (2005) Engaging the vascular component of the tumor response. Cancer Cell 8(2):89–91
Gao Q, Wang XY, Qiu SJ, Yamato I, Sho M, Nakajima Y, Zhou J, Li BZ, Shi YH, Xiao YS, Xu Y, Fan J (2009) Overexpression of PD-L1 significantly associates with tumor aggressiveness and postoperative recurrence in human hepatocellular carcinoma. Clin Cancer Res 15(3):971–979
Gide TN, Silva IP, Quek C, Ahmed T, Menzies AM, Carlino MS et al (2020) Close proximity of immune and tumor cells underlies response to anti-PD-1 based therapies in metastatic melanoma patients. Oncoimmunology 9(1):1659093. https://doi.org/10.1080/2162402X.2019.1659093
Golden EB, Demaria S, Schiff PB, Chachoua A, Formenti SC (2013) An abscopal response to radiation and ipilimumab in a patient with metastatis non-small cell ling cancer. Cancer Immunol Res 1(6):365–372
Goodman BD, Mannina EM, Althouse SK, Maluccio MA, Cardenes HR (2016) Long-term safety and efficacy of stereotactic body radiation therapy for hepatic oligometastases. Pract Radiat Oncol 6(2):86–95
Gustafson MP, Bornschlegl S, Park SS, Gastineau DA, Roberts LR, Dietz AB, Hallemeier CL (2017) Comprehensive assessment of circulating immune cell populations in response to stereotactic body radiation therapy in patients with liver cancer. Adv Radiat Oncol 2(4):540–547
Herfarth KK, Debus J, Wannenmacher M (2004) Stereotactic radiation therapy of liver metastases: update of the initial phase-I/II trial. Front Radiat Ther Oncol 38:100–105
Höchst B, Schildberg FA, Sauerborn P, Gäbel YA, Gevensleben H, Goltz D, Heukamp LC, Türler A, Ballmaier M, Gieseke F, Müller I, Kalff J, Kurts C, Knolle PA, Diehl L (2013) Activated human hepatic stellate cells induce myeloid derived suppressor cells from peripheral blood monocytes in a CD44-dependent fashion. J Hepatol 59(3):528–535
Ibrahim J, Nguyen AH, Rehman A, Ochil A, Jamal M, Graffeo CS, Henning JR, Zambirinis CP, Fallon N, Barilla R, Badar S, Mitchell A, Rao R, Acehan D, Frey AB, Miller G (2012) Dendritic cell populations with different concentrations of lipid regulate tolerance and immunity in mouse and human liver. Gastroenterology 143(4):1061–1072
Kamran SC, Palma D, Katz MS, Zietman AL (2021) Technical giants but biologic infants: defining a more sophisticated role for local therapy in metastatic disease. Semin Radiat Oncol 31(3):200–204
Kelly RJ, Ajani JA, Kuzdzal J, Zander T, Van Cutsem E, Piessen G et al (2021) Adjuvant nivolumab in resected esophageal or gastroesophageal junction cancer. N Engl J Med 384(13):1191–1203
Kruse N, Neumann K, Schrage A, Derkow K, Schott E, Erben U, Kühl A, Loddenkemper C, Zeitz M, Hamann A, Klugewitz K (2009) Priming of CD4+ T cells by liver sinusoidal endothelial cells induces CD25low forkhead box protein 3^- regulatory T cells suppressing autoimmune hepatitis. Hepatology 50:1904–1913
Larkin J, Chiarion-Sileni V, Gonzalez R, Grob J-J, Rutkowski P, Lao CD et al (2019) Five-year survival with combined nivolumab and ipilimumab in advanced melanoma. N Engl J Med 381(16):1535–1546
Lee JC, Mehdizadeh S, Smith J, Young A, Mufazalov IA, Mowery CT et al (2020) Regulatory T cell control of systemic immunity and immunotherapy response in liver metastasis. Sci Immunol 5(52):eaba0759

Li W, Li L, Hui L (2020) Cell plasticity in liver regeneration. Trends Cell Biol 30(4):329–338
Liu X, Yao J, Song L, Zhang S, Huang T, Li Y (2019) J Immunother Cancer 7(1):204
Luke JJ, Lemons JM, Karrison TG, Pitroda SP, Melotek JM, Zha Y et al (2018) Safety and clinical activity of pembrolizumab and multisite stereotactic body radiotherapy in patients with advanced solid tumors. J Clin Oncol 36(16):1611–1618
Marks KB, Yorke ED, Jackson A et al (2010) Use of normal tissue complication probability models in the clinic. Int J Radiat Oncol Biol Phys 76(3 Suppl):S10–S19
Mehlen P, Puisieux A (2006) Metastasis: a question of life or death. Nat Rev Cancer 6(6):449–458
Mendez Romero A, Wunderink W, Hussain SM, De Pooter JA, Heijmen BJ, Nowak PC et al (2006) Stereotactic body radiation therapy for primary and metastatic liver tumors: a single institution phase I-II study. Acta Oncol 45(7):831–837
Moseman EA, Liang X, Dawson AJ, Panoskaltsis-Mortari A, Krieg AM, Liu Y-J, Blazar BR, Chen W (2004) Human plasmacytoid dendritic cells activated by CpG Oligodeoxynucleotides induce the generation of CD4+CD25+ regulatory T cells. J Immunol 73:4433–4442
Onderdonk BE (2019) Multi-site SBRT and sequential pembrolizumab: treated metastasis control and immune-related expression predict outcomes. In: Oral presentation at the 2019 American Society for Radiation Oncology (ASTRO) Research Workshop: Treatment of Oligometastatic Disease. 2019 Jun 13–14, Washington, DC
Pérez-Romasanta LA, Portillo EG-D, Rodríguez-Gutiérrez A, Matías-Pérez Á (2021) Stereotactic radiotherapy for hepatocellular carcinoma, randiosensitization strategies and radiation-immunotherapy combination. Cancers (Basel) 13(2):192
Prieto J, Melero I, Sangro B (2015) Immunological landscape and immunotherapy of hepatocellular carcinoma. Nat Rev Gastroenterol Hepatol 12(12):681–700
Protzer U, Maini MK, Knolle PA (2012) Living in the liver: hepatic infections. Nat Rev Immunol 12(3):201–213
Reck M, Mok TS, Nishio M, Jotte RM, Cappuzzo F, Orlandi F et al (2019) Atezolizumab plus bevacizumab and chemotherapy in non-small-cell lung cancer (IMpower150): key subgroup analyses of patients with EGFR mutations or baseline liver metastases in a randomised, open-label phase 3 trial. Lancet Respir Med 7(5):387–401
Rokni MB, Pointer KB, George J, Luke JJ, Chmura SJ, Redler G (2021) Radiation treatment planning study to investigate feasibility of delivering immunotherapy in combination with ablative radiosurgery to ultra-high DoSes (ICARUS). J Appl Clin Med Phys 22(3):196–206
Rusthoven KE, Kavanagh BD, Cardenes H, Stieber VW, Burri SH, Feigenberg SJ et al (2009) Multiinstitutional phase I/II trial of stereotactic body radiation therapy for liver metastases. J Clin Oncol 27(10):1572–1578
Sapisochin G, Barry A, Doherty M, Fischer S, Goldaracena N, Rosales R, Russo M, Beecroft R, Ghanekar A, Bhat M, Brierley J, Greig PD, Knox JJ, Dawson LA, Grant DR (2017) Stereotactic body radiotherapy vs. TACE or RFA as a bridge to transplant in patients with hepatocellular carcinoma. An intention-to-treat analysis. J Hepatol 67(1):92–99
Sasaki A, Nakamura Y, Togashi Y, Kuno H, Hojo H, Kageyama S et al (2020) Enhanced tumor response to radiotherapy after PD-1 blockade in metastatic gastric cancer. Gastric Cancer 23(5):893–903
Sato H, Demaria S, Ohno T (2021) The role of radiotherapy in the age of inmunotherapy. Jpn J Clin Oncol 51(4):513–522
Schmid S, Diem S, Li Q, Krapf M, Flatz L, Leschka S et al (2018) Organ-specific response to nivolumab in patients with non-small cell lung cancer (NSCLC). Cancer Immunol Immunother 67(12):1825–1832
Scorsetti M, Comito T, Tozzi A, Navarria P, Fogliata A, Clerici E et al (2015) Final results of a phase II trial for stereotactic body radiation therapy for patients with inoperable liver metastases from colorectal cancer. J Cancer Res Clin Oncol 141(3):543–553
Shi F, Shi M, Zeng Z, Qi RZ, Liu ZW, Zhang JY, Yang YP, Tien P, Wang FS (2011) PD-1 and PD-L1 upregulation promotes CD8(+) T-cell apoptosis and postoperative recurrence in hepatocellular carcinoma patients. Int J Cancer 128(4):887–896

Sia J, Szmyd R, Hau E, Gee HE (2020) Molecular mechanisms of radiation-induced cancer cell death: a primer. Front Cell Dev Biol 8:41

Thomas HR, Feng M (2021) Stereotactic body radiation therapy (SBRT) in hepatocellular carcinoma. Curr Hepatology Rep 20:12–22

Topalian SL, Hodi FS, Brahmer JR, Gettinger SN, Smith DC, McDermott DF et al (2019) Five-year survival and correlates among patients with advanced melanoma, renal cell carcinoma, or non-small cell lung cancer treated with nivolumab. JAMA Oncol 5(10):1411–1420

Tumeh PC, Hellmann MD, Hamid O, Tsai KK, Loo KL, Gubens MA et al (2017) Liver metastasis and treatment outcome with anti-PD-1 monoclonal antibody in patients with melanoma and NSCLC. Cancer Immunol Res 5(5):417–424

Uhrig A, Banafsche R, Kremer M, Hegenbarth S, Hamann A, Neurath M, Gerken G, Limmer A, Knolle PA (2005) Development and functional consequences of LPS tolerance in sinusoidal endothelial cells of the liver. J Leukoc Biol 77:626–633

Vanpouille-Box C, Alard A, Aryankalayil MJ, Sarfraz Y, Diamond JM, Schneider RJ, Inghirami G, Coleman CN, Formenti SC, Demaria S (2017) DNA exonuclease Trex1 regulates radiotherapy-induced tumour immunogenicity. Nat Commun 9(8):15618

von Oppen N, Schurich A, Hegenbarth S, Stabenow D, Tolba R, Weiskirchen R, Geerts A, Kolanus W, Knolle P, Diehl L (2009) Systemic antigen cross-presented by liver sinusoidal endothelial cells induces liver-specific CD8 T-cell retention and tolerization. Hepatology 49:1664–1672

Wong TC-L, Lee VH-F, Law AL-Y, Pang HH, Lam K-O, Lau V, Cui TY, Fong AS-Y, Lee SW-M, Wong EC-Y, Dai JW-C, Chan AC-Y, Cheung T-T, Fung JY-Y, Yeung RM-W, Luk M-Y, Leung T-W, Lo C-M (2021) Prospective study of stereotactic body radiation therapy for hepatocellular carcinoma on waitlist for liver transplant. Hepatology 74:2580–2594

Xia S, Guo Z, Xiongfei X, Yi H, Wang Q, Cao X (2008) Hepatic microenvironment programs hematopoietic progenitor differentiation into regulatory dendritic cells, maintaining liver tolerance. Blood 112:3175–3185

Wu K, Kryczek I, Chen L, Zou W, Welling TH (2009) Kupffer cell suppression of CD8+ T cells in human hepatocellular carcinoma is mediated by B7-H1/programmed death-1 interactions. Cancer Res 69(20):8067–8075

Yu J, Green MD, Li S, Sun Y, Journey SN, Choi JE et al (2021) Liver metastasis restrains immunotherapy efficacy via macrophage-mediated T cell elimination. Nat Med 27(1):152–164

Zhang M, Xu S, Han Y, Cao X (2011) Apoptotic cells attenuate fulminant hepatitis by priming Kupffer cells to produce interleukin-10 through membrane-bound TGF-β. Hepatolog 53 (1):306–316. PMID: 21140375. https://doi.org/10.1002/hep.24029

Zhu AX, Finn RS, Edeline J, Cattan S, Ogasawara S, Palmer D, Verslype C, Zagonel V, Fartoux L, Vogel A, Sarker D, Verset G, Chan SL, Knox J, Daniele B, Webber AL, Ebbinghaus SW, Ma J, Siegel AB, Cheng AL, Kudo M, KEYNOTE-224 investigators (2018) Pembrolizumab in patients with advanced hepatocellular carcinoma previously treated with sorafenib (KEYNOTE-224): a non-randomised, open-label phase 2 trial. Lancet Oncol 19(7):940–952

Index

A
Ablation, 7–10, 27, 42, 59, 63–67, 70, 97, 179, 265, 269, 271, 272, 285, 308, 309, 315–318, 355
Angiogenesis, 55, 56, 88–96, 98, 109–112, 118, 119, 123, 124, 154, 233, 234, 259, 271, 290, 292, 295, 299, 327, 329, 330, 337
Antigen, 86, 88, 92, 95, 96, 110, 111, 114, 116, 123, 234, 235, 265, 271, 272, 287, 293–295, 297, 327, 350–352, 356
Autophagy, 180, 260, 332–334, 336–341

B
Biliary tract cancer (BTC), vii, 2–4, 8, 14, 16, 18, 229–246
Biomarker, 4, 5, 20, 49–59, 96, 117, 118, 121, 123, 127, 141–144, 151, 174, 196, 197, 215, 217, 218, 220, 242–246, 257, 269, 352, 357, 358

C
Cancer immunity, 109–111
Cancer metabolism, 141, 147
Cancer stem cells (CSCs), 93, 139–156, 327, 329, 333, 335, 339
Cellular senescence, 173, 175–178, 180–182
Checkpoint inhibitors, 116, 267, 357
Cholangiocarcinoma (CC), 3, 6, 14–19, 146, 230, 231
Cirrhosis, 2, 3, 6, 42–44, 46–48, 54, 57, 58, 60, 63, 64, 87, 88, 109, 113, 118, 168–171, 176, 196, 230, 256–258, 260, 261, 269, 297, 311, 326
Clinical trials, 56, 59, 69, 70, 96, 118, 181, 242, 243, 261–269, 287, 289, 298–299, 317, 357, 358
Co-delivery, 295
Cytokines, 56, 70, 87–93, 95–97, 99, 114, 118, 123, 125, 126, 169, 174, 176, 197, 233–235, 258–260, 266, 271, 272, 287, 288, 294, 297, 330, 337, 351

D
Diagnosis, 2, 4, 6–8, 19, 20, 24, 27, 42, 45, 51, 53, 68, 70, 71, 117–118, 127, 144, 196, 217, 240, 308
DNA damage, 150, 156, 173, 175–177, 179, 215, 327, 339, 356
Drug resistance, vii, 70, 112, 145, 152, 154, 232, 261, 296, 325–341

F
Fibrosis, vii, 46, 49, 70, 87, 89, 91, 94, 100, 113, 114, 118, 168–172, 175, 176, 181, 182, 197, 203, 208, 219, 255–273, 293, 297, 314

G
Gene therapy, 288, 289, 295–298

H
Hepatic fibrosis, 88, 170, 196, 218, 257, 258
Hepatitis B virus (HBV), 4, 5, 9, 42, 43, 45–53, 56, 58, 59, 86, 114, 118, 140, 171, 257, 270, 271, 284
Hepatocellular carcinoma (HCC), 2, 42, 86, 108, 140, 168, 196, 256, 284, 308, 326, 355

Interdisciplinary Cancer Research, https://doi.org/10.1007/978-3-031-44528-6

Hepatocytes, 45, 50, 54, 58, 87–89, 91, 95, 109, 114, 115, 140, 143, 148, 152, 168–170, 174–176, 178, 179, 202, 256–260, 284, 292–294, 350, 351

I
Immune cells, 86–90, 94–98, 109, 110, 112, 114, 115, 153, 170, 176–178, 234, 259, 261, 270, 273, 285, 286, 293, 298
Immune checkpoint, 13, 96, 116, 122–124, 285
Immune checkpoint inhibitors (ICIs), 63, 68–70, 119, 122, 123, 127, 140, 236–243, 245, 246, 261, 265–270, 272, 273, 287, 356–358
Immune system, vii, 85–99, 110, 114, 125, 176, 233, 235, 260, 265, 271, 287, 293, 351, 356
Immunosurveillance, 110, 111, 116, 121, 177, 179
Immunotherapy, vii, 2, 62, 68, 70, 71, 96, 97, 108, 116, 119, 121–122, 140, 229–246, 255–273, 285, 287–288, 292, 293, 298, 317, 318, 337, 351–359
Immunotolerance, 86, 88, 95–96
Inflammation, 86–92, 94, 99, 109, 110, 113, 115, 118, 122, 125–126, 148, 150, 153, 154, 168–173, 175, 197, 215, 233, 235, 257, 260, 271, 273, 351, 355
Interdisciplinary communication, 60
Ionizable lipids, 291–295

L
Lipid nanoparticles (LNP), 288–299
Liposomes, 124, 290, 294, 297–299
Liver, 2, 42, 86, 108, 140, 168, 196, 230, 256, 284, 308, 326, 350
Liver cancer, vii, 2, 3, 20, 41–71, 85–99, 108, 112–116, 118, 127, 140, 150, 171, 172, 257, 269, 270, 283–299, 308, 309, 326, 333, 337, 341, 357, 358
Liver-directed therapies (LDTs), 256, 261, 271–273
Locoregional therapy, vii, 2, 63, 68, 140, 270, 307–318

M
Macrophages, 70, 71, 87, 89, 90, 92, 97, 111–116, 118, 121, 123–126, 153, 169, 173, 177, 178, 181, 218, 233, 234, 245, 259, 260, 268, 272, 286, 287, 297, 298, 329, 338, 351, 352, 354, 357
Metabolism, 53, 54, 57, 71, 94, 109, 110, 112, 126, 141, 145–147, 149–154, 156, 170, 172, 200, 201, 205, 212, 215–218, 220, 260, 327, 338, 339
Microenvironment, 87, 89, 90, 95–97, 108–119, 121, 122, 124, 126–128, 141, 143, 148, 154, 175, 197, 257, 259, 261, 267, 271, 286, 334, 336, 355
mRNA vaccines, 289, 293, 294, 299

N
Nanomedicine, vii, 283–299
Nanotechnology, 285
Neoplasms, 235
Non-alcoholic fatty liver disease (NAFLD), vii, 3, 44, 47, 49, 54, 69, 116, 140, 148–150, 168–182, 193–220, 257
Noncoding RNA, 50, 329–337, 341

O
Outcomes, 2, 4, 8–11, 15, 21–24, 56, 59, 61, 62, 66, 71, 88, 96, 97, 147, 154, 180, 235, 236, 239, 243, 259, 260, 267, 271, 309, 310, 312, 314, 315, 317, 355, 357, 359

P
p53 restoration, 177

R
RNA silencing, 289

S
Signature-based metabolic profiling, 193–220
Small interfering RNA (siRNA), 13, 288–290, 292–293, 295, 296, 299
Sorafenib, 10, 55, 97, 108, 140, 217, 256, 285, 310, 326
Stereotactic body radiotherapy (SBRT), vii, 68, 272, 349–359
Systems biology, 218, 219

T
Targeted therapy, 2, 14, 15, 68, 145, 146, 231, 285
Therapeutics, 2, 10, 13, 176, 289, 290, 295, 298

Transarterial chemoembolization (TACE), 7–10, 14, 42, 59, 60, 63, 66, 67, 70, 119–120, 231, 264–266, 269–273, 285, 286, 308–314, 316–318
Transarterial radioembolization (TARE), 8, 10, 14, 66, 67, 70, 231, 272, 308–315, 318
Transcriptional regulatory units, 197
Tumor, 2, 50, 86, 108, 176, 197, 230, 256, 284, 308, 326, 350
Tumor microenvironment (TME), vii, 70, 86, 87, 89, 90, 94, 96–98, 107–128, 153–154, 229–246, 256, 267, 268, 271, 272, 286, 287, 338, 352, 356, 357

V

Vaccine, 13, 70, 96, 246, 285, 289, 293–295

MIX
Papier aus verantwortungsvollen Quellen
Paper from responsible sources
FSC® C105338

If you have any concerns about our products,
you can contact us on
ProductSafety@springernature.com

In case Publisher is established outside the EU,
the EU authorized representative is:
Springer Nature Customer Service Center GmbH
Europaplatz 3, 69115 Heidelberg, Germany

Printed by Libri Plureos GmbH
in Hamburg, Germany